PRINCIPLES OF GEOLOGY

Principles of Geology

BY

JAMES GILLULY *U. S. Geological Survey*

AARON C. WATERS *The Johns Hopkins University*

A. O. WOODFORD *Pomona College*

ILLUSTRATIONS BY

Robert R. Compton Stanford University

1952 **W. H. Freeman & Company**

SAN FRANCISCO, CALIFORNIA

Preface

THIS book attempts to summarize some of the knowledge that geologists have won from the study of the earth. A subject so large must be treated very briefly if it is to be presented between the covers of a single book; we have chosen to concentrate on the analysis of processes that are at work upon and within the earth, rather than to present a catalog of descriptive facts and terms. We have felt, too, that the student is entitled to know something of the kind of evidence on which geologic conclusions are based, even though its presentation takes valuable pages that might be used to put forth more facts.

Some teachers will regret our brief treatment of many of the standard topics usually found in textbooks of physical geology. We can only hope that the loss will be balanced by the new material included covering many phases of the science in which rapid advances have been made in recent years, and more particularly by the emphasis on leading the student through approximately the same sequence of reasoning that was used in the historical development of the subject. We believe that the student may retain more of the basic principles on which geology is based if he knows how a geologic map is made, and if he is introduced to Werner's and Desmarest's divergent views on the origin of basalt, than if he is instructed too minutely on the purely technical terminology of landscape morphology or rock classification. It is our hope, too, that such a presentation carries with it an understanding of the intrinsic uncertainties of indirect evidence, upon which so much of geology depends.

Geology, as we know it, could hardly exist without the foundation of stratigraphy, which gave the dimension of time to the science. Accordingly, we have outlined a little of the development of stratigraphy instead of leaving it entirely for a later course in historical geology.

We are indebted for assistance in the preparation of this book to many persons, only a few of whom can be mentioned here. The contribution of Robert R. Compton goes far beyond that indicated on the title page; in addition to preparing the illustrations, he wrote one chapter of the book and acted as critic on all the others.

The staff of W. H. Freeman and Company gave its unfailing help and encouragement and relieved us of many bothersome details.

Special thanks are due our colleagues S. E. Clabaugh, John Shelton, George A. Thompson, Roger Revelle, Walter Munk, John C. Crowell, Arthur D. Howard, C. Melvin Swinney, Robert Sharp, D. I. Axelrod, W. C. Putnam, Cordell Durrell, George Tunell, M. N. Bramlette, and George Bellemin, who have read certain chapters and have generously aided us with constructive criticism and new ideas.

Specific credit for illustrative material is given in the captions for individual figures. More generally, we wish to acknowledge here the kindness of the U. S. Geological Survey, the Geological Survey of Canada, and the U. S. Air Force for opening photographic files to us. Individuals who also allowed us to make selections from large photographic collections include Eliot Blackwelder, Robert C. Frampton, Howard A. Coombs, John Shelton, and Arch Addington.

Miss Margaret Ellis and Mrs. Priscilla Feigan typed the manuscript and helped in other ways.

Finally, the three of us are greatly indebted to our families who patiently served as "guinea pigs" for our ideas and as good-humored critics of our literary eccentricities.

December 23, 1950   JAMES GILLULY

                                                        AARON C. WATERS

                                                        A. O. WOODFORD

Contents

Appendixes

1. *Introduction*

The Earth's Riddles

SINCE the dawn of civilization, men have been filled with curiosity about the earth on which they live. Why does a volcano erupt? What makes rain? What force causes the earthquake? What is the source of the water that bubbles up in a spring?

As man's curiosity led him to seek solutions to such riddles, he often found that he was faced with new ones even more baffling. How did sea shells become entombed in the rocks of high mountain ranges? Why does one stream have quicksand on one bank and solid rock on the other? What controls the beautiful geometric forms of snowflakes and other crystals? Why does one well yield water in abundance, whereas another dug to the same depth is dry?

It would be interesting to know how early man attempted to solve such riddles. By comparison, we of today have signposts along the way, for reasoning man has built up the method of investigation and the compilation of knowledge that is known as *geology*—the science of the earth.

To some of earth's riddles, geologists have won the final answers. To others, the suggested answers are still tentative; and to still others, only the faintest glimmerings of light that may ultimately illuminate the way to final solutions have, so far, been discovered. Progress in geology has not been at a uniform rate. There have been periods, following fundamental discoveries, when outbursts of fruitful activity quickly revolutionized some of geology's theories and methods. At other periods there has been little advance. At times geologists even followed the wrong trail, and progress in some branches of the science came to a dead end. Then more information and new skills were acquired until, finally, the accumulated dogmas were overthrown and a new start made.

The first roots of geologic knowledge are lost in antiquity. The early Greeks and some of the peoples of other early civilizations made progress in geologic study, but their ideas were largely based on untested speculations, and little has survived. The modern science known as geology is of comparatively recent origin—the word itself is less than 200 years old.

Despite its youth, however, geology has already done much to stimulate and unshackle the thinking of mankind. The demonstration that sea shells and other fossils* entombed in the rocks are but the remains of animals and plants that lived in the geologic past routed dogmas that had warped men's thinking for centuries. From the detailed study of the biological relationships of living and fossil organisms, coupled with geologic investigation of the sequence and changes of fossil assemblages with time, the doctrine of evolution emerged. This doctrine has profoundly influenced modern philosophical and scientific thought.

Evidence, well documented, that the landscapes about us are not static but are slowly changing, has not failed to stimulate the imagination of thinking men. The wheat farmer tilling the cold, wind-swept plains of Alberta is curious about the shells turned up by his plow, and becomes amazed when told that scientific comparison of these shells with living marine organisms shows that his farm was once the bottom of a warm, shallow sea.

Who can deny the thrill that comes with the realization that less than 20,000 years ago the site of Chicago lay under a sheet of ice such as enshrouds Antarctica today? Or that the green, well-watered hills of Scotland's Midland Valley were once the site of shifting sand dunes similar to those of the modern Sahara? Yet, preserved in the rocks and soils along Lake Michigan's shore and in sandstone quarries near the city of Glasgow are the proofs —as clearly recorded as are the deliberations of the Roman Senate preserved in the writings of Seneca and Cicero.

The science of geology has brought to mankind new conceptions of time, just as astronomy has revolutionized ideas of space. The rocks record events, some of which date back at least 1,800 million years, and throw into sharp perspective the short lapse of human history, as compared with that of the earth as a whole. There is a fascination in reading from the rocks the evidence on which we may reconstruct events of millions of years ago. It is this fascination that has led men to develop the science of geology—the attraction that will cause them to undertake deeper exploration of the earth's riddles.

Minerals, Wealth, and Politics

Man's interest in the minerals and rocks of the earth's crust ceased long ago to be that of mere curiosity. There are sound practical reasons for his inves-

---

* Fossils are the remains or imprints of animals and plants of the geologic past, naturally preserved by burial under sediments.

tigations. Our modern civilization makes many uses of the minerals and rocks that compose the earth's crust. Industry is almost wholly dependent on them. From minerals we obtain the iron, copper, aluminum, and other metals that make an industrial civilization possible. Our chief sources of power are the mineral fuels, coal and petroleum. In recent years, we have learned how to release stupendous amounts of energy from radioactive minerals.

Even our individual desires and needs are closely tied to the mineral industries. The bricks in our houses, the salt that seasons our food, the material that paves our highways, the gold and silver ornaments and precious stones with which we adorn ourselves—all have been won from mineral deposits in the earth's crust. Man's avid search for the gold and silver, the copper and gem stones that pleased his vanity and brought him security and wealth began early in the annals of civilization. With possession of the minerals, he sought to refine and improve them and to discover new uses to which they could be put. As a result, the arts and crafts in metal and stone were born; and these, in turn, expanded into the vast industries we know today.

On the international scene, the power and wealth of a nation is largely determined by its supplies of useful minerals, its authority over the areas that contain them, and its skill in discovering and utilizing them. In this age of political readjustment between nations, we know that the vast accumulation of petroleum in Iran and Arabia is a potent force in world politics. We shall be wiser in world affairs if we know how petroleum occurs, how it is discovered, and how its quantity may be estimated.

Without the economic urge to find and exploit the mineral wealth hidden in the earth, many of the great forward steps in geology would never have been made; for geology is the science of the mine and the quarry, of the oil field and the placer.

The Study of Geology

Although geology is a complex and varied subject, it is also a stimulating and interesting one. Relatively few of its problems are so simple that they can be solved directly by one method of approach. Many even require supplementary investigations with the techniques of other sciences. Geologists are constantly taking over from chemistry, biology, physics, and engineering new methods, data, and theories that can be adapted to their needs. Geologists, in turn, have contributed data and ideas to these bordering sciences. Progress in one science advances all the others.

Because of the complexity of its problems, geology has not advanced so far as has physics or mathematics. The geologist cannot move a volcano to the laboratory to observe the growth of its cone, nor can he spread a bed of coal on the laboratory table to watch for millions of years its development. Yet, these are the simpler phenomena of geology. Factors of size and time make experimental study of many geologic processes difficult and often impossible. It is not possible to put a lava flow in a calorimeter and measure its output of energy. Faced with these apparently insurmountable difficulties, geologists have had to devise ingenious, indirect methods for getting the answers to many of their questions. Despite these inherent difficulties in subject matter, however, geologists have been outstandingly successful in predicting where to drill for oil or other mineral deposits, and in arriving at verifiable solutions of complex scientific problems.

Geologists, if they would be successful, must develop resourcefulness and imagination. They must be able to make sound decisions on the basis of incomplete, and even partially conflicting, data. In deciding where to drill an oil well or where to develop a gold placer, the geologist must, in many instances, evaluate and coordinate several kinds of evidence. His fundamental guides, of course, are the data from geologic mapping and other geologic techniques. He may also need to consider results from geophysical exploration, data regarding production of other wells or placers, and miscellaneous additional evidence drawn from engineering, economics, chemistry, physics, and many other sources.

These very factors of complexity and diversity, together with the newness of the science, combine to make geology a vigorous, rapidly expanding field. A student who selects geology as his profession has a wide choice of what he will learn and do. For his first two years of training, he will study more chemistry, physics, and engineering than geology. A sound elementary knowledge of these basic sciences is essential for many advanced geology courses. The student will learn something of ordinary laboratory techniques, but will soon find that his main laboratory is not a building lined with bottle-filled shelves and machinery. The geologist's laboratory is the bold cliffs of high mountain peaks, the walls of deep canyons, and the slopes of desert ranges. A part of his education will be spent in strenuous hiking and climbing in some mountainous area, perhaps far from civilization, where he will map the rocks and their structures and collect other geological data. Such "field work" is essential to geological training.

Upon completion of his training, the geologist will find many opportunities open to him. He may work for an oil company and travel the earth in search of new petroleum deposits. He may direct exploration to find new bodies of ore in a mine. He may have the responsibility of estimating accurately the reserves of ore in the ground beneath a mining property or the amount of oil that can be recovered from a partially developed oil field. He may be called upon to decide which of several small mines is the best prospect for development and investment.

As an employee of a federal or state geological survey, the geologist may map rocks and mineral deposits, investigate conservation problems such as soil erosion and mineral depletion, or classify public lands as to their minerals, soils, water, and other natural resources.

Or the geologist may teach at a college or university, training future geologists, and at the same time, engaging in efforts to discover new principles or to unify and correlate old ones. Other opportunities for research are open in government work, in various research institutes, and in industrial laboratories.

In time of war, the geologist can serve by giving authentic information on problems of terrain, by discovering and assisting to develop critically short mineral supplies, and by selecting targets in enemy territory which, when demolished, will put an end to some vital industry of the enemy nation. He may sit at peace conferences and advise on the mineral resources of various nations and their resulting industrial potential.

There are also opportunities in commerce for the geologist. His geological knowledge can be used to good purpose in the development of a cement plant, or in the operation of a stone quarry, a brick yard, or a sand-and-gravel pit.

Whatever path the geologist takes, it is likely to lead to widespread travel, for the whole earth is the field for his investigations.

The Branches of Geology

Geology is such a large and varied field that only the briefest résumé of its subject matter can be outlined here. In general, this book emphasizes *physical geology* (also called *dynamical geology*). Physical geology is concerned with the physical processes that operate on and within the earth—the processes that have given the rocks of the earth's crust their composition and structure, and the forces that have shaped the landscapes we see on its surface. Many separate geologic sciences contribute to the broad field of physical geology. Among the more important are *mineralogy,* the science of minerals; *petrology,* the science of rocks; *geodesy,* which is concerned with measuring the form and size of the earth; *structural geology,* which seeks to interpret the structures to be seen in the rocks in terms of the dynamical forces that have produced them; and *geomorphology,* which deals with the origin of landscapes and with the changes that are constantly occurring in them.

Touched much more lightly in this book is the broad field of *historical geology*—the science that traces the evolution and development of the earth and of its animal and plant inhabitants with time. Historical geology is based, first of all, upon physical geology, but also draws extensively upon *paleontology*, the science that deals with the study of animals and plants of the geologic past, and on *stratigraphy*, the science that is concerned primarily with the order and sequence of the rocks that make up the earth's crust.

Another great field considered only briefly in this book is *economic geology*—the application of the science of geology to the uses of man, as illustrated in the finding and recovery of valuable mineral deposits from the rocks of the crust, the search for and recovery of oil and water from wells, or the study of the depletion of soils by erosion.

The subdivisions we have enumerated are not independent sciences. For example, physical geology could never have attained its present development without the concurrent progress in paleontology. In a broad view of geologic science practically every branch contributes in some measure to all the others.

2. The Earth's Broad Pattern

To BEGIN a study of the earth, we shall need to know something about its dimensions and form. How large is it? What is its shape? How high are its mountains, and how deep are its seas? What is the pattern of its rivers and ridges? And by what methods have men tried to find the answers to such questions?

The Plumb Bob

Some of the answers may be gained through observations made with the aid of an ordinary plumb bob. The plumb bob is one of the oldest and most useful instruments of civilization. The early Egyptians used it in recovering land boundaries after the annual floods of the Nile. Today, just as did our ancestors, we use plumb bobs in all surveying; in the construction of buildings, bridges, roads, and tunnels; in the making of all maps; and in measuring the size and shape of the earth.

The Earth's Gross Size and Shape

Early Measurements

The ancient Greeks noticed that the earth casts a round shadow upon the moon during eclipses. They also observed that the surface of any large body of water is curved, for only the upper part of a ship's mast is visible at a distance, and the ship appears to rise gradually out of the sea as it approaches. From these observations, the Greeks correctly inferred that the earth is, roughly, spherical. It is much easier for us to arrive at the same

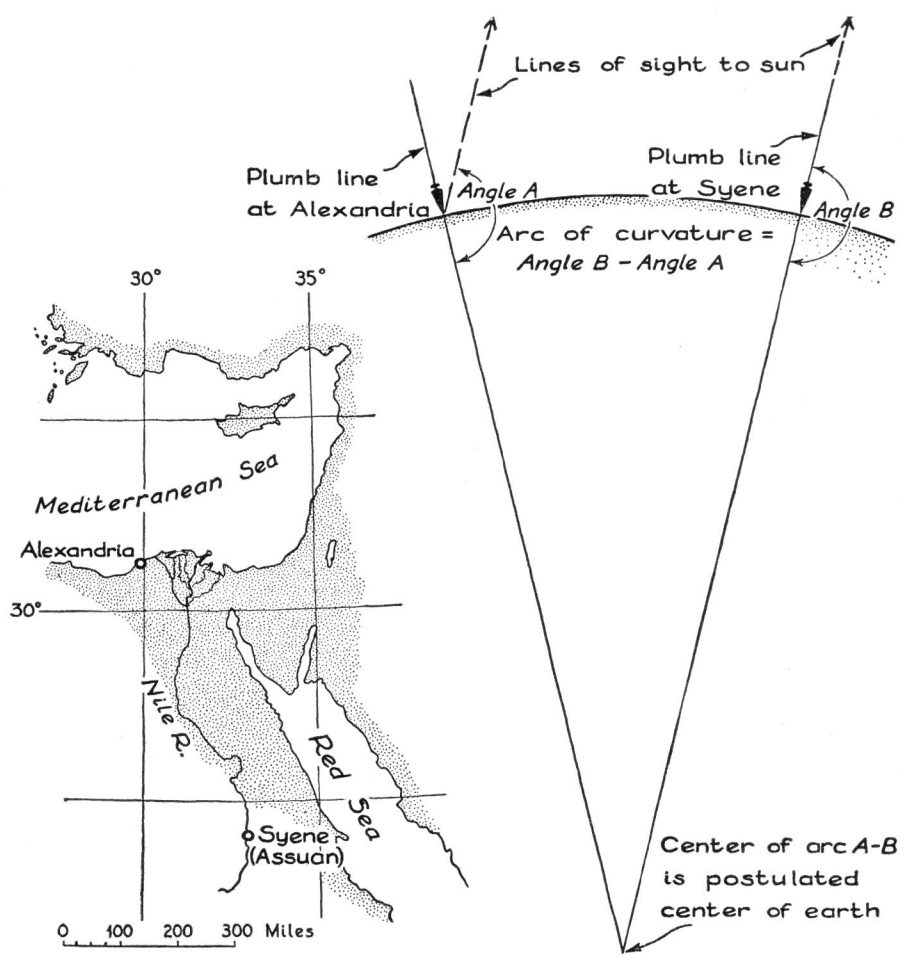

FIGURE 2-1. *Eratosthenes' method for measuring the size of the earth. Note that Syene does not lie due south of Alexandria, so that the distance between Syene and Alexandria was not measured along a meridian as he assumed. His result was therefore too large.*

conclusion today. In modern airplanes, we can circle the globe in a few days. The curvature of the earth can be seen plainly in photographs taken from stratosphere balloons and rockets.

A few of the early Greeks made even more penetrating interpretations. More than 2,000 years ago, Eratosthenes, a Greek geometer and astronomer, first measured the curvature of the earth's surface. Then, assuming the earth to be spherical, he computed its dimensions. Though measuring techniques have been greatly refined, his reasoning is still used in modern geodesy.

Eratosthenes learned that in southern Egypt, at Syene (now Assuan), the sun shines vertically down a well only at noon on the longest day of the year. On such a day, he measured the angle between the plumb line and the edge of the shadow cast by the sun at noon in a well at Alexandria (Fig. 2-1). Alexandria lies 5,000 stades (the ancient Egyptian stade equals about 600 feet) north of Syene. On the premises—

a. that the sun is so distant that its rays to Syene and Alexandria are parallel;

b. that Alexandria lies due north of Syene so that a plane through Alexandria, Syene, and the center of the earth also includes the noon sun;

c. that the plumb line points directly toward the center of the earth; and

d. that the earth is a sphere—

the angle between the plumb line and the shadow at Alexandria is equal to the arc of the earth's curvature between the two points (Fig. 2-1). On these assumptions, the circumference of the earth is given by solving the equation:

$$\text{Circumference} = \frac{360°}{\text{Angle of sun's rays to vertical at Alexandria}} \times 5{,}000 \text{ stades}$$

Eratosthenes' earth was too large, but only 14 per cent larger than the figures now accepted. About a century later, Poseidonius, the Greek philosopher, applied the same method to another arc but was not so favored with compensating errors. The size of the earth deduced from his measurement was a quarter too small, and led to Columbus' error in mistaking America for India.

We shall see how, with refined modern measurements, the simple assumption that the earth is a sphere—which was satisfied by the early crude measurements—has had to be successively refined as succeeding measurements have become more and more accurate. It would, indeed, be difficult to find a better example of scientific method, and of successive changes in theory made necessary by improved observations, than is afforded by the history of investigation of the figure of the earth.

Modern Measurements

In the Seventeenth and Eighteenth centuries, as the expansion of navigation and the accurate surveying of land boundaries became more important, Eratosthenes' method came into wider use. Arcs of meridian (North-South lines), equivalent in length to one degree of geographic latitude or to some simple fraction thereof, were measured at many different localities. It was found that a degree of latitude is longer near the poles than at the equator. Or, stated in another way, if we hold to the assumption that the earth is a

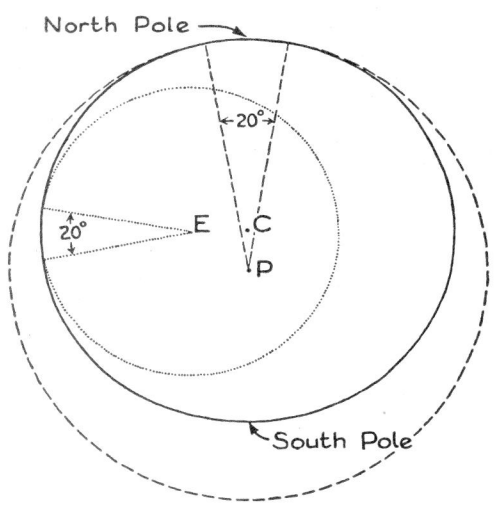

FIGURE 2-2. *Diagram showing how unequal circular arcs measured in different latitudes require an ellipsoidal figure for the earth.*

sphere, its radius, as determined by observations near the pole, is notably greater than that deduced from observations near the equator. This is shown greatly exaggerated in Figure 2-2. Here the dashed circle, with the center P, illustrates the size of the earth based on observations made near the pole. The dotted circle, with the letter E at its center, illustrates the size as determined from observations made near the equator. Note the discrepancy in size. These measurements indicate that the earth is not a perfect sphere. The inconsistency between the polar and equatorial measurements vanishes if we modify our assumption that the earth is a sphere and assume instead that it is slightly flattened at the poles—or, in technical words, that the earth is an ellipsoid. The solid line in Figure 2-2 is an ellipse with its center at C. This line fits the data from both the polar and the equatorial observations.

Thus, the simplest model of the earth that conforms with modern measurements is an oblate ellipsoid—the solid figure that is obtained by revolving an ellipse about its shorter axis. In the ellipse indicated by a solid line in Figure 2-2, the short axis would be a straight line connecting the North Pole with the South Pole. In the drawing, the flattening at the poles is greatly exaggerated, for the earth actually does not depart greatly from a sphere. The measurements now accepted and used internationally as the basis for official mapping, are:

Equatorial radius	6,378,388 meters	(3,963.5 miles)
Polar radius	6,356,912 meters	(3,950.2 miles)
Difference	21,476 meters	(13.3 miles)

In determining the size of the earth, Eratosthenes did not concern himself with mountains and other irregularities of its surface. He regarded the earth as a smooth-surfaced sphere. Actually, this is a reasonably valid assumption, for the irregularities of the earth's surface in comparison to its size are not so great as those on the surface of a billiard ball. But to a Tibetan, living in the shadow of towering Mount Everest, such a comparison must seem a great oversimplification.

How far does the earth's surface actually depart from perfect smoothness? To answer this question, we must find the difference in height between its highest and lowest points. This requires a kind of surveying different from that used by Eratosthenes, for now we are concerned with measuring the differences in altitude between various points on the earth's surface. The results of such surveys have been assembled in the form of *topographic* and *hydrographic* maps.

Maps and Mapping

Maps are the shorthand summary used by the student of the earth in presenting his data and observations. The earth's crust is complex. The intricate patterns of land and water, the forms of hills and valleys, the labyrinths that men have dug in mining are all so complicated in form that a true picture of them cannot be given by words alone. A map, however, condenses in intelligible form the findings regarding them.

Map Scales

A blueprint of a machine part or a dress pattern may be thought of as a map. Most of these are drawn as *"full-scale" maps. An inch on such a map corresponds with an inch on the object it represents.*

Few geographic and geologic maps, however, are full size. Most of them are *scale drawings.* In such *"reduced-scale" maps, an inch on the map may correspond to 10 inches, 1,000 inches, 1,000,000 inches, or whatever unit of reduction the map maker deems desirable* to show the features he wishes to portray. If he decides to reduce the length of objects on the map to 1/10 of their true length on the ground, he plots on a "1/10 scale." The fraction is the ratio of reduction and simply means that 1 inch on the map equals 10 inches on the ground. Many of the newer maps of the *Topographic Atlas of the United States* are drawn on a scale of 1/24,000—1 inch on the map corresponds with 24,000 inches or 2,000 feet on the ground.

Limitations of Maps

On a full-scale drawing, it is possible to show, for example, the head of a nail 1/10 inch across in full size. If the nail were to be correctly represented on a 1/200 scale, however, it would have to be drawn as only 1/200 of 1/10, or 1/2,000 of an inch across. Such a point is too small to be visible. Thus, if such nails were to be shown at all on the 1/200 scale, they would have to be shown diagrammatically. Their positions might be indicated, but

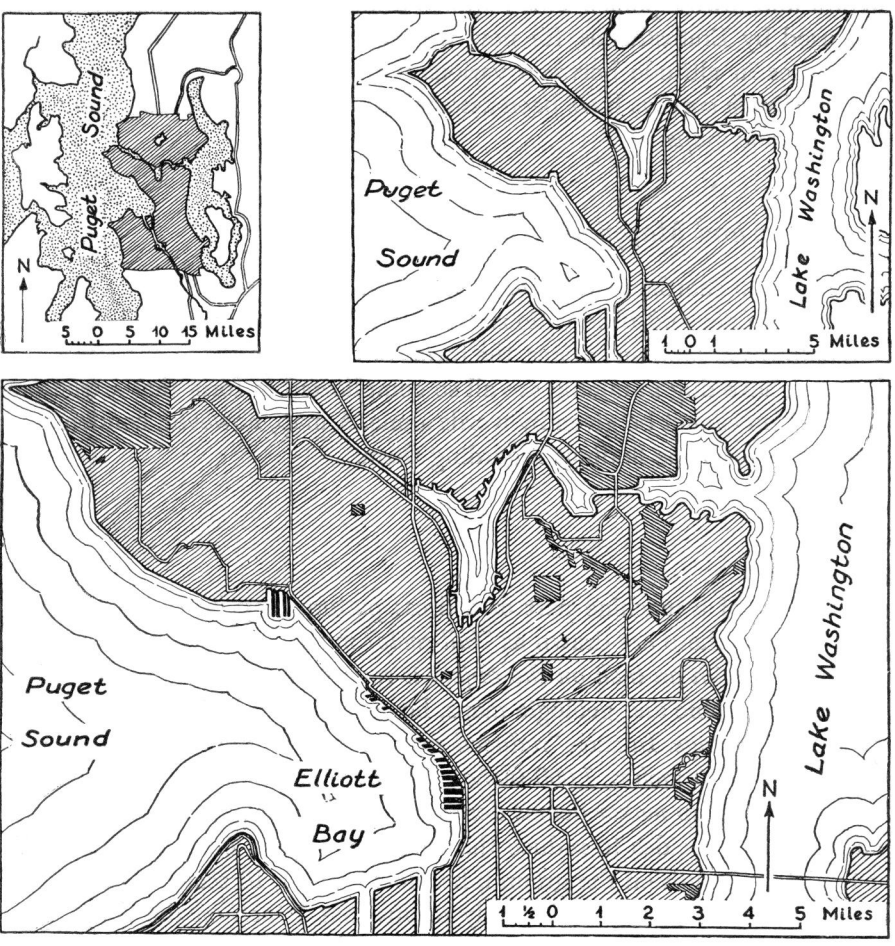

FIGURE 2-3.   *Maps of Seattle Harbor on different scales. Note that the smaller scale maps show much less detail, though all contain about the same number of lines per square inch. (After maps of U. S. Geological Survey and Chamber of Commerce of Seattle.)*

their size must be greatly exaggerated if they are to be seen. This limitation of reduced-scale maps must constantly be remembered by the map user.

All maps are generalizations, drawn to perform a particular service or function. Therefore, all represent selections of data chosen to illustrate the particular purpose and are commonly exaggerated with respect to other data. A navigator's chart emphasizes the features useful to navigation; a good road map stresses highway junctions and, in doing so, may distort the distances between them.

The maps of Seattle Harbor shown in Figure 2-3 illustrate one effect of map scale. Details such as the docks in Elliott Bay can be shown only on the larger scale map.

Whatever the scale, limitations in drawing and printing make it almost impossible for maps to be more accurate than 1/100 inch in the location of points. It is difficult to make a legible pencil mark less than 1/100 inch

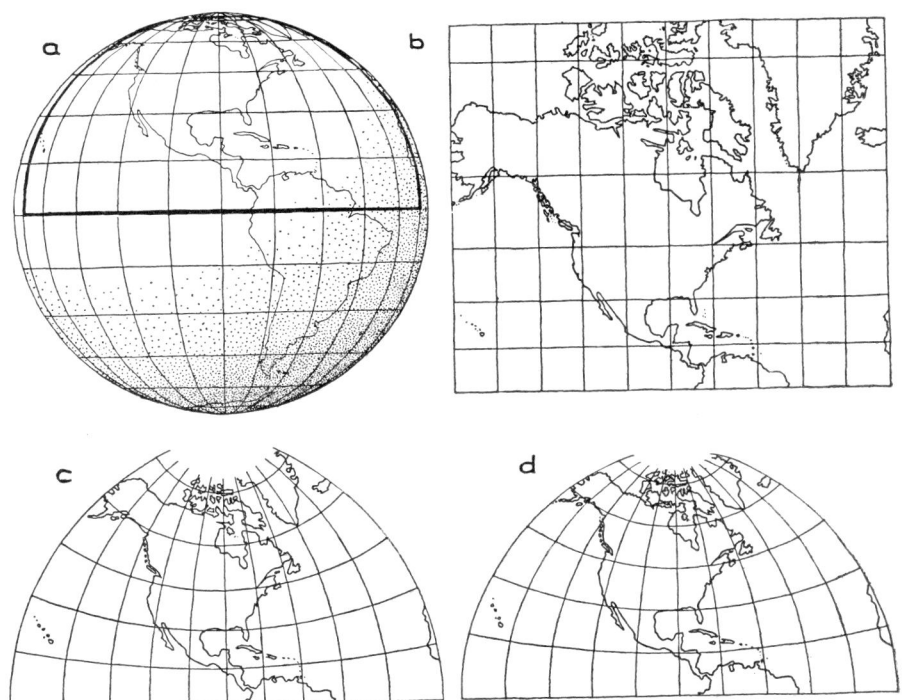

FIGURE 2-4. *The part of the world outlined in* a *is reproduced in three common map projections:* b, *the Mercator;* c, *the stereographic; and* d, *the polyconic. By comparing the size and shape of the longitude and latitude grid, it can be seen that each projection distorts the earth's surface to a varying degree —a necessary evil in transferring a curved surface to a flat one.*

across. A map of the United States reduced to a scale so that it could be printed on this page would be about 1/25,000,000; on it points less than 8 miles apart could not be shown as distinct without distortion. A thin line representing the Mississippi River on such a map would scale at least 4 miles in width.

Maps of the earth—the summaries of geographical knowledge—have still another limitation: They must depict, on a *flat* surface, the *curved* surface of the earth. It is impossible to do this without distorting the distances between points or the angles between intersecting lines. Most maps are compromises between these evils. Some of the distortions that result from various methods of plotting are shown in Figure 2-4.

Topographic Maps

The maps described thus far may be called *planimetric maps;* they show the relative positions of points but do not indicate their elevations above or below sea level. The *relief* of an area is the difference in altitude between the highest and lowest points within it. Although, by means of skillful shading, a so-called *relief map* may give an impression of the relative steepness of the slopes in an area, such a map cannot be used to determine accurately the actual differences in elevation between any two points. It is impossible to read height accurately from such a map.

To meet this difficulty, geodesists have devised *topographic maps,* which are designed to show the elevations as well as the positions of points. They portray the three-dimensional form of the land surface, its *topography.*

A topographic map depicts a three-dimensional surface—one having length, breadth, and varying height above a reference plane or *datum* (usually *mean sea level*)—on a two-dimensional piece of paper. On such a map, lines called *contours* are drawn to portray the intersections of the ground surface with a series of horizontal planes at definite intervals above (or below) the datum plane (Fig. 2-5). A convenient way of visualizing contours is to regard each of them as the shoreline if the level of the ocean should rise to exactly the height that the contour represents. The 100-foot contour line on a map, therefore, shows the position of the shore if sea level should rise 100 feet. Every point on it is exactly 100 feet above the present sea level. Thus, each contour shows the course of a level line of definite elevation as it would appear if traced across the country shown on the map.

In order to understand something about how topographic maps are made,

FIGURE 2-5. *Relief model and topographic (contour) map of the volcano Vesuvius. (After* Il Vesuvio *sheet, Inst. Geografico Militaire.)*

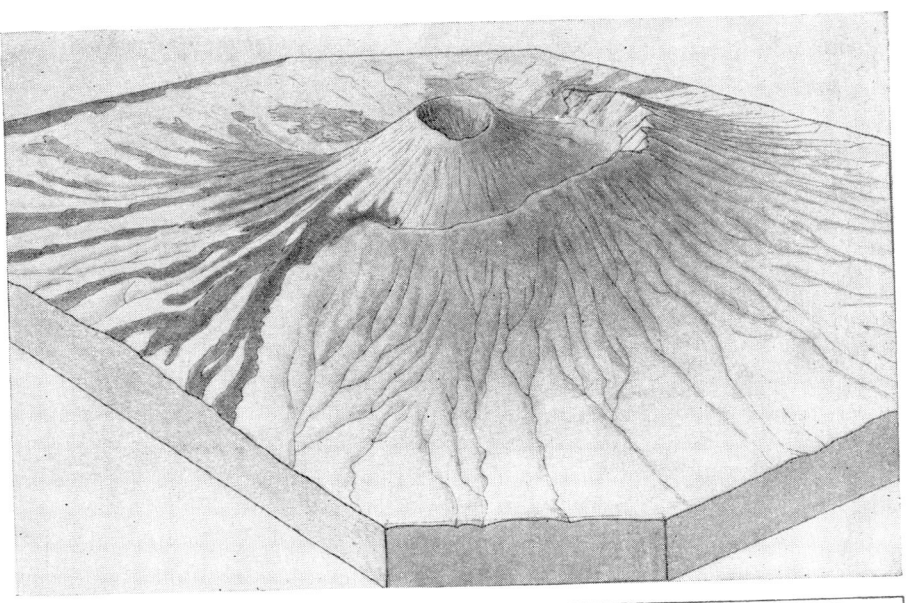

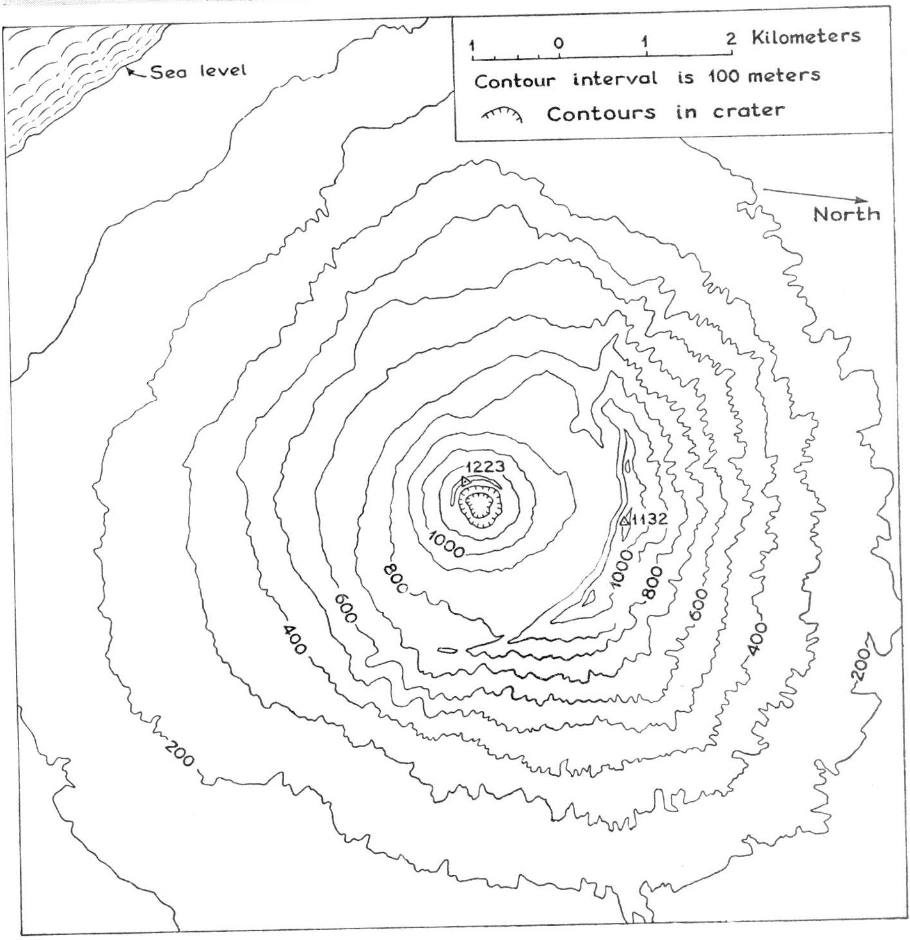

1 0 1 2 Kilometers
Contour interval is 100 meters
Contours in crater
Sea level
North
1223
1132
1000
800
600
400
200
1000
800
600
400
200

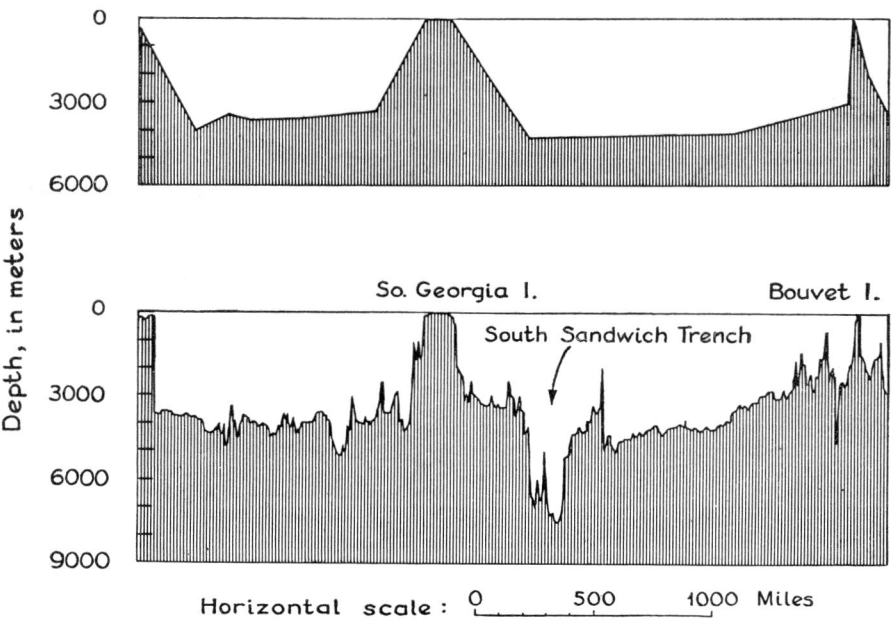

FIGURE 2-6.   *Two profiles of the bottom of the Atlantic near the Southern Antilles.*
The upper *profile shows the interpretation based on a few wire soundings;*
the lower *that measured by sonic soundings. Note not only the greater*
detail but that the great South Sandwich Trench was completely missed by
the wire soundings. (After Stocks *and* Wüst, *redrawn from H. U. Sverdrup,*
M. W. Johnson, and R. H. Fleming, The Oceans, *Copyright 1942 by*
Prentice-Hall, Inc.)

the student is referred to Appendix I on "Techniques of Topographic Mapping," especially to the discussion of triangulation.

Hydrographic Maps

Hydrographic maps do for sea areas what topographic maps do for land areas. They not only depict the outlines of the water bodies but, by soundings (measurements of depth made by vessels at sea), they also show something of the topography of the bottom.

Although a few of the major irregularities of the ocean bottom became known during the Nineteenth Century, it was generally believed that the ocean floors were relatively much smoother than the surface of the land. With the advent of sonic methods of sounding (see Appendix I), however, it soon became evident that at least a few parts of the ocean floors are highly

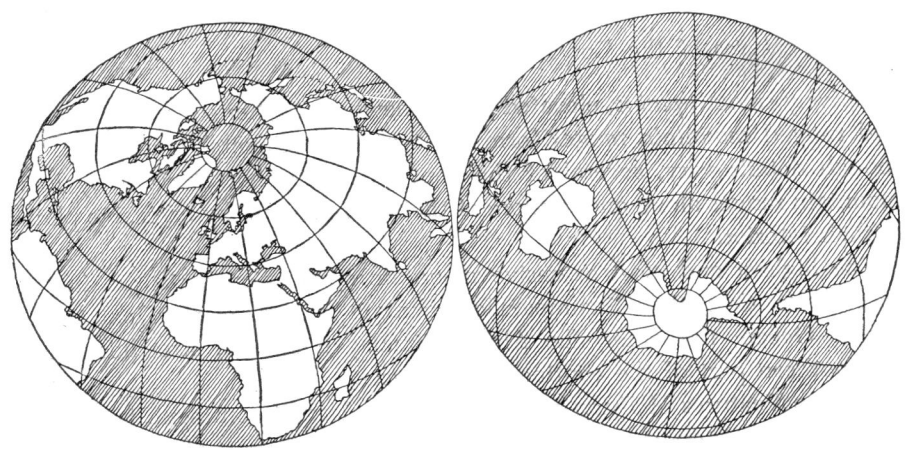

FIGURE 2-7. *The land and water hemispheres.*

irregular. Figure 2-6 shows the change in the interpretation of the bottom configuration of an area in the South Atlantic with the added data from sonic sounding.

Irregularities of the Earth's Surface

Land and Sea Areas

The major irregularities of the earth's surface are, of course, the continents and the ocean basins. The maps shown in Figure 2-7 summarize our knowledge of the distribution of these major features. Perhaps the most obvious fact is that most of the earth's surface is covered by the ocean. Using the best maps and charts available, the areas of land and sea have been carefully determined as follows:

Area of the sea	361,059,000 sq. km.	70.8%
Area of the land	148,892,000 sq. km.	29.2%
Total	509,951,000 sq. km.	100.0%

The land and sea are not uniformly distributed over the earth's surface. The surface of the earth can be divided into two hemispheres (Fig. 2-7), one containing four-fifths of all the land, whereas the other is nearly nine-tenths covered by water.

Let us now look at the topography of the earth's surface as it is recorded on maps of land and sea.

Relief Features of the Continents

Relatively smooth *plains,* elevated and somewhat rougher surfaced *plateaus,* and rugged *mountain ranges* form the typical continental landscapes. Although details vary, several of the continents are characterized by lofty mountain ranges along one or more of their margins and by extensive plains in their interior.

The highest and most extensive mountain systems on earth are in two great belts. One, the Circumpacific mountain system, is marginal to the Pacific Ocean; the other, the Alpine-Himalayan system, stretches across the southern part of the continent of Eurasia.

Surface Features of Eurasia. Eurasia, the largest continent, is also the most diversified topographically. Europe, the western part of this vast continent, can be divided roughly into three great topographic divisions: (1) The northern highlands, which include rugged mountains rising 4,000 to 5,000 feet above sea level in the Scandinavian peninsula, and lower and more rounded ranges in the British Isles and northwestern France. (2) The great central plain, which includes the vast plains of Russia, and which extends southwestward along the Baltic and North seas across Germany and into France. (3) The Alpine system of mountain ranges and bordering highlands. This includes the lofty ranges of southern Europe, such as the Alps, Pyrenees, Apennines, Carpathians, and Caucasus, and also extensive areas of lower highlands that form the typical landscapes of Spain, central and northeastern France, Germany, Czechoslovakia, and Yugoslavia.

The mean altitude of Europe is slightly less than 1,000 feet. Its highest point is Mount Elbrus, which rises 18,465 feet above sea level; its lowest point is the Caspian Sea, whose surface is 86 feet below sea level.

Asia is much more diversified than Europe. It has both the highest elevation and the lowest depression of any continent. Mount Everest in the Himalayas towers 29,002 feet above the sea. The Dead Sea lies 1,294 feet below sea level.

Asia is wrinkled by scores of lofty mountain ranges. Many of them meet in the Pamir knot of northwestern India, which is often called the "Roof of the World." Among the great mountain chains that radiate from the Pamir knot are the Himalayas, Kuenlun, Hindu Kush, Karakorum, and Altyn Tagh ranges. Many other great mountain chains also diversify the face of Asia. Among the most prominent are the Elburz Mountains of Iran; the Tian Shan, Great Altai, and Khingan ranges, which stretch from west China northeastward into Mongolia and Siberia; and the prominent mountain chains that form the island arcs off the east coast of Asia, such as the Kurile, Japanese, Philippine, and East Indies archipelagos.

Between the great ranges of central Asia lie many high plateaus; among them are the Tibetan Plateau, the Tsaidan and Tarim desert basins, the great Plateau of Mongolia, and the Iranian Plateau.

Vast expanses of plain—the tundras and steppes of Asiatic Russia—occupy the area northwest of the central Asiatic plexus of mountain ranges. Other extensive plains form parts of China and India.

Surface Features of Africa. In general, Africa may be described as a vast plateau bordered in part by narrow coastal plains and diversified by numerous short, irregular mountain ranges. Unlike most of the continents, Africa has no long continuous mountain systems, although the Atlas chain, along the northwestern coast, can be followed with minor interruptions for nearly 1,500 miles. Most of this chain consists of low mountains, but a few peaks rise more than 13,000 feet above sea level.

Extending from Abyssinia southward to Lake Nyassa is a mixed group of irregular ranges and elevated plateaus diversified by two linear depressions —the *Rift Valleys* of east Africa. Lakes Tanganyika, Albert, and Edward lie in the western Rift Valley; the eastern Rift contains Lake Rudolf and part of Lake Victoria. The highest peaks in Africa (Mount Kilimanjaro, 19,710 feet; and Ruwenzori, 16,815 feet) are in this area.

The extreme southern and southwestern edge of Africa is bordered by the relatively low Cape Mountains.

Surface Features of North America. North America is mainly a broad central plain lying between marginal mountain systems. Toward the north this plain widens to include the plains of the Canadian prairie provinces and, on the northeast, the low rolling surface broken by inconspicuous mountains known as the Canadian Shield. Far to the north, in the islands of the Arctic, Ellesmere Land and Grant Land, stand other mountains, higher than the Appalachians but little known. The Appalachians, which border North America on the east, form a low but continuous series of small ranges and parallel ridges. The much wider and higher Cordilleran mountain system lies west of the central plain. In the United States, the Cordilleran system is nearly 1,000 miles wide and includes the Rocky Mountains, the Cascade-Sierra Nevada chain, and, near the Pacific, a series of lower and somewhat discontinuous Coast Ranges. Between these mountain chains lie broad plateaus. The principal ones are the Colorado Plateau west of the southern Rockies, the Columbia Plateau between the Cascades and the Rockies, and the Interior Plateau of British Columbia between the Rockies and the Coast Range.

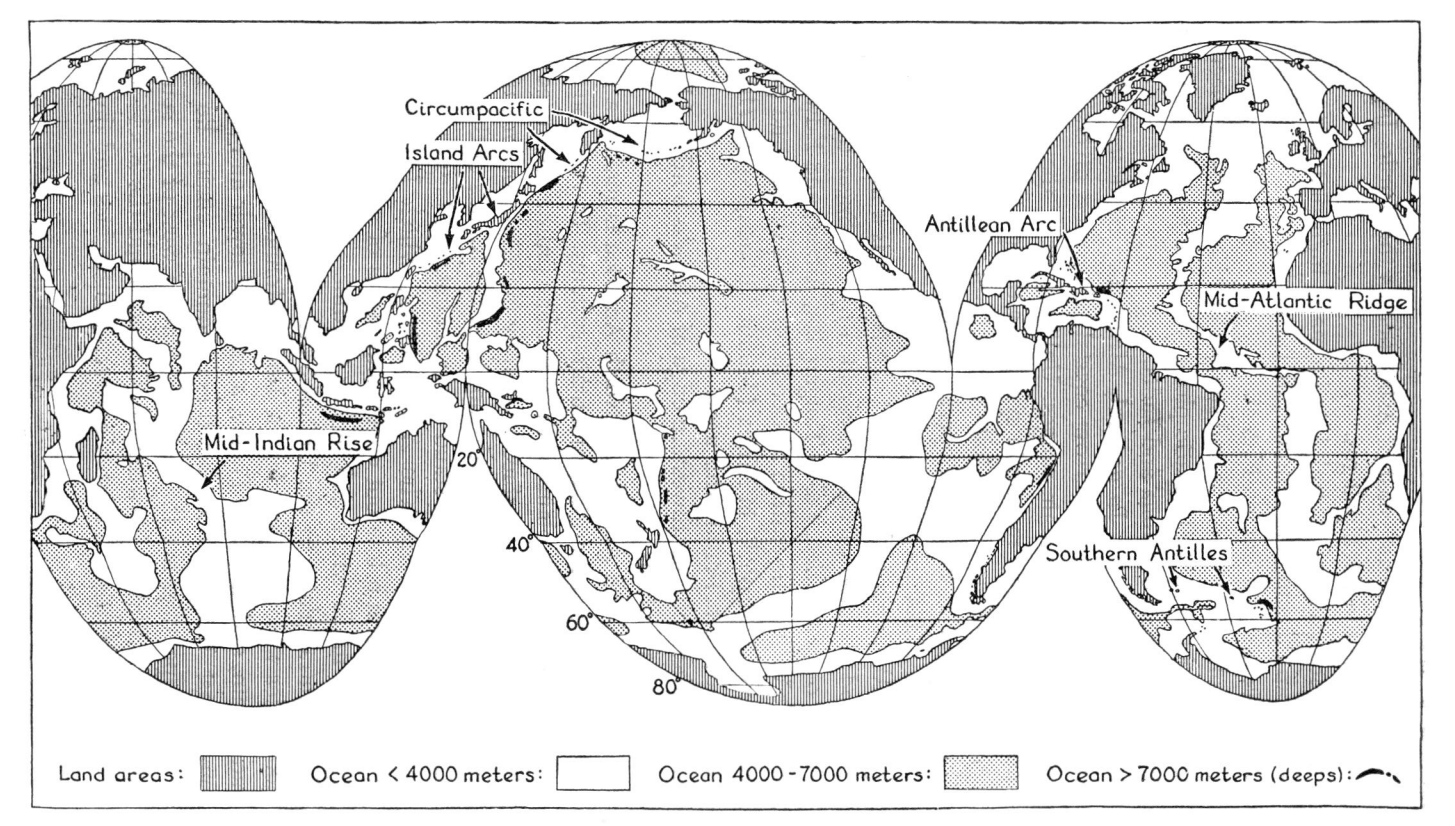

FIGURE 2-8. *Map of the world showing the major relief features of the ocean floor. Note the Mid-Atlantic Ridge, the Mid-Indian Rise, and the island arcs with their associated deeps. (From H. U. Sverdrup, M. W. Johnson, and R. H. Fleming, The Oceans, Copyright 1942 by Prentice-Hall, Inc.; Goode Base Map used by permission of the University of Chicago.)*

In Mexico, a high plateau, the Mesa Central, lies between two mountain ranges which merge into a high volcanic tableland south of Mexico City.

The highest point on the North American continent is Mount McKinley in Alaska, 20,300 feet, and the lowest is Death Valley in California, 276 feet below sea level. The mean elevation of the continent is approximately 1,300 feet.

Surface Features of South America. South America is similar in topography to North America. A long and high mountain system culminating in the Andes forms the continent's western margin. On the east, from southern Brazil northward, a low and irregular series of highlands lies near the Atlantic Coast. The interior of the continent between these two mountain systems is a vast plain which, in Argentina, extends unbroken to the shores of the Atlantic.

The Andes are surpassed in height only by the Himalayas. Several of the Andean peaks rise more than 21,000 feet above the sea; the highest is the Argentine volcano, Aconcagua, 22,867 feet above sea level. At their northern end, the Andes split into three distinct chains. The easternmost extends into Venezuela and thence connects with the island arcs of the West Indies.

Between the parallel ranges of the Andes lie high intermontane plateaus. The Pacific coastline is bordered, especially in the southern part of the continent, by a low and rather discontinuous Coast Range.

Surface Features of Australia. Australia is a broad, low plateau. It is diversified on the east by the Dividing Mountains, a low series of ranges extending from northern Queensland to Tasmania. Other low ranges are found north of Adelaide and in parts of western Australia, but most of the continent has little contrast in elevation. The highest peak is Mount Kosciusko in southeastern Australia, which rises only 7,323 feet above the sea.

Surface Features of Antarctica. The continent of Antarctica is only partially explored. It consists of a great interior dome of ice rising to elevations of 6,000 to 10,000 feet, through which project peaks of some of the higher mountain ranges. Most of the continent appears to be an irregular plateau buried beneath glaciers. Mount Erebus, a volcano overlooking the Ross Sea, rises to a height of over 13,000 feet.

Relief Features of the Ocean Floors

The outstanding relief features of the ocean floor are portrayed in Figure 2-8. Just as the continents are diversified with mountains, plateaus, and plains, the sea floor also has its larger topographic features. Among these are the *continental shelves* and *slopes*.

The continental shelves are the submerged edge of the continents. Off most coastlines the bottom descends gradually from the shore to a depth of about 200 meters, or perhaps somewhat less, before breaking off abruptly to the much steeper continental slope. The shallow continental shelves include more than 7 per cent of all the marine areas of the earth. They range in width from zero to as much as 800 miles (off the Siberian coast in the Arctic Ocean), and average about 30 miles in width for all coastlines. The surface of the continental shelves thus has an average seaward slope of about 20 feet per mile, though most of them are somewhat irregular and contain local ridges and depressions.

At depths that may be only a few meters or as many as 300 meters, the gently sloping surface of the continental shelf changes abruptly to the *very much steeper continental slope*. Off mountainous coasts, the continental slope may descend abruptly to the ocean floor at the rate of 300 feet or more per mile. Off wide coastal plains, its slope is generally about half this figure.

The continental slope off many shores is cut by so-called "submarine canyons" (Fig. 16-29). Some of these are deeper and more precipitous than the Grand Canyon of the Colorado, and form notches in the slopes recognizable to depths as great as 12,000 feet or more. Some head off the mouths of major streams such as the Congo and Hudson; others, equally impressive, head at the shoreline far from the mouth of any stream. These perplexing features will be discussed in Chapter 16, Part 2, as their origin presents one of the most fascinating of geologic puzzles.

On the ocean floor itself, there is a diversity of topographic forms. Some are submarine mountains or other elevated tracts; some are depressions in the sea floor. Depressions may be grouped into the more-or-less rounded *basins,* the elongated *troughs* with gentle side slopes, and the similar but steeper-sided *trenches.* Any one of these depressions that has a floor more than 7,000 meters (23,000 feet) below sea level is called a *deep.*

Long, relatively narrow, elevated tracts are called *ridges;* or, if they are broader and larger, *rises.* Small, but significant submerged topographic eminences are the *sea mounts,* which are isolated steep-sided elevations; and the *banks,* which are shallower and broader, flat-topped elevations.

The Floor of the Atlantic Ocean.   One of the greatest mountain ranges on earth, the *Mid-Atlantic Ridge,* almost bisects the Atlantic Basin from north to south. Its southern portion faithfully reflects the great bend of the coasts of South America and Africa. Throughout its length, the summit of the Mid-Atlantic Ridge rises approximately 10,000 feet or more above the floor of the Atlantic on either side. Most of its summit is less than 3,000 meters below sea level, and a few peaks—among them the Azores, Saint Paul's Rocks off the Brazilian coast, Ascension, Saint Helena, Tristan da Cunha, and Bouvet—project as islands above the surface of the sea.

The Walfisch Ridge, a range as high as the Alps and three times as long, branches from the Mid-Atlantic Ridge near its southern end and joins it to the west coast of Africa. The Rio Grande Rise, a lower and more gently sloping range, lies between the Mid-Atlantic Ridge and the coast of southern Brazil. In the North Atlantic, the Mid-Atlantic Ridge merges with the generally shallow bottom of the Norwegian Sea between Scandinavia and Greenland.

Two striking mountain arcs are present in the Atlantic. The Antillean Arc curves through the islands of the Lesser Antilles from the Venezuelan coast to the Florida shelf and separates the Caribbean Sea and the Gulf of Mexico from the Atlantic Basin proper. In the South Atlantic, the arc of the Southern Antilles is, similarly, convex eastward. It joins southern South America to the Antarctic continent through the South Georgia, South Hebrides, and South Sandwich islands. The only deeps in the Atlantic—the Puerto Rico Trough and the South Sandwich Trench, each more than 8,000 meters deep —lie on the convex sides of these arcs.

The Floor of the Indian Ocean. The Mid-Indian Rise, extending from India to Antarctica, resembles the Mid-Atlantic Ridge in lying medially in the ocean. It is broader and not so sharply marked off from the adjacent basins. Only about half its length reaches within 3,000 meters of the surface. West of the Mid-Indian Rise lie several other ridges, some of which branch from it. Others extend southeastward from the African coast; Madagascar lies on one of these. South of Australia, the Indian and Pacific basins are separated by a moderate rise that joins Australia to Antarctica west of the Ross Sea. The area between Australia and southeastern Asia contains generally shallow seas, but the complex bottom topography of the East Indies includes several deep but relatively small basins and narrow trenches. This area also contains many curving island arcs, the archipelagos of the East Indies. The Sunda Trench, south of Java, is the only deep in the Indian Ocean. Here, as in the Atlantic, the deep is on the convex side of an island arc.

The Floor of the Pacific Ocean. The Pacific Ocean floor is not well known. Apparently much of it is relatively flat. A sprawling, poorly defined rise extends from Antarctica to the coast of South America and across the southeastern Pacific to Central America. A considerably shallower rise between New Zealand and Australia reaches far to the northwest off the east coast of Australia and the north coast of New Guinea. This rise is notable for its steep eastern slope and for bordering several more-or-less isolated deep basins. East of the Tonga and Kermadec islands, it drops off abruptly to the Tonga Deep (more than 9,000 meters). Another great deep (9,400 meters), south of New Britain, is exceptional in that it lies on the concave side of the arc that connects New Britain and Bougainville islands. The Hawaiian Islands are perched on a rather straight ridge, trending northwest. A steep-sided ridge almost at right angles to the Hawaiian Ridge has been discovered recently. It crosses the Hawaiian Ridge northwest of the Hawaiian Islands. Northwest of Hawaii, the ocean floor appears highly irregular, though chiefly deeper than 5,000 meters.

The most striking features of Pacific Ocean topography are the great island arcs that festoon its northern and western sides. Among them are the Aleutian, Kurile, Japanese, Ryukyu, and Philippine arcs. These island-crowned arcuate ridges separate generally shallow seas—Bering, Okhotsk, Japan, Yellow, and East China seas—from the Pacific Basin proper. Several comparable arcs branch southward from Japan through the Bonin and Marianas islands. These are not bordered by shallow seas; the Philippine Basin, on their concave side, is as deep as much of the main Pacific.

As in the Atlantic and Indian oceans, great deeps are common on the convex sides of the Pacific arcs. Among them are the Aleutian, Kurile, Japan, Bonin, Philippine, Ryukyu, and Marianas trenches. The only other known deeps in the Pacific Ocean are next to the Andean arc off the Chilean coast (Atacama Deep, 7,635 meters) and between New Zealand and Antarctica (Byrd Deep, 8,590 meters).

The Floor of the Arctic Ocean.   The Arctic is, of course, the least known ocean. In a sense, it is a land-locked mass of water, for the greatest depth connecting it with the Atlantic is only 1,500 meters (between Greenland and Spitsbergen). The Bering Strait, its connection with the Pacific, is only 55 meters deep. The Arctic is notable for the very wide continental shelves that border Siberia, Europe, and northern Canada. The sparse soundings that have been made in the Arctic have not yet revealed any great deeps. The greatest depth reported, 5,440 meters, north of Alaska, may even be in error, as indicated by later soundings in approximately the same area.

The Earth's Maximum Relief

Our brief review of the topography of the continents and ocean basins permits some generalizations about the earth's relief. It is in the great ocean deeps of the western Pacific that we find the maximum departure of the solid earth from sea level. Here, depths of more than 30,000 feet have been recorded by soundings. The greatest depth so far known is in the Philippine Deep, which descends to more than 34,000 feet, or approximately 6.5 miles below sea level.

FIGURE 2-9. *Graph showing the percentage of the earth's surface lying between various levels above and below the sea. The bars at the left represent the percentages lying between the respective contours at intervals of 1,000 meters. Note the flattening of the respective contours at intervals of 4,000 and 5,000 meters depth. (Redrawn from H. U. Sverdrup, M. W. Johnson, and R. H. Fleming, The Oceans, Copyright 1942 by Prentice-Hall, Inc.)*

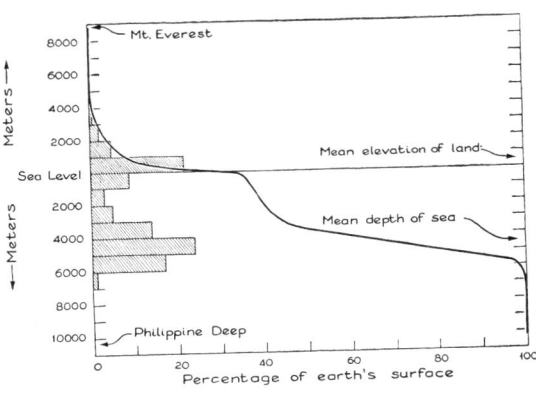

The greatest height of land above sea level is Mount Everest, which rises 29,002 feet or nearly 5.5 miles. Great as these distances are by human standards, they dwindle into insignificance when considered in relation to the radius of the earth. If the largest circle that can be drawn on this page represents the earth, a moderate weight pencil line would, on the same scale, include within its width all the irregularities of the earth's surface, ranging from Mount Everest to the Philippine Deep. Thus, we see that, in relation to its size, the earth actually is about as smooth as a billiard ball.

From the best maps and other data available, careful estimates have been made of the total areas of the land surface that lie between various altitude limits. These are summarized in the accompanying graph, Figure 2-9. This graph brings out two important facts: (1) The highest mountain peaks and the extreme abyssal depths of the sea are very small in area; (2) within the extreme range represented by the Philippine Deep and Mount Everest, the altitudes are not evenly distributed by area but there are two altitude ranges especially prominent. One is the range between 4,000 and 5,000 meters (13,-120 to 16,400 feet) below sea level, which embraces nearly one-fourth of the earth's surface. The other lies in the range from 200 meters below sea level to 500 meters above sea level and includes approximately one-fifth of the earth's surface. The submerged portion of this second level, extending from a depth of 200 meters to the sea coast, is the continental shelf. At its outer edge, the sea floor descends rapidly in the continental slope to the ocean floors proper, which lie at the first altitude range mentioned, approximately 4,000 to 5,000 meters below sea level.

What are the reasons for these two dominant levels in the architecture of the earth? Why is the boundary between continental shelf and ocean floor so marked? These questions are among the great problems of geology. Their answers, so far as answers are possible, cannot be given in a sentence, but they are among the principal themes of this book. A part of the answer will be suggested by a more detailed discussion of certain aspects of the earth's relief.

Facts, Concepts, Terms

THE EARTH'S GROSS FORM:
> Sphere? Oblate ellipsoid? Measuring an "arc of meridian"

MAPS AND MAP-MAKING
> Choosing the scale of a map
> Limitations of maps
> Topographic maps
>> Triangulation (see Appendix I)
>> Portraying relief by contours
> Hydrographic maps and charts
>> Sonic sounding (see Appendix I)

MAJOR RELIEF FEATURES OF CONTINENTS
> Plains, plateaus, mountain systems

MAJOR RELIEF FEATURES OF THE OCEANS
> Continental shelves, continental slopes, ocean floors
> Basins, troughs, trenches, deeps
> Ridges, rises, island arcs

RATIO OF EARTH'S MAJOR RELIEF TO EARTH'S RADIUS: 12 to 4,000 miles

Questions

1. How many lines of evidence can you adduce suggesting that the earth is roughly spherical and not, for example, shaped like a doughnut?
2. What assumptions underlie Eratosthenes' measurement of the earth? Neglecting observational errors, what are the reasons for doubting their validity?
3. If, along a single meridian, we imagine plumb lines to be suspended at latitudes 10° N, 30° N, 60° N, and 80° N, will their projections toward the interior of the earth meet at its center? At any single point? At points successively further from the points of observation as these are taken in pairs from low to high latitudes? Draw a diagram to support your statement.
4. Eratosthenes' reasoning is used in measuring geographic latitudes, which are defined as the angle between the plane of the equator and the plumb line at the point considered. A degree of geographic latitude is longer in high than

in low latitudes. What conclusion does this suggest regarding the shape of the earth?

5. If we imagine rays drawn from the center of the earth to a single meridian, say that of Greenwich, at equal central angles of 10°, how would the lengths in miles of meridianal arcs marked off by these rays compare in high and low latitudes?

6. The angles between such rays are angles of *geocentric latitude* (measured from the equator). Is a degree of geocentric latitude longer or shorter than a degree of geographic latitude near the poles?

7. What is a contour?

8. Figure 2-9 presents the relation between altitude and area of the earth's surface. What would you expect the shape of such a graph to be if the altitude distribution (between the present extremes) were completely random?

9. What broad features of the Atlantic bottom are not paralleled in the Pacific?

10. What aspects of the Antilles arc resemble those of the border of the western Pacific? How is this arc related to the Circumpacific mountain belt? What is its relation to the Alpine-Himalayan chain?

11. What features of the continental borders of the Atlantic differ from those of the Pacific, broadly considered?

12. How does the Pacific Basin differ from those of the Indian and Atlantic oceans?

13. What is the usual relation between oceanic deeps and island arcs?

Suggested Readings

1. Bucher, W. H., *The Deformation of the Earth's Crust*, Princeton University Press, Princeton, N. J., 1933. (An excellent outline of the earth's relief, including also a different interpretation of the graph of Fig. 2-9.)

2. Raisz, E. J., *General Cartography*, McGraw-Hill Book Company, New York, 1948.

3. Gravity, Isostasy, Strength

FROM OUR study of the earth's relief, we found that Mount Everest rises approximately twelve miles above the level of the Philippine Deep. Although twelve miles may seem insignificant in comparison with the 4,000-mile radius of the earth, nevertheless the irregularities of the earth's surface pose some fascinating problems. To one curious about the earth on which we live, several implications about the difference in level between its mountains and its ocean deeps are of lively interest.

An engineer knows that there are practical limits to the height of any building or tower. Even if we used the strongest rocks known, it would be quite impossible to construct a vertical tower as high as Mount Everest. The sheer weight of five and one-half miles of rock would crush the stone blocks at the base. Long before the tower could be built to such a height, it would come crashing down under its own weight. How, then, is Mount Everest supported at such a high elevation? Why does not the floor of the Pacific Ocean crush in and fill the Philippine Deep?

To gain some insight into how the great irregularities of the earth's surface are sustained, we must begin by studying the force that draws all objects toward the earth's center. This force is *gravity*, one manifestation of the universal force of gravitation.

Gravity

Water in a pond and coffee in a cup both assume a *level surface*. If we hang a plumb bob above either liquid, we find that *the plumb line is exactly perpendicular to the liquid's surface*. Both the plumb bob and the liquid are responding to the force of gravity. Each has approached as close to the center of the earth's mass as its nature will permit—the liquid by flattening out so that it is as near the earth as possible, the plumb bob by bringing its mass as close to the earth as the string from which it hangs permits. *The earth evidently has an attraction for each particle of the liquid, and for each particle of the plumb bob.*

The Law of Gravitation

The attraction of the earth for a plumb bob, for water, and for all other objects in the universe illustrates the universal *Law of Gravitation,* first formulated by the great English scientist Isaac Newton (1642-1727). This generalization is usually stated:

Every particle in the universe attracts every other particle with a force that is directly proportional to the product of their masses, and inversely proportional to the square of the distance between them.

The mathematical statement of this law is as follows:

$$F \propto (\text{read "varies as"}) \frac{M' \times M''}{D^2}$$

In this expression, F is the force of attraction, M' is the mass of one body, M'' that of a second body, and D the distance between the bodies.

As a numerical example of the way in which the law operates, consider the earth's attraction for a plumb bob that weighs one pound at sea level. At sea level, the distance of the plumb bob from the earth's center is approximately 4,000 miles. If this distance were doubled, the attractive force would be only one-fourth as great, for the denominator would be $8,000^2$ instead of $4,000^2$. In other words, if it were possible to suspend the plumb bob in space 8,000 miles from the earth's center (4,000 miles above the earth's surface), it would weigh only 1/4 pound.

Gravity and a Level Surface

In making a topographic map, or in setting the floor joists of a building, what do we mean when we speak of a level surface? The surface of a pond is level, and it appears to be a plane but, as we know, in large bodies of water such surfaces are curved. A level surface is thus not a flat plane, but a surface that is, at every point, exactly perpendicular to the direction of the plumb line at that point. The direction of the plumb line is the vertical—it defines the *zenith,* the point directly overhead.

What Eratosthenes did in measuring the size of the earth was to determine the curvature of the level surface of the earth by measuring the angle be-

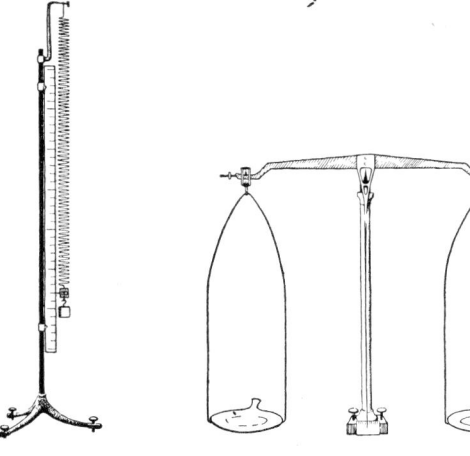

FIGURE 3-1. *Two common weighing instruments: spring scale,* left; *beam balance,* right.

tween two such vertical lines; and then, knowing the distance between the vertical lines at the surface, he calculated the size of the hypothetical, perfect sphere he considered the level surface to bound.

There is another method, involving careful *measurement of the exact force of gravity at various points on the earth's surface,* by which we can determine some interesting things about the size and form of the earth.

Weighing a Mass

When we measure the force of gravity, we are, in reality, weighing a mass, for *the weight of any mass is the measure of the force of attraction between the earth and itself.* There are two common methods of weighing—the spring scale and the beam balance (Fig. 3-1).

A spring scale measures weight by the amount of stretching of its spring; it is calibrated by marking the stretching produced by the earth's pull on a series of standard weights. By international agreement, such calibrations are based on *the standard kilogram mass* preserved at the International Bureau of Weights and Measures at Sèvres, France. At Sèvres, we calibrate our spring scale so that the pointer reads exactly one kilogram when the standard kilogram mass is placed on the scale. A second weight that brings the pointer to the same mark weighs, of course, exactly one kilogram and will counterpoise the standard kilogram mass when the two are placed in opposite pans of the beam balance. But, we shall find some interesting changes in these relations if we move our equipment from Sèvres to some other point on the earth's surface.

Effect of Altitude. If, for example, we take scale, weights, and balance to a pass in the Alps, two miles above sea level, we shall find that on the spring scale our kilogram weight will no longer weigh one kilogram but slightly less, though it will still exactly counterpoise the standard kilogram mass in the beam balance. Of course, all our equipment is identical with that at Sèvres, but the distance to the center of the earth has been increased by two miles. Reference to Newton's Law shows why a change in elevation above the sea results in a change in the force of gravity. At Sèvres, nearly at sea level, the distance to the earth's center is approximately 4,000 miles; at the Alpine pass, this distance is roughly 4,002 miles. The product of the masses of the earth and our kilogram weight must be divided, at Sèvres, by $4,000^2$ and, at the Alpine pass, by $4,002^2$. This is a difference of nearly 0.1 per cent.

We see immediately that the distance from the earth's center must be considered when we weigh on a spring scale. It affects the force of attraction because it affects the denominator in the expression of Newton's Law.

Effect of the Earth's Rotation. Factors other than altitude also affect the force of gravity on objects at the earth's surface. Because the earth rotates on its polar axis, there is a tendency for objects near the equator to be thrown off, just as mud is thrown from a spinning automobile wheel, or sparks from an emery grinder. The force causing this tendency is the *centrifugal force of rotation*. It is very small compared to the force of gravity; hence, objects do not leave the equator and fly off into space. Nevertheless, the centrifugal force of rotation is not negligible and must be considered in our studies of the form of the earth (Fig. 3-2). As Newton showed long ago, this force should produce an equatorial bulge and a polar flattening in the figure of the earth. Thus is explained the difference between the polar and equatorial diameters actually found by measuring arcs of meridian—Eratosthenes' method.

The force of gravity is greater at the poles than at the equator because the earth is not a perfect sphere; the surface at the poles is 13 miles closer to the earth's center than at the equator. Even if the earth were a perfect sphere, an object at the equator would weigh less than the same object transported to one of the poles. The centrifugal force of rotation is greatest at the equator and steadily diminishes to zero at the pole (Fig. 3-2). The actual difference in the force that results from both of these factors is approximately 0.5 per cent. Thus, if a polar bear weighing 1,000 pounds in his natural habitat were transported to an equatorial zoo, he would there weigh only 995 pounds.

Effect of Local Variations in Rock Density. Still another factor produces variations in the weight of an object at different points on the earth's surface. This factor is the local distribution of rock masses of different density. Let us imagine that the gold buried at Fort Knox, Kentucky, as part of the United States Treasury reserve, formed a single block of pure metal. A cubic foot of gold weighs more than 1,200 pounds. A moment's consideration of the inverse-square relation tells us that a sensitive device for weighing should show a greater weight for an object immediately on top of the gold than it would if the object were weighed on the roof of one of the great limestone caverns not far away, where much empty space is immediately beneath (Fig. 3-3). Thus, even though the latitude and altitude of two localities are identical, the force of gravity at each point must vary with the *density* (mass per unit volume) of the immediately underlying rocks. As we shall see in later chapters, we know the rocks just below the earth's surface do vary considerably in density from place to place, though, of course, not to the extreme degree of our artificial example at Fort Knox.

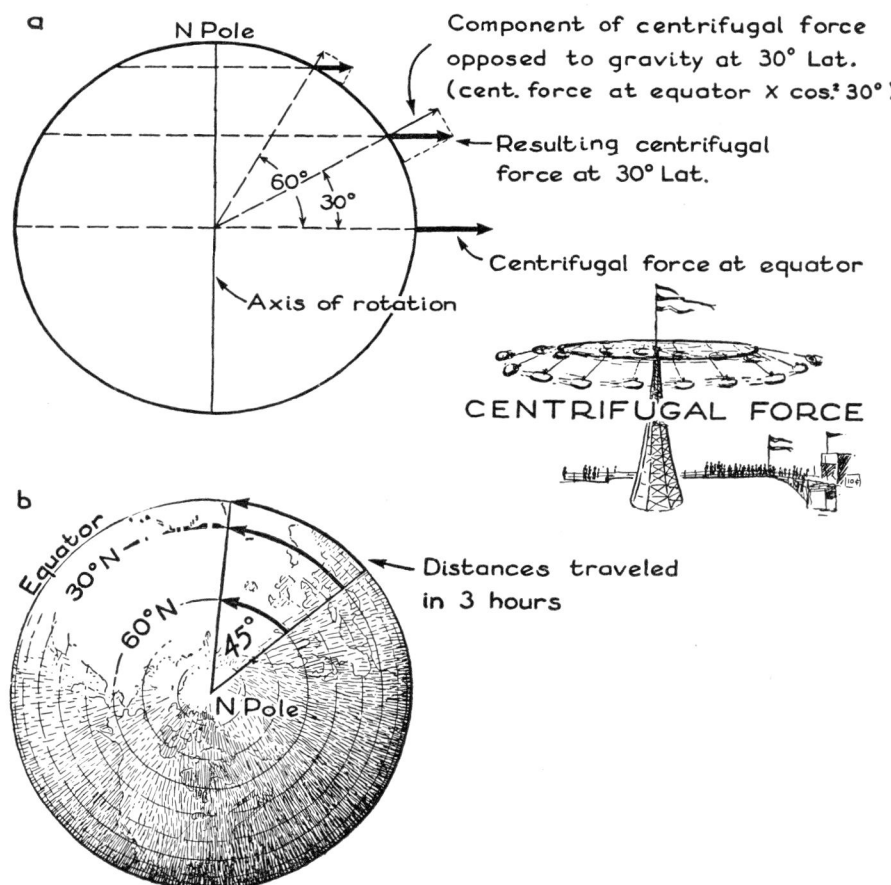

FIGURE 3-2. *Diagram showing a familiar demonstration of centrifugal force and (a) how the centrifugal force due to rotation varies from equator to pole, or, stated differently, (b) how the linear velocity varies from equator to pole.*

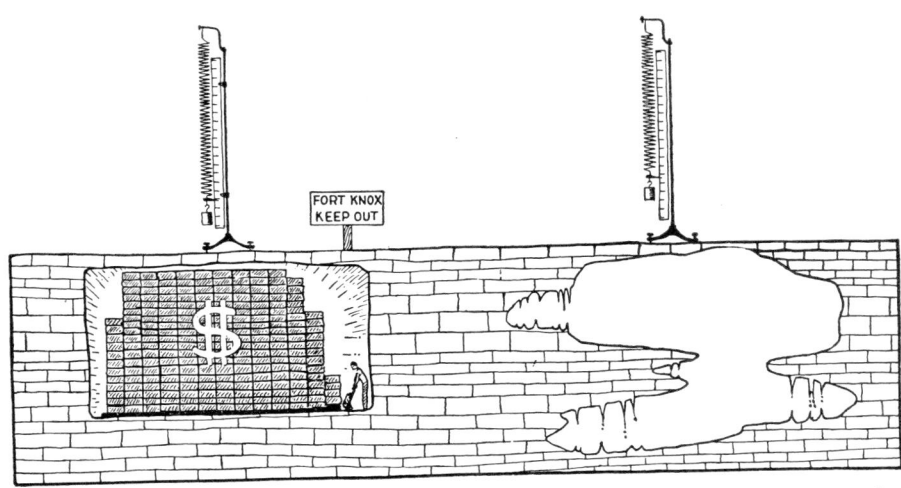

FIGURE 3-3. *The effect of differences in density of nearby masses upon weight.*

The Gravity Pendulum

In summary, we can see that three factors—(1) altitude, (2) latitude, and (3) variations in density of the nearby rocks—affect the force of gravity at any point on the earth's surface. But the effect of each of these factors is small. In order to learn much of value from studies of the variation in the force of gravity from one place to another we need a very sensitive instrument for detecting small variations in weight—far more sensitive than the crude spring scale of which we have spoken. Fortunately, we have precisely this kind of instrument in the *gravity pendulum*.

A free-swinging pendulum—that is, one which is not driven by clockwork or other external means—oscillates to and fro because of the force of gravity. If we suspend a heavy weight from a string, pull back the weight and release it, the weight falls toward the center of the earth following the arc that the suspending string allows. It is being pulled along its path by the force of gravity. But the *inertia*—the resistance that an object offers to any change in its motion—of the weight carries it past the lowest point on its swing, and causes it to rise against the force of gravity along the arc that the suspending string allows. It continues to rise until its inertia is counterbalanced and eventually overcome by gravity. Then it falls back again toward the lowest point, repeating the process again and again. The pendulum oscillates back and forth through the lowest point, rising in each oscillation to a little less height because of friction with the air. Gradually, the oscillations die down until, ultimately, the pendulum loses its motion and hangs straight down perpendicular to a level surface. It has become a plumb bob.

We borrow from physics the law that governs the *period of oscillation* (the time for one complete to-and-fro movement) of a pendulum:

The period varies inversely with the square root of the local acceleration of gravity, and directly with the square root of the length of the pendulum.

Newton demonstrated that this relationship explains why even good pendulum clocks show systematic gains or losses in time when moved about from place to place. We have seen that the *force of gravity* (the local acceleration of a freely falling body) varies from place to place on the earth's surface. A pendulum clock that keeps good time in Paris would systematically lose time when taken to a point high in the Alps because the force of gravity (and hence the acceleration of the falling pendulum) is less at high altitudes than at low. From this effect of the force of gravity upon the period of a pendulum, we can measure gravity by means of a pendulum of known weight and length. To determine the force of gravity, we count the number of oscillations of the pendulum in a given time and calculate the force from the law for the period of oscillation.

Modern gravity pendulums are so constructed that they are almost frictionless. They are hung on knife-edge, jeweled bearings and are swung in chambers from which most of the air has been evacuated. The number of oscillations is counted by precision chronometers that can time the swing to less than 1/10,000 second. With such equipment, geodesists are able to measure the force of gravity with an accuracy of a few parts per million.

Ordinary gravity pendulums cannot be used on a ship at sea because of the rocking motion of the waves, but the Dutch geodesist, F. A. Vening-Meinesz, has devised an adaptation of the gravity pendulum that can be used in a submarine submerged beneath the zone of strong waves. Thus, in recent years, we have obtained thousands of measurements from the land and sea areas of the earth. From these measurements and from study of the angles between plumb lines at different points, geologists and geodesists have been able to make important deductions concerning the interrelations of gravity, topography, and the density of materials within the earth's crust.

Some of the most interesting of these interrelationships were discovered as a result of attempts to explain what appeared to be systematic errors in the location by triangulation of points on the earth's surface. Triangulation (described in Appendix I) is the method of locating a third point by sighting on it from each of two points of known position. But a point can also be located without recourse to triangulation by determining its latitude and longitude astronomically, as is done in navigation. Astronomic determinations of latitude and longitude are, of course, made with reference to the "vertical through the point," that is, the plumb line. They are, in essence, determinations of the local zenith. If the plumb line invariably pointed directly to the earth's center, determinations of position made by triangulation should coincide exactly with determinations made by astronomic methods. The plumb line, however, does not everywhere point directly to the earth's center. According to the law of gravitation, a plumb bob is attracted by a given mass one mile away with a force one hundred times greater than it is attracted by an equivalent mass ten miles away. We ought, then, to find the plumb line deflected toward a nearby mountain. This is, indeed, the fact.

Consider the situation in a deep Norwegian fjord. A narrow arm of the sea lies between massive cliffs 4,000 feet high. A plumb line suspended near one side of the fjord will be deflected toward the nearby mountain mass. Also, the sea's surface will be tilted slightly upward toward the mountain mass and downward toward the center of the fjord (Fig. 3-7, p. 43).

Similarly, at the edge of a continent, the surface of the sea must be tilted upward slightly toward the continent by the attraction of the continental mass. Such tilts are very small, generally only a few seconds of arc, but they lead to appreciable errors in determining the position of a point by astronomic methods. Also, from consideration of such distortions of the water surface, we see that the concept of an earth with the figure of an oblate ellipsoid, though a closer approximation to the truth than the concept of a sphere, does not precisely represent, in detail, the shape of the earth's level-surface—that is, the surface which is everywhere at right angles to the plumb line.

Isostasy

Even the largest mountains, however, have very small masses compared with the mass of the earth as a whole. Despite their being close at hand, we can expect their effect in deflecting the plumb line to be small. To calculate the theoretical deviations operating at a given station by the law of gravitation requires considerable mathematical labor, for usually one has to calculate the effects of many irregular masses lying in different directions and at different distances from the observing station. When this is done for a large series of stations, however, an exceptionally interesting relation appears, showing that mountain chains are not mere loads of rock heaped on the surface of the earth, but that they are sustained in some other way.

From such studies it has been discerned that mountain masses do not actually deflect the plumb line sideways as much as it would be deflected if the mountain were actually a load on top of the earth's crust. The same conclusion is reached from consideration of measurements of the force of gravity with the gravity pendulum. If a mountain were really a load heaped on the crust of the earth, the force of gravity (after correction for the effect of added altitude) should be greater on the mountain than on an adjacent plain because of the added gravitational pull of the mass of the mountain. The results of many investigations with the gravity pendulum show, however, that over large areas there is little relationship between topography and the force of gravity, when this is corrected for elevation and masses above or below sea level.

From these relations, geologists and geodesists have inferred that the major irregularities of the earth's crust are sustained, not as loads borne up by the strength of the earth's crust, but by a process of flotation upon a heavier, plastic interior. Everyone knows that an oak plank will float lower in water than a pine plank; it is denser and must sink deeper to displace its own weight of water (Archimedes' principle). The idea is that *the excess mass of high areas,* such as mountain chains and continents, *is compensated for by a deficiency in density of the material of which the elevated tracts and their roots are composed.*

Mountain chains and continents float high because of their lower density. Plains and sea floors stand lower because either they are composed of rocks of higher density than the mountains, or else, if they are made of the same stuff as the mountains, this material is thicker under the mountains than under the adjacent lowlands (Figs. 3-5 and 3-6, pp. 40 and 42). Such a condition of flotational equilibrium between blocks of the earth's crust is called *isostasy* (Greek: "equal standing"). The word isostasy is also used to indicate a *tendency* for blocks of the earth to approach a condition of balance (flotational equilibrium) with adjacent blocks; that is, a tendency for heavy blocks to sink and light ones to rise, in the earth's plastic interior.

The theory of isostasy is an important concept that we shall refer to again and again. It will be well, therefore, to explain more fully the kind of geodetic evidence on which the theory is based by describing an actual example. There is, perhaps, no better example than the first study of this kind ever made—the analysis of the deflections of the plumb line discovered in the course of the survey of the land mass of India. It was the investigation of these deflections that first led to the discovery of the condition of isostatic equilibrium in the earth's crust.

The Trigonometrical Survey of India

In the middle of the Nineteenth Century, the Trigonometrical Survey of India was organized under Sir George Everest (for whom Mount Everest is named) to carry out an extensive plan of precise surveying in order to establish "control points" for mapping that great subcontinent. The method used for determining the relative position of the control points was triangulation (see Appendix I). The triangulation was done very carefully and gave highly accurate distances between stations. Scores of points in India were located by this method, and their latitudes and longitudes were also carefully determined by astronomical observations.

If the latitude and longitude of one point in a triangulation net is known, it is, of course, a relatively simple matter to calculate the latitudes and longitudes of all the other points in the net with respect to the known one. We know the amount of curvature of the earth (Chap. 2), and, hence, the distance along the meridian that will be subtended by an arc of one degree. By the process of triangulation, we can determine the distance between two stations with a high degree of accuracy, and also determine the exact compass direction between the stations. It is, therefore, easy to translate measured distances, determined by triangulation, into terms of degrees of arc, and hence to calculate the latitude and longitude of each point in the net. Such calculations are, of course, not dependent on astronomic observations at the new station. But if an astronomic determination of the latitude and longitude of the new station is made independently, it can be used as a check on the triangulation.

In the survey of India, it soon became apparent that, for some stations, determinations of their relative positions made by triangulation did not agree with those made by astronomic methods. At first, it was thought that perhaps errors had crept into the triangulation, but when checked the triangulation results were consistent. Also, several of the apparent "errors" were far too large to be accounted for by inaccuracies in surveying.

Two of the stations, Kaliana and Kalianpur (Fig. 3-4), are among the stations discussed by Archdeacon Pratt, a British cleric who interested himself in the problem and who, in seeking an explanation for the apparent "errors," discovered the isostatic relationship. Now Kaliana is on the Indo-Gangetic plain, immediately beside the towering Himalayas. Kalianpur lies far to the south, near the center of the Indian peninsula (Fig. 3-4). Archdeacon Pratt expected that, at Kaliana, the plumb line would be appreciably deflected northward toward the Himalayas, and, therefore, that the distance between the two stations, in terms of degrees of curvature of the earth's surface, would prove to be greater when calculated from the triangulation results than when based on astronomic determinations. This expectation was borne out. The difference in latitude between the two stations as determined by the two methods was:

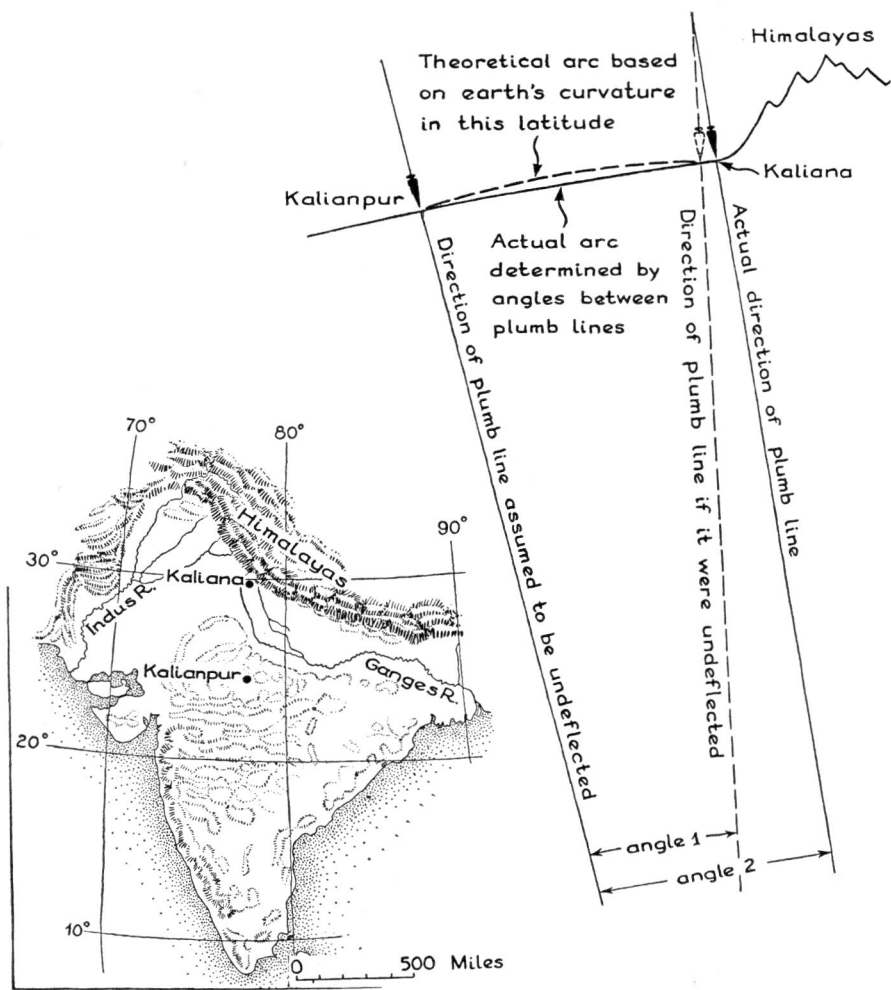

FIGURE 3-4.   *The effect of the Himalayan Mountains on the computed distance between Kaliana and Kalianpur. Note the difference in the arc distance between the two cities produced by the shifting of the plumb bob, so that angle 1 is greater than angle 2. Relations greatly exaggerated.*

Difference in latitude, determined by triangulation $= 5°23'42.29''$
Difference in latitude, determined astronomically $= 5°23'37.06''$
Discrepancy   .   .   .   . $5.23''$

The discrepancy of 5.23 seconds of arc corresponds to a distance of about 500 feet and is far greater than can be explained by surveying errors.

Archdeacon Pratt was interested to find out whether the sideways gravitational pull of the Himalayas on the plumb bob accounted entirely for this discrepancy. Although no accurate topographic maps of the Himalayas were available, enough was known about the height and position of their principal peaks for Archdeacon Pratt to compute the approximate mass of the Himalayas above sea level and their average distance northward from each station. Assuming that the mountains constitute an additional load on an otherwise uniform crust, he could then compute how much the plumb line should have been deflected northward at each station and the difference in astronomic latitude that should result from the deflection. He found that, on these assumptions, the plumb line at Kaliana should have been deflected 27.853 seconds to the north, and at Kalianpur it should have been deflected 11.968 seconds. The difference between these is 15.885 seconds, over three times the 5.23 seconds difference actually found by the Trigonometrical Survey. The difference between Pratt's calculated results and the figure actually found was far too great to be explained by errors in triangulation, or by errors in Pratt's estimate of the volume and mass of the Himalayas.

Pratt's Theory of Isostasy

Pratt saw that some of his assumptions were not in conformity with fact. One of the assumptions implicitly accepted was that the density of material below sea level was identical, whether the material lay beneath the plain of India or beneath the Himalayas. He had assumed that the mass of the Himalayas, from their highest peaks to sea level, was a load heaped on a crust considered to be uniform in density below sea level. Pratt saw that the discrepancy might be explained if the rock below the Himalayas were of less mass (lower density) than that to the same depth beneath an equivalent area of the plain of peninsular India. He suggested that both the Himalayas and the peninsula are "floating" on a deeper layer of still denser earth material, and that the heights to which their surfaces rise above the surface of this deeper layer are inversely proportional to the average densities of the material composing the two "blocks" (Fig. 3-5). In other words, the high-standing mass of the Himalayas is "compensated" or balanced by a corresponding deficiency of mass extending far below sea level beneath the Himalayas.

A simple illustration of Pratt's idea is shown in Figure 3-5. In the upper diagram, blocks of four different metals, each of which weighs exactly the same amount and each of which has the same cross section, are shown floating in a pan of mercury, a liquid of very high density (13.6 gr./cm.^{3}). The metal blocks have different lengths because their densities vary. The antimony block (density 6.6 gr./cm.³) is longer than the lead block (density 11.4 gr./cm.³) because its volume must be about twice that of the lead block in order to weigh the same amount. The blocks sink in the fluid until they displace enough mercury to equal their weight. Therefore, the antimony and lead blocks will sink to the same depth in the mercury, but, since the antimony block has nearly twice the volume of the lead block, it will project much farther above the surface than the lead. Similarly, zinc rises higher above the surface than iron, but all four metal blocks project into the mercury to the same depth—exactly deep enough to displace their weight

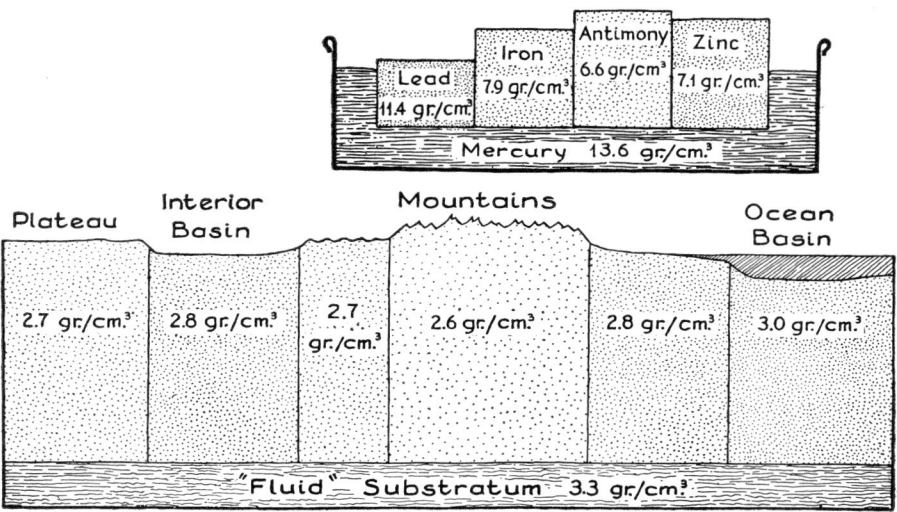

FIGURE 3-5. *Pratt's hypothesis of isostasy. The densities shown in the lower diagram were not specified by Pratt but conform roughly to some modern estimates. (Modified in part from W. Bowie, Isostasy, E. P. Dutton, 1927.)*

of that fluid. Pratt assumed that mountains, plains, and ocean floors showed essentially the same relations (lower diagram in Fig. 3-5). Mountains are like the antimony block. They project higher above the surface of the "fluid" substratum because they are composed of material less dense than that which underlies a plain.

Pratt's suggestion was the first formulation of the theory of isostasy, and his scheme of "compensating" for differences in elevation above sea level by variation in density of the blocks below sea level is known today as the *Pratt Theory of Isostasy*. Pratt did not use the word isostasy; the condition of equilibrium which he discovered was named by Dutton some thirty-four years later.

Airy's Theory of Isostasy

The same volume (1855) of the *Transactions of the Royal Society* that contains Pratt's discussion also contains a brief contribution by G. B. Airy, the Astronomer Royal of Great Britain. Airy accepted most of Pratt's reasoning, saying that this conclusion should have been anticipated because it could be shown that if masses as great as plateaus and high mountains were loads on a solid earth, no rocks could be strong enough to sustain them. The rocks beneath would break and flow out laterally until a state of balance was restored. The only possibility, then, is to regard mountain chains as floating masses, as Pratt did.

Airy, however, suggested a different mechanism of flotation. He saw no reason to believe that the density of the material immediately beneath a mountain is any different from that directly beneath a plain. If both blocks are of the same density, but of unequal thickness, the difference in surface elevation may still be readily explained. The thick block (mountain block) will float higher above the surface, but it will also sink deeper into the heavier "fluid" below. The height of the mountain block is compensated by a "root" of light material that projects deeper into the underlying "fluid" layer and displaces it, just as an iceberg one hundred feet thick will float higher in water than one fifty feet thick, although it also projects deeper into the water. Airy's view is called the *Roots of Mountains Theory of Isostasy*. This theory accounts for the "errors" in the India survey just as well as Pratt's theory, and accords much better with what we know and can infer about the composition of the materials beneath the surface. Therefore it is more widely accepted by geologists.

A simplified view of the "Roots of Mountains" theory is shown in Figure 3-6. In the upper diagram, several blocks of copper, all with the same density but with different weights because they are of different lengths, are seen floating in a pan of mercury. The longest blocks rise higher above the surface and also sink deeper into the mercury. The lower diagram shows how blocks of the earth's crust, all composed of the same kind of material, may, nevertheless, float to varying heights above sea level provided the blocks are of different thicknesses.

The diagrams shown in Figures 3-5 and 3-6 are, of course, greatly oversimplified. Good evidence will be developed in later chapters to show that the earth's crust is not divided into simple blocks free to move past one another along frictionless boundaries. Also, the substratum is not a fluid, although it may respond by plastic flow to heavy loads of long duration and so act essentially like a very viscous fluid to these large stresses. We know, furthermore, that the strength of the earth's crust is considerable; loading must reach a certain amount before this strength is overcome and the rock beneath gives way and begins to flow.

The point to be emphasized here is that both Pratt and Airy concluded

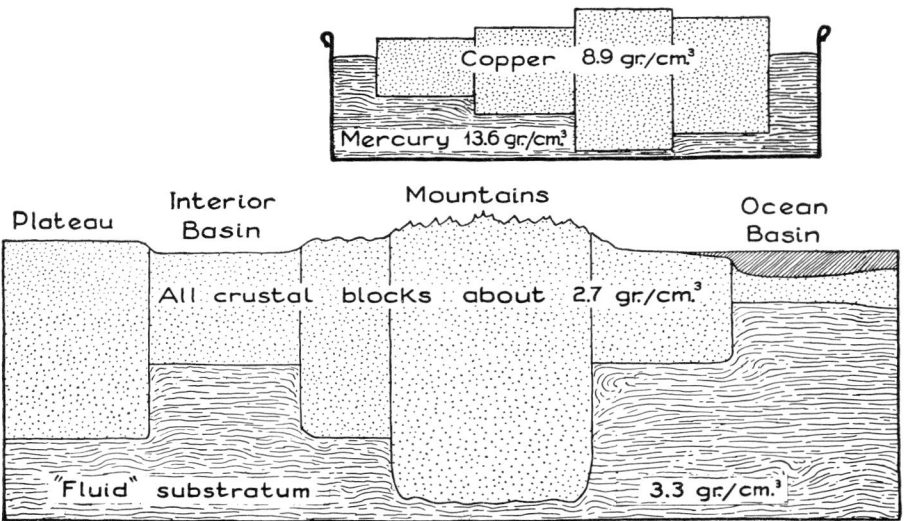

FIGURE 3-6. *Airy's hypothesis of isostasy. The densities shown in the lower diagram were not specified by Airy but conform to some modern estimates. (Modified in part from C. R. Longwell, Geog. Review, 1925.)*

that the Himalayas do not stand at their present height because their huge pile is a load supported by the strength of the earth's crust, but because it is, in some way, sustained by flotation. The general validity of this idea also finds strong support in studies of the intensity of the force of gravity as determined from observations with the gravity pendulum.

The Geoid and the Spheroid

We have seen how the modern measurements of arcs of meridian at different latitudes forced the modification of Eratosthenes' "Figure of the Earth" from a sphere to an oblate ellipsoid. But a moment's thought over the facts just discussed must show that still further refinement is necessary if we are to deal with the question in detail.

Though nearby mountain ranges do not deflect the plumb line as much as one might expect—a fact that we have already explained by the concept of isostasy—they nevertheless do deflect it. Thus, even the surface of the sea cannot be an oblate ellipsoid, at least near to the shores, for it is warped upward by the gravitational pull of the adjacent lands. This warped surface of the sea, which is, of course, *everywhere at right angles to the plumb line, is called the* GEOID (Fig. 3-7). It may be thought of as *the surface of the ocean, and as the surface to which the ocean would stand in a system of narrow sea-level canals cut through the continents.*

The geoid, or in other words the level surface corresponding to sea level over the whole earth, obviously does not correspond to any regular mathematical figure; unlike the ellipsoid, for example, it cannot be formed by rotating an ellipse on its axis. Hence, it offers great mathematical difficulties to geodesists who are primarily concerned with precise comparisons of measurements. To get around this difficulty, geodesists have devised a theoretical figure of the earth which is called the *spheroid*. The spheroid is really the

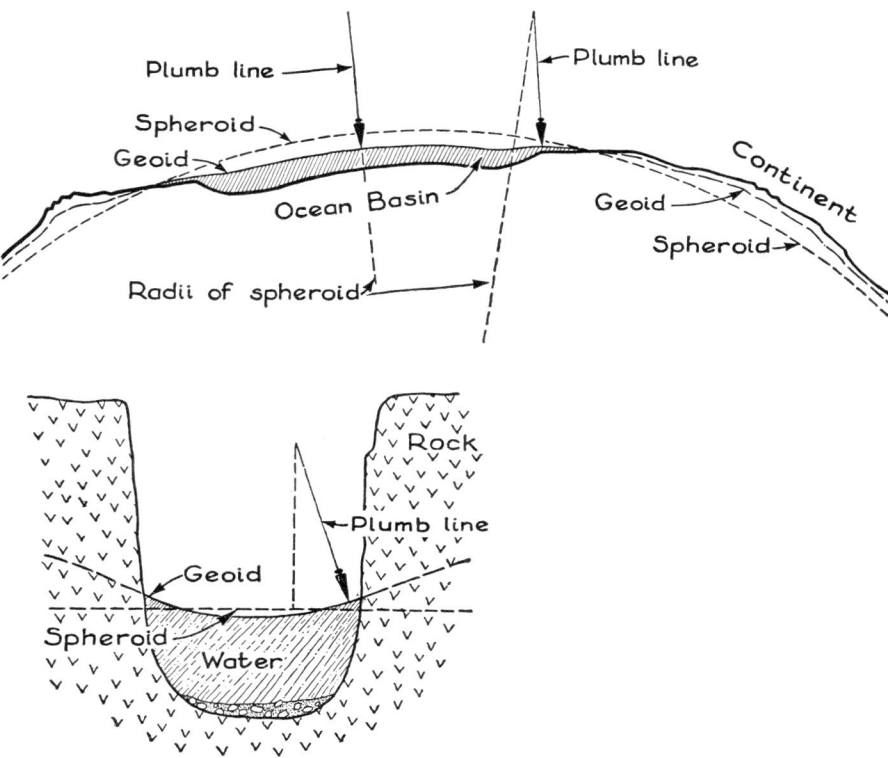

FIGURE 3-7. *The relations between spheroid and geoid, tremendously exaggerated to show the qualitative effect of irregular topography. Above, broad view of their relations as between continental and oceanic segments. Below, the hypothetical relation in a Norwegian fjord.*

world-average form of the geoid. The spheroid will lie very slightly above the geoid in oceanic areas and somewhat below the geoid in continental areas. These relations are illustrated on a greatly exaggerated scale in Figure 3-7. When we recall that the oceans cover 71 per cent of the earth, it should not be surprising that the spheroid, or "world-average geoid," is only slightly different from the ellipsoid.

Gravity Measurements and Isostasy

To every point on the theoretical figure of the earth, the spheroid, there corresponds a theoretical value of the force of gravity which depends only on the latitude of the station. That is, the theory takes account of the centrifugal force of the earth's rotation. By swinging a pendulum it is possible to measure the actual force of gravity. It is clear that, in order to compare the measured and the theoretical values of gravity on a common basis, we must allow for several factors. In the first place, nearly every continental station lies higher than sea level and hence is above the spheroid. It is farther from the earth's center of mass than the theoretical point on the spheroid beneath. The effect of this is to make the measured value less than the theoretical. On the other hand, the material between the ground surface and the spheroid adds its local attraction to the measured value. This should make it too great as compared with the theoretical value. Also, there are the disturbing effects of nearby topographic features. Obviously, many complex measurements and computations are necessary before the measured and theoretical values can be compared (Fig. 3-8).

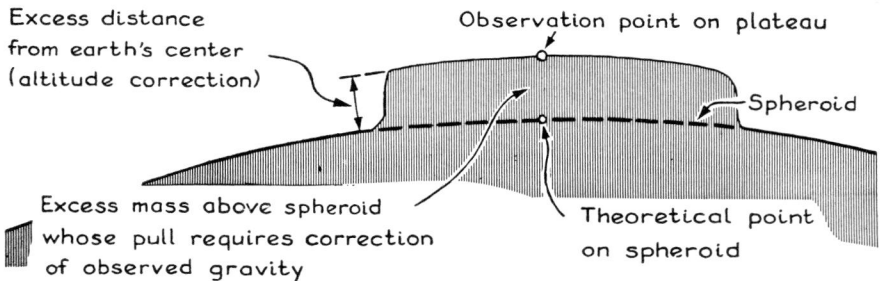

FIGURE 3-8. *The factors involved in correcting observed gravity for comparison with the theoretical value by the Bouguer method. Height above the spheroid taken alone diminishes the observed value as compared with the theoretical, but the added attraction of the mass between spheroid and observation point increases it. When both corrections are made, most land stations yield negative anomalies and most sea stations positive. This fact constitutes strong evidence for isostasy.*

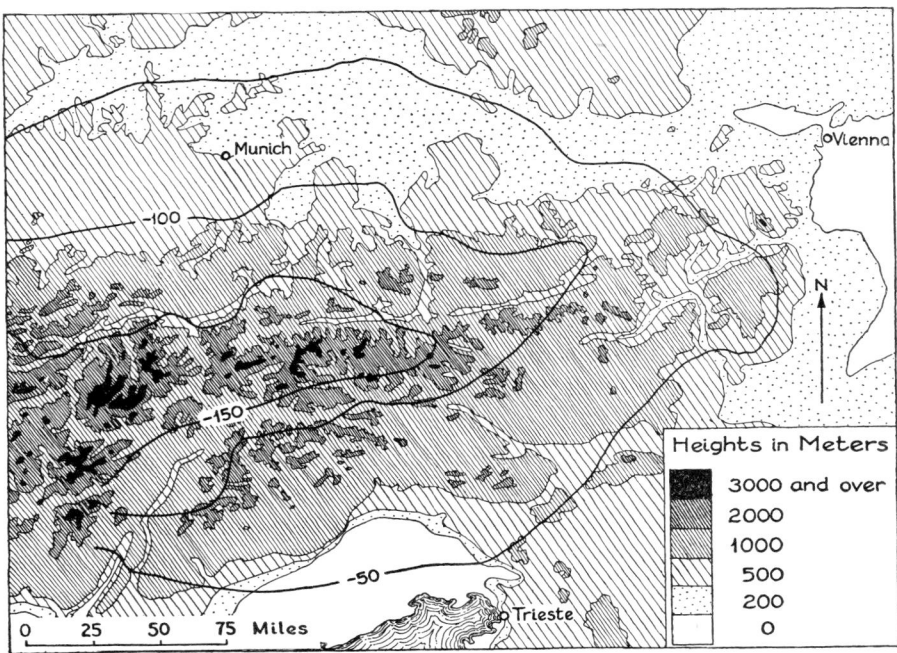

FIGURE 3-9. *The topography and Bouguer gravity anomalies of the Alps. Note the general relation of topography with lines of equal gravity anomaly, the negative anomalies being greatest in the mountainous areas. (From Paavo Holopainen, 1947.)*

When this work is done, however, as it has been for many thousands of stations on land and many hundreds at sea, a most significant fact is disclosed. By convention the theoretical value is subtracted from the measured value as corrected for all these disturbing influences. The difference is called *the gravity anomaly*. Now, gravity anomalies that are computed after correcting for elevation of the station and for the mass between the station and sea level (called *Bouguer anomalies* after the French geodesist who first suggested this method) are chiefly negative on land areas and in general more highly negative the higher the station above sea (Fig. 3-9). Conversely, the sea stations have slightly positive anomalies as a general rule.

What this means can be easily seen. When we allow for the gravitational pull of the masses between sea level and the station, our corrected value is too low. But in making this "correction" we have assumed *uniform density of material below the spheroid*. As the anomalies are in general greater the higher the station, it is clear that one of two factors has influenced our result: Either (1) the material *between the station and the spheroid* is not so dense and hence has not so great an attraction as we assumed it would have, or

(2) the material *below the spheroid* is less dense beneath high land than beneath low. But the material between the station and sea level cannot be far from the average rock in density; as we shall see in later chapters this is quite well known. It must, then, be the second factor that has brought about our anomalies. But this is precisely the conclusion reached from plumb-line observations: *The excess mass of the rocks above sea level is compensated for by a deficiency of mass below sea level.*

Conversely, the positive anomalies at sea stations must be explained by the error involved in assuming that the oceans are underlain by material of the same density as that beneath the continents. If we replace this assumption with another—that the oceanic segments of the earth are underlain by *more dense material than the continents*—the anomalies are greatly reduced.

Thus, both the deflections of the plumb line and the relations between gravity measurements and the earth's relief indicate the same thing: The larger segments of the earth are roughly in floating equilibrium (isostatic balance), one with another. Large areas of highland stand above large areas of lowland because either: (1) They are composed of less dense materials than underlie the lowlands; or (2) if they are made of the same material, it is thicker beneath the highlands. In later chapters we shall show that certain geologic relations can be explained only by accepting the idea of isostatic balance. We shall also show that, for some areas of limited size, isostatic equilibrium is far from complete—locally, indeed, there is good evidence indicating that masses of rock of the size of small mountain ranges may actually be held lower or higher than their equilibrium position by the strength of the crust. The maximum size of the masses that can be so held out of their equilibrium position is still vigorously debated by geologists. But there must indeed be a maximum limit, for considerations of strength and scale compel us to conclude that the great earth blocks comprising the continents and the ocean basins, and even blocks of the size of the major mountain chains, are incapable of supporting themselves except essentially in flotational equilibrium, one with another.

Strength

The preceding discussion of isostasy assumes that the earth has a plastic core capable of buoying up the crust by flotation. Airy emphasized the point that a "solid" or rigid core, no matter how "strong" we think the hardest rocks are, could not be strong enough to support the earth's crust; therefore, the core must behave as though it were essentially a fluid mass.

What does this observation mean? A piece of rock we pick up is strong and does not behave at all like a fluid. We have also seen that the earth's surface is irregular in detail, with great mountain ranges towering thousands of feet above the sea and deep troughs submerged thousands of feet beneath it. The great mountain cliffs do not seem to be flattening out of their own weight. As far as we can see, they are made up of rigid and strong rocks capable of maintaining the topographic relief.

How can we reconcile this paradox between what we observe and the evidence that large blocks of the crust rest in flotational (isostatic) equilibrium upon a plastic interior? Must we conclude from these evidences of surface strength and of bodily weakness that the earth has a strong solid crust floating on an interior that is liquid? As we shall see in later chapters, there are cogent reasons to doubt that the subcrust is really liquid. Perhaps a part of the answer may be found if we consider carefully just what is meant by "strength."

Definition of Strength

> The strength of a body is defined as the force (load) per unit area that is required to deform that body permanently—that is, to break it or make it yield continuously.

Thus, we say that a block of granite an inch square that breaks under a load of 30,000 pounds has a *compressive strength* of 30,000 pounds per square inch. A steel cable with a cross section of one square inch that breaks when a weight of 150,000 pounds is suspended from its end has a *tensile strength* of 150,000 pounds per square inch. Each solid substance requires a definite force per square inch before it will rupture or flow.

Fluids have no strength. Water has no strength; it will flow into, and fill, all available spaces of a vessel into which it is poured. Tar has no strength; a piece of iron placed on it will eventually sink to the bottom. It is merely more viscous than water. Indeed, part of the technical definition of a fluid is that it yields continuously under the slightest load or stress.

Solids, too, can be made to flow—that is, yield continuously without rupture—under particular conditions of temperature and pressure. *But solids, unlike liquids, require that a definite threshold force per square inch must be applied before the strength of the material is overcome and continuous yielding begins.*

Effect of Temperature

Whether a particular solid will break or will flow under an applied force often depends on the temperature prevailing while the force is applied. At

FIGURE 3-10. *Quarrying the state of Texas. (Reproduced by permission, from M. King Hubbert, Am. Assoc. Petrol. Geol. Bull., 1945.)*

red heat, a bar of iron is still solid, but it will flow under relatively small stresses that would not cause it to flow at room temperature—a fact that is utilized in forging iron.

As shown in Chapter 5 and subsequent chapters, there is good reason to believe that many solid rocks buried deep within the earth's crust have yielded continuously by flowage in response to the heat and pressure of the earth's interior. The same rocks at the surface deform only by breaking. Such flowage of solids should not be confused with liquid flow; the rock does not melt, and a definite *stress* (equivalent to the strength of the material under the prevailing environment) must be applied before continuous yielding begins. The general rule is that *any particular solid is weaker at high temperature than at low.*

Effect of Size, or Scale

In everyday life, we seldom think of the effect of scale upon the strength of materials, although, as we observed earlier, engineers have to consider scale in providing strength for massive structures. The earth is a large structure. How are we to think of rock strength in huge masses? A dramatic illustration of the effect of scale has been presented by M. King Hubbert[*] in the quarry operation illustrated in Figure 3-10.

Suppose we are to quarry a single block of granite the size of the state of Texas. It is roughly 1,200 kilometers across, and we want to hoist a piece one-quarter as thick as it is broad—that is, 300 kilometers thick. Grant that we have a quarry crane capable of hoisting it. The rock is assumed to be flawless and to have the average crushing strength of granite, about 30,000 pounds per square inch. Will it be strong enough to allow itself to be hoisted without disintegrating? Obviously, it is impossible to test such

---

[*] Hubbert, M. K., "Strength of the Earth," Am. Assoc. Petrol. Geol. Bull., vol. 29, 1945, pp. 1630-1653.

a problem directly, but we can investigate the properties of such a block by using a scale model.

What would be the properties of a model reduced to a size suitable for our experiment? To obtain a convenient size for our model of the state of Texas, we could reduce the true length of 1,200 kilometers to 60 cm (about 2 feet). Our model would then be on a scale of 5/10 millionths of the original. On this scale, the 300-kilometer thickness would be reduced in the model to 15 cm. (about 6 inches). The force of gravity does not concern us because both the original and the model are at the earth's surface. We can also use material of the same density as the original, about 3 gm/cm³. What other factors must be considered?

If the model is to act the same as the original—that is, if its mechanical behavior is to be identical—there must be the same ratio of strength to size in the model as in the original. In order to obtain the same ratio of load in our model, we must use material whose strength is 5/10 millionths that of granite. The strength of the material in our model, then, will be

$$\frac{5}{10,000,000} \times 30,000 \text{ pounds per square inch, or .015 pounds per square inch.}$$

In Hubbert's words:

> It is difficult to envisage a solid of this weakness. A crushing strength of .015 pounds per square inch is the same as 1 gram per square cm, which, for the density of 3 grams per cubic cm, would be the pressure at the base of a column 1/3 of a cm high. Any column higher than this would collapse of its own weight. Yet the size of the reduced block would be such that its thickness would be about 15 cm (6 inches) and its total weight about 180 pounds. The pressure at its base would be about 45 grams per square cm or 45 times the crushing strength of the materials.
>
> Consequently, if we tried to lift such a block in the manner indicated in the figure, the eyebolts would pull out; if we should support it on a pair of saw-horses, its middle would collapse; were we to place it on a horizontal table, its sides would fall off. In fact, to lift it at all would require the use of a scoop shovel. That this is not an unreasonable result can easily be verified by direct calculation upon the original block (the State of Texas). For it, too, the pressure at its base would exceed the crushing strength of its assumed material by a factor of 45. The inescapable conclusion, therefore, is that the good State of Texas is utterly incapable of self-support!

Effect of Confining Pressure

The strength of a substance is increased when it is placed under pressure from all directions simultaneously—that is, when it is under *confining* (hydrostatic) *pressure*.

Laboratory experiments show that the crushing strength of the limestone from Solenhofen, Germany, is six times as great under a confining (hydrostatic) pressure of 10,000 atmospheres as it is under ordinary laboratory conditions. Its crushing strength under ordinary conditions is 25,000 pounds per square inch. Under a confining pressure equivalent to 10,000 atmospheres (147,000 pounds per square inch), its strength is 150,000 pounds per square inch. Such a pressure is equal to the weight of a column of granite about 22 miles high. Presumably, then, it is also equal to the hydrostatic pressure at a depth of 22 miles within the earth.

We see, therefore, that pressure due to gravity is a great factor in the earth's strength. In short, there is nothing really inconsistent in the apparent incongruity of an earth with "rigid and strong" rocks at its surface capable of maintaining relief features of considerable size yet so "weak" that it reacts almost like a liquid to large differential loads equivalent to the weight of a mountain chain. Our chief difficulty lies in visualizing the way in which a substance must act in very large masses or under different pressure-temperature conditions from those familiar to us. Some of these difficulties will be lessened when we learn (in Chaps. 4 and 5) more about the actual minerals and rocks that comprise the earth's crust.

Facts, Concepts, Terms

GRAVITATION

Law of gravitation $F \propto \dfrac{M' \times M''}{D^2}$

Mass, weight, and inertia
Density
Relation of a level surface to the plumb line
Influences modifying force of gravity on the earth's surface:
    Centrifugal force
    Altitude
    Variations in density of nearby rocks
Free period of a pendulum
Deflection of the plumb line (warping of level surface) by mountains

ISOSTASY

    Evidence from deflection of the plumb line
    Pratt isostasy
    Airy isostasy

The geoid
The spheroid
Evidence from gravity anomalies

STRENGTH
Solid and fluid
Effect of temperature
Effect of scale or size
Effect of confining pressure

Questions

1. When we sight along a telescope accurately adjusted to a level position, are we sighting parallel to (1) the geoid, (2) the spheroid, or (3) the ellipsoid? Where, in general, would you expect all three of these surfaces to be most nearly parallel? (Refer to Chap. 2.)

2. How would you expect the plumb line to be deflected from the line at right angles to the spheroid at Denver, just east of the Rockies? If surveys failed to show any such expected deflection (geoid and spheroid parallel) what might you suspect as the cause?

3. The statement is common in geologic writing that the hydrostatic pressure at depths of a few tens of miles is equal to the weight per unit area of the overlying rocks. How can this statement be justified?

4. There is a very large negative anomaly of the type described in this chapter at Seattle nearly at sea level. Within less than 20 miles east or west of Seattle, also at sea level, measurements of the force of gravity show practically no anomaly. What suggestion can you offer to explain this fact?

5. What change in the rate of swing of a gravity pendulum would be noticed between a station at the surface and a second station at the bottom of a 5,000-foot mine? Between a station at Seattle (see Question 4) and another station 20 miles west of Seattle?

6. In a certain valley in Turkestan the plumb lines are actually deflected *toward* the center of the valley instead of away from it toward the mountains. Suggest a possible explanation of this anomalous behavior.

7. Account for the fact that rocks which are brittle and break under stresses in the laboratory have obviously been folded and have flowed under stresses in the earth.

8. Why does not the continental mass spread out over the ocean floor?

Suggested Readings

1. Daly, R. A., *Strength and Structure of the Earth*, Prentice-Hall, New York, 1940. (Especially the Introduction and Chap. 1.)

2. Hubbert, M. K., "Strength of the Earth," *Am. Assoc. Petrol. Geol. Bull.*, vol. 29, 1945, pp. 1630-1653.

3. Poynting, J. H., *The Earth*, Cambridge University Press, Cambridge, England, 1913.

4. *Minerals*

THE EARTH'S crust is not a homogeneous substance. If we examine its surface during a walk along a stream or a drive along the highway, we seldom fail to find a variety of rocks and soils which differ in color, in coherence, in density, and in other characteristics. If we pick up a fragment of rock or a handful of soil and examine it more closely, we find that it, too, is a mixture of different substances (Fig. 4-1). However, the individual particles that

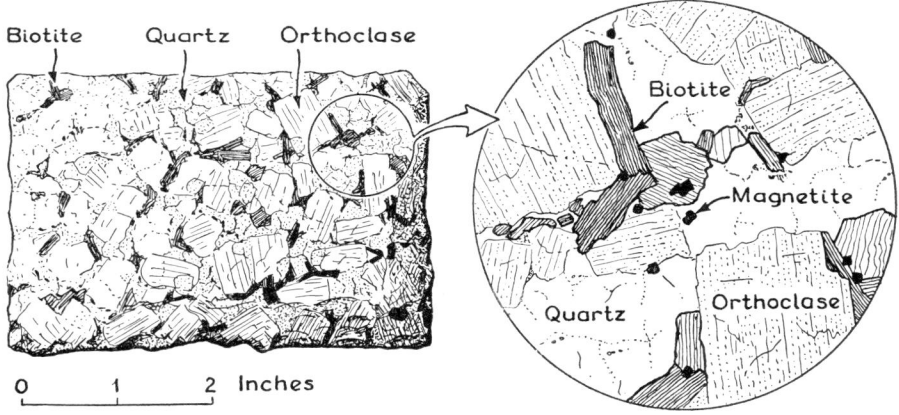

FIGURE 4-1. *The mineral constituents of the common rock granite. Note that the mineral magnetite is visible only when the rock is greatly magnified* (right).

make up the rocks and soils are not mixtures. Each is a distinct, homogeneous substance with definite chemical and physical characteristics. Some may be hard, transparent particles that resemble bits of broken glass; some may be dull, earthy grains; some may be tiny, elastic flakes that flash brilliantly in the sun. Each of these distinct, homogeneous substances is a mineral. *Rocks and soils are aggregates of minerals.* Hence, if we are to understand the origin and classification of rocks, we must learn something about the various minerals that compose them.

Definition of Mineral

If we are to learn to recognize different minerals, and to discriminate one from another, we must define their real nature precisely. Stated formally:

A mineral is a naturally occurring substance with a characteristic internal structure, and with a chemical composition and physical properties that are either fixed or that vary within a definite range.

Minerals, then, are natural substances, found ready-made out-of-doors. The synthetic products that a chemist makes in a laboratory are not minerals, despite the corruption of the word in advertising. A druggist who tells you that a certain pharmaceutical preparation is "rich in vitamins and minerals" is using the term "mineral" in an entirely different way than it is used by a geologist. He is not prescribing a diet of mud and rocks.

To say that minerals have definite chemical and physical properties or properties that vary within certain definitely fixed limits is merely to point out that all particles of a single kind of mineral are alike in their physical and chemical characters, whether one particle comes from Brazil and another from the United States, or whether one may have crystallized in the shell of a snail and another from the water of a hillside spring.

The most definitive characteristic of a mineral is its internal structure, the core of our definition. To understand what we mean by internal structure, and, hence, the true nature of minerals, we must digress into a discussion of the fundamental structure and properties of matter itself. Here, the sciences of chemistry and physics unite with geology to arrive at a solution.

Structure of Matter

Atoms

The English chemist, John Dalton, in 1805 advanced the hypothesis that all matter is made up of individual particles which he called *atoms*. Later work of many physicists and chemists has verified Dalton's prediction. All solids, liquids, and gases are made up of atoms.

Atoms are extremely minute particles. One hundred million of them placed side by side would make a row only an inch long. Yet, a century after Dalton's work, it was discovered that the atom, small as it is, is composed of particles still smaller. Only three of the various subatomic particles that have been discovered are important in discussing the chemical behavior of minerals—the *proton*, the *neutron*, and the *electron*.

Every atom has a nucleus that is very small and dense. The nucleus con tains one or more protons and, nearly always, one or more neutrons. There is an important difference between these two kinds of particles: *The proton has a unit of positive electrical charge,* whereas *the neutron is electrically neutral.*

The part of the atom that surrounds the nucleus is largely empty space, but within it the electrons of the atom revolve in orbits. Compared with the diameter of the nucleus, this outer part of the atom is very large—about 10,000 times as great. *Each electron has a unit of negative electrical charge. Each atom contains the same number of electrons as protons and, therefore, is electrically balanced or neutral.* Electrons are very much lighter than protons or neutrons; they weigh only 1/1,845 times as much.

The orbits along which the electrons revolve about the nucleus are arranged in successive shells. The simplest atom, that of hydrogen, consists of a nucleus containing a single proton with a single electron revolving about it. Helium has two protons and two neutrons in the nucleus, and it contains two electrons revolving in an electron shell close to the nucleus. In more complex atoms, the inner electron shell containing two electrons—the helium structure—is retained, but additional electrons, sufficient to match the number of protons contained in the nucleus of the atom, are added in one or more successive shells farther out from the nucleus.

The size of an atom, however, is not determined solely by the number of electron shells it contains. The calcium atom, which has 20 electrons distributed in 4 electron shells, is nearly identical in size to the sodium atom, which contains 11 electrons distributed in 3 shells. Both are very much smaller than the potassium atom which has 19 electrons distributed in 4 shells. This matter of atomic diameter is highly significant in minerals, as we shall show later (p. 64 and Fig. 4-7) in describing how one kind of atom can substitute for, or replace, another atom within the atomic lattice-work that makes up a crystal.

Elements and Compounds

From the chemist's point of view, the number of protons contained in the nucleus of an atom is fundamental, because the chemical characteristics of an atom are determined entirely by the number of positive electrical charges in its nucleus. The number of these electrical charges, of course, equals the number of protons in the nucleus. All atoms having the same number of protons in the nucleus belong to one species of matter, called an *element,* or elementary substance. About ninety elements have been recognized in nature, and the atomic physicists and chemists have recently produced a few new ones by artificial means.

Each element has been assigned a definite *symbol,* such as H for hydrogen and Pb for lead—a convenient shorthand in writing chemical formulas and equations. The elements, together with their symbols, atomic numbers (number of protons in the nucleus), and atomic weights are given in Appendix II.

An elementary substance, such as hydrogen, oxygen, or iron, is made up entirely of atoms of a single kind, whereas a *chemical compound,* such as water, salt, or mica, is composed of two or more kinds of atoms bound together by electrical charges. We shall now consider the nature of these electrical bonds that hold diverse elements together in compounds.

For some reason, *those elements are most stable whose outermost electron shell contains 8 electrons.* Such elements—for illustration, the gases argon, neon, and xenon—are never found combined with other elements in the form of chemical compounds and, hence, are called the *inert gases.* The atoms of other elements may contain from 1 to 7 electrons in their outer shell.

The number of electrons in the outer shell is a major factor in determining the relative ease with which elements combine to form compounds. For example, the elements sodium (Na) and chlorine (Cl) are highly active chemically. Sodium has 1 electron in its outer shell, chlorine has 7. Sodium has only to lose an electron and chlorine only to gain one for both to attain the stable grouping of 8. This is just what does happen when sodium combines with chlorine to form *sodium chloride* (NaCl), the substance we know as table salt. An electron is transferred from the sodium atom to the chlorine atom. But, the loss of an electron leaves the sodium atom no longer electrically neutral; it now has one more positively charged proton in the nucleus than it has negatively charged electrons in its electron shells. Hence, the atom has one unbalanced positive charge. Similarly, the chlorine atom, by gaining an electron, acquires one unbalanced negative charge; there is one more electron in its electron shells than there are protons in its nucleus. Such a charged atom, with the number of its protons either more or less than the number of its electrons, is called an *ion.* As unlike charges of electricity attract, and like charges repel, if the positively charged sodium ion is free to move, as in a solution or a gas, it is drawn to the negatively charged chlorine ion and the two may be held together to form a molecule of sodium chloride. *Molecules* are distinct groups of two or more atoms tightly bound together. The compound, sodium chloride, has very different properties from either of the two elements that combined to form it.

Two atoms also combine to form stable outer shells of 8 electrons without actual transference of one electron to another. This happens by the "sharing" of one or more electrons in such a way that, if the shared electrons are counted as belonging to each of the bonded atoms, each atom achieves a stable shell of 8. Thus, two chlorine atoms with 7 electrons in their outer shells may unite to form a chlorine molecule by sharing two electrons, as shown in the following structural diagram:

$$:\ddot{C}l. \quad + \quad :\ddot{C}l. \quad \longrightarrow \quad :\ddot{C}l:\ddot{C}l:$$

chlorine atom   chlorine atom   chlorine molecule

The chemist's concept of molecules thus helps to explain the formation of chemical compounds and also the structure of gases and liquids in which most chemical reactions take place. The molecular concept of matter is also useful in describing the structure of some solids. With most solids, however, a somewhat different view of the structure of compounds, a view in which the idea of molecules is not strictly applicable, seems to be needed. Most minerals are solids of this kind, as has been proved by many studies within the past twenty-five years. Some of these studies throw much light on the question of how elements combine, as well as on the internal structure of minerals.

Structure of Crystals

Geometrical Form.   Everyone is familiar with crystals of certain minerals: halite (or rock salt), garnet, quartz (also called rock crystal), ice. Some crystals occur in beautiful geometric forms with surfaces bounded by smooth planes called *crystal faces* (Fig. 4-2). Even the early Greeks noticed that some minerals, such as garnet and quartz, commonly have characteristic crystal forms.

Perfect crystals are rare. Most snowflakes fall as beautiful, six-sided, perfect crystals (Fig. 13-2); but frost on a windowpane shows much less perfect ones, and the granules of ice that form on the surface of a freezing pond may show few, if any, crystal faces. This is also true for other minerals. Relatively few are completely bounded by plane faces and some show none at all. Nevertheless, study of the common imperfect crystals, together with the relatively rare perfect ones, permitted mineralogists (as geologists who specialize in minerals are called) to make sound deductions about the internal structure of minerals and the fundamental properties of matter long before physicists and chemists had proved the atomic theory.

The first important step in the analysis of minerals by their crystal faces was made by Nicolas Steno (1631-1687), a Danish physician who lived in Florence, Italy. Steno, one of the outstanding figures in the history of geology, also made fundamental contributions to our knowledge of the origin and structural relations of rock strata.

FIGURE 4-2.   *Common minerals showing good crystal form: a, epidote; b, ortho-clase; c, garnet; d, pyrite. (Photos by Alexander Tihonravov.)*

Constancy of Interfacial Angles.   Steno showed, with the crude instruments at his disposal, that, if crystal faces are found on a specimen of quartz, they always meet in characteristic angles regardless of the size and gross shape of the crystal. An Italian student, Guglielmini, showed in 1688 and 1705 that similar relations held for other minerals. He also noted that the angles characteristic of one species of mineral differed from those of another.

Thus, in halite ($NaCl$) the angle between adjacent surfaces is always a right angle; this means the crystal may be a cube, a rectangular, boxlike figure, or any other right-angled parallelepiped. In quartz, as Steno had found, the angles between the long crystal faces that form the sides of the crystal are always 120°.

Steno's and Guglielmini's methods were refined and extended by later workers until thousands of similar measurements on many kinds of mineral crystals had been made. From them, mineralogists long ago reached the conclusion that the internal structure of each kind of mineral is unique. They reasoned that the constancy of the interfacial angles in different specimens of the same mineral, regardless of the size and shape of the crystals, could mean only that each mineral is built up of minute particles regularly packed together in a definite geometric pattern. The pattern of packing determines the angles between faces and is identical in all specimens of a particular mineral. The size of the specimen depends merely on the number of such particles it contains.

Optical Properties. Other studies of minerals from a very different viewpoint fortified this conclusion. Among the most important were the studies of the effects that crystals produce on light transmitted through them. The Dutch physicist, Christian Huygens (1629-1695), discovered the phenomenon of "double refraction" (Fig. 4-3) while studying the transmission of light through the mineral calcite. One can easily observe this phenomenon by placing a fragment of transparent calcite above a dot on a sheet of paper. Instead of one dot, two are seen. If the calcite is revolved slowly, one dot describes a circle about the other. More than 100 years later, physicists showed that this phenomenon could be explained in terms of a theory of light. According to this theory, the light ray which penetrates the crystal is broken into two rays which deviate slightly from each other as they travel through the crystal.

Furthermore, in contrast to ordinary light which vibrates in all directions at right angles to the line of propagation, each of the two rays that pass through the crystal vibrates in only one plane. Light so modified that it vibrates in only one plane is said to be *polarized.*

These discoveries opened the way for an important new technique in studying and identifying minerals. William Nicol, who taught natural philosophy (today, we call it physics) at Edinburgh, showed in 1829 that transparent fragments of calcite could be cut and glued together in such a way as to eliminate one of the two polarized rays. Others adapted these *Nicol prisms* to microscopes, making it possible to study the effects that

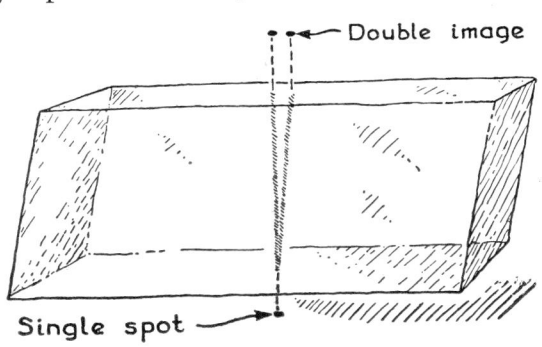

FIGURE 4-3. *A fragment of clear calcite showing double refraction.*

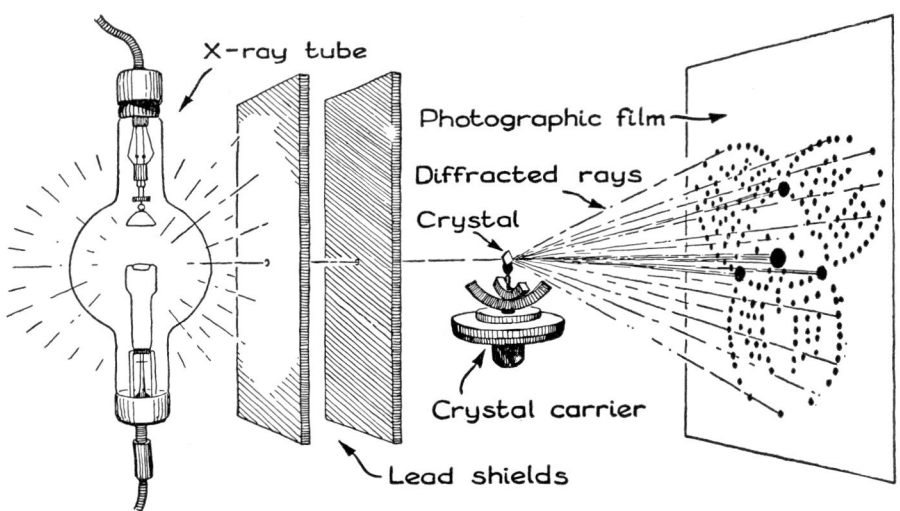

FIGURE 4-4. *Sketch of apparatus used in obtaining a refraction pattern. The crystal is calcite.*

crystals produce on transmitted polarized light. Discussion of these effects is beyond the scope of this book, but it should be emphasized that the *optical properties* of minerals, determined by means of the *petrographic microscope* (a microscope equipped with two Nicol prisms), are precise and diagnostic. By them, most minerals can be quickly identified. The petrographic microscope is the geologist's most efficient instrument for mineral and rock study.

But more important for our present purpose is the fact that the petrographic microscope gives us clues to the fundamental nature of matter and light. The behavior of light in minerals is systematically related to the angles between the crystal faces. This suggests that light is influenced by very minute, systematically arranged particles within the crystal, and strengthens the conclusion from interfacial angles that minerals are made up of submicroscopic particles systematically packed together.

X-ray Studies.   Final proof that the atoms in crystals are arranged in a geometrical latticework came in 1912.

A young Munich student, W. Friedrich by name, was interested in the theory on wave properties of X-rays in relation to the structure of crystals, as expounded by M. Laue, a specialist in the physics of light. When a series of closely spaced parallel lines are scratched on a mirror surface, the light reflected from the mirror is broken into the colors of the spectrum. Laue reasoned that if X-rays are like light but much shorter in wave length, and if crystals really are composed of submicroscopic particles geometrically packed in parallel planes, the surface of a crystal might act toward X-rays much as the ruled mirror surface does toward light.

In order to test this idea, Friedrich and a fellow student, P. Knipping, using a crystal of copper sulfate on an apparatus much like that illustrated in Figure 4-4, developed the first "Laue X-radiogram." Later tests showed the same results as the first one, and gave definite and conclusive proof of patterns of crystal structure, confirming Laue's reasoning on the wave properties of X-rays. The experiments also proved beyond debate the inference of mineralogists regarding the internal structure of crystals. Also, a versatile new tool had been made available for the study of minerals. By X-ray studies the geometrical arrangement of the atoms within a crystal, its *internal structure,* could now be worked out (Figs. 4-5, 4-6, and III-2). The technique could even be applied to grains so small that they can hardly be seen with the petrographic microscope. By X-ray studies, it also became possible to measure accurately the volume occupied by atoms of almost all the elements represented in the crystal, since the distances between similar planes in the geometrical pattern could now be measured. From the spacing and arrangement of the different kinds of atoms, it was possible to analyze more closely the way in which the atoms in crystals are held together.

The Chemical Bond in Crystals

X-ray studies show, for example, that halite (NaCl) has the structure illustrated in Figure 4-5. We have already seen that by the transference of the one lone electron in the outer shell of a sodium atom to the almost filled outer shell of a chlorine atom, both atoms can achieve the stable arrange-

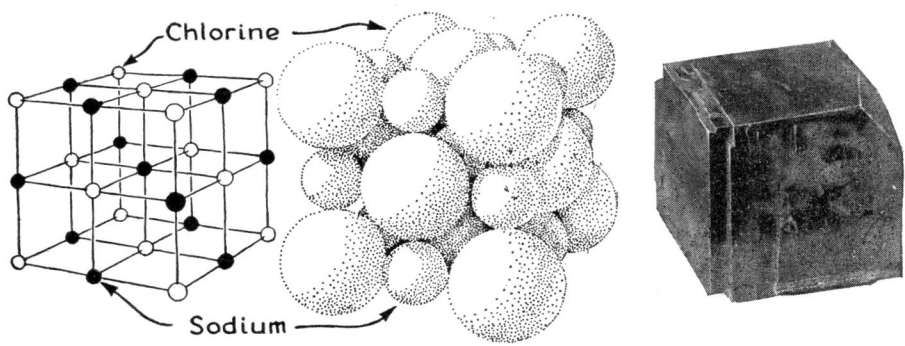

FIGURE 4-5.   *The cubic form,* right, *and internal structure of halite. The lattice diagram,* left, *shows the relative position of Na and Cl nuclei, and the packing arrangement of the ions is depicted in the* center. (*Photo from the Smithsonian Institution.*)

ment of 8 electrons in the outer shell. In acquiring this stable arrangement, each atom sacrifices its electrical neutrality and acquires a charge. It becomes an ion. In a liquid or gas, two ions with unlike charges, such as a sodium ion and a chlorine ion, might be drawn together to form a molecule. In a crystal, the geometrical packing of the particles necessitates a fixed arrangement of the ions, an arrangement that must satisfy the electrical forces set up by the attraction of ions of unlike sign and the repulsion of those with the same sign. Figure 4-5 shows how this is accomplished in halite (NaCl). Each positively charged sodium ion is equidistant from, and at the center of, 6 symmetrically placed chlorine ions. Each negatively charged chlorine ion is similarly surrounded by 6 symmetrically placed sodium ions. Therefore, we say that halite is an *ionic crystal* held together by the electric charges on its symmetrically arranged ions. Most minerals are held together by similar chemical bonds, although, in general, the internal structure is far more complex and not so easily visualized as in halite.

Some minerals—diamond, for example—are held together in another way.

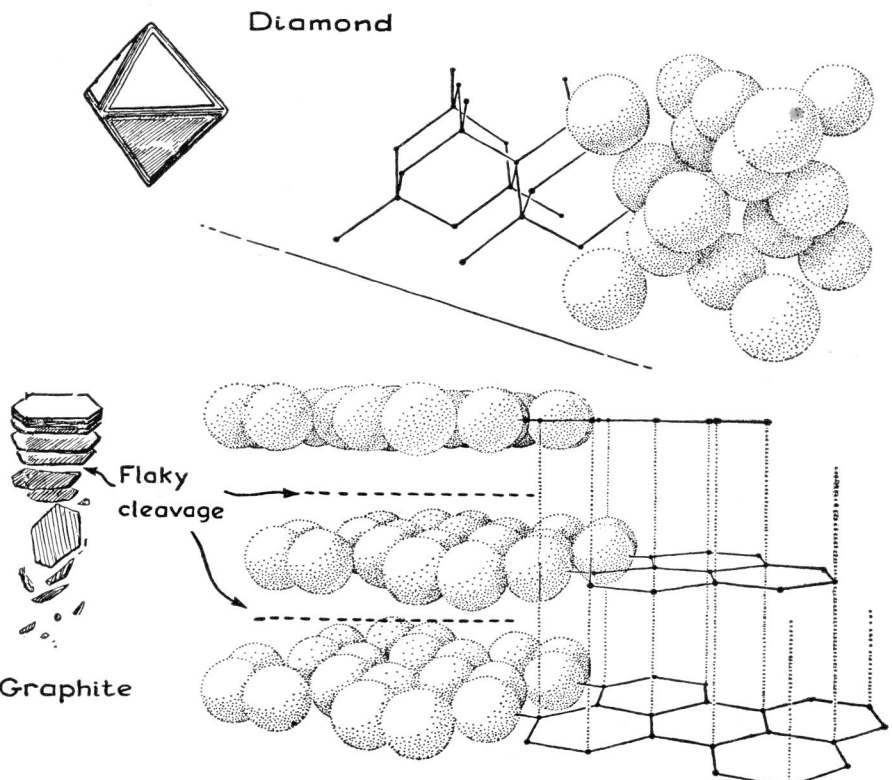

FIGURE 4-6. *The crystal forms and internal structures of diamond and graphite.*

Diamond is composed entirely of carbon; it is one of the crystalline forms of this element. Carbon atoms have 4 electrons in the outer shell. In crystals of diamond (Fig. 4-6), each carbon atom is linked with four others. This linking allows each of the 4 outer electrons in each carbon atom to be "shared" with an adjacent carbon atom. Thus, the carbon atoms in diamond achieve stability; each may be considered to have a complete outer shell of 8 electrons, though every electron is actually shared with a neighboring carbon atom. In terms of atomic structure, each electron may be thought of as oscillating between the orbits of the two neighboring carbon nuclei.

This close bonding of the atoms in diamond is very strong; hence, diamond is the hardest substance known. In this type of crystal, there are no ions—the atoms retain electrical neutrality but are selectively packed in positions so that electrons may be "shared."

It is readily seen that in either ionic crystals, such as halite, or in atomic crystals, such as diamond, there is no particular part of the crystal that can be considered a molecule. The chemist's concept of molecular association of atoms, applicable to gases and liquids, fails to apply to most minerals.

The Chemical Composition of Minerals

Some minerals, like diamond, sulfur, copper, and gold, are elements. Others, like ice (H_2O), quartz (SiO_2), calcite ($CaCO_3$), and kaolinite ($H_4Al_2Si_2O_9$), are compounds whose compositions can be expressed by simple chemical formulas. Nevertheless, as is indicated in the definition of mineral, many minerals vary within certain limits in the percentage of the various elements they contain, and the composition of such a one cannot be expressed by a simple formula. The fundamental distinction among minerals rests not on their chemical composition but on their differing internal structures.

Perhaps the most striking illustration of this is found in the contrast between diamond and graphite. Both are pure carbon; their chemical composition is identical but they occur in different kinds of crystals (Fig. 4-6). Diamond is the hardest known substance, graphite is soft and greasy. Most diamonds are transparent, graphite is opaque. Diamond is used as an abrasive and cutting tool, graphite as a lubricant because it yields fine flakes that glide readily over one another.

Isomorphism

Certain elements may replace one another within a compound so that a range of chemical composition occurs within a single mineral species. Such replacement is called *isomorphism*. The elements that substitute for one another may be chemically similar, but in many of the commonest minerals, the plagioclase feldspars, for example, they are not. Perhaps as simple an isomorphous series as any is the olivine group of minerals. The formula of this group is written $(Mg,Fe)_2SiO_4$, meaning that different specimens of olivine may have chemical compositions intermediate between the two "end members"; that is, they range from pure Mg_2SiO_4 to pure Fe_2SiO_4. It is only the iron (Fe) and the magnesium (Mg) that vary; the proportions of silicon and oxygen remain constant. The intermediate members of the series are regarded as *solid solutions* of the two end members because they are homogeneous crystals and not merely mixed aggregates of two minerals.

Mechanism of Substitution in Isomorphism

The application of X-ray studies to minerals has revealed much of the mechanism by which substitution of one element for another can take place in crystals. The controlling factor in isomorphism is not the number of electrons in the outer shell of the atoms of the two elements concerned, as might have been thought, but their atomic (or ionic) diameters. These are measured in units of length called *Ångströms*. An Ångström is 0.00000001 centimeter long. In olivine, Fe and Mg can readily substitute one for the other, for not only does each contain 2 electrons in its outer shell, but the ionic diameters are very nearly the same.

In many mineral groups, sodium (1 electron in the outer shell) readily substitutes for calcium (2 electrons in the outer shell) because their ionic diameters are very similar (0.98 and 1.06 Ångströms, respectively). But sodium cannot substitute to nearly the same extent for potassium, although each has 1 electron in its outer shell, because the potassium ion is much larger (1.33 Ångströms) than the sodium ion. This discovery was at variance with earlier chemical theory because, since sodium and potassium are very similar in chemical properties, it had been thought they would substitute readily for one another, as indeed they do in aqueous solutions. In crystals, however, a small amount of potassium may be substituted for sodium—the internal structure is warped to take care of the difference in ionic diameters—but, when the substitution exceeds a certain amount, the warping is evidently too great for the structure to remain stable; it breaks up into interlocking crystals of two distinct minerals. Such warping of the internal structure during substitution is illustrated in Figure 4-7, which shows the result of substituting calcium ions for about half of the iron and magnesium ions in the internal structure of olivine, giving the slightly different structure of monticellite, a mineral of closely related chemical composition.

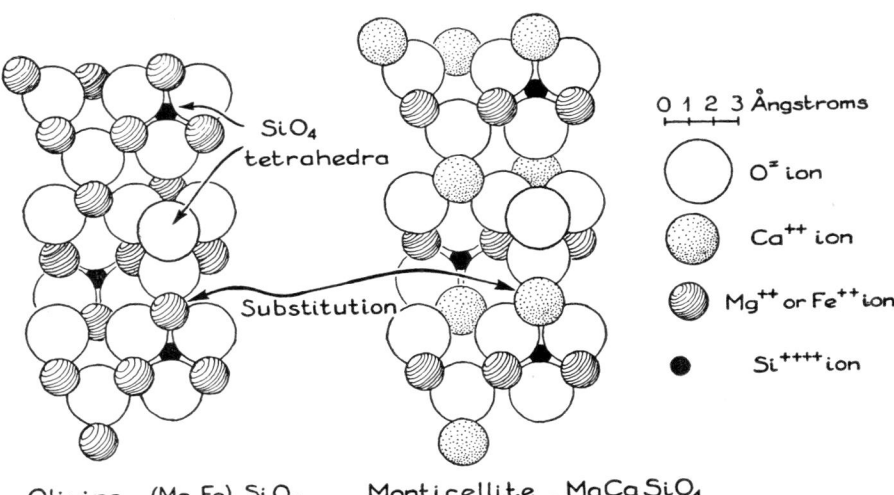

Olivine — $(Mg,Fe)_2 SiO_4$ Monticellite — $MgCa SiO_4$

FIGURE 4-7. *Olivine and monticellite, showing how substitution of the larger calcium ion for ferrous or magnesium ion requires an expansion of the crystal structure. (After W. H. Bragg, 1928.)*

Clearly, an element with one electron in its outer shell cannot be substituted for another with two electrons in its outer shell without destroying the electrical neutrality of the structure; therefore, a second, concurrent substitution is required to maintain neutrality. For example, in the plagioclase series of solid solutions, the change from pure albite ($NaAlSi_3O_8$) to pure anorthite ($CaAl_2Si_2O_8$) takes place by the simultaneous substitution of Ca (2 electrons in outer shell) and Al (3 electrons) for Na (1 electron) and Si (4 electrons). Hence, chemical neutrality is attained, since $3 + 2 = 4 + 1$. The ionic diameter of sodium is nearly the same as that of calcium; the ionic diameter of aluminum is sufficiently near to that of silicon so that the internal structure is not warped enough to become unstable. These slight changes in structure and composition, however, are sufficient to produce small variations in the optical properties.

In many minerals, similar substitutions of one element for another are numerous, and there may be several end members instead of only two. The composition of such minerals can seldom be expressed by a simple chemical formula in which the ratios of the elements are expressed in whole numbers. The formula for the common mineral hornblende

$$(Na, K)_{0.5-2.0} Ca_{3-4} Mg_{3-8} Fe^{2+}_{2-4} (Al, Fe^{3+})_2 (OH)_4 Al_{2-4} Si_{12-14} O_{44}$$

can only be understood in terms of such substitutions between several different end members.

Identification of Minerals

About 2,000 minerals have been recognized and described but only about twenty of them are abundant constituents of the earth's crust. Most of these can be readily identified on sight by anyone who will make a careful study of their ordinary physical properties. See Appendix III, which describes the methods used in identifying minerals, and lists the properties of 30 common minerals.

Facts, Concepts, Terms

MINERAL, ROCK, SOIL

ATOMS, IONS,

ELEMENTS, COMPOUNDS, AGGREGATES

INTERNAL STRUCTURE OF MINERALS
>   Crystal form
>   Constancy of interfacial angles
>   X-ray studies

CHEMICAL BOND IN MINERALS
>   Ionic crystals
>   Atomic crystals
>   Isomorphism or "solid solution"
>   Role of ionic diameter in isomorphism

Questions
(Based in part on Appendix III)

1. Why does the idea of molecules fail to apply to minerals?
2. What factors control the substitution of one element for another in isomorphous mineral series?
3. Can you suggest a reason why the streak of a mineral is often more characteristic than the color of the mineral in bulk?
4. Sketch the internal structure of graphite and halite, and explain why graphite has only one good cleavage, whereas halite has three.
5. Explain how the difference in internal structure of diamond and graphite accounts for their differences in such physical properties as cleavage, hardness, and specific gravity.

6. Why is the specific gravity of quartz definite (2.65), whereas that of pyroxene is variable (3.2 to 3.6)?

Suggested Readings

1. Bragg, W. L., *The Atomic Structure of Minerals*, Cornell University Press, Ithaca, N. Y., 1937.
2. Pauling, L., *General Chemistry*, W. H. Freeman and Company, San Francisco, 1949.

5. Rocks

"The Present is the Key to the Past"

GEOLOGY, like any science, has certain principles and generalizations that help to systematize and interpret the data collected by observation and experiment. The inquiring student should look critically into the validity of these generalizations.

A basic principle in geology, often called the *Uniformitarian Principle,* was proposed by James Hutton of Edinburgh in 1785, and was popularized by the British geologist, Charles Lyell, in 1830. This principle of "uniformity in the order of nature" may be stated as follows:

> "The present is the key to the past," or, applied more specifically to our present subject: Rocks formed long ago at the earth's surface may be understood and explained in accordance with processes now in operation.

The Uniformitarian Principle assumes the *uniform operation of physical laws throughout the geologic past.* It assumes, for example, that in the geologic past, just as today, water collected into streams and carried loads of mud and silt to the sea, that marine organisms lived and died in the ancient seas, and that their shells were buried in the sands and mud accumulating on the sea floor. Hence, if features identical to those we now see in process of formation along streams and beaches can be recognized in and among the solid rocks, it is reasonable to infer that these features were produced by processes of the kind we now see in operation.

The Uniformitarian Principle, like any other scientific "law," rests on the circumstance that all the facts known conform to it. Long and careful searches have failed to find good evidence for ancient conditions totally unlike any existing today. Yet, like most scientific laws, this one must be interpreted carefully and rather broadly. In applying the principle that "the present is the key to the past," we must keep in mind that, although there is good evidence to believe geologic processes have always operated in the same way, they have not necessarily always operated at their present rate or intensity. From evidence developed in Chapter 13, we shall see that the climates of the world were colder some 15,000 years ago and that glaciers were much more widespread than they are now; but there is every reason to believe that the glaciers of that time formed, moved, eroded, and deposited in precisely the same way that glaciers do today.

In the following discussion it will be apparent that some applications of the Uniformitarian Principle may be made to the problems of the origins of rocks.

Three Great Classes of Rocks

The walls of many deep roadcuts show that the surface soil and the unconsolidated mantle rock beneath it form only a thin veneer. At depths of a few feet, we pass through them and enter rock. The deepest wells that have been drilled extend downward into the rocks about 20,000 feet—less than one-tenth of 1 per cent of the earth's radius. Materials now at greater depths cannot be examined, but evidence presented in other chapters indicates that the rocky shell extends to depths of at least several miles, and that some rocks now at the surface were once buried deeply.

Let us briefly summarize a few of the conclusions geologists have reached about rocks, leaving the evidence upon which these generalizations are based until later. Rocks are aggregates of minerals. Three great classes are recognized. The *sedimentary rocks* have been formed at the surface of the earth, either by the accumulation and cementation of fragments of rocks, minerals, and organisms, or as precipitates from sea water and other surface solutions. The *igneous rocks* have formed from molten material that solidified on cooling. The *metamorphic rocks* have been formed by the transformation, *while in the solid state,* of pre-existing rocks beneath the earth's surface through the agencies of heat, pressure, and chemically active fluids.

Rocks are difficult to classify because they grade one into the other. Even the three broad classes of rocks intergrade, and not every rock can be placed unequivocally in one class or the other. Nevertheless, we can classify most rocks into these three great classes and also into smaller divisions, just as it is possible to classify people into tall and short, even though we know there is every gradation between.

Sedimentary Rocks

Everyone has observed how rills formed on a hillside during a downpour of rain spread sheets of mud, sand, and gravel at the base of steep slopes. Similarly, every stream, whether it be a small brook or a great river, carries unconsolidated debris downstream. Most of the rock waste is dropped in sand bars or in beds of silt and gravel in the slack-water parts of the stream course, but it is carried further by later floods and most of it eventually reaches the ocean and accumulates in a delta at the stream mouth or is scattered over the beaches and sea floor by ocean waves and currents.

So commonplace and easily observed are the transportation and deposition of debris by running water that even the Greek philosophers learned to recognize water-borne deposits. They reasoned that beds of gravel and sand clinging to the walls of steep valleys high above the reach of present-day floods were the deposits of former streams. They saw in the clam and oyster shells protruding from some sandstone ledges on inland mountain ridges evidence of former higher-standing seas.

It was a more difficult step—and one probably not made by the ancients —to conclude that a hard, well-consolidated sandstone containing scattered fossil shells and exposed in the gorges of a mountain range is merely the *cemented,* shell-strewn sand of an ancient beach or sea floor that has been warped upward to its present height. In the Middle Ages, churchmen concluded that such fossils, many of which are not closely similar to the shells of animals now living in the sea, were not organic remains but "sports of nature," perhaps put in the rocks by the devil to confuse mankind. Even the accurate and relatively enlightened German scholar, Georg Agricola (1494-1555), in whose words and woodcuts the late medieval Saxon mines and miners are still preserved for us, described only the leaves, wood, bones, and fish skeletons embedded in the rocks as organic remains. To him the fossil shells were "solidified accumulations from water" (whatever that may mean).

The restraints of tradition and authority were not thrown off until the geological pioneers of the late Seventeenth and Eighteenth centuries collected and compared the shells in hard rocks with those in unconsolidated sands near the seashore and, in the years between 1790 and 1815, made maps showing the surface distribution of strata of sandstone, limestone, and other rocks, demonstrating that soft sandstones may grade laterally into hard rock and that in some localities both types contain the same varieties of fossil shells.

Today, by modern tools such as the petrographic microscope, it is only the work of minutes to trace the steps whereby loose sand like that on the floor of the sea has been transformed to solid rock (Fig. 5-1). It is easy to see that the individual grains of the fossiliferous sandstone (Fig. 5-1c), when magnified by the microscope, are of the same shapes and composed of essentially the same minerals as the sand grains of a modern beach (Fig. 5-1a). The fossil shells of the sandstone, even though different from those of animals

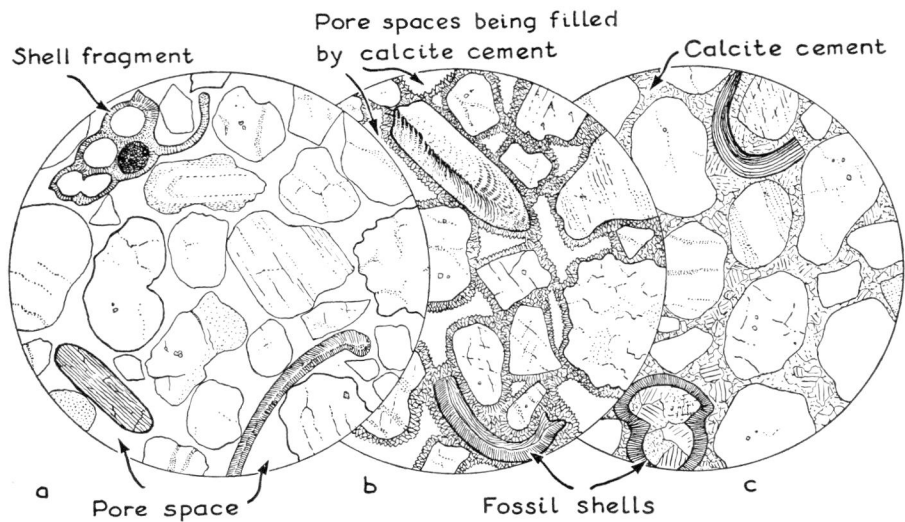

Shell fragment

Pore spaces being filled
by calcite cement

Calcite cement

a

Pore space

b

Fossil shells

c

FIGURE 5-1. *Cementation of sand, as seen under the microscope:* a, *loose sand from an Oregon beach;* b, *partially cemented sandstone from a Brazilian coral reef; and* c, *completely cemented sandstone from Ohio.*

living today, show under the microscope structures so similar as to compel the belief that they are the remains of organisms that lived in the past. Thus, the loose beach sand and the firm sandstone differ only in *cementation*. The voids or pores between the grains of the fossiliferous sandstone have been almost completely filled with mineral matter. By the filling of these voids, unconsolidated sand has been made into solid rock.

The process of natural cementation is slow, but partially cemented rocks are common and, since all intermediate stages have been observed, the inference that sandstone is cemented sand is logical. By digging deeply into some modern beaches, we find that certain layers of sand below the surface, where they are not disturbed by wave action, are coated with a thin film of mineral substance that evidently has precipitated from the solution surrounding them. Although this film may not have grown thick enough to bind the sand grains firmly, complete cementation will surely occur if the process keeps on for a sufficiently long time. More striking examples may be seen along certain stream beds in localities that have hot, dry summers. Here, the newly deposited gravel brought down by a spring flood may, under exceptional conditions, be completely cemented in a single season by the calcite and other minerals precipitated out of the water as the stream dwindles and finally disappears under the hot summer sun. Thus, even modern sediments in process of accumulation show many stages in cementation from unconsolidated debris to solid rock. Yet cementation may also be extremely slow. Some ancient sedimentary rocks are only partially cemented and contain abundant open voids (Fig. 5-1b).

Streams, waves, and currents are not the only agents that deposit sediments. Glaciers and winds also move and deposit detritus which may then be cemented into rock. Organisms are also sedimentary rock builders. Off the shores of northern Australia and islands of the East Indies are great reefs many miles long, made up of the shells of corals, clams, marine snails, and a wide variety of other organisms which live in the shallow, clear water of these warm seas. Even today, the living shells are being cemented together by lime-depositing plants and the interstices filled with microscopic shells and limy (calcite-rich) mud. Such accumulations of cemented shells form the common sedimentary rock called limestone. Limestones that show all the characteristics of these underwater reefs are also found on land far above the reach of present seas. Even the New Guinea natives recognize that some of the limestone in hills near the shore has been formed in the same way as the reefs accumulating offshore today. Similar reef limestones, interstratified with other marine sedimentary rocks, are found in west Texas, the Balkans, and many other places thousands of feet above the sea and hundreds of miles from it. The conclusion that these rocks were once deposited in ancient seas as "coral reefs" of the kind that we now see in the southwest Pacific seems logically inescapable.

Coal is another organic sedimentary rock; the well-preserved cell structures and other plant characteristics visible under the microscope prove that it is made from accumulations of plant remains.

Still other sedimentary rocks, such as the salt deposits of the Bonneville flats west of Great Salt Lake, Utah, are residues from the evaporation of saline lakes—identical beds can be seen in process of forming in parts of the lake a few miles farther east.

Most sedimentary rocks form distinct layers or *strata*. This *stratification* (also called *bedding*) generally results from variations in the supply of sedimentary detritus during deposition, or from changes in the velocity of currents that are laying down the material, or from still other factors. Even a few minutes' observation of a sand bar in a small stream shows that such variations have taken place recently. A smooth-walled pit dug a few inches into the sand bar usually reveals stratification identical with that of many sandstones.

Visits to an ocean beach during and after a storm reveal comparable changes in the coarseness of the beach material. A pit dug by a child in the sand commonly shows layers of varying coarseness recording such variations in current and wave strength.

The strong waves and currents of a heavy storm may partially erode an earlier deposited layer of sand and mud on the sea floor and sweep a sheet of coarse gravel over it, as has been proved by samples dredged from the same spot before and after storms. Even the accumulation of shells in an offshore reef is sometimes interrupted by a fall of ash and pumice from a nearby volcano or by mud swept far out to sea during unusually heavy floods in the rivers of the adjacent land. By such interruptions and accidents during deposition, distinct sedimentary strata are formed; and because such changes vary in intensity and frequency, some strata are thin, others many feet in thickness.

Laws of Sedimentary Sequence

Observations of strata now accumulating make possible the following rather obvious generalizations, which are useful in interpreting ancient sedimentary rocks:

> **In any pile of sedimentary strata that has not been disturbed by folding or overturning since accumulation, the youngest stratum is at the top and the oldest at the base.**

In other words, the order of deposition is from the bottom upward. This is the *Law of Superposition*, first clearly stated in 1669 by Nicholas Steno.

> **Water-laid sediments are deposited in strata that are not far from horizontal, and parallel or nearly parallel to the surface on which they are accumulating.**

Though the bottom surface of a stratum may conform roughly to the irregularities of the base, its top must be nearly horizontal. This is the *Law of Original Horizontality*, also stated by Steno.

Many applications of these generalizations will appear in subsequent chapters; that they are not insignificant truisms may be realized from the fact that in most mountain ranges the strata of sandstone, limestone, and other sedimentary rocks are no longer horizontal. Instead, they are steeply tilted or even overturned. Because we know that these rocks were once sheets of sand, shells, and gravel deposited in nearly horizontal layers and then cemented together, their present distorted attitudes show that great forces must have been at work in the region. Thus, a simple structure like the stratification of a sediment gives us the clue to read the record of great changes of energy in the earth—changes that have deformed and mashed once horizontal sheets of rock into fantastically complex patterns.

Classification of Sedimentary Rocks

Sedimentary rocks are classified and named on the basis of their *texture,* that is, the *size and shape of their constituent particles;* and on their *composition,* that is, *the kinds of material that compose the particles and cements.* Two general subdivisions may be recognized: *Clastic sedimentary rocks* are composed mainly of fragments of rocks and minerals which have been transported to the site of deposition and cemented there; *organic and chemical sedimentary rocks* are composed dominantly of the cemented shells of organisms, or of precipitates from aqueous solutions such as sea water.

Hundreds of different kinds of sedimentary rocks have been described and named but most of them are comparatively rare. For an elementary knowledge of geology, learning to recognize the common ones listed in Table IV-1 of Appendix IV will suffice. In addition, this appendix contains a list giving full descriptions of each of the individual rock groups. The student should turn to the appendix at this point and read the descriptions of *sandstone* and *limestone*—two rocks that have been given more than passing mention in the preceding pages.

The rock descriptions have been relegated to an appendix for convenience of use in the laboratory. The descriptions should be studied with a specimen of the rock at hand, so that one can note and compare its properties with those listed in the description. In making such a comparison, do not be dismayed if specimen and description do not correspond exactly, for, as we remarked earlier, rocks vary widely and they grade into one another.

The brief notes on the origin of each of the rocks listed in the appendix are intended only to present summaries of conclusions reached by geologists from evidence outlined in this and later chapters. To give all of this evidence here would involve lengthy presentation of indirect as well as observational data. The rock tables and lists in the appendix might be said to constitute a résumé of the subject as a whole, for rocks are the basic documents of geology and the sedimentary rocks are the "dated" documents.

Igneous Rocks

At active volcanoes, also, we can see rocks being made. Molten material rises to the surface and flows for a time in lava streams, but eventually cools and hardens into *igneous rock.*

Two subdivisions of the igneous rocks are recognized. The lavas and solid fragments erupted from volcanoes are called *volcanic rocks.* They are composed in large part of microscopic mineral crystals and glass. In a few places volcanic rocks can be seen to grade into rocks composed of much larger crystals. These coarse-grained igneous rocks are called *plutonic rocks,* and are widespread at the earth's surface, but have not been observed in process of formation. As we shall see, there is good reason to believe that the plutonic rocks were not spewed out to the surface like the lavas, but solidified deep underground. Subsequently, the roof rock that covered many of these buried masses was eroded away, thus exposing the plutonic rock at the surface.

Volcanic Rocks

In January 1938, a white-hot stream of molten lava issued from a fissure near the base of the volcano Nyamlagira, in central Africa, and poured quietly downward into a forested plain below. For two years and four months the lava continued to emerge until more than 500,000 cubic yards of molten rock had devastated an area of over twenty-five square miles. Finally the flow ceased, the fissure froze over, and the lava congealed into the black slaggy rock that we call basalt. Similar but smaller flows have been observed at many other volcanoes.

Some volcanoes erupt explosively, blowing vast quantities of volcanic "ash" (fine bits of volcanic glass and pumice, see p. 607) and broken rock fragments into the air. Around volcanoes such as Mount Katmai in Alaska, Bandai-San in Japan, or Mount Pelée in the West Indies, large areas have more than once been buried under fifty feet or more of hot ash and rock. Such loose debris, erupted from Vesuvius in Italy, has been known to consolidate into firm, coherent rock within a generation.

That rocks were made by volcanic action was well known to the peoples of early civilizations because of the many active volcanoes in the Mediterranean countries and in Persia. Early writings contain a wealth of information on Vesuvius in Italy, Santorin in the Aegean Sea, Etna in Sicily, and Erebus in ancient Persia. The eruption from Vesuvius that overwhelmed Herculaneum and Pompeii in 79 A.D. was but one of a series of volcanic disasters that impressed on early peoples both the awesome power of volcanic eruptions and the characteristics of the lavas and explosive products that come from them.

Nevertheless, the ancients did not recognize that volcanic rocks are widely distributed over the surface of the earth in areas far removed from volcanoes now active. It is one thing to stand on the brink of a fissure and watch liquid lava emerge, roll down a slope, and congeal into a mass of basalt, and another to recognize a basalt flow that was extruded millions of years ago, and has since been detached from its parent cone or fissure by erosion, or buried under a later accumulation of sedimentary rock. From their manner of formation, it is common for volcanic and sedimentary rocks to be inter-stratified. In the Samoan Islands, flows of basalt have been seen to spread over reefs in which limestone was forming; today, deposits of coral and shells are collecting on the upper surface of the congealed lava. The great flow at the base of Nyamlagira covers older volcanic material, but earlier flows in the same region spread out over a plain underlain by lake and river deposits and, in turn, were partially buried under new deposits from the lakes and rivers. Scarcely any thick pile of sedimentary rocks is completely free from volcanic interlayers.

It is not surprising that flows of lava and beds of volcanic ash interstrati-fied with sedimentary rocks were considered to be hardened sediments by the early geologists, just as they often are by the uninitiated today. It was not until after much careful field study that reliable criteria were put forth distinguishing buried lava flows from sedimentary deposits. Indeed, fifty years of violent controversy took place before the volcanic origin of basalt was definitely established.

The Controversy over the Origin of Basalt

The interpretation of geological phenomena is often influenced by the philosophy of the worker. The history of geology is replete with examples of unsuccessful attempts to fit the features seen in the field into the precon-ceived notions of the observer. One of the classical examples of the conflict between theoretical and field interpretations was the controversy concern-ing the origin of basalt, a controversy that raged from about 1775 to 1822.

Some of the hills of Saxony near the famous mining academy of Freiberg (Bergakademie) are composed chiefly of sedimentary rocks, but interstrati-fied with them are a few layers of a hard, dark-colored rock long ago named basalt. The basalt is more resistant to erosion than the sedimentary rocks with which it is associated and appears in picturesque colonnaded cliffs at or near the summits of many of the hills.

In 1775, the Stolpen, one of the basalt-capped hills, was visited by Abra-ham Gottlob Werner, professor of mining and mineralogy at Freiberg, a scientist who was destined to wield great influence on the development of geology ("geognosy," as he called it). From his observations on this and later visits, Werner wrote, in 1787, that the hill showed ". . . not a trace of volcanic action, nor the smallest proof of volcanic origin. . . . After further more matured research and consideration, I hold that no basalt is volcanic, but that all these rocks, as well as the other Primitive and Floetz* rocks, are of aqueous origin."

---

* The terms "Primitive" and "Floetz" refer to two subdivisions in Werner's classification of rock strata. These terms are not used today.

Following up his idea that all basalts were precipitated from the ocean, Werner developed a "system." With all the precision that he applied to the classification and organization of the mineral specimens in the laboratory collections at Freiberg,* he proceeded to divide the crust of the earth into a series of "Universal Formations." These, he taught, were all precipitated from a primeval ocean, and could be definitely recognized all over the world, each formation having the same character and occurring in the same order no matter in what country it might be found.

Werner's personal charm was great, and he attracted large numbers of able students whom he fired with great zeal. They came to believe that the "Universal System" of the Master would unlock the geologic history of every country.

Some of them were rudely awakened; in fact, Wernerism received its first really serious setback from two of Werner's own students, D'Aubuisson and von Buch. Both of these young men became disillusioned with Werner's "system" after visiting the Auvergne region in central France. The Auvergne has had no eruptions within historic times but it contains almost perfectly preserved craters, lava flows, and other volcanic phenomena.

The Auvergne had been studied several years before by Nicholas Desmarest (1725-1815), a hard-working government official who in his spare time published several excellent papers on geology. Among these, his treatises on the Auvergne volcanoes have won for him the title of "Father of Volcanology."

How different was the approach of this thorough observer from that of the dogmatic Werner to essentially the same problem! In his first journey into the Auvergne in 1763, Desmarest found a cliff of basalt. Searching at the base of the cliff, he noticed that the soil beneath the flow had apparently been burned and hardened. He also noticed that the basalt grades into and contains masses of *scoria*, a coarsely frothy basaltic rock filled with small, spherical holes. Scoria is common at the base, and even more abundant in the upper part, of basalt flows; it has been observed to form when rising bubbles of steam are caught by the congealing of the sticky lava around them. When he first visited the Auvergne, Desmarest had never seen moving lava from an active volcano. By careful observation and by reasoning, however, he established two of the criteria now generally used in the recognition of ancient basalt flows—the baking of the ground beneath the flow and the presence of scoria formed by the congealing of bubble-filled lava. Desmarest, however, did not proceed forthwith to establish a "system." His

---

* Because of the errors he promulgated about the structure of the earth's crust and the origin of basalt, the remarkably effective work that Werner did in mineralogy is often neglected or forgotten. The science of mineralogy in his day was a chaos of jumbled terminology and haphazard description; he reduced some of this to order, and provided the original impetus that has led Germany to excel in this branch of geology.

curiosity was aroused, but he drew no certain conclusions. He decided to examine and map the outlines of the entire flow. This work disclosed similar features at many other points along the base of the flow. He also traced the flow to its source at the base of a round, steep-sided hill which still retains the characteristic form of a volcanic cone.

Still not entirely satisfied, Desmarest decided to plot on a map the distribution of all the different kinds of rocks of the Auvergne. By carefully following the junctions between the lavas and other kinds of rocks, and by plotting these junctions or *contacts* on maps, he proved that the volcanic history of the Auvergne was very long. Some eruptions had been followed by long quiet periods during which streams cut valleys in the flows and removed much of the ash from the cones. Then, these newly cut valleys were filled and obliterated by renewed volcanic activity. Eventually he traced out three main cycles in the volcanic history. His map, one of the first geologic maps produced, is a monument to his good judgment and ability to interpret field relations. Desmarest's maps will stand comparison with modern maps of similar volcanic districts.

Thus, nearly 200 years ago, the failure of Werner's speculations to withstand the test of Desmarest's rigorous field observations showed geologists that the ultimate worth of a theory can only be found in the field. Ironically, although most of Desmarest's reports were published before Werner tried to compress all the vagaries of volcanic action, sedimentation, and erosion into a "Universal System," they remained almost unnoticed for many years. While Werner's teachings and ideas were sweeping over Europe, Desmarest took no part in the controversy. When asked for his opinion on the origin of basalt, he would reply: "Go to the Auvergne and see."

Eventually D'Aubuisson and von Buch, eager to establish the "Universal System" of their teacher in other countries, visited the Auvergne. How great was their disillusionment as they followed, step by step, the same evidence and lines of reasoning that had led Desmarest to his conclusion! Here in the Auvergne, D'Aubuisson and von Buch saw the burned and hardened ground on which the lava had flowed, and the scoriaceous tops and bottoms of the flows where rising gas bubbles had been trapped in the congealing lava. All these features could have been seen at Stolpen had Werner taken his students there on field trips instead of merely lecturing about the "Universal System" and giving them specimens of basalt to study in the laboratory. Even more devastating for the Wernerian doctrines—D'Aubuisson and von Buch perceived that the lavas of the Auvergne capped not only sedimentary rocks (as at Stolpen) but nearby they also rested on granite, which supposedly occupied quite a different position in the "Universal System." The reports of D'Aubuisson and von Buch did much to overthrow the theory that basalt was a precipitate in the "Universal Ocean," although other students of the great teacher, still content to work in the laboratory instead of examining rocks in the field, continued to promulgate the Wernerian doctrines.

The controversy over the origin of basalt had happy results for the development of volcanic geology. Interest in ancient volcanoes spread wide and far and bore abundant fruit, particularly in the British Isles, an area uniquely rich in varied and spectacular volcanic phenomena, notwithstanding the fact that none of the British volcanoes has been active in historic time.

Plutonic Rocks

As the details of lava streams and deposits of volcanic ash became better understood, the British geologists turned their attention to the conduits through which the molten rock, or *magma*, as such natural silicate melts are called, had reached the surface. In the Midland Valley of Scotland and in parts of the Hebrides, erosion has swept away most of the lava streams and cones and has cut deeply into the foundation beneath. Here, as in northwestern New Mexico and in central Oregon, erosion has displayed in clear and diagrammatic fashion the conduits that fed the volcanoes. Many of these feeders are simple, pipelike masses (*volcanic necks*) or filled fissures (*dikes*); but some connect with other subterranean igneous masses much more complex in form and origin.

Soon a wide variety of different kinds of igneous bodies formed by the solidification of magma underground had been mapped and described. These are called *intrusive igneous bodies* in contrast to the *extrusive bodies* formed by magma erupted to the surface. The rocks that have solidified within them are called *plutonic rocks,* in contrast to the *volcanic rocks* that have formed at the surface.

Observations of hundreds of different intrusive masses have shown that plutonic rocks are characterized by coarser mineral grains than comparable lava flows. This has led to the inference that the slower-cooling underground affords time for the mineral grains to grow larger. In large intrusions, some of which are many miles in diameter, the grains are easily visible, and the minerals can be readily identified without the microscope. Laboratory investigations prove that heat is transmitted by solid rock less readily than it is carried off by rising air (convection currents) above a molten lava flow. Hence, the roof rock acts as a blanket which permits only slow cooling of a magma emplaced below the earth's surface.

Rounded gas bubbles are absent or at least very rare in plutonic rocks, apparently because of the considerable pressure of the roof which keeps the gas in solution in the magma. The gas collects into bubbles only when the pressure is released, as when magma is suddenly extruded from depth to a point upon or near the surface.

Although some intrusive bodies can be traced continuously into recognizable lava flows so that their origin from a common magma is obvious, most are separate bodies. These may never have had any direct outlet to the earth's surface. That these masses were likewise formed by crystallization from a magma is proved by several lines of evidence: They commonly harden, or even completely recrystallize, the rocks along their contacts; tongues and stringers from them penetrate into cracks in the adjacent rock in the manner of a liquid; they commonly show chilled borders of fine-textured rock similar to that of lava flows; and they display, with minor

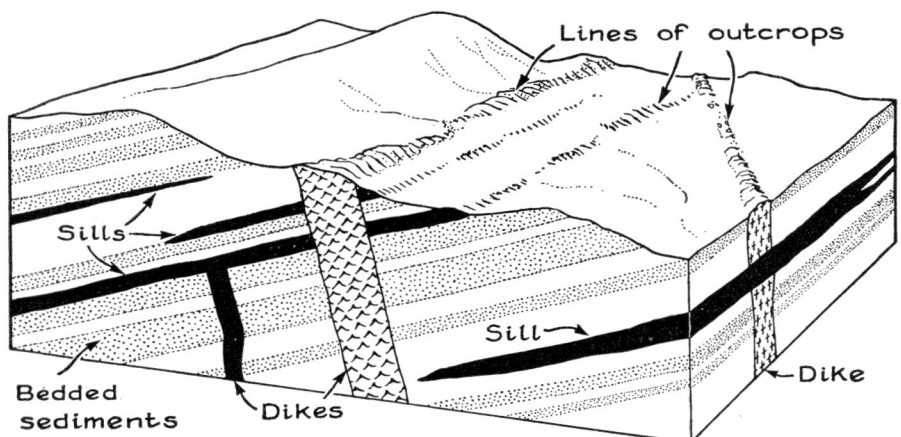

FIGURE 5-2. *Sills are concordant, dikes are discordant tabular intrusions.*

exceptions, the same mineral and chemical compositions as surface flows.

The varied forms assumed by intrusive igneous rock bodies are systematically described in Chapter 17, but one form, called a *sill*, is worthy of special comment here. Most intrusive bodies cut across the bedding of the enclosing sedimentary rocks, hence they are called *discordant*, but a sill is a rock mass congealed from magma forced *concordantly* between sedimentary strata. The magma has spread out like the grease squirted between metal surfaces by a grease gun (Fig. 5-2). A sill might thus occur low in the sequence and yet be much younger than strata hundreds of feet above it. Here, then, is a possible source of error in applying our generalization about the sequence of strata (p. 73), for how can we tell an intrusive sill from a lava flow that has been buried under a later accumulation of sedimentary rocks?

Contacts of Rock Masses

The key to this question, as to many others in geology, lies in the interpretation of the contacts between rock bodies. A mass of a single kind of rock, whether sandstone, basalt, or any other, does not extend indefinitely; somewhere it must end against another. These common boundary surfaces of adjacent rock masses are called their *contacts*. There are two general kinds: *abrupt*, with a definite surface of junction; and *gradational*, in which there is no sharp boundary but an intermediate zone of greater or less thickness in which the transition from one rock mass to the other takes place.

In interpreting the relative age of two rock masses in contact, the following generalizations are useful. They apply not only to problems of sequence among igneous rocks, but to all kinds of rock bodies:

Of two rock masses in contact, that one is younger which contains within it fragments or inclusions of the other.

Thus, a sill might be expected to enclose, at some places near its upper surface, fragments torn from the overlying stratum at the time of intrusion. On the other hand, fragments of a buried lava flow are likely to be found as inclusions in an immediately overlying sedimentary stratum, since loose

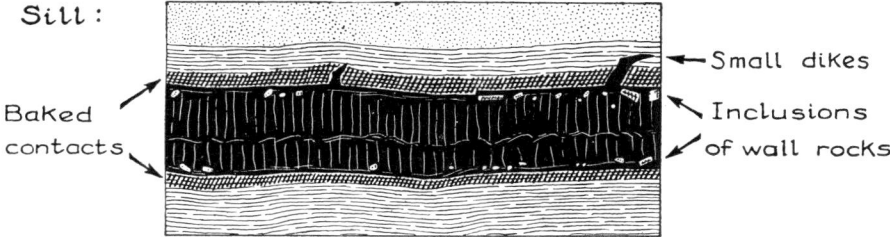

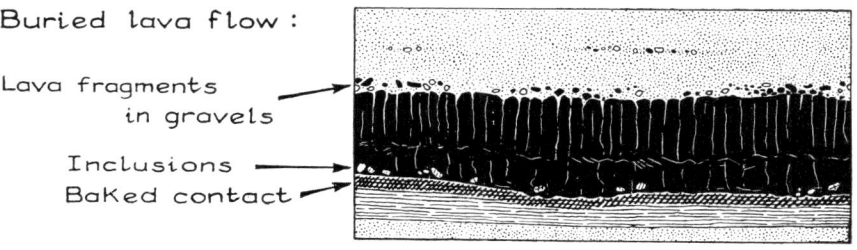

FIGURE 5-3. *Criteria for distinguishing between a sill and a buried lava flow.*

pieces of lava and scoria would probably be picked up and mixed with the overlying detritus by the moving agent that deposited the sediment (Fig. 5-3).

If a rock sends tongues and branches into another, it is younger than the rock it penetrates.

The sedimentary strata above most sills, or for that matter the older rock in contact with any igneous intrusion, commonly contain tongues or dikes of igneous rock formed by the magma penetrating fissures or other openings in the older rock and solidifying there.

Although this rule applies mostly to igneous rocks, sediments may also, in places, penetrate adjacent rocks. Thus, cracks and openings in a lava flow may contain debris identical with that of an immediately overlying sedimentary rock. Presumably, some of the unconsolidated detritus filtered into the

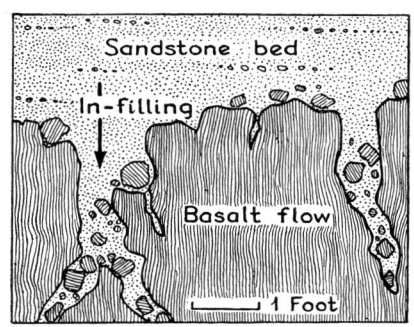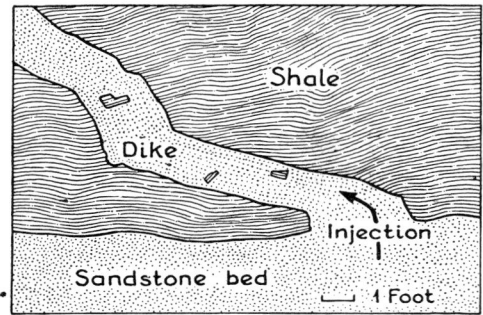

FIGURE 5-4. *Sandstone dikes; left, near Stanford University, California;* right, *Santa Monica Mountains, California.*

cracks during deposition of the sediment and cemented there into rock (Fig. 5-4, *left*).

The generalization must be used with care, however, when applied to sedimentary rocks. For example, an underlying sandstone may send dikes upward into overlying shale (Fig. 5-4, *right*). This is an exception to the rule as stated, though of course the rise of the sand into dikes actually occurred after the formation of the overlying shale. What apparently happened was that underlying, unconsolidated sand was forced into cracks opened in overlying rock. Presumably cementation of the sandstone was delayed until after the shale had been laid down and consolidated (Fig. 5-4, *right*).

If an igneous rock bakes or alters another rock with which it is in contact, it is younger than the rock it bakes.

So stated, this is an obvious truism, but to recognize such contact alteration is not always easy. Some rock masses are exceptionally well cemented or impregnated with mineral matter along their contacts because of later percolating waters. Along other contacts, rocks have been discolored by water stains until they may look like chilled zones. Generally, careful observation or even microscopic study is needed to distinguish these spurious features from actual baking or chilling.

We have already seen how Desmarest noted the alteration of the rock beneath the Auvergne lavas. A sill will alter not only the strata that form its floor, but also those of its roof. A moving lava flow has no roof, and lava congeals before any considerable sedimentary cover can be deposited upon it.

In general, the criterion of alteration by a hot magma can be more readily applied to intrusive rocks than to lava flows. The interior of a lava flow commonly remains molten long after a thick rind of congealed lava has formed on its surface. Many flows have been seen to move forward beneath this rind; during movement the still-liquid interior breaks up the solidified crust and rolls blocks of it under the front of the advancing flow. Such blocks are commonly so cool that they fail to bake the material beneath—or at least baking is so slight that it is not readily recognized. But intrusive masses of igneous rock invariably alter their walls, at least to some degree.

The Intrusive Nature of Granite

As careful geologic mapping in many areas of volcanic and shallow plutonic rocks disclosed more and more about the contact relations of intrusive igneous rocks, a spirited discussion arose over the origin of granite, a common, coarse-grained rock composed chiefly of feldspar and quartz. The origin of granite is still a topic of lively discussion but many points have been clarified by careful study of granite contacts.

Geologic mapping in many different countries shows that granite and the similar rock, granodiorite (p. 608), are among the most abundant rocks in the accessible part of the earth. They are found in great monotonous bodies hundreds of square miles in extent. At many places, eroded granites are covered with sedimentary rocks that contain pebbles of the underlying granite and are therefore younger than the granite. Other bodies are covered by younger lavas and beds of volcanic ash. These relations persuaded some early geologists to regard granite as the earth's "original crust," formed when the earth solidified from a molten state, though, as we have seen, Werner considered it to be the first precipitate out of a universal ocean.

The idea that all granites are old, however, was not left unchallenged. In many localities, geologists began to note that granitic dikes, tongues, and complicated offshoots of considerable size penetrate the overlying rocks, indicating that the granite had injected the roof rocks as a molten mass. Only in scale of the injection features—in miles or thousands of feet rather than in feet or inches—were the contacts different from those of small intrusive masses. Furthermore, the invaded rocks at such contacts were altered, and, in places, so thoroughly recrystallized that it was difficult to tell whether they had originally been sedimentary or volcanic rocks. Such changes in the wall rocks suggested the effect of high temperatures. More detailed field studies also showed that most of the great expanses of granite, far from being as monotonous as had been supposed, actually embrace several distinct intrusions of granite and related plutonic rocks, the younger ones penetrating the older. Also, along many contacts, fragments of the adjacent older

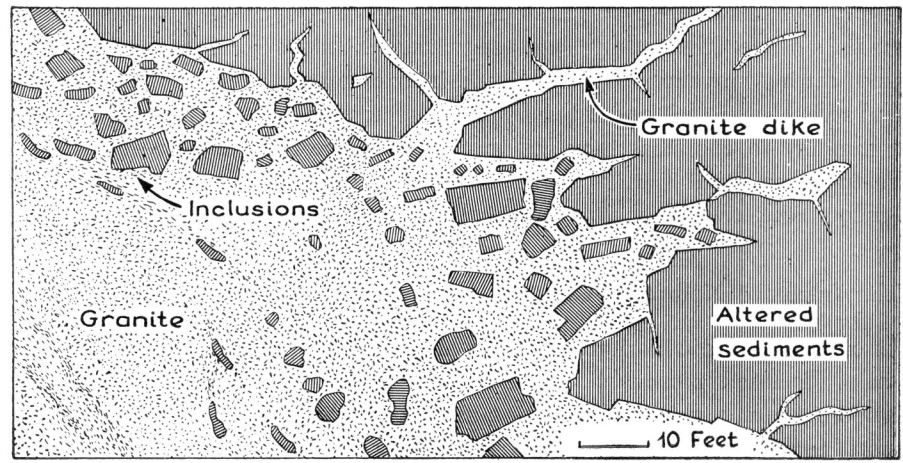

FIGURE 5-5. *Intrusive relations at a granite contact.*

rock are strewn through the margin of the granite in large numbers as though they had been pried off by the intruding granite magma. (Fig. 5-5).

Today, no remnants of the hypothetical "original crust" have survived the results of detailed mapping. At some points along their contacts, most granite masses can be seen to have invaded and altered rocks of sedimentary or volcanic origin, although elsewhere younger sedimentary and volcanic rocks may have been deposited upon the eroded granite surface.

Thus, many granites were shown to be igneous rocks, congealed at considerable depth. Diked contacts and other injection phenomena also indicated that some masses were not single intrusions but had been formed by successive invasions separated by intervals of inaction. This recalls the Auvergne, where Desmarest demonstrated three distinct effusive periods separated by long quiet interludes. Apparently the emplacement of large igneous bodies, whether at the surface or far beneath it, is a slow and some-what intermittent process.

If we left the subject of the origin of granite here, however, we would be greatly oversimplifying a complex problem. Not all granite contacts are either clearly erosional or clearly intrusive. Many of them are gradational, the granite appearing to fade gradually into rocks of undoubted sedimentary or volcanic origin. Also, certain rocks, definitely granites in composition and texture, contain faint, nebulous patterns that resemble stratification, outlines of pebbles, or other structures found in sedimentary or metamorphic rocks. The origin of these transitional contacts and of granites with ghostlike foreign structures is a fascinating problem that has not yet been completely solved. It will be further discussed in Chapters 17 and 19.

Classification of Igneous Rocks

With the petrographic microscope, much information can be obtained about the minerals in an igneous rock, including their kinds, shapes, sizes, patterns of arrangement, order of crystallization, and the alterations brought about in them either by hot gases from the magma or by outside agencies such as air and atmospheric moisture. Great variation in so many different features yields infinite possibilities for classification, and so literally hundreds of kinds of igneous rocks have been discriminated and given separate names. How-ever, the great bulk (approximately 95 per cent) of the igneous rocks can be lumped into about 15 major groups (see the Appendix, Table IV-2). This simplified list suffices for our purpose; we shall not need names for the numerous varieties within each group, or for the rare groups not included. Such a simplified classification of the igneous rocks can be based on the minerals and textures visible to the eye, thus dispensing completely with the microscope. The classification outlined in Table IV-2 is widely used in field work where laboratory equipment is not immediately available.

Comparison of mineral composition and the bulk chemical composition of thousands of igneous rocks shows that *the kinds and amounts of different minerals depend, in general, on the chemical composition of the original magma. Magmas rich in silica yield, on cooling, large amounts of feldspar and quartz; magmas low in silica form rocks rich in the ferromagnesian minerals* such as pyroxene, olivine, and amphibole.

Textures of Igneous Rocks

The size of the mineral grains in an igneous rock depends chiefly on the rate of cooling, although bulk chemical composition of the magma plays a part.

It has been inferred from field observations, and confirmed by laboratory experiment, that a high content of water and other volatile substances promotes the growth of larger crystals. Field observations clearly indicate that when magma congeals slowly in a large subterranean mass, the minerals commonly grow large enough to be readily identified with the eye. If the magma forms a lava flow, however, rapid cooling in contact with the air prevents the growth of large crystals, and, instead, the rock congeals to mixtures of microscopic mineral grains and glass. If cooling is even more rapid, the magma may congeal into a glass containing almost no crystals.

These differences in *degree of crystallinity* and in *size of crystals* determine the *texture* of the igneous rock. A rock is said to have a *granular texture* if its mineral grains have grown to a *size large enough to be seen and identified without the aid of lens or microscope.* In different rocks, the average size of the grains may vary from about 0.5 mm. to more than 1 cm. in diameter, but the common coarse granular rocks such as granite have grains averaging from about 3 mm. to 5 mm. in size. In rocks with *aphanitic texture,* the constituent minerals are less than 0.5 mm. in diameter; they are commonly mere specks, too small to identify without the microscope. In many aphanitic rocks, the microscope shows a considerable residue of uncrystallized glass between the minute crystals, but others are completely crystalline. Rocks with *glassy texture* are almost entirely volcanic glass, though even they may contain a few scattered crystals.

Pyroclastic texture is the texture of rocks formed of volcanic ash and rock fragments blown out of a volcano and cemented together in the manner of sedimentary rocks, or else welded together by heat from the chemical reactions that take place between the gases in the eruption cloud.

Many igneous rocks contain crystals of two widely different sizes and therefore appear spotted; these rocks are said to have a *porphyritic texture* and are often loosely called "porphyries" (p. 609). Because porphyritic texture is most common in small intrusive bodies or in lavas, it has been attributed to *a change in the rate of cooling while the magma was crystallizing.* The process is hypothetically explained as follows: A large body of magma underground may cool to the temperature where one or more minerals begin to crystallize. Because cooling is slow, the crystals of these minerals grow to considerable size. If, when the magma is perhaps half crystallized, a fissure opens in the roof of the chamber, some of the magma with its suspended crystals may escape to form a lava flow at the surface. The still-liquid portion of the magma quickly freezes and surrounds the large crystals, or *phenocrysts,* with a *groundmass* of crystals of aphanitic size. The phenocrysts were formed underground, the aphanitic groundmass at the surface. Such a lava has a *porphyritic aphanitic* texture. Rocks with *porphyritic granular* texture—large crystals in a granular groundmass of finer grain—are common in some intrusive bodies. *Porphyritic glassy* texture is seen in some lava flows. Rarely, conditions other than a change in the rate of cooling may locally produce porphyritic rocks.

With the five basic textural distinctions—granular, aphanitic, glassy, porphyritic, and pyroclastic—we can build a table of the major groups of igneous rocks such as is presented in Appendix IV, Table IV-2. Following the table are descriptions of the different rock groups. The student should now turn to the descriptions of *granite* and *basalt,* given on p. 608 and p. 607. The descriptions of the other rocks in the table can be more advantageously studied in the laboratory where specimens are available.

Metamorphic Rocks

We have seen that the sedimentary origin of many rocks can be confidently inferred from their stratification, water-worn pebbles, fossils, or other characteristic features. Other rocks display textures and mineral compositions that, by analogy with those of rock masses of known volcanic or plutonic origin, permit us to classify them just as confidently as igneous rocks.

There remains a third great group of rocks in which diagnostic features of igneous or sedimentary origin either are absent or have been so obscured and altered by the growth of new minerals and new textures that they are scarcely recognizable. Thus, in places, we find rocks showing rounded pebbles and stratification much like that of normal conglomerates, but the pebbles are ellipsoidal instead of spherical; the matrix between them con-

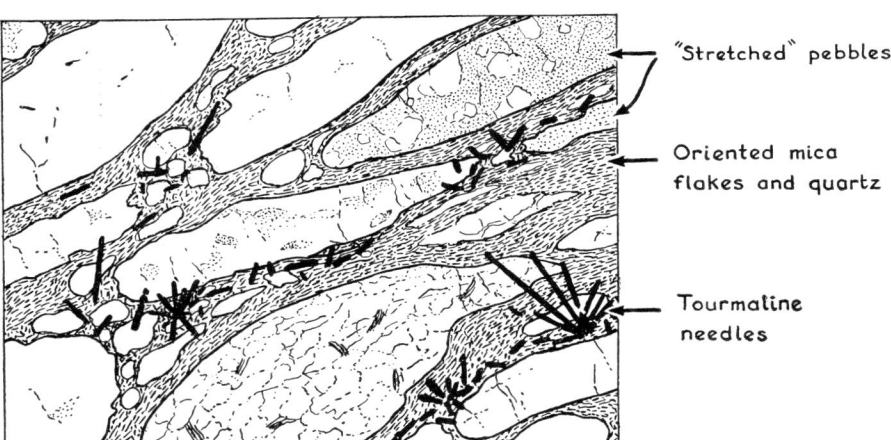

FIGURE 5-6. *Metamorphosed conglomerate with tourmaline needles cutting the other constituents. (Enlarged about four times.)*

sist not of sand and clay as in ordinary conglomerate, but of lustrous foils of mica and quartz; and, cutting across two or more adjacent pebbles and the intervening matrix, are long delicate needles of tourmaline (Fig. 5-6). It is inferred that the rock was once a conglomerate and that the tourmaline could not have been present when the conglomerate was deposited but must have grown in the solid rock long after deposition. Apparently the original round pebbles have been squeezed and stretched, and the sand and clay originally present in the conglomerate matrix have been altered to mica and quartz. Similarly, igneous features such as phenocrysts or flow banding may locally be recognized in their metamorphosed equivalents.

Still other rocks show no trace whatever of the textures of known sedimentary and igneous rocks. Yet, in places, they grade imperceptibly into rocks in which partially obliterated igneous or sedimentary structures still survive.

From such relations, the conclusion that this last great group of rocks represent altered or *metamorphosed* rocks is well established. We therefore recognize this third group as *metamorphic rocks*. There can, from the nature of their origin, be no sharp boundary between metamorphic rocks on the one hand and the igneous or sedimentary rocks from which they were derived on the other.

The survival of relic features such as pebbles or phenocrysts in metamorphic rocks must mean that the original rock did not melt in attaining the metamorphic condition. Instead, the alteration appears to have been accomplished mainly by recrystallization while the rock remained essentially solid.

Foliated Metamorphic Rocks

Most metamorphic rocks have a banded or layered structure called *foliation*. The layers may be coarse bands ¼ inch or more thick, as in gneiss (see p. 612), or layers thinner than a sheet of paper, as in slate. The rock usually splits readily along these planes. The foliation appears to record widespread pervasive movement within the rock mass, during which most of the original minerals were broken, streaked out, and recrystallized into new minerals. Because movement is essential to their development, the foliated rocks are called *dynamic metamorphic* rocks.

The Origin of Slate

When geology was emerging as a science, the coarse-grained, faintly foliated rocks were usually called gneisses and classified with the granites, and the fine-grained, well-foliated ones with the sedimentary rocks. For example, the metamorphic foliation in common slate, a rock much used for roofing, flag-stones, and other building purposes, was erroneously considered to be strati-fication. Geologic study of European and American slate quarries disclosed, however, that most slates have two distinct layered structures. One can be seen to cross and displace the other. The older of these is parallel with other interlayered rocks like limestone or quartzite, and with alternations in grain size, color, and composition; it is the true sedimentary stratification. The rock, however, does not break or split along the stratification planes as in normal sedimentary rocks except when the younger structure, the true meta-morphic foliation, happens to coincide with it. When foliation and stratifica-tion do coincide, recognizable but distorted fossils are found locally on the

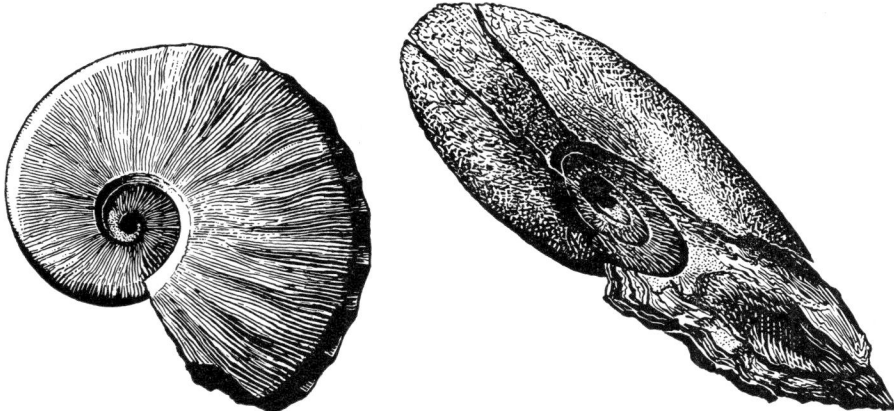

FIGURE 5-7.   *The fossil at the left was collected from unaltered limestone in Idaho; that at the right, the identical species, occurred in slates in the Inyo Moun-tains, California.*

foliation surfaces. Where these fossils can be compared with like specimens from otherwise little-changed sedimentary rocks, they are seen to have been greatly thinned in the direction at right angles to the foliation and are stretched out parallel to the foliation (Fig. 5-7). If the foliation cuts the stratification at a high angle, however, it is almost useless to look for fossils, because the rock will rarely split along the stratification planes where the fossils occur. Furthermore, each foliation plane, where it crosses the strati-fication, commonly breaks and offsets the stratification planes by a small (usually microscopic) distance (Fig. 5-8), showing that it is a structure younger than the bedding. Thus, a fossil may be broken and displaced by so many microscopic shears as to be unrecognizable.

In most slate quarries the foliation is nearly parallel throughout, although the strata themselves may be thrown into folds and become highly distorted. The foliation cuts through the folds, intersecting the strata at all angles.

From the relics of stratification and the occasional battered fossils, it is clear that most slates were derived from fine-grained sedimentary rocks, such as shale. This is confirmed by chemical analyses which show no essential difference between shale and most slates, except that the slates contain less chemically combined water than the shales. Field and laboratory studies show, however, that not every slate is derived from shale; some were formed from other fine-grained rocks such as volcanic tuff.

Only by the use of the petrographic microscope and by X-rays can we unravel the true nature of the foliation. The minerals of slate are exceedingly

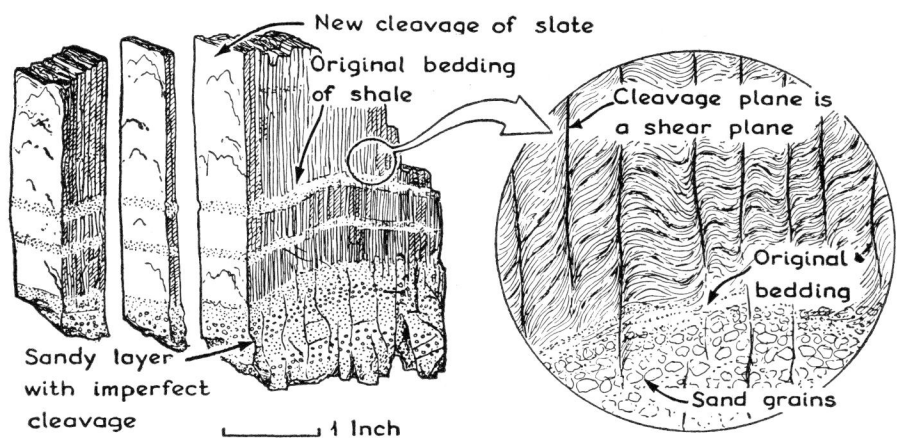

FIGURE 5-8. *Fragments of slate, left, showing relics of original bedding. The enlargement shows small offsets along the cleavage surfaces.*

fine grained—indeed, most are ultramicroscopic. X-ray studies show, however, that slate is made up largely of tiny flakes of minerals akin to muscovite, the white mica. These flakes are nearly parallel. Since mica has only one perfect cleavage which is parallel to the platy form of each grain, the cleavage of all the grains in slate lies nearly parallel. Slate splits so readily into thin parallel sheets because of this cleavage parallelism. Evidently slate is merely shale that has been sheared and heated sufficiently to cause the original clay minerals composing it to recrystallize into tiny micaceous flakes. Movement along the innumerable surfaces of foliation, as shown by offsets of the bedding, has granulated the original minerals and has lined up the newly formed mica, which continued growing during and after movement.

In some areas, slate may be seen to pass by imperceptible gradation into more coarsely crystalline rocks such as mica schist. In these more highly metamorphic rocks, every trace of the original stratification and of fossils has been obliterated. Yet, in chemical composition they are identical to the slate into which they grade. Although they may contain minerals different from those in slate, the new minerals in the transitional rocks can be seen to have grown from the micaceous minerals of the slate, gradually replacing and obliterating them just as the micas in slate replaced the minerals of the shale. From such mineral relations and transitions in rock type, we conclude that even these rocks, showing no trace of original sedimentary structures, were, in fact, derived from the sedimentary rock shale. They are merely more thoroughly metamorphosed.

Recrystallization at Igneous Contacts

One of the simplest kinds of metamorphism is the recrystallization of rocks by an igneous intrusion. In many places where, for example, pure quartz sandstone containing marine shells has been invaded by a mass of gabbro or some other plutonic rock, shells cannot be found in the rock close to the contact. Instead, the sandstone near the intrusion contains sheaves of a light-colored mineral called wollastonite. These relations indicate that the shells have broken down into new products according to the following reaction:

SiO_2	$+$	$CaCO_3$	$\longrightarrow$	$CaSiO_3$	$+$	CO_2
sand grains		shells		(wollastonite)		(carbon
(quartz)		(calcite)				dioxide)

The wollastonite formed by the reaction has no resemblance to either the quartz grains or the shells from which it was derived. Most fossiliferous sandstones, however, contain many substances besides quartz—usually a little clay, some limonite or ferromagnesian minerals, and various other impurities. These generally join in the reaction, yielding, not a simple mineral like wollastonite, but a more complicated compound, usually a member of an isomorphous series such as amphibole, garnet, or pyroxene.

Where wollastonite actually does appear, however, we can make certain additional inferences about the conditions of metamorphism because the wollastonite reaction has been carefully investigated over a wide range of temperatures and pressures in the laboratory. The results of this investiga-

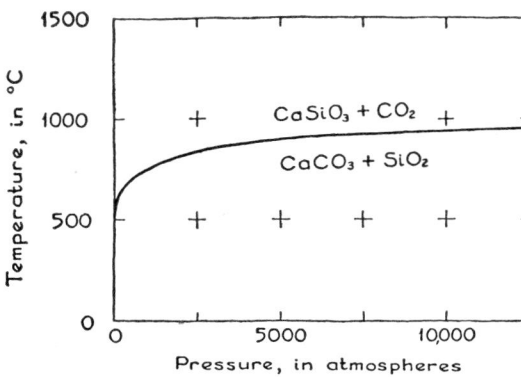

FIGURE 5-9. *All points on the curve represent conditions at which wollastonite will begin to form from calcite and quartz. (After V. M. Goldschmidt.)*

tion are plotted in Figure 5-9. From this graph, we infer that if a rock containing wollastonite was formed at depths equivalent to a pressure of 3,000 atmospheres, the temperature attained by the metamorphosed rock was at least 850° C. This illustrates what is meant when certain minerals are called *"geologic thermometers."*

Locally, *contact metamorphism,* as the kind of metamorphism caused by the heat of igneous masses is called, is greatly complicated by the addition of various substances from the igneous mass. That such additions have taken place is demonstrated by the finding of rocks of different mineral and chemical composition along the contacts—for example, calcium-iron silicates taking the place of calcite in limestone near a granite contact. These different minerals are attributed to hot gases and aqueous solutions escaping from the magma and carrying large amounts of iron, silica, sulfur, and other substances into the wall rock. Here they combined with the minerals already present to produce new minerals.

Recrystallization of the Stassfurt Salts

Deep wells show that the temperature in the outer crust of the earth increases, on the average, about 1° Centigrade for every 100 feet of depth. Therefore, if the rocks are buried to great depths, some of their constituent minerals may become unstable at the higher temperatures.

An interesting example of such *geothermal metamorphism* is afforded by the well-known salt deposits at Stassfurt, Germany. These salt deposits were studied in 1900–1905 by the famous Dutch chemist van't Hoff. Because of the associated sedimentary rocks, he reasoned that the salts had been formed from a shallow body of sea water, cut off from the ocean by a bar or other obstruction, and slowly evaporated in a desert climate.

Van't Hoff was puzzled by one feature of the salts, however. Although their bulk chemical composition is consistent with their being the residue left from the evaporation of sea water, the actual minerals of the deposits are quite different from those obtained by evaporating sea water to dryness.

Patiently and thoroughly, van't Hoff worked out the physical chemistry of the salts by many laboratory experiments, and, ultimately, produced many of the minerals found in the Stassfurt salts by holding sea water at temperatures well above normal surface temperatures but below boiling during the period of evaporation. Van't Hoff then announced that the average annual temperature of the earth's crust at the time when the salts were laid down must have been much greater than at present—perhaps close to the temperature of boiling water!

Geologists, though accepting van't Hoff's experimental data, could not agree with the conclusion. Fossil fish and mollusks are found in strata associated with the Stassfurt salts. It is inconceivable that the marine animals of that ancient time had lived in hot seas. From the relations shown on a detailed geologic map of the Stassfurt area, it is obvious that, after the Stassfurt salts had been laid down in an arm of an ancient sea, they were buried under several thousand feet of younger sedimentary rocks. The rise in temperature because of this deep burial could have caused the already precipitated salts to recombine into new minerals stable in the hotter environment. The textures of the salt, as seen in thin sections under the petrographic microscope, give abundant evidence of postdepositional recrystallization of the solid salt beds. Thus, van't Hoff's conclusion as to the temperature of mineral formation was correct; his inferences as to the climate were absurd because he neglected the field evidence.

Salt beds are particularly susceptible to metamorphic changes; they recrystallize completely at temperatures that are still too low to cause notable changes in the silicate minerals of which most rocks are composed. In sandstones associated with the Stassfurt salts fossil shells are completely unaltered. Therefore, the rise in temperature was clearly not enough to cause calcite to combine with quartz and form wollastonite. It would have required much deeper burial to metamorphose the sandstones and shales at Stassfurt; only the sensitive salt beds were affected.

Metamorphic Processes

Even these brief descriptions—and scores of additional examples could be given—show that many minerals are chemically stable only within a limited range of pressure and temperature. If brought into a part of the earth's crust where the temperature is higher, or where crushing and shearing take place, or where hot fluids attack them, many minerals break down and form other minerals stable in the new environment. Metamorphic rocks result from such transformations. In some, the alteration affects only a few minerals of the original sedimentary or igneous rock, and the metamorphic rock retains in the surviving unchanged minerals or textures clear evidence of its origin; but more commonly, complete transformation has occurred.

From relations revealed by geologic maps, we know that nearly all metamorphism takes place deep within the crust of the earth—far below the depths we can reach in mines and wells. Therefore, in general, we must infer the reactions that take place from the resultant products, although we can also get valuable clues from experiments made possible by laboratory equipment designed to imitate conditions deep within the earth. Unfortunately for our purpose, knowledge of the chemistry of solids, and particularly of solids under shearing stress, is still meager as compared with the vast amount of information chemists have obtained on reactions in solutions.

Significance of Foliated Metamorphic Rocks

Where large areas of metamorphic rocks—particularly the foliated rocks called crystalline schists—are found at the surface, deep erosion has taken place.

Such a generalization is based on observed relations of metamorphic and nonmetamorphic rocks in many parts of the world. We infer from these relations that the textures and minerals characteristic of foliated metamorphic rocks were formed at great depths, under conditions of high pressure and temperature. Thus, we also interpret other rocks of like mineral composition and texture as having been under like conditions—even where the local structural relations do not permit direct proof of deep burial. We reason that a thick covering must have been eroded from any area where crystalline schists are now seen at the surface.

Classification of Metamorphic Rocks

Since any rock may be recrystallized by metamorphism and there are several different ways in which metamorphism may take place, the number of different kinds of metamorphic rocks is large. It is possible, for example, to find at least five different metamorphic rocks all with the chemical composition of basalt—and indeed, all derived from basalt—yet each differing from the others in texture, mineral composition, and general appearance. A greater variety of minerals appears in metamorphic rocks than in either sedimentary or igneous rocks. Therefore, metamorphism is best studied with the petrographic microscope, aided by the principles of physical chemistry and, of course, primarily by many field observations. Certain of the more common groups of metamorphic rocks, however, can be roughly distinguished by sight, as indicated in Appendix IV, Table IV-3. (See, for example, the descriptions of *gneiss* and *slate*, p. 612 and p. 611.)

Metamorphic rocks, like sedimentary and igneous rocks, are classified on the basis of texture and composition. The principal textures to learn for determinations made without the aid of the microscope are:

Gneissose Texture.   Coarsely foliated; individual folia 1 mm. or more, even up to several centimeters thick. The separate folia are commonly lenslike or wavy, and generally differ in composition; feldspars, for example, commonly alternate with dark minerals. Mineral grains are coarse, easily identified.

Schistose Texture.   Finely foliated, forming thin parallel bands or flat lenses. Individual mineral grains are distinctly visible. The minerals are mainly platy or rodlike—chiefly mica, chlorite, and amphibole. Equidimensional minerals like feldspar and pyroxene are not abundant.

Slaty Texture.   Very fine foliation, producing almost rigidly parallel planes of splitting due to the parallelism of microscopic and ultramicroscopic grains of platy minerals such as mica.

Granoblastic Texture.   Unfoliated, or only faintly foliated. Mineral grains are large enough to be visible. Corresponds roughly to the granular texture of igneous rocks.

Hornfelsic Texture.   Unfoliated. Mineral grains commonly microscopic to ultramicroscopic, though a few may be visible. Corresponds to the aphanitic texture of lavas, from which it can be distinguished only with difficulty.

Facts, Concepts, Terms

SEDIMENTARY ROCKS, IGNEOUS ROCKS, METAMORPHIC ROCKS

UNIFORMITARIAN PRINCIPLE

CEMENTATION OF SAND; BURIAL AND PRESERVAL OF FOSSILS

LAW OF SUPERPOSITION; LAW OF ORIGINAL HORIZONTALITY

TEXTURE OF ROCKS

VOLCANIC AND PLUTONIC ROCKS
Magma

ORIGIN OF BASALT
Importance of field studies

ORIGIN OF GRANITE
INTERPRETATION OF CONTACTS
Fragments of one rock in another

Penetration of one rock by another

Alteration of one rock by another

RECRYSTALLIZATION OF ROCKS BY METAMORPHISM

ORIGIN OF FOLIATION

Relation of foliation to relic sedimentary and igneous features

GEOLOGIC THERMOMETERS

CRYSTALLINE SCHISTS AS INDICATORS OF DEEP EROSION

Questions
(See Appendix IV)

1. Some of the lavas from Mauna Loa in Hawaii congeal on a slope of approximately 10°. How do you reconcile this with the Law of Original Horizontality?

2. All geologists who have examined the Palisades of the Hudson agree in calling the mass a sill. What features would you expect to find here different from those present in the rocks of Keeweenaw Point, Michigan, most of which are interpreted as lava flows?

3. Wollastonite was recognized as an indication of thermal metamorphism long before laboratory determinations of its temperature of formation were made. What kind of evidence was probably drawn on to establish this?

4. If you observed two intersecting sets of parallel structures in a rock, how would you decide which was the older?

5. The surface temperatures of lavas are considered as high as or higher than those of most intrusive magmas in the part of the crust accessible to us. Yet evidences of contact metamorphism are widespread along intrusive contacts and trivial along volcanic vents. Can you suggest a reason?

6. Sands and muds are both highly porous when deposited. The component grains of sands are chiefly equidimensional; those of muds platy. Can you suggest why consolidated sandstone, even when cemented, is commonly more permeable than shale, which has little or no interstitial cement?

7. Some limestones, otherwise quite normal, contain scattered well-formed crystals of feldspar. What evidence would you seek to determine whether these are remnants of volcanic ash falls or were developed in place by mild metamorphism?

8. How do you distinguish limestone from sandstone? basalt from limestone? phyllite from shale?

9. In examining a contact between granite and an overlying rock, what would you look for to tell whether the granite had intruded the overlying rock, or had been eroded to form it?

10. Discuss the origin and significance of porphyritic texture.

11. What holds the sand and other mineral fragments together in a sandstone?

What holds the mineral grains together in a granite? What holds the fragments together in welded tuff?

12. Why are fossils rare or absent in slate and other metamorphic rocks?

Suggested Readings

1. Geikie, Sir Archibald, *The Founders of Geology,* Johns Hopkins Press, Baltimore, 1901. (A fascinating account of the history of geology.)
2. Adams, F. D., *The Birth and Development of the Geologic Sciences,* Williams and Wilkins, Baltimore, 1938.

6. *Climates, Weathering, and Soils*

Climates

THE WORD "climate," from the Greek for "incline," preserves one of the oldest scientific discoveries, that year in and year out weather is largely controlled by the tilting of the earth's axis of rotation to the plane of its path around the sun (Fig. 6-1). The Greeks noticed that, in general, the climates are colder at higher latitudes; they correlated this with the obvious fact that the air is generally warmer when the sun is high above the horizon. As the sun rises higher in summer than in winter (a fact that indicates tilting of the earth's axis), it was natural to attribute seasonal weather differences to this tilting. Modern studies of *weather* (the day-to-day variation) and *climate* (the long-period weather, including seasonal changes) show that many other factors besides tilting of the earth's axis are involved. Weather and climate involve not only heat and cold, but also such features as wind and calm, the direction and intensity of the winds, cloudiness, and precipitation either as rain or snow. All of these aspects are closely interrelated, and depend on such factors as heat supplied to the atmosphere by the sun's radiation, the movements of the atmosphere, and the distribution of land, sea, and mountains. A full treatment of these interrelations is the subject of a separate earth science—meteorology, the study of the atmosphere—but any geologic study requires some acquaintance with them. Not only do geologic processes differ greatly in different climates, but the clear evidences of drastic climatic changes in the geologic past cannot be appreciated without some understanding of the factors involved. We begin our study with an analysis of the nature and movements of the atmosphere.

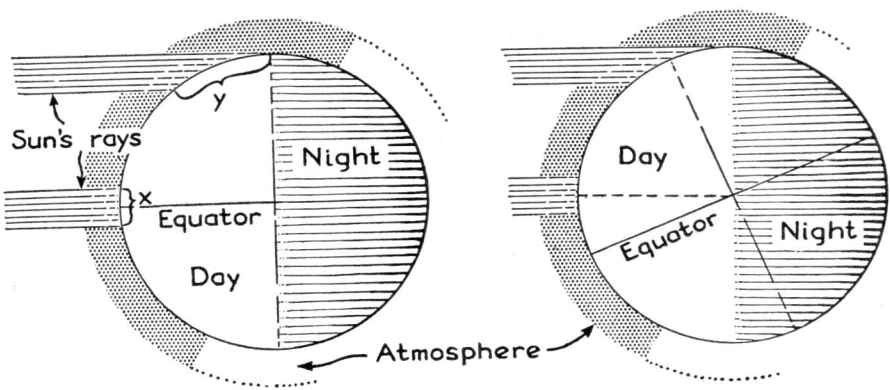

FIGURE 6-1. *The variation of average temperature with latitude*, left, *and season,* right. *Note that the sun's rays are much more concentrated near the equator* (y *is greater than* x), *and that here they are less absorbed by the atmosphere. As the earth tilts*, right, *the incidence of the sun's rays at any one place changes, producing seasonal temperature variations.* (*After G. T. Trewartha*, Introduction to Weather and Climate, McGraw-Hill Book Co., 1937.)

The Atmosphere

The atmosphere is an envelope of gas around the earth. At the earth's surface the air contains about 79 per cent nitrogen, 21 per cent oxygen, 0.03 per cent carbon dioxide, small but variable amounts of water vapor, and small amounts of chemically inert gases.

Air Pressure

As in all gases, every portion of the atmosphere exerts pressure against every adjacent portion. Such pressure is one of the evidences that gases are made up of particles in constant motion, colliding with each other and rebounding. The atmospheric pressure can be measured by the height of the column of mercury in an inverted evacuated tube (Torricelli's barometer, 1643).

One of the first discoveries made with the aid of Torricelli's device was that the pressure of the air is less at high elevations than at low. The pressure at any elevation results from the interplay of two tendencies: (1) that of the gas to expand equally in all directions, and (2) that of each particle to fall toward the earth's surface because of gravity. The barometric pressure, actually a measure of the weight of the air particles above the barometer, decreases rapidly with height and has only about half its sea-level value at a height of 17,500 feet, thus indicating that only half the mass of the atmosphere is above that elevation, even though the atmosphere extends upward for several hundred miles.

Movements in the Atmosphere

The energy to drive all atmospheric motion is ultimately derived from the sun's radiation. The actual motion is in response to differences in the gravitational attraction of the earth upon air masses of different densities, but the density differences are due to changes in temperature set up by the sun's radiation.

Over a long time the amount of heat received from the sun must be balanced by the heat lost to outer space, or the earth would be either heating up or cooling down. True, climatic fluctuations are not negligible (see Chap. 13) but the fossil record strongly indicates that these are minor; roughly, the heat budget of the earth (amount received minus the amount lost) must be nearly balanced. Much of the sun's radiation striking the outer atmosphere is reflected back into space; it does not affect the climate or temperature of the earth's surface. Of the radiation entering the atmosphere, part is radiated back into space as heat waves, part is reflected, and a part is absorbed, raising the temperature of the land, air, and water.

Measurements and calculations of heat received at the earth's surface indicate that the amount varies with latitude, much as the ancient Greeks thought. Losses by reflection and back-radiation into space also vary with latitude because a warm body radiates more than a cold one, but these differences are not so great as those in the amounts received. Thus, in low latitudes, more heat is received than lost, and in high latitudes more is lost than received (see Table 6-1). As the equatorial zones are not becoming

TABLE 6-1. *Heat Budget and Heat Transport from Lower to Higher Latitudes (Sverdrup)*

LATITUDE (DEGREES)	HEAT RECEIVED (GRAM CALORIES PER SQ. CM. PER MINUTE)	HEAT LOST (GRAM CALORIES PER SQ. CM. PER MINUTE)	SURPLUS OR DEFICIT	HEAT TRANSPORT ACROSS PARALLEL (PER CM. LENGTH OF THE PARALLEL OF LATITUDE)
0 (Equator)	0.339	0.300	+0.039	0.00
30	0.297	0.283	+0.014	1.07
60	0.193	0.245	−0.052	1.20
90 (Pole)	0.140	0.220	−0.080	0.00

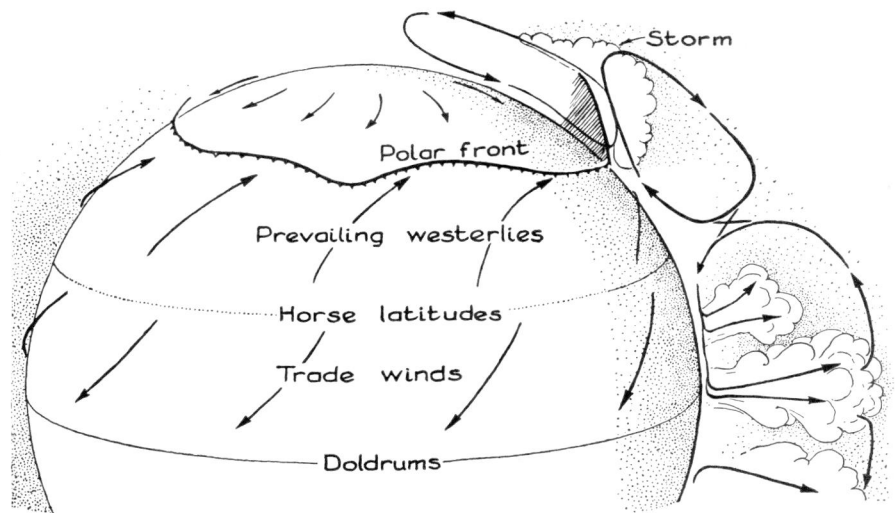

FIGURE 6-2. *Atmospheric circulation in the northern hemisphere.* (*From U. S. Dept. of Agriculture,* Climate and Man, *1941.*)

hotter and the polar cooler, the net heat transported from the tropics to higher latitudes must compensate nearly exactly for these differences, as shown by the last column of the table. Most of this transport is by movements of warmed air (winds) but some is by ocean currents (see Chap. 16, Part I).

Consider first the condition in the tropics: Air heated near the equator must expand (become less dense). A column of air in the tropics thus weighs less than a column of cooler air of equal height in high latitudes. Such a system is unstable, and, since no gas has rigidity, these pressure differences cannot be sustained. The warm equatorial air column rises and flows poleward, while the denser cool air from higher latitudes flows equatorward beneath it. This establishes a convective circulation (Fig. 6-2) like that seen in a pot of soup slowly simmering on the stove.

But flow cannot be directly toward the equator because, due to its inertia, the air is deflected by the rotation of the earth.

Effect of the Earth's Rotation

The equatorial circumference of the earth is about 25,000 miles; a point on the equator thus moves eastward at the rate of more than 1,000 miles per hour. At latitude 40° the eastward velocity is about 750 miles per hour, but at latitude 45° only about 700. An air mass moving with the earth (on a windless day) has the velocity appropriate to its latitude. If it later moves southward from a point in north latitude toward the equator, its speed is less than that of the ground at the lower latitude, and to an observer there results a northeast wind. In general, if an observer stands with his back to the wind, he finds that in the northern hemisphere all winds are deflected to his right (Fig. 6-2), and those in the southern hemisphere to his left, because of the earth's rotation.

One result of these two factors—the equatorial heating and the earth's rotation—is a prevailing wind called the *trade wind*, blowing obliquely toward the equator in each hemisphere. In the northern hemisphere this is the *northeast trade;* in the southern, the *southeast trade* wind. Between lies the equatorial belt of calms.

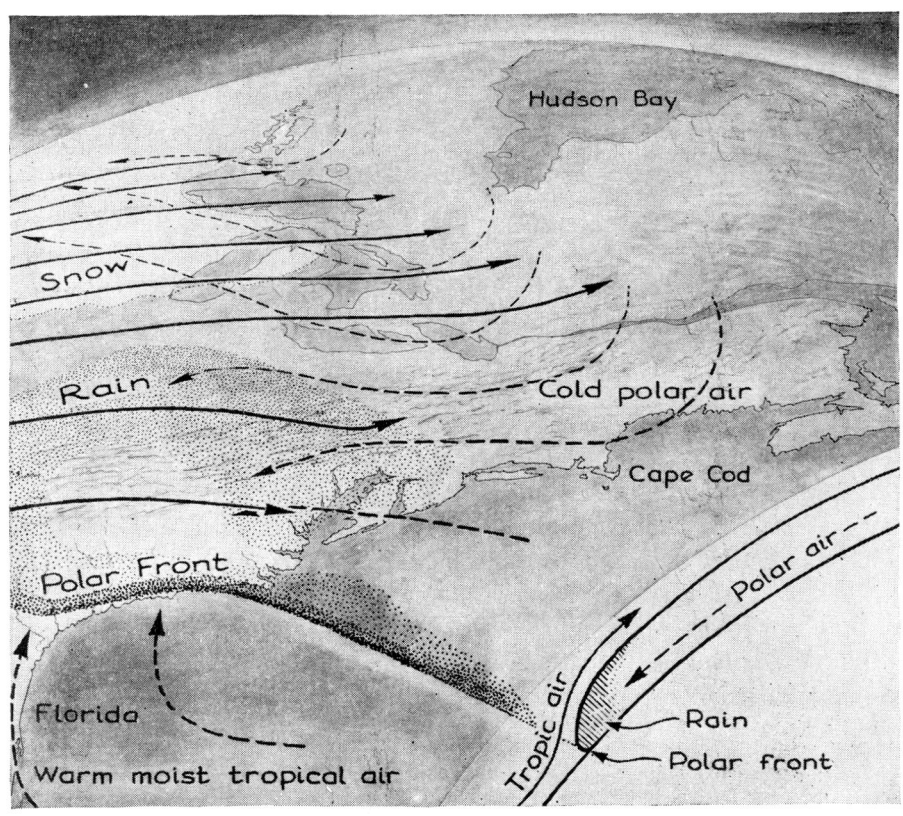

FIGURE 6-3.   *Air movements over eastern North America causing the storm of January 20, 1937. At the lower right is a cross section showing the general relations.* (*Modified from U. S. Dept. of Agriculture,* Climate and Man, 1941.)

If the circulation were a single pair of simple circuits, one in each hemisphere, trade winds would blow all the way from the poles. Actually two (or three) circuits appear in each hemisphere, the upper air coming down in middle latitudes and moving along the earth's surface both toward the equator and toward a pole.

Air Masses

On a smaller scale, air masses, commonly 500 to 1,000 miles across, move slowly in the lower levels of the atmosphere in response to regional or local differences of pressure. Thus, a meteorologist speaks of a cold, dry, polar air mass penetrating beneath a warm, moist, tropical air mass. The advancing margin of such a polar mass is called a "cold front." Rain is likely to occur behind a cold front. Figure 6-3 illustrates the storm of January 20, 1937, in the eastern United States. A large polar mass, twisting southeast (dashed arrows) across North America, insinuated a thin advancing wedge beneath a mass of warm moist air moving north and northeast (solid arrows) from the Gulf of Mexico. The lower part of the water-saturated tropic air mass was lifted above the cold heavy mass, just as air rises when a liquid flows into a pan. In rising into a region of lower pressure, the air expanded and cooled. As cold air can hold less water vapor than warm, the moisture condensed as a torrential rain, bringing devastating floods to the Ohio River Valley and snow to the northern Midwest.

Effect of Mountain Ranges

Mountain ranges profoundly modify local climates. Moist oceanic winds lose much of their water content as rain or snow when they rise over a mountain range. Such an ascent has the same effect as rise over a cold front. For example, the windward sides of the Hawaiian Islands may have 50 to 200 inches of rainfall a year; the sheltered sides 5 to 10. Similarly, the deserts of Turkestan are behind the Himalayas, whose seaward slopes receive more rain. The air that pours down the inland slope of a mountain range is compressed and heated, so that precipitation is usually prevented.

Climate Types

If one considers rainfall alone, climates may be called *humid* if the mean annual precipitation exceeds 20 inches, *arid* if less than 10 inches, and *semi-arid* if between 10 and 20 inches. From another point of view, *maritime climates* are characterized by relatively uniform temperatures, moist air, fog and cloudiness, rainy winter, high wind velocities, and freedom from dust.

In *continental climates,* both daily and seasonal temperature ranges are greater, windless days more frequent, the winter rainfall less, and the air drier and often dust-filled. The extreme continental climate is the arid or desert climate, which is much windier than other continental types, partly because of the great daily ranges in temperature. Many deserts lie in "rain-shadows" behind high mountains, but some of the greatest—the Sahara, Australian, and southwest African deserts—extend to the ocean shores in the subtropical high-pressure belts.

Weathering and Soils

The principal significance of the different climatic types to the student of geology lies in their influence on soil formation and on erosion. Each climatic environment places its own stamp upon the soils developed there, and each influences, through its control over vegetation, amount of rainfall, and evaporation losses, the geological processes involved in molding the details of the earth's surface. Erosion, the process of removal of rock waste, will be discussed in later chapters; here, we will examine the influence of several different climatic environments upon the weathering of different rock types.

The most familiar example of *weathering* is the etching and discoloration of the surface of an unpainted board left out-of-doors. Rock, exposed on the earth's surface, also decays and leaches, but much more slowly. If the product of rock decay is merely broken and discolored, it is called *mantle rock;* but if it is loose and porous enough for plants to find a foothold, it is called *soil.*

Soil is more common than rock at the earth's surface. Almost all outcrops of rock are less firm—more easily crumbled and broken—than is the same rock at a depth of 20 or 100 feet. Many rocks that are black or steel gray where penetrated in mines, wells, or deep quarries are yellow or brown in outcrops. In some, the yellow color is a mere stain on or near cracks, but in others it is more pervasive and is accompanied by drastic changes in mineral composition of the rock. That the changes result from weathering is shown by observations on building stones. For example, the exposed faces of the sandstone used in the older buildings at Stanford University turned yellow in 5 to 10 years, and, where exposed to repeated wettings from garden sprinklers, began to crumble in 20 to 30 years.

Analysis of Weathering

Weathering results chiefly from exposure to the air or to the action of substances derived from the air, the most important of which are water, oxygen, and carbon dioxide. Most studies of weathering have been made on one of two very different materials: (1) soils clearly derived from immediately underlying rocks or (2) building or monumental stones weathered superficially after years or centuries of exposure under various conditions. Such studies have shown that weathering usually includes both *mechanical disintegration* and *chemical decomposition*. *Disintegration* includes only loss of coherence, with little or no changes in composition. It does not include abrasion and removal of the constituents; such movements are part of erosion. *Decomposition* or *chemical weathering* is the term for change in chemical (and mineralogical) composition. The complex silicates that make up the bulk of most crystalline rocks alter into other substances, such as hydrous silicates, hydrous oxides, and carbonates.

Examples of Mechanical Weathering

Frost Action.   Wherever water freezes, the 9 per cent expansion is an effective disruptive force. The expanding of ice to break rock is called *frost action* or *wedging*. Water freezes in the soil and in cracks or pores of rocks. The colder the climate, the deeper the freeze. Thawing and repeated freezing add to the effect, however, so that frost action is more effective in cold temperate zones than under extreme arctic conditions.

In soil, especially, because of the high water content and the complexity of the pore systems, freezing increases the volume considerably. After thawing, the soil is open and spongy. To break rocks, freezing water must be confined. As the water in a crack begins to freeze at or near the ground surface, it may be completely confined in the lower part of the crevice in the rock. Over a long time, rocks become extensively shattered by frost action. Many high mountain peaks are covered with rubble formed in this way.

Plants and Animals as Aids to Weathering.   Organisms assist in the breakdown of rock to soil. Their action is in part disintegration, and in part chemical effects. The roots of growing plants powerfully wedge the soil aside, raising and breaking slabs of rock (or concrete), and widening cracks. Burrowing animals move and mix the soil effectively. Charles Darwin, in his last published work (1881), showed that English earthworms may spread their casts over the ground to the average thickness of 0.1–0.2 inch per year. The soil is thus loosened, aerated, and subjected to the chemical action of the worms' digestive processes. This mixing probably accounts for the uniformity of the humid, and especially the forest or meadow, soils. There are practically no earthworms in arid regions.

Most plant tissues are carbohydrates (compounds of carbon, hydrogen, and oxygen). Microscopic organisms, such as molds and bacteria, in the absence of oxygen, change leaves, fruit, and wood on or in the soil into the dark organic substance known as *humus*. In the presence of oxygen, the organic substances are destroyed and CO_2 and H_2O are formed from them. Some H_2CO_3 is formed by both processes and contributes to chemical weathering. Living or dead, land plants and animals are aids to weathering.

Other Agents of Disintegration.   Frost action and wedging by plants are the most effective agents of disintegration, but many other processes play a role.

Chemical decomposition is one cause of disintegration, and possibly a major one. Feldspars swell as they weather into clay (see p. 108), and many other minerals decompose to substances that occupy more volume. In a granite composed of feldspar, quartz, and mica, this swelling of the rotting feldspar may disintegrate the massive rock to a pile of granitic sand. Alternate swelling and shrinking of the clay as it is wet by the rains and heated by the sun aid the process.

Rain may also wash soluble salts into cracks in rocks and soils where they crystallize. The growing crystals may exert enough force to break and disintegrate the rock still further. Telephone poles set in the Bonneville salt flats west of Salt Lake City show this effect to a striking degree. During the wet season the pole acts as a wick, drawing the salt brine up into the wood by capillarity. Here, the brine evaporates, precipitating salt crystals so abundantly that, after a few years, the pole up to about 18 inches above ground is swelled to twice its normal size, and the wood fibers are completely shattered.

A forest fire often shatters and spalls the edges from exposed rock ledges and boulders. Rock is a poor conductor of heat, so the interior is not appreciably warmed by the fire. Consequently, the highly heated outside expands and pulls away from it. Lightning also breaks and shatters rocks.

It is possible that the drastic day-to-night changes of temperature that occur in deserts may cause enough expansion and contraction to weaken or even break some minerals and rocks. However, specimens of granite have been placed in electric ovens and run through the range of temperatures characteristic of deserts many thousands of times without showing any appreciable loss in strength. Doubtless, the swelling effects of decomposing minerals and the expansion of growing salt crystals are far more important in producing disintegration in deserts than are the diurnal temperature changes.

Examples of Chemical Weathering

We will now consider a few details of the chemical processes whereby rock-forming minerals are decomposed into hydrous silicates, oxides, and carbonates. Our first example is the chemically simple rock, limestone.

Weathering of Limestone. Limestone is almost wholly calcite, but generally contains a few particles of clay. Calcite dissolves very slightly in pure water, sending some Ca^{2+} and CO_3^{2-} ions into solution. As indicated in chemical notation:

$$(1) \quad CaCO_3 \rightleftarrows Ca^{2+} + CO_3^{2-}$$

calcite   calcium   carbonate
             ion   ion

In water containing dissolved carbon dioxide (as all rain water does), the calcite becomes much more soluble. The water and carbon dioxide react to yield carbonic acid and this in turn furnishes ions of hydrogen and bicarbonate. Thus:

$$(2) \quad H_2O + CO_2 \rightleftarrows H_2CO_3 \rightleftarrows H^+ + HCO_3^-$$

water   carbon   carbonic   hydrogen   bicarbonate
          dioxide   acid   ion   ion

But the hydrogen ion thus formed reacts with carbonate ion from the calcite to form more bicarbonate ion:

$$(3) \quad H^+ + CO_3^{2-} \rightleftarrows HCO_3^-$$

hydrogen   carbonate   bicarbonate
ion   ion   ion

The arrows in these equations are the chemist's shorthand for a reversible process. In solutions the combination and dissociation of ions are constantly proceeding at rates governed by the abundance of the several ionic species present. For example, at equilibrium in reaction (1) as much calcite is being precipitated from solution as is being dissociated into ions in that solution. But if the ions on one side of the reaction are selectively removed, they obviously cannot furnish as many products on the other side as they formerly did—the reaction will take place in the direction of yielding more of the ions being removed.

When calcite is dissolved in water containing carbon dioxide, the carbon dioxide enters into new combinations with the water, as in equations (2) and (3), so that more of the products on the right side of equation (1) are formed than would be in pure water; that is, more calcite is dissolved. Thus, in the weathering of limestone the most important reaction is the slow solution of the calcite in percolating water that has absorbed carbon dioxide from the air. Calcite is dissolved not only at the surface, but along every crack and fissure into which the solutions penetrate. The clay in the limestone accumulates on the surface as the underlying rock slowly dissolves. A few inches of clay soil may thus represent the residue of tens or scores of feet of dissolved limestone. Where underground circulation is vigorous (Chap. 14), the limestone may be extensively dissolved, resulting in the formation of great subterranean caves.

Weathering of Granodiorite. The individual minerals of granodiorite weather very differently from calcite and from each other. In the chemical reactions given below we list only the products, rather than the ionic reactions through which they are largely formed.

1. Quartz: Persists almost unchanged. Quartz dissolves only with extreme slowness.

2. Orthoclase: $2KAlSi_3O_8$ + H_2CO_3 + nH_2O $\longrightarrow$ K_2CO_3
   orthoclase   carbonic   water   potassium carbonate
                 acid   (readily soluble)

   + $Al_2(OH)_2Si_4O_{10} \cdot nH_2O$ + $2SiO_2$
   clay mineral   soluble silica,
                              or finely
                             divided quartz

3. Plagioclase: $CaAl_2Si_2O_8 \cdot 2NaAlSi_3O_8$ + $4H_2CO_3$
              anorthite   albite   carbonic acid
                      plagioclase

   + $2(nH_2O)$ $\longrightarrow$ $Ca(HCO_3)_2$ + $2NaHCO_3$
       water   calcium bicarbonate   sodium bicarbonate

   + $2Al_2(OH)_2Si_4O_{10} \cdot nH_2O$
   clay mineral

4. Biotite: $2KMg_2Fe(OH)_2AlSi_3O_{10}$ + O + $10H_2CO_3$
           biotite   oxygen   carbonic acid

   + nH_2O $\longrightarrow$ $2KHCO_3$ + $4Mg(HCO_3)_2$ + $Fe_2O_3 \cdot H_2O$
       water   potassium   magnesium   "limonite"
             bicarbonate   bicarbonate

   + $Al_2(OH)_2Si_4O_{10} \cdot nH_2O$ + $2SiO_2$ + $5H_2O$
   clay mineral   soluble silica   water
                        or quartz

5. Hornblende: Alteration similar to that of biotite, with similar products, but goes to completion more readily.

It is seen, then, that *the complete weathering of granodiorite in a humid temperate climate leaves residual quartz and newly formed aluminum-silicate clay, stained yellow with hydrated ferric oxide ("limonite")*. Sodium and calcium, and much of the magnesium and potassium have gone off as ions in the water draining from the soil. Some magnesium and potassium ions may remain in the clay minerals, held there by feeble electrical forces. As potassium is an important plant food, its retention is agriculturally significant.

It is noteworthy that, despite the differences in composition of the under-lying rock, soils derived from limestone and granodiorite are both rich in clay. The clay minerals are among the stablest under surficial conditions, both in temperate and in arctic regions.

From this analysis of the processes of weathering we can see that disinte-gration and decomposition take place together, each aiding the other. In moist temperate regions, which are the areas that have been most studied by soil scientists, decomposition appears to be much the more important. Disintegration by frost action, however, is widespread, and in cold regions it may predominate.

Exfoliation

The splitting away of successive scalelike layers of rock from an exposed or soil-covered surface of massive rock (not a platy rock such as shale or schist) is called *exfoliation*. The separated sheets or plates may be flat or curved, paper thin or many feet thick, a fraction of an inch or hundreds of feet long. Two types of exfoliation may be distinguished—small thin flakes on massive rocks of all kinds; and giant plates, usually on granite or granodiorite.

Closely fitted blocks of dolerite, sandstone, granite, and other medium- or fine-grained rocks frequently develop numerous thin concentric layers of partially weathered material (Fig. 6-4). During weathering, the material

FIGURE 6-4. *Exfoliation in granodiorite. Note that the rounded forms of weath-ering are guided by pre-existing fractures. (Photo by Eliot Blackwelder.)*

FIGURE 6-5. *Granite dome showing coarse exfoliation, Sierra Nevada. The sheet-like slabs of rock are tens of feet thick. (Photo by G. K. Gilbert, U. S. Geological Survey.)*

expands as hydrated clay minerals are formed from the original minerals. Probably the outermost shell weathered and expanded first, pulling away from the fresher rock beneath. The shells formed successively inward in similar fashion. In each shell, chemical weathering preceded the mechanical process of exfoliation.

The giant curved plates of fresh rock (Fig. 6-5) that have split away in forming the Yosemite domes in California, Stone Mountain in Georgia, and Sugar Loaf at Rio de Janeiro, indicate that exfoliation may take place without much weathering. These domes are all in plutonic rocks that originally consolidated beneath a thick cover of overlying rocks. Perhaps unloading by erosion of thousands of feet of the cover has permitted upward expansion and the formation of cracks parallel to the surface.

Colors of Soils

Most soil colors are due to iron minerals or organic matter. Iron exists in two states: ferrous and ferric. A ferric ion may be formed from a ferrous ion by the losing of one electron, a process called *oxidation*. When hydrated ferric oxide ("limonite") forms by chemical decomposition of a ferrous silicate such as biotite, the reaction is made possible by the addition of oxygen from the air. The finely divided "limonite" stains the resulting soil yellow. If the soil is repeatedly dried out, as in a climate with a warm dry season, the "limonite" may lose its water and change to red hematite; the soil then becomes red.

Iron is reduced to the ferrous state as easily as it is oxidized. A pale greenish or a dark gray soil usually indicates that the iron has been reduced by plant residues or other reactions with organic matter. Ordinarily this occurs where the air is excluded because pores have been filled with water. Thus, a black soil may indicate swamp conditions; the iron has been reduced and the soil is rich in humus. Colors of soils give clues to the conditions of weathering and to the abundance of organic matter.

Residual Soil and the Soil Profile

The surface changes in a fairly uniform rock brought about by weathering are well illustrated in the Sierra Nevada of California. The canyon walls expose hard gray granodiorite, composed chiefly of plagioclase, with smaller amounts of quartz, orthoclase, biotite, and hornblende. The rolling forest-covered uplands between the canyons have a red-brown soil. The surface layer, called by soil scientists the "A horizon," is a sandy loam (a mixture of quartz sand, silt, minute clay particles, and decomposed plant residues). In the fine-grained reddish matrix are embedded many irregular quartz grains, alike in size and shape to those in the granodiorite. At about a foot depth, the clay content increases. This sandy clay is the subsoil or "B horizon." At two to four feet, the soil becomes paler and sandier. The sand grains include not only quartz, but abundant feldspar, stained but readily recognized by its flashing cleavage faces, and also many micaceous flakes. These are not jet black, as is the mica of the granodiorite, but iron-leached pearly yellow scales sometimes mistaken for gold. Still deeper, this material grades imperceptibly into stained and crumbly material in which the texture and characteristic minerals of the granodiorite can be vaguely recognized. The transitional material below the zone of high clay content (B horizon) is called the "C horizon." See Figure 6-6, *left*.

The conclusion seems inescapable that the surface material is developed by weathering of the underlying granodiorite. Such a soil, still in place on its source rock, is called *residual*.

FIGURE 6-6. *Soil profiles on Sierra Nevada granodiorite,* left, *and Kentucky limestone,* right.

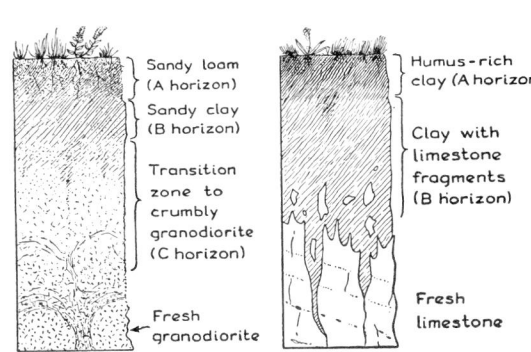

The A horizon of the residual Sierra soil is looser and sandier than the compact B horizon. Apparently some clay formed by decomposition of feldspar near the surface has been washed downward from the A horizon, giving the B horizon an extra portion.

In contrast with the residual soil of the Sierra granodiorite is that of certain parts of Kentucky where the bedrock is limestone instead of granodiorite, and the climate, though also temperate, is more humid. The A horizon consists of black humus-rich clay; the B horizon of light-gray limy clay which passes downward into white limestone (Fig. 6-6, *right*).

Most soils show characteristic "profiles" composed of two or more distinct layers or soil horizons, of which many kinds have been distinguished from various climatic environments by soil scientists. Comparison of various soils is facilitated by comparison of their respective soil horizons.

Soils Developed on Transported Material

Many soils have developed not on bedrock, but on unconsolidated material deposited by streams, glaciers, or other agents of erosion. Such materials have generally been at least partially weathered before deposition. After deposition the loose material allows easy entry of air and water, and so weathers quickly to form typical A and B soil horizons.

The soils of the Great Valley of California afford an example. The residual soil of the Sierra Nevada, developed from granodiorite, has been washed into streams and transported to the Great Valley. On debouching from the mountains, some streams have deposited hundreds of feet of soil and rock debris at the mountain base. In places, such sediments, even some gravels, have been left undisturbed long enough to have rotted into brown soils with a characteristic soil profile. The brown silty A horizon, two or three feet thick, contains scattered quartz pebbles and numerous groups of closely spaced angular quartz grains that are probably relics from decayed granodiorite pebbles. The compact B horizon, one or two feet thick, has been formed in part by the filtering down of minute clay grains from the A horizon. Such compact clayey B horizons, especially when cemented by iron oxide, calcium carbonate, or other bonding materials, are called *hardpan*. The C horizon consists of the relatively unaltered gravel or other parent material below the B horizon.

Another example of soil developed on transported material is that in eastern Massachusetts derived from rock flour (very finely ground fresh rock), boulders, and clay transported by glaciers in the geologic past (Chap. 13). Here, the A horizon consists of three or four inches of dark humus-rich soil overlying sterile rock flour and fragments.

Climatic Factors in Weathering

In Moist Temperate Climates. As we have seen, the weathering of limestone and granodiorite produces mainly minute grains of one or more of the clay minerals, with more or less quartz sand, iron oxide, flakes of altered mica, and perhaps some partially decomposed feldspar.

For this climate, a rough, but not precise, rule is that the more iron and the less silica in a rock, the more rapid its weathering. Thus, basalt weathers readily, but quartzite is almost immune. Limestone weathers rapidly in this climate.

In Dry and Cold Climates. From the relatively few investigations reported it appears that disintegration is the chief weathering process in the driest deserts, such as those of southwest Africa. It is also dominant in polar latitudes and on the high mountain peaks of the temperate and torrid zones. In arctic regions, frost action predominates; in the subtropical deserts, shattering of unknown origin prevails, perhaps caused chiefly by slight chemical changes, facilitated by expansion of the surface layer by decomposition, and by solar heating. Obvious chemical changes are minor and fine-grained soils rare. Vegetation is sparse; hence, carbonic and other soil acids are of less than usual importance. In subarctic regions such as Finland, however, chemical weathering of a special type, producing very siliceous soils, appears to predominate, especially where vegetation is fairly abundant and its decomposition slow and steady.

Chemical weathering of another sort, producing soils rich in calcium carbonate and high-silica clay minerals, is predominant in semi-arid regions such as the western interior of the United States. The calcium carbonate accumulates because, during most of the year, the soil water rises to the surface through small connected (capillary) pores, and releases its dissolved carbonate by evaporation, rather than draining away through subterranean pores, cracks, and channels. A striking contrast between moist and semi-arid weathering is furnished by limestone, which dissolves to leave pock-marked valleys in humid regions but stands out as bold ridges, almost like quartzite, in a semi-arid environment like that prevailing over most of Arizona and Nevada.

In the Tropics. The rain forests of the Congo and Amazon basins grow in soils which have not been adequately studied, but which appear similar to the aluminum-silicate clay soils of moist temperate regions.

The grass- and tree-covered savanna lands characterizing wide belts north and south of the tropical rain forests are largely underlaid by yellow and red-brown soils called *laterites*. The characteristic dark-brown upper portions of laterites dry to bricklike hardness and are sometimes used as building materials. Typical laterite areas are found in India, Nigeria, Central America, and Brazil.

Because laterites offer very different problems than do the products of weathering in temperate climates, and because they illustrate one way in which a chemical element such as iron or aluminum may be concentrated into valuable ores, we will discuss them in some detail.

Laterites vary considerably in mineral composition, but are typically aluminum and iron hydroxides and oxides mixed with some residual quartz. A rare variety called *bauxite* is almost pure $Al_2O_3 \cdot nH_2O$ and, hence, valuable as the ore of aluminum. In laterites, practically all the silicon of the original silicates has been leached out by rain waters, along with the easily soluble sodium and potassium, and the acid-soluble calcium and magnesium. The chemical aspects of this are difficult to understand. Silica is normally soluble only in water containing many more OH^- (hydroxyl) ions than hydrogen ions (an alkaline solution). However, calcium and magnesium are not soluble in such a solution, but in those with excess hydrogen over hydroxyl ions (acid solutions). Yet all three of these elements have clearly been leached from the soils.

Most laterite regions have marked wet and dry seasons. It has been suggested that during the warm dry seasons organic acids are so completely oxidized (to carbon dioxide and water) that the dry soil contains no acid-producing materials because the carbonic acid goes into the atmosphere as a gas. But the soil may retain a very small amount of material which would, when dissolved, give an alkaline solution. Then, at the onset of the first rains, silica is carried off in the temporarily alkaline soil solution before the decay of new vegetation again restores the carbonic acid supply.

Residual laterite soils are characterized by a pale *zone of leaching* just above the parent rock, and a dark-brown *concretionary zone* at or near the surface (Fig. 6-7). Each zone is usually a few feet or a few tens of feet thick, locally deepening to hundreds of feet. The concretionary zone is a concrete-like mass, composed chiefly of either dark-brown "limonite" or of numerous limonitic nodules (*concretions*) of pea or marble size, more or less well cemented into a solid mass. Possibly the concretionary zone should be called the "B_1 horizon," and the zone of leaching the "B_2." The uppermost parts of some lateritic soils are crumbly or even sandy (because of unaltered quartz) and perhaps represent the A horizon.

In laterites some elements needed by plants have been leached away, and others, such as phosphorus and iron, are precipitated as iron phosphates or other compounds so insoluble as to be practically unavailable to plants. Therefore, pineapple plants grown on the lateritic soil of the Hawaiian

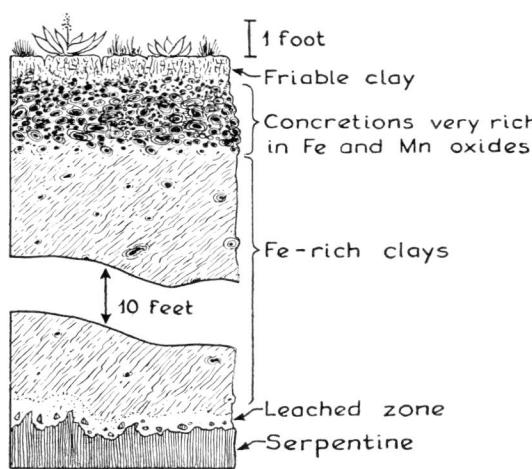

FIGURE 6-7. *Profile of Cuban laterite. The concentration of iron in these soils is so great that they are locally mined as iron ore. (Data from H. H. Bennett and R. V. Allison, 1928.)*

Islands are sprayed with soluble iron salts, and small quantities of soluble phosphates are applied frequently.

What happens to the silica leached out during lateritization? No definite answer can yet be given because the subtropical distribution of even so abundant a substance as silica is little known. Extensive siliceous crusts have been reported from Angola, just south of the Congo Basin, and from other parts of tropical and subtropical Africa, but laterite is not definitely known near them. When the total amount of silica in certain rocks is compared to the amount in the streams that drain them, it appears probable that silica is more rapidly removed in tropical climates. It has been shown, for instance, that the relative solubility of the silica in certain igneous rocks in tropical British Guiana is about twice as great as the world average for all rocks in all climates.

Rates of Weathering

The rate of weathering varies greatly with different rocks and different climates. In less than eighty years the Edinburgh inscription on marble in memory of Joseph Black, the discoverer of carbon dioxide, was rendered illegible by the action of carbon dioxide and other acid-forming substances. On the average, about nine millimeters of rock per hundred years has dissolved from faced limestone in Edinburgh. Furthermore, there has been marked disintegration of the limestone by frost action. In small Scotch towns, with less coal smoke, the marble in tombstones has been dissolved more slowly, but the effects are notably greater on limestone than the barely perceptible roughening of slate slabs in one or two hundred years.

Europe, southeast Asia, and Central America are dotted with ruined stone castles or temples. For many of these structures the dates of building and abandonment are known quite accurately. A few ruins have been studied by soil scientists or archeologists. Kamenetz fortress in southern Russia was built in 1362 and remained in use until 1699. Thereafter, through neglect, the stone blocks began to disintegrate and decompose. The top of one limestone tower developed a soil 4 to 16 inches thick in the 231 years between 1699 and 1930. This soil was much like that on undisturbed, similar limestone in the vicinity.

Other towers and castles under the European climatic environment have weathered more slowly, especially if constructed of sandstone, granite, or other high-silica rocks. Most observers consider that the rate of chemical weathering is probably most rapid in moist tropical climates, even though the evidence from dated structures is inconclusive. The arched roofs of Angkor in Cambodia, Indo-China, stand intact after seven centuries of neglect. Though the jungle crowds close, and clumps of green plants grow in the crannies, the endless vistas of sculptured walls indicate only slight weathering of the rock. In a somewhat less luxuriant coastal jungle southeast of Vera Cruz, Mexico, a slab of basalt, definitely dated by archeologists, lay in moist soil for at least 400 years without blurring of the Mayan inscription,

FIGURE 6-8. *Cleopatra's Needle, in Egypt, left, and in Central Park, New York City, right. (Photos by courtesy of The Metropolitan Museum of Art.)*

whereas other inscriptions at the same place, with an almost identical history, have become illegible.

Weathering proceeds most slowly in a dry warm climate. Inscriptions carved between 3 B.C. and 79 A.D. above the doors of sepulchers in the crumbly sandstone cliffs of northwest Arabia could still be read in 1876. Forty centuries ago, not far from the site of Assuan Dam in Upper Egypt, a surface of Syene red granite was smoothed by men's tools and dated by an inscription. This surface, though exposed to direct sunlight, is still firm. At various dates, ranging from 2850 to 313 B.C., colossal statues from this quarry were set up at Luxor and elsewhere in the somewhat moister atmosphere of Middle and Lower Egypt, and blocks of the same rock were used to face pyramids near Cairo. The average rate of exfoliation for all the Syene cut granite studied by D. C. Barton, an American visitor to Middle and Lower Egypt in 1916, was about 0.5 to 1.0 centimeter per 5,000 years. In general, the older monuments showed the greater effects, but the maximum measured exfoliation was 1.0 centimeter in the youngest blocks studied (313 B.C.).

Two obelisks of the Syene granite, each bearing many deep-cut hieroglyphics, and each now called Cleopatra's Needle, stood about 3,500 years at Heliopolis and Alexandria in Egypt with only slight weathering. One, removed to London, has weathered appreciably but not disastrously. The other needle, brought to New York about 1880 and set up in Central Park where it is exposed to frost as well as frequent wetting, exfoliated so extensively that by 1950 (despite the application of shellaclike preservatives in the nineteen-twenties and thirties) part of the pictured story was completely erased (Fig. 6-8). More weathering had taken place in 70 New York winters than in 50 times 70 Egyptian years.

Facts, Concepts, Terms

CLIMATE

HEAT BUDGET

AIR PRESSURE; AIR MASSES
    Movements of the atmosphere
    Trade winds; cold front

WEATHERING

DISINTEGRATION OR MECHANICAL WEATHERING
    Frost action, root wedging, and related processes

Questions

1. If air flows from places of high barometric pressure to those of low, why does it not always rise toward mountain peaks from the surrounding plains?
2. Why is the equatorial belt cloudier than the high-pressure belt in middle latitudes (the so-called "horse latitudes")?
3. The local winds that blow down from mountains onto the adjacent lowlands are locally called "Föhn" in the Alps, "Chinooks," in northwestern United States. Why are such winds warm enough sometimes to melt six inches of snow in one night when they are flowing down from snow-covered peaks?
4. Why is the general air drift near the equator westward and irregular, whereas in the latitude of Cape Horn strong winds blow nearly incessantly from the west?
5. In much of Illinois, Iowa, Indiana, and nearby states, deep road cuts expose several soil profiles each with well-developed A and B horizons. In broad terms, what does this superposition of profiles indicate as to the geologic past?
6. Western Nebraska is semi-arid, with dry summers; eastern Iowa considerably more moist and with considerable summer rainfall. Assuming the soils are derived from similar parental rocks, how should the characteristic soil profiles of the two areas differ, if at all?
7. Why does calcium accumulate in the A horizon of some soils of Nevada, whereas it is practically absent from this horizon in soils developed on limestone in Kentucky?
8. A soil in which the A and B horizons are distinctly developed is called a "mature soil" by the soils scientists. Few of the soils of extreme west Texas

are mature, whereas those of eastern Texas are characteristically mature. What possible explanations of this fact can you suggest?

Suggested Readings

1. *Climate and Weather for Flight in Naval Operation Zones,* Aviation Training Div., U. S. Navy, 1944. (Especially Chap. 1, "Climatic controls," and Chap. 2, "Air masses and fronts of the world.")
2. *1941 Yearbook of Agriculture,* House Doc. 27, 77th Cong., "Climate and Man." (Especially "Climate and soil," pp. 266-291; and "Flood hazards and flood control," in part, pp. 531-557. Soil profiles; soil map of U. S.; examples of U. S. storms described in terms of air mass behavior.)
3. Jenny, Hans, *Factors of Soil Formation,* McGraw-Hill Book Co., Inc., New York, 1941. (A thorough treatment of the formation of soils by weathering. Five principal factors are recognized: time, character of rock, topography, climate, and organisms.)
4. Geikie, Archibald, "Rock weathering as illustrated in Edinburgh church yards," *Proc. Roy. Soc. Edinburgh,* vol. 10, pp. 518-532, pl. 16, 1880.

7. Erosion

THE LOOSENING and transporting away of rock debris by moving agents operating at the earth's surface is called *erosion*. Weathering of rock is essentially a static process; if there were no natural agents to remove the clays, sands, and other products of weathering, all surface rocks would eventually be deeply buried beneath a thick mantle of their own decomposition products—a common situation on flat ground in areas with a humid climate. On sloping ground, however, the products of weathering are constantly being transported away by rainwash, rills, and streams. On gentle slopes the soil cover is generally a few feet deep, but on steeper slopes, weathering products are generally removed nearly as rapidly as they are formed. On many steep mountain slopes the erosional agents are so active that they even scour and remove fresh rock before it has been appreciably softened by weathering.

As the products of weathering are removed by erosion, new rock surfaces are exposed to attack by the weathering agents. Thus a cycle of change is set up: Weathering *prepares* the rock for transport by decomposing and disintegrating it. Erosional agencies *transport* the weathered material to a different locality (for example, to the ocean), and there deposit it in layers (strata) which may be ultimately *cemented* into new rock. This rock may then be raised above sea level, exposed to the air and the rains, and again weathered into soil, to start a new cycle.

The interrelations of these processes, and of the products produced by them, are shown in Figure 7-1.

Agents of Erosion

Erosion is a dynamic process. Only moving agents are capable of picking up rock debris and transporting it to a new location. Although many different agents—as varied in their nature as earthworms, lightning, and avalanches of snow—can transport a minor amount of rock waste, the great bulk of erosional debris is transported by four main agents: (1) the *wind;* (2) *wave*

FIGURE 7-1. *The cycle of rock change.*

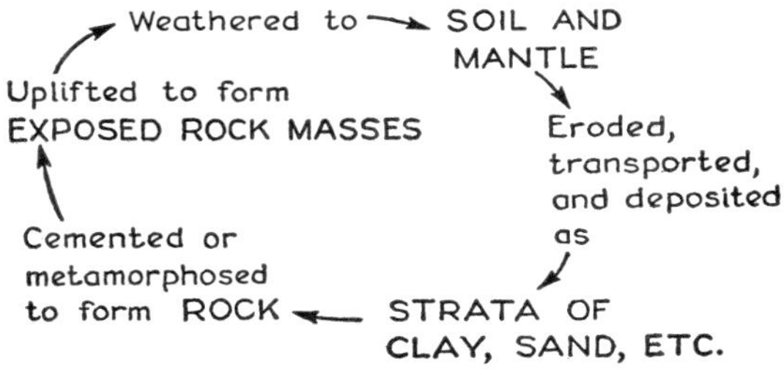

```
                      Weathered to ──► SOIL AND
                     /                   MANTLE
          Uplifted to form                  \
          EXPOSED ROCK MASSES          Eroded,
               ▲                       transported,
               |                       and deposited
               |                       as
          Cemented or                    /
          metamorphosed                /
          to form  ROCK ◄── STRATA OF
                              CLAY, SAND, ETC.
```

and currents in the ocean and, to a lesser extent, in lakes and other small water bodies; (3) *glaciers;* and (4) *running water.* Another important process, because it modifies and increases the erosional effects of running water, is the slow, or occasionally rapid, *downslope movement,* under the action of *gravity,* of solid or semi-liquid masses of mud, soil, and rock debris. On hills and mountainsides such downslope movements deliver a steady increment of rock waste to the streams at the foot of the slopes. The streams then have material to transport.

To evaluate the effectiveness of the different agents in eroding the land requires analysis of many factors. Because of the size of the earth, the complexity of the erosional processes, and the difficulty of measuring accurately the amount of debris in transit at any one time, it is impossible to get quantitative figures on either the rate at which erosion is lowering the land masses (the *rate of denudation*), or on the amount of debris transported by any one particular agent of erosion. Yet, despite the incomplete data, fairly reliable estimates as to the relative importance of the different agents of erosion, and a rough but useful approximation of the rate of denudation of the land areas can be made.

Wind

When the Great Plains area east of the Rocky Mountains was settled, the natural grass sod was plowed under, and the light soils were excessively tilled in order to grow corn and wheat. Even before settlement, parts of this area had undergone considerable wind erosion. The farmers who raced madly into the "Cherokee strip" to stake out homesteads, when the Indian Territory of what is now Oklahoma was opened to settlement, were greeted by blinding dust storms. As more and more land was put under the plow, these "black dusters" became bigger and more frequent. In the early 1930's a succession of dry seasons led to the terrible drought years of 1933 and 1934. Pulverized by tillage and parched by drought, the soil had lost not only its original protective grass cover, but much of its cohesiveness. Thus, the stage was set for one of the most spectacular and destructive events that has occurred in modern agriculture.

On May 12, 1934, strong winds lifted vast quantities of dust from the fields of Kansas, Oklahoma, Texas, Colorado, and other Plains states high into the air and drove it swiftly to the east as a gigantic dust storm. Sweeping swiftly out of the Plains as a blinding, choking mass of particles so dense as to blacken the sky and change day into darkness, the seething dust clouds rolled eastward across the well-watered lands east of the Mississippi River. Here, where little dust could be gleaned from the forest and grass-covered landscape, the clouds lost some of their density, but they still retained enough solid material to blot out the sun as they swirled around the skyscrapers of New York City and out over the Atlantic. Dense, dirty-brown clouds of dust engulfed ships 300 to 500 miles from shore and filtered even into well-protected holds, blanketing the cargoes.

Measurements of the amount of dust in the air, reported by observers from widely scattered points over central and eastern North America, indicate that more than 100 tons of dust per square mile fell in the areas covered by the dust cloud. Since the area affected embraced approximately two-thirds of the North American continent and much of the western Atlantic Ocean, it appears that more than 300,000,000 tons of soil was removed from the Great Plains during this storm and strewn over the lands and sea to the east.

Wind strongly winnows the material it transports. Only the finest particles are swirled high in the air and carried far. Most of the material moved by wind is not dust, but consists of coarser grains of silt and sand that roll or skip along the surface of the ground (Chap. 15). Studies by soil conservationists indicate that in the Plains country about three-quarters of the soil moved by the wind does not rise into the air but drifts along the surface for a few yards, or at most a few miles, and then accumulates in ditches, around clumps of vegetation, and against fences, buildings, or other obstructions. Thus, for every ton of air-borne dust there are generally two or three tons of coarser debris piled in drifts of sand and silt within or near the source.

The dust storm of May 12, 1934, was followed by many others. How a vast area of the southern Great Plains was converted into a barren "Dust Bowl" with attendant economic disruption and enforced mass migration of people is now an episode of American history. Because of the great havoc wrought, many of the later dust storms were carefully investigated. One of the most completely studied storms was a much smaller one that swept out of the Dust Bowl on February 7, 1937. Most of the land to the northeast was blanketed with snow so it was easy to collect and measure the dust that settled. At Ames, Iowa, 34.2 tons of material per square mile was deposited by this storm, 14.9 tons per square mile fell at Marquette, Michigan, and 10 tons per square mile in southern New Hampshire.

More rain and better methods of farming have again brought prosperity to the Dust Bowl, but wind erosion on a minor scale still continues, and grimly threatens a major revival if a series of drought years returns.

Although the great dust storms of the 1930's were largely due to changes in the soil cover induced by man, wind erosion is constantly at work on all surfaces of the continent, and is particularly potent in deserts. In the western Sahara, strong southward-blowing winds (the *Harmattan*, a local name for the Trade Winds) sweep across the desert for about six months of each year. In exceptionally stormy years they are reported to deposit as much as a foot of silt, dust, and sand along the edge of the desert in northern Nigeria. At times, terrific dust storms also sweep northward across the Mediterranean into Europe. In 1859 such a dust storm covered most of Europe, and deposited 80 to 100 tons of dust per square mile in southern Germany.

Oceanic Waves and Currents

Turbulence in the water envelope of the earth (oceanic waves and currents), like turbulence in the air (winds), is an important agent of erosion. Anyone who has watched the waves of the open ocean crash against the coast need not be told that here is a powerful agent capable of picking up and transporting sand, gravel, and other rock waste. Houses, breakwaters, and seaside roads may be destroyed overnight by the waves, cliffs are undermined, ships driven against the rocks and broken into matchwood, and boulders and sand rolled about like ninepins in the swirling water along the shore.

Sand drifted along the coast during a storm may completely block the mouth of a river, or may be swept far out into deep water. Even on a relatively quiet day, the sand along an ocean beach is in constant agitation: each oncoming wave, as it strikes the beach, brings sand rolling along the bottom with it, and as the wave ebbs the sand is rolled back seaward. During severe storms huge boulders may be dashed back and forth in the breakers, and even hurled against a sloping rock beach with such violence as to catapult high in the air with the spray. Occasionally, shore-driven rocks are thrown over seawalls and come crashing through the roofs of buildings.

Obviously the shore of the open ocean is a zone of vigorous erosion. Most ocean coasts are cliffed because the waves rapidly undermine the shore. Along many shorelines, such striking changes take place during storms that valuable beach property ordinarily must be protected from erosion (or in some places from unwanted deposits of new sand or other beach materials) by breakwaters.

Glaciers

Probably the Nisqually glacier, which is easily reached by highway in Rainier National Park, has been visited by more people than any other in North America. The Nisqually is a pygmy among glaciers; less than 5 miles long and less than 1,000 feet thick, it does not compare with the much larger Emmons and Winthrop glaciers in the more remote northeastern section of Rainier National Park, not to mention the 60-mile-long ice streams of Alaska or the vast moving ice sheets that blanket Greenland and Antarctica.

Dwarf though it is among its fellows, no visitor can examine the front of the Nisqually glacier without being impressed by the tremendous erosive power of moving ice masses. At the glacier front very little ice is visible—most of it is obscured by blocks of rock riding on the moving ice mass or pushed forward in front of it. Some of these blocks are fragments of andesite, each weighing several tons. By matching the rock with outcrops far up the valley, it is obvious that they have been torn from parent rock ledges upstream and carried to their present resting place. Some have been rafted along on the back of the glacier, some were frozen into its body and dragged forward with it, and some were pushed in front of it. Smaller particles of rock and soil are mixed with the larger boulders. As the ice melts, they accumulate in hummocky piles in front of the glacier. Thick accumulations of loose rock also rest on top of the glacier, especially at its edges, because, here, boulders, loosened by frost action from the cliffs above, are constantly rolling down, and snowslides and streams of meltwater also add debris to the glacier's surface.

The Nisqually glacier has been slowly shrinking in recent years. Downstream, below the front of the glacier, lie irregular piles of debris left by the ice as it retreated upstream. The rock floor, where bared by the melting ice, is scored and polished by the blocks of rock and finer material that has been dragged across it by the moving glacier.

Meltwater pouring from tunnels in the Nisqually glacier is a dirty milky white. Examination shows that it is cloudy, not from clay, but from pulverized rock flour obviously made by the crushing and milling of rock particles against one another and against the bedrock floor as they were rolled beneath the tremendous weight of the moving glacier and dragged forward by it.

Relative Importance of Different Agents

Spectacular as the results of wind, ocean, and glacier appear, they are of relatively little importance as compared with the movement of rock debris by running water.

In the years since the middle 1930's, rainwash has almost completely obliterated the drifts of silt and sand made during the dust storm of May 12, 1934. Only in the driest deserts are features due to wind erosion conspicuous, and even here the presence of numerous gullied slopes, dry stream beds, and sheets of water-deposited detritus shows that the rare rainstorms are able to put their stamp upon the desert landscape.

Oceans are limited in scope and effect. The attack of the sea upon the land is chiefly confined to the shoreline and to the shallow submerged margin of the continents. Here they have undoubtedly played a part in extending the upper of the two dominant surface levels we noted in Chapter 2 (p. 25), but stream deposits have doubtless played an even greater role.

Again, compared to running water, glaciers are quantitatively unimportant as erosional agents. Streams are practically universal—there is scarcely any part of the continents that is not subject to at least some rainwash each year. Glaciers, on the other hand, are confined to high mountain ranges and to polar regions. The combined area of all the glaciers constitutes only 0.03 per cent of the total land surface. Their contribution to the denudation of the continents is insignificant when compared to the work of streams.

Streams, aided by downslope movements, are the great levelers. Although individual rivers during stage of flood may produce sudden and spectacular feats of erosion, the quantitative importance of streams is due to the fact that rain falls on all parts of the continental masses, and in coursing off them to the sea it steadily removes rock debris from the surface. Let us turn our attention to an analysis of the way in which streams perform their work.

Analysis of Stream Erosion

Source of the Runoff

"From whence springs the water of the rivers?" is a question to which reliable answers have been won only in comparatively modern times. The early philosophers held that rainfall was totally inadequate to account for the vast flow of water in rivers like the Seine, Rhône, and Nile, and that the earth's surface was too impervious to allow the percolation of rain water into the soil and rocks from which it could be returned in the form of springs. Although there were occasional dissensions, the general belief up to the middle of the Seventeenth Century was that the water rising in springs and flowing to the ocean in streams could not possibly be the runoff from rain and snow.

The modern concept of hydrology (the science that treats of water) began with the work of a Frenchman, Pierre Perrault (1608-1680). Perrault made measurements of the rainfall in a part of the Seine River basin over a three-year interval and also measured the amount of water flowing in the Seine. From the available maps, he then estimated the area of the river drainage basin above the point where he had measured its flow. These measurements indicated that the Seine carried off only one-sixth of the water that fell

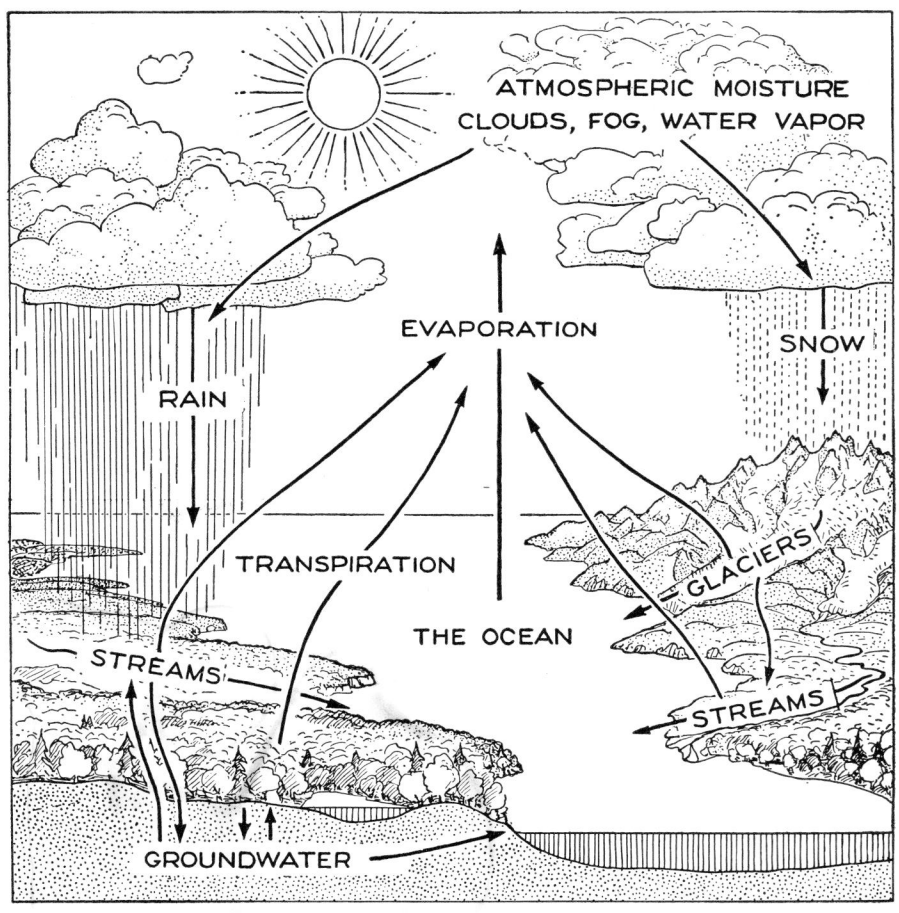

FIGURE 7-2. *The hydrologic cycle.*

within its drainage basin as rain and snow. Perrault thus proved the fallacy of the old assumption that the streams carry more water to the sea than falls on their drainage basins. In fact, the problem of the streams was now re-versed—the question was no longer, What is the source of the water? but, What has become of the vast volume of water, equivalent to five times the flow of the Seine, that has been precipitated on the drainage basin but is not being carried out by stream flow?

A few years after Perrault's measurements, Edmund Halley, an English astronomer, showed by experiment that the moisture evaporated from the surface of the oceans is entirely adequate to supply the water returned as runoff.

Since the time of these pioneers a wealth of hydrologic observations has become available. The flow of water in the major streams of practically all countries has been accurately determined by stream gaging. Such measure-ment, generally carried on by federal or state agencies, has many purposes. Precise data on river flow are valuable to navigation, irrigation, hydroelec-tric power, flood control, and many other interests. Similarly, the accurate recording of rainfall and snowpack has many uses; and data are available from thousands of stations scattered all over the world. Numerous tests have been made on the rate of evaporation from bodies of water under different climatic conditions, and on the rate of loss of water into the atmosphere through transpiration by plants. Rates of underground percolation of water and measurements of the amount of water stored in underground reservoirs have been obtained from thousands of observations on springs and wells. From these measurements and observations we can fairly accurately deduce the relations between rainfall and runoff, and can answer the question of what happens to the water that falls as rain (Fig. 7-2).

The Hydrologic Cycle

Of the moisture precipitated on the land as rain and snow: (a) Some evapo-rates at once from the surface of the ground and from the vegetation on which it falls; (b) some is absorbed by the roots of growing plants and is quickly transpired back into the atmosphere through their leaves, although a little enters into the plant tissues and is trapped there until the plant is dead and destroyed; (c) some seeps into the soil and rock where it may be temporarily stored underground until it is (1) brought to or near the surface by capillarity and evaporates; (2) delivered to the roots of growing plants; or (3) oozed to the surface at a lower elevation, reappearing either in springs or as ground-water runoff (*effluent seepage*) pouring into streams along their banks and bottoms; and (d) some runs off in surface rills, brooks, and rivers and is returned directly to the ocean. It is the last part, the *run-off*, that we are concerned with in the study of stream erosion.

The hydrologic cycle is depicted graphically in Figure 7-2. The sun is the source of energy that operates this great system of waterworks. Solar energy raises the temperature of the oceans and the lands, and evaporates water from their surfaces. It also stimulates the growth of plants and leads to the transpiration of water vapor into the atmosphere from their pores. Atmospheric moisture gained in these ways is wafted inland and raised to high altitudes by the winds. It thus acquires potential energy of position. When the moisture falls as rain, this energy of position is converted to kinetic energy—the water has become a moving agent, capable of acquiring and moving a load. On reaching the surface of the earth, that part of the rain that constitutes the runoff flows downslope in response to gravity, moving obstacles in its path and sweeping soil and rock debris along in its currents.

In studies of the relation of runoff to precipitation it is convenient to define *runoff* as *the total discharge of water by surface streams*. Thus defined, the runoff includes not only that part of the precipitation that flows directly from the rains across the surface of the land, but also the increments of ground water that enter the streams from springs and effluent seepage.

It is also convenient to distinguish that part of the precipitation returned to the air by evaporation and transpiration as the *evapo-transpiration factor*. Hydrologists, for convenience, usually simplify the relations within the hydrologic cycle by assuming that evapo-transpiration can be determined by subtracting runoff from precipitation. The equation

$$\text{Precipitation} = \text{Runoff} + \text{Evapo-transpiration}$$

is only approximately correct, however. It neglects such locally important (though generally unmeasurable) factors as the seepage of water through the soils and rocks directly into the ocean. For the land masses as a whole such seepage is not great, but in a few areas of highly permeable rocks, as in the Hawaiian Islands, it may locally dispose of a large part of the total rainfall. Other very minor factors are the amounts of water locked up in the tissues of plants and animals, and the amounts chemically combined in minerals during weathering.

Variable Factors Affecting the Runoff

The ratio of runoff to precipitation is not everywhere 1:6 as Perrault determined for the basin of the Seine. Several variable factors affect the ratio.

Careful studies of many different hydrologic basins in the United States, made by the Water Resources Branch of the U. S. Geological Survey, have revealed great variations in the ratio of runoff to precipitation. From these measurements some of the factors may be summarized as follows:

Temperature. Evaporation is, of course, much greater in warm regions than in cold. High temperatures also favor growth of vegetation, and hence increased loss by transpiration. In humid regions a close correlation has been found between the amount of runoff and the mean annual temperature. Even in arid and semi-arid regions, if large groups of data are considered, not simply those from a single drainage basin, there is also a reasonably close accord. The data relating temperature to runoff in the United States has been summarized by Langbein, a hydrologist of the U. S. Geological Survey, in Figure 7-3. Thus, for an average annual precipitation of 40 inches, more than half the rainfall (21.5 inches) runs off where the mean annual temperature is 40° F.; but the runoff decreases to 10.2 inches in a mean annual temperature of 60° F., and to less than 3 inches in a mean annual temperature of 80° F.

Slope. Slope obviously exerts an important control on runoff. Steep slopes of bare rock in mountainous regions shed practically all of the rain

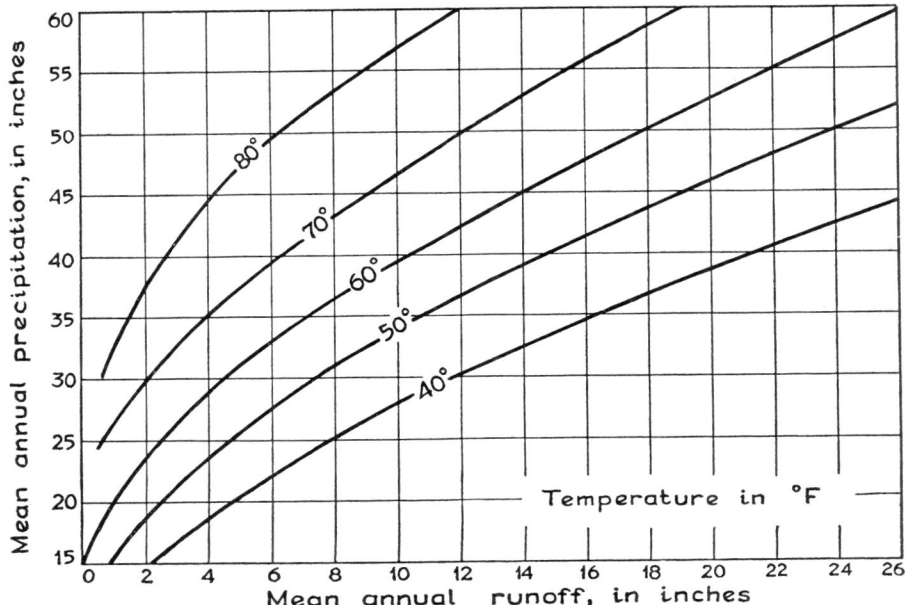

FIGURE 7-3. *The variation of runoff with rainfall and temperature.* (*After W. B. Langbein*, Annual Runoff in the United States, 1949.)

that falls on them. On flat ground the rain may remain on the surface in shallow puddles until it is evaporated or absorbed by the soil and plants.

Permeability of the Ground. Different soils and rocks have highly different permeabilities. Most of the rain falling on the porous ash of a recent volcanic cone sinks directly into the ground and becomes a part of the ground-water circulation, but nearly all will run off on a landmass of similar slope that is underlain by relatively impermeable shale.

The ability of a soil to absorb rainfall or snowmelt depends on other physical factors as well as permeability. Water-logged soil, although its permeability may be high, has reached its maximum capacity of absorption; frozen ground is similarly nonabsorptive. The absorption of a highly permeable soil also may be limited because of a nonpermeable subsoil or bedrock a few feet below. Permeability, then, is not alone an accurate measure of the rate of infiltration. Hydrologists use the term *infiltration capacity* to define *the maximum rate at which the soil, in a given physical condition, can absorb falling rain.* Artificial tests of infiltration capacity made by playing sprinklers on enclosed plots of soil from which the runoff is caught and measured show, as would be expected, that the rate of absorption is high at the beginning of a rain, then diminishes rapidly until a fairly constant value is reached. The high initial rate usually changes to the slower uniform rate within a half hour after the rain begins to fall.

Amount and Duration of the Rainfall. Distribution of rainfall through the year greatly influences the runoff. Rain uniformly distributed in many small showers may be largely evaporated or absorbed by the ground and plants before it reaches flowing streams. During violent rainstorms, on the other hand, percolation is too slow to lead much of the water underground; most of it courses rapidly across the surface and into stream channels. Rapid melting of the winter snowpack by warm rains or winds is one of the most common causes of river flooding.

If the annual rainfall is high (80 inches or more), most of it is likely to run off, even though the rain may be uniformly distributed. Here, the underground reservoirs are constantly replenished, and the pores within the soil and rock are filled with water almost to the surface of the ground. In deserts, on the other hand, most of the scanty rainfall is absorbed by the parched soil, or evaporated, and there is little runoff except immediately after violent storms.

Vegetation. Vegetation retards the runoff. Tangled stalks of grass or the mulch of decaying leaves and twigs in a forest absorb rain like a blotter. Earthworms and other burrowing animals that live in the plant-rich soil aid percolation by opening tunnels in the soil.

Very striking increases in runoff have been observed where the forest cover has been burned off, or where the natural sod has been plowed under, as in our Dust Bowl. Figure 7-5 (p. 136) summarizes some of the data for a few such areas.

These variable factors result in marked differences in runoff in different stream basins. In the western United States runoff varies tremendously with differences in temperature, infiltration capacity, and other local factors. The range is from less than 0.25 inch in the southwestern Arizona desert to more than 80 inches on the western slopes of the Olympic and Cascade mountains of Washington and Oregon.

Amount of Water Available for Erosion

The rivers of the United States, as determined by stream gaging, deliver water to the oceans at a rate of 1,800,000 cubic feet per second (about 330 cubic miles per year). Roughly one-third of this amount is carried by the Mississippi River.

The average annual rainfall of the United States, determined from thousands of rain-gaging stations, is 30 inches. The average runoff for the country as a whole is 8.6 inches. This gives a ratio of runoff to precipitation of approximately 1 to 3.5, as compared with the ratio of 1 to 6 determined by Perrault for the basin of the upper Seine.

Data for runoff and precipitation of other continents is less complete than for the United States. L'vovich, a Russian hydrologist, has summarized the available data on runoff for the various continents. His results are given in Table 7-1. The table shows that Australia, with a runoff of only 3.0 inches, is the driest of the continents, and that South America, with 17.7 inches of runoff, is the wettest. The world average, as given by L'vovich, is 10.5. Other hydrologists have reached slightly different, and in general somewhat lower, figures (around 7.5 to 9 inches) for the average world runoff. Despite these uncertainties, we can make a fairly reliable estimate of the total amount of water available for erosion: The average annual precipitation over the land areas of the earth is about 40 inches per year. This means that each year approximately 35,000 cubic miles of water falls on the land in the form of rain and snow. According to different estimates, the rivers return to the oceans from 7.5 to 10.5 inches per year. Using an average figure, this means that in round numbers *approximately 8,000 cubic miles of water courses off the lands and into the seas each year. This is the amount of water that is available for the erosion of the lands.*

TABLE 7-1. *World Distribution of Runoff (from Langbein, after L'vovich).*

CONTINENT (OR OTHER AREA)	AREA (THOUSANDS OF SQUARE MILES)	RUNOFF (INCHES)
Europe (including Iceland)	3,734	10.3
Asia (including Japanese and Philippine Islands)	16,321	6.7
Africa (including Madagascar)	11,510	8.0
Australia (including Tasmania and New Zealand)	3,075	3.0
South America	6,941	17.7
North America (including West Indies and Central America)	7,893	12.4
Greenland and Canadian Archipelago	1,499	7.1
Malayan Archipelago	1,012	63.0
TOTAL AREA AND AVERAGE RUNOFF	51,985	10.5

Energy Released by the Runoff

On the average, the lands stand about one-half mile above sea level (see p. 25 and Fig. 2-9). Therefore, the continental runoff descends from an average altitude of one-half mile to sea level in flowing from its source to the ocean. This is an average figure; runoff falling on a plain near sea level may descend only a few feet to reach the ocean; the meltwater from the snows on the summit of Mount Everest falls over five miles. A tremendous amount of energy is developed by the world's streams during their descent to the sea. Some qualitative idea of the amount can be gained by imagining all the continental runoff to be concentrated in one gigantic river that is pouring over a huge waterfall equivalent in height to the average altitude of the landmasses. In other words, imagine 8,000 cubic miles of water per year cascading down a waterfall one-half mile high! One could easily compute the amount of power available at such a waterfall, but the resulting figure is so large in human terms as to be utterly meaningless.

Yet this is the amount of energy available for the erosion of the lands by streams. It is easy to see why running water is the great leveler—much more important than all the other agents of erosion combined. Fortunately for mankind, very few streams work at anywhere near their full erosional capacity. Because weathered material ready for transport is moved downslope to the streams only slowly, many rivers carry little load and dissipate

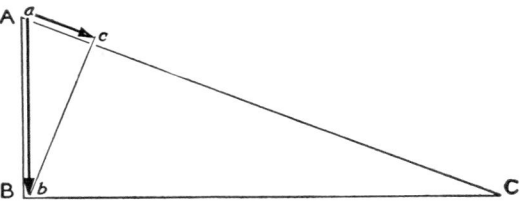

FIGURE 7-4. *Resolution of gravity for a stream of uniform slope.*

most of their energy in work other than erosion. Otherwise, the continents would be largely denuded of soil.

Mechanics of Stream Erosion

In dealing with influences on stream flow it is essential to keep in mind the *Law of Conservation of Energy,* and the distinctions between potential energy (energy of position), kinetic energy (energy of motion), and heat energy. The *Law of Conservation of Energy* requires that, although one form of energy may be transformed into another, for example, kinetic energy into heat energy when a falling rock strikes the ground, the sum of all the different kinds of energy does not change during the transformation. In other words, the total amount of energy remains constant, and no energy is gained or lost.

Flowing streams are powered by gravity. It is convenient to use a simple graphical representation, the "resolution of forces" (Fig. 7-4), to determine the component of the force of gravity effective down the particular slope on which a stream is flowing. Consider a mass of water poised at the top of a uniform slope (point *a* of Fig. 7-4). Its total energy is measured by the potential energy of vertical fall, and hence the final velocity after movement down the slope would, if there were no friction, be the same as for direct vertical fall, though the time required to attain the velocity would be greater because the component of gravity operating parallel with the slope is less. The acceleration of gravity (an increase in velocity of 32 feet per second at the end of each second) is represented by the hypotenuse (*a-b*) of the right triangle *abc*. Under frictionless flow, the mass of water would move down the slope AC with an acceleration *ac* (an increase of velocity of about 5 to 6 feet per second in a second for the slope shown in Fig. 7-4) and would travel the distance *ac* in the same time that it would fall freely through the vertical distance *ab*. Even with the much flatter slopes characteristic of typical streams, the acceleration for frictionless flow would often be as much as 0.5 feet per second in a second. As there are 3,600 seconds in an hour, the stream, after an hour's frictionless flow down an ordinary slope, might attain the astounding velocity of 1,800 feet per second—more than 1,200 miles per hour!

No such velocities are even remotely approached in natural streams. Actual streams are slowed by friction of the moving particles against one another and against the land surface. They commonly attain and maintain velocities of only a few feet per second (generally less than 5 miles per hour); if their velocity changes appreciably, it usually decreases downstream. We conclude that the energy of fall is mostly converted to heat energy by friction. Some of the energy is also used up in the transportation of eroded rock particles from upstream.

The actual quantity of mud, sand, and gravel eroded varies widely for different rivers. For example, compare the crystal-clear Columbia River at Wenatchee, Washington, with the Missouri River (often called the "Big Muddy") near Saint Louis, Missouri. The amount of water carried by each is about the same, but the Columbia, though a roily turbulent stream filled with violent rollers and eddies, contains very little silt and mud. It is flowing on smooth bedrock and coarse boulders, hence almost no fine debris is available in its bed to be picked up and carried. By contrast, the Missouri is laden with mud and silt. It flows across the Great Plains, where the soil cover is deep and the vegetation scanty; consequently, abundant fine-grained rock waste is available for transport. Much of the energy of flow of the Columbia is used up in internal friction within its boils and eddies, whereas in the Missouri much of the available energy is used to transport the abundant rock waste. But even the muddy Missouri is transporting far less material than it might. Only rarely does the Missouri carry more than 20,000 parts per million (2%) by weight of solid material. This is a high load for a big river, though it is exceeded by the Colorado, the Yellow River of China, and several others. The Mississippi, generally considered a very muddy stream, carries only about 5,000 parts per million (0.5%). Small streams in semi-arid regions with a thick soil mantle, however, may carry enormously greater loads—up to 30 per cent or more by weight—especially during floods. A sample from the San Juan River in Colorado when it was in flood contained over 75 per cent by weight of red silt and sand. There is, of course, a complete gradation from such highly loaded streams to mudflows and landslides.

The evidence indicates, however, that nearly all streams are working far below their potential capacity as erosional agents. If they used most of their energy in carrying rock waste, few slopes could retain a cover of soil.

Rate of Denudation

What, then, are the actual figures on the rate of denudation? At the present time, measurements on the rate at which different drainage basins are being lowered are so few and incomplete, and quantitative data so hard to obtain, that estimates can be only tentative.

The total load of debris being carried by a stream is difficult to measure. Although reliable figures covering a period of more than a year or two are available for only very few rivers, there are enough data on hand to indicate the general magnitude of erosion. A carefully controlled series of measurements of the silt load of the Missouri River above Kansas City made during 1930 and 1931 indicates that this river is lowering its drainage basin about 1 inch in 650 years, or 1 foot in 7,800 years. The Colorado River, which, in the ten-year period preceding September 30, 1935, carried an average annual load of 250,000,000 tons of silt and sand past the Grand Canyon gaging station, is certainly lowering its drainage basin more rapidly than the Missouri; but the Columbia, at least above the junction with its muddy tributary, the Snake, is undoubtedly denuding its basin much more slowly than the Missouri. Each year the Mississippi River brings to the Gulf of Mexico about 730,000,000 tons of dissolved and solid material.

From such measurements it is estimated that *the rate of denudation for the entire United States is about 1 foot in 9,000 years.* If this rate could be maintained, and if there were no compensating upward movements of parts of the earth's crust, all the landmasses would be eroded to sea level in about 23 million years. This is a long time by human standards, but, as shown in the next chapter, it is rather short in terms of the total length of geologic time.

Changes in the Rate of Erosion

Because of certain variables not taken into account in the measurements, our figure of 1 foot in 9,000 years for the average rate of denudation must be modified considerably in dealing with specific erosional problems. For some kinds of problems the figure is too high; for others it is much too low. For example, can this figure be used as the average rate of erosion for all of geologic time? There are many reasons for believing that present-day streams carry much higher loads than the average for the geologic past. The activities of civilized man contribute vastly to present stream loads. Rivers in populous areas carry a vastly increased load because they are polluted with industrial and municipal wastes. Enormously increased amounts of silt and mud are also being washed into our streams from cultivated fields; before human settlement, the original forest and grass cover surely impeded erosion and greatly lessened the amount of soil washed to the sea. Such an increased rate of erosion has been proven recently by the Soil Conservation Service of the U. S. Department of Agriculture in making extensive tests of the amount of soil removed under varying conditions of tillage. The bureau caught and measured carefully the amount of soil and water running off

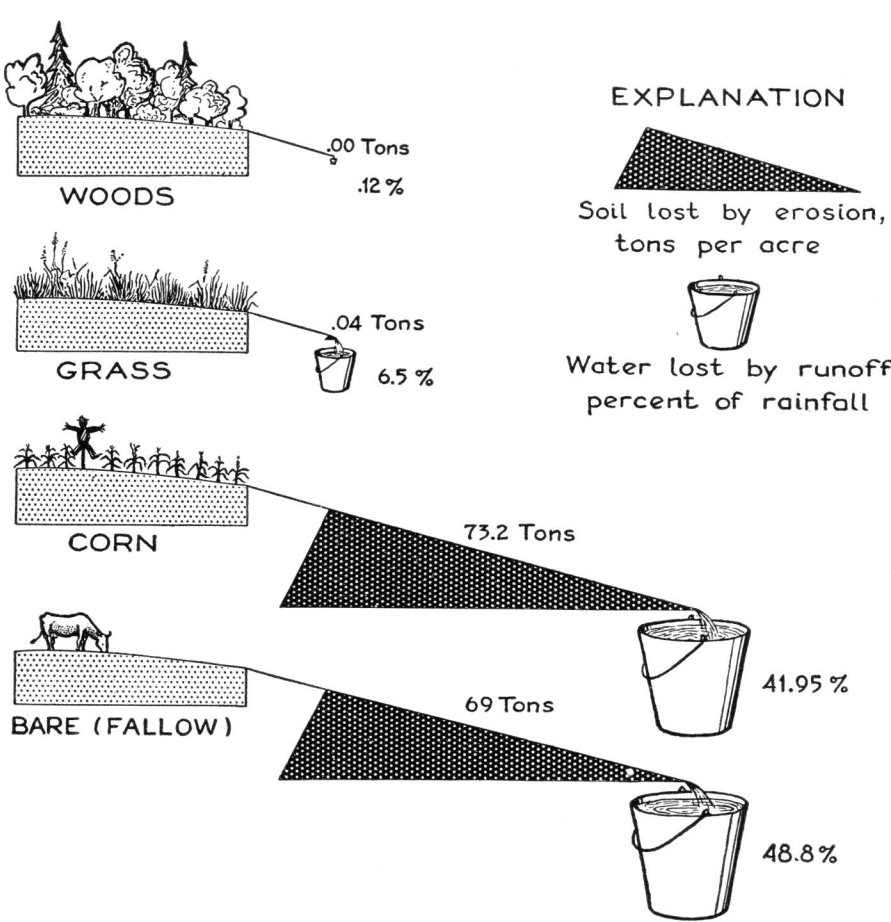

FIGURE 7-5. *Results of soil erosion tests on plots of ground with different vegetative covers. (Redrawn by permission from H. H. Bennett,* Soil Conservation, *McGraw-Hill Book Co., 1939.)*

from plots of ground with identical areas and slopes but with different vegetative covers. A few results from such tests are given in Figure 7-5.

There are also some reasons to think, as shown in later chapters, that during most of the geologic past the lands probably did not stand as high as now; hence the energy available for erosion may have averaged considerably less.

These factors cannot be precisely evaluated and, thus, the rate of erosion during the whole of geologic time cannot be fixed quantitatively. Certainly, the rate of 1 foot in 9,000 years is much too high—perhaps two or three times too high. In the next chapter we shall see how an ingenious attempt to measure the age of the oceans by their content of salt failed because the amount of dissolved salt carried to the sea by the rivers of the geologic past could not be accurately estimated—the age, as determined from the present rate, was many times too low.

Can we apply the rate of 1 foot per 9,000 years to the problem of soil erosion? The soil conservationist wants to know how rapidly the topsoils of American farms are being gullied and rendered unusable by erosion. In many sections of the country productive topsoil is rapidly disappearing, and once-prosperous farms are now barren wastes of deeply eroded gullies and exposed bedrock. It is estimated that about 282 million acres of farmland in the United States have been so seriously eroded as to make farming unprofitable. Still another 775 million acres have lost enough topsoil to affect their productivity; and another billion acres of land show some evidence of damaging erosion. These figures do not include the large acreage of the Dust Bowl, in which the land has been damaged by wind erosion.

The soil conservationist has difficulty in reconciling this striking evidence of damaging erosion with the figure of 1 foot in 9,000 years for the rate of denudation. Most of our productive soils are at least 1 foot thick; how can so much damage have been accomplished in the relatively short time since the sod of the frontier was plowed into farmland? Only after intensive investigations of the kind represented in Figure 7-5 on thousands of soil plots scattered all over the United States did the answer become clear. Most of the soil removed from the fields by rainwash is almost immediately redeposited; only a part is transported directly to the sea. In the words of Hugh Bennett, an American soil scientist with the Department of Agriculture:

> . . . the sediment entering the oceans represents merely a fraction of the soil washed out of the fields and pastures. The greater part is piled up or temporarily lodged along lower slopes, often damaging the soil beneath; or it is deposited over rich, alluvial stream bottoms or in channelways, harbors, reservoirs, irrigation ditches, and drainage canals . . . Available measurements indicate that at least 3,000,000,000 tons of solid material is washed out of the fields and pastures of America every year.

Bennett's figure for the tonnage lost from the fields is nearly three times the amount of soil delivered to the oceans by streams. Evidently, the figure of 1 foot per 9,000 years for the rate of denudation, though a reasonable estimate of total lowering of the continent, does not take into account the *local* erosion and redeposition of soils of primary interest to the soil conservationist.

Geologic Evidence of Erosion

Most of the data on which this chapter is based have been accumulated during the last one hundred years, and much of it within the last fifteen or twenty. Since the earliest days of the science, however, geologists have been aware of the vast changes in the landscape brought about by running water. Their observations, that indicated an immense amount of erosion of the land surfaces, were not concerned with the amount of silt being carried to the sea in streams, nor even the spectacular effects occasionally produced by catastrophic river floods, but with the etched surface of the earth itself (Figs. 7-6 and 7-7).

On the walls of the Grand Canyon of the Colorado River (Fig. 7-6) appear horizontal layers of limestone, shale, and sandstone. The limestones, particularly the thick Redwall limestone which occurs 1,000 feet or more below the canyon rim, form great cliffs. These cliffs can be traced in and out of the short tributary canyons that score the walls of the Grand Canyon. The cliffed limestone lies everywhere at the same level, and is exposed on both walls of the canyon and its tributaries, and in all the isolated buttes within the canyon that rise high enough to intersect it. Clearly, the limestone must once have been continuous across the canyon; the Colorado River has cut down through the horizontal layers, and eroded away the portions of them

FIGURE 7-6.   *The Grand Canyon of the Colorado River. (Photo by L. Noble, U. S. Geological Survey.)*

FIGURE 7-7. *Erosion of tilted sedimentary rocks. The drawing shows the rocks as they would appear if sliced vertically at the lower boundary of the photograph. The dashed lines represent only a small part of what erosion has removed. (Photo by U. S. Air Force.)*

that formerly filled the area now occupied by the canyon. At the bottom of the Grand Canyon the river is cutting into granite and metamorphic rocks. From their mode of origin, these rocks must have been formed far below the surface of the ground.

In many mountain ranges, strata of sandstone, shale, and limestone have been tilted from their original horizontal position. The edges of these tilted beds, laid bare by erosion, can be followed in ridges like those shown in Figure 7-7. Obviously, the deeper and older beds in the sequence could only have been exposed to view through the removal by erosion of the younger overlying beds that once covered them. In several mountain chains such series of tilted beds are exposed in belts many miles wide; to strip away all of the overlying beds and to expose the deepest layers in some of these ranges have required the removal of 5, 10, or even more miles of overlying rock.

The study of this phase of erosion, however, can best be deferred until we have learned about geologic maps, for the most striking proofs of deep erosion have come from the mapping of the surface geology of mountain ranges.

Other Implications of Erosion

There are other interesting implications that follow from our study of erosion. Streams, glaciers, and waves all move material downhill; it ultimately comes to rest at or slightly below sea level. Can this be one factor in accounting for the upper of the two dominant levels of the earth's surface (Chap. 2, Fig. 2-9)?

What isostatic response (Chap. 3) can we expect in the earth's crust as a result of the unloading of mountainous areas by erosion, and the loading of the continental shelves and other low areas by deposition? Answers to these questions must be deferred to later chapters, but the point emphasized here is that erosion is a process that is continuously operating to upset the condition of isostatic balance in the earth.

Facts, Concepts, Terms

EROSIONAL AGENTS
>   Wind, streams, glaciers, waves, and currents

RELATIVE IMPORTANCE OF THE AGENTS OF EROSION
>   Streams are the great levelers

THE HYDROLOGIC CYCLE
>   Source of the energy that powers the cycle
>   Factors affecting the runoff
>   Precipitation = runoff + evapo-transpiration

AMOUNT OF WATER AVAILABLE FOR EROSION

ENERGY GENERATED BY THE RUNOFF

STREAMS ARE POWERED BY GRAVITY

RATE OF DENUDATION

GEOLOGIC EVIDENCE OF EROSION

ISOSTATIC IMPLICATIONS OF EROSION

Questions

1. Why can a stream move larger particles than wind moving with the same velocity?
2. Which of the erosional agents listed in this chapter could cut a valley to depths of 1,000 feet or more below sea level? Why?
3. Point Barrow, Alaska, and Yuma, Arizona, have about the same rainfall (5 inches per year). Yet Yuma lies in a parched desert whereas the country around Point Barrow is largely swamp. Explain.
4. In a temperate humid climate, which will erode more rapidly: a basalt cone composed of loose cinders or a hill of the same size composed of clay? Why?
5. What factors influence the amount of runoff?
6. In view of the discussion of infiltration, in which area would you expect the streams to be spaced closer together, assuming similar slope, vegetation and precipitation: (1) on limestone, (2) on shale, (3) on granite? Give reasons.
7. In which area would you expect them to be most closely spaced, assuming uniform granite bedrock: (1) steep slopes, (2) gentle slopes, (3) nearly flat terrain?
8. On aerial photographs of heavily brush-covered country it is often possible to trace boundaries between different kinds of rock despite the fact that the ground itself cannot be seen. What factors permit this?
9. Most stream junctions are at grade, that is, the bed of the tributary and of the main channel are identical in elevation at the point where they join. Why?
10. If the kinetic energy of water is consistently being converted to heat, why does a swift stream not become appreciably warmer downstream?

Suggested Readings

1. Meinzer, O. E., editor, *Hydrology*, vol. 9, *Physics of the Earth*, National Research Council, McGraw-Hill, New York, 1942.
2. Bennett, H. H., *Soil Conservation*, McGraw-Hill, New York, 1939.

8. *Geologic Maps, Fossils, and Time*

Geologic Maps

No TRAVELER can miss the contrast between a lava flow of black basalt and the prevailing light-colored granodiorite of the Sierra Nevada, or, half way around the earth, between a similar basalt flow and the gleaming white coral rock of Samoa. To show the outlines of such a flow on a map we must trace the contacts between the flow and the other rock. This is done by plotting the contacts of the flow on the map in their proper relation to valleys and hills, as shown by the contour lines and by other features on the map such as streams and roads. The result is a *geologic map*.

> A geologic map shows the distribution of the different rock masses that underlie the ground surface plotted accurately to scale in relation to the topographic features and other control points on the map.

Such maps are economically useful. If basalt is needed to surface a road in the Sierra Nevada, or to build a breakwater in Samoa, we know from our geologic map where it can be quarried. From the contours we can also estimate the thickness of the flow, and thus calculate the approximate tonnage of basalt.

From geologic maps, in conjunction with topographic maps, we can determine much of the size and shape of rock bodies hundreds or thousands of feet below the surface.

The economic power of this geologic tool is especially great in locating oil accumulations, water supplies, coal, iron ores, and other valuable substances hidden beneath a cover of soil and rocks. Though no surface indications of these valuable materials may be apparent, geologic mapping often reveals where they can be found by tunneling or drilling. The accuracy of such geological predictions has been proved again and again by actual tests.

The geologic map is also the principal tool in deciphering the history of a mountain range, the course of evolution of fossil organisms, the changes of local climates, or, indeed, the history of the earth itself.

Difficulties of Geologic Mapping

The distribution and relationships of rock bodies are not commonly so obvious as those in the High Sierra or on Samoan beaches. Indeed, in a fertile agricultural area like the lower Mississippi Valley, any systematic arrangement of the rocks is not readily seen. In such farming areas the rocks are nearly everywhere masked by soil or alluvial deposits. Unweathered rock is exposed only in stream banks and deep ravines, or in road cuts, quarries, and other artificial excavations. To make a geologic map in such regions is not easy and often requires digging pits and trenches in critical spots, or drilling holes and examining fragments of the rocks penetrated by the drill.

Early Geologic Maps

The principles of geologic mapping were developed chiefly in western Europe, where the rocks are seldom well exposed. Among the earliest geologic maps that showed the relations of sedimentary strata over a considerable area are two maps of the region around Paris, France, published jointly by the French naturalists Georges Cuvier and Alexandre Brongniart in 1810 and 1822. At about the same time (1815) William Smith, an English surveyor, published a geologic map of England which was one of the milestones in the development of geology.

Long before the work of Cuvier and Brongniart, French scientists knew that the rocks near Paris are gently tilted layers of limestone, clay, gypsum, and sandstone. These rocks could be seen in many natural and artificial exposures, notably in the pits dug by the makers of pottery and porcelain in their search for plastic clay. As early as 1782, Lavoisier took time from his epochal discoveries in chemistry to demonstrate that quarry after quarry near Paris showed the same succession of clay, limestone, gypsum, impure limestone, and sand, from the base upward, with a siliceous limestone forming a cap at the top.

Cuvier and Brongniart went much farther than Lavoisier. They demonstrated that certain characteristic fossils were present in some of the different layers, that some layers changed in character both laterally and vertically, and that the different strata near Paris could be grouped into units called

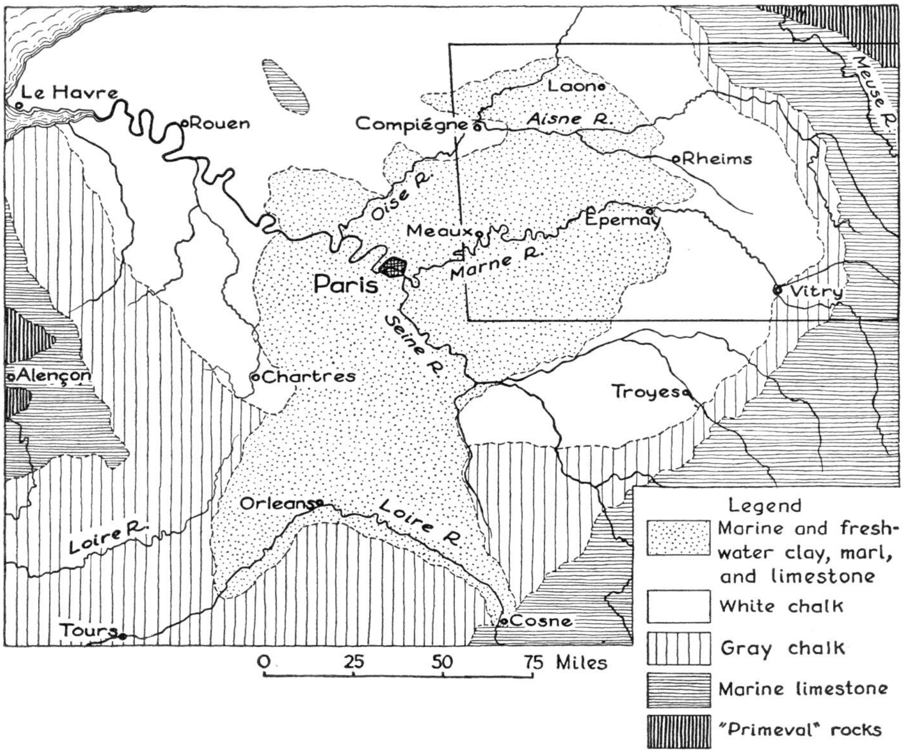

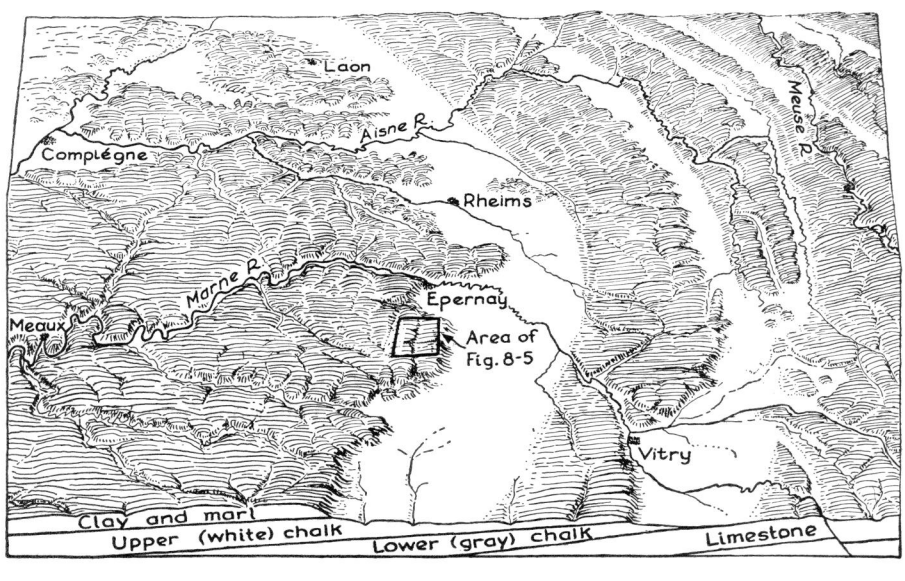

Legend
- Marine and fresh-water clay, marl, and limestone
- White chalk
- Gray chalk
- Marine limestone
- "Primeval" rocks

(Top map labels)
Le Havre
Rouen
Compiégne
Laon
Aisne R.
Meuse R.
Rheims
Oise R.
Meaux
Epernay
Marne R.
Paris
Vitry
Seine R.
Alençon
Chartres
Troyes
Orleans
Loire R.
Loire R.
Tours
Cosne

0   25   50   75 Miles

(Bottom block diagram labels)
Laon
Aisne R.
Compiégne
Meuse R.
Rheims
Marne R.
Epernay
Area of Fig. 8-5
Meaux
Vitry

Clay and marl
Upper (white) chalk
Lower (gray) chalk
Limestone

formations, whose thickness and physical characteristics are sufficiently distinctive to allow them to be traced continuously for long distances.

Succession of the Rocks Near Paris. East of Paris one enters a series of low hills and narrow lowlands (Fig. 8-1) that curve in a broad arc, partially encircling the city. These hills have steep slopes on the side away from Paris and more gentle slopes toward the city.

The lowland just beyond the first group of curving hills, though largely covered with soil, contains scattered rock *outcrops* (natural exposures) and also man-made excavations, all of which expose chalk, a variety of limestone (see p. 602).

The chalk occurs in a lowland that almost completely encircles Paris. On the side of the lowland farthest from Paris, the rock is gray, with many layers of pale green sandstone; closer to Paris the chalk is a white, very porous rock containing numerous potato-shaped nodules of black flint scattered along thin layers within it. Such layers may be separated from one another by 15 to 20 feet of massive chalk containing no flint nodules. The layering in both the gray and white chalk inclines gently downward toward Paris; hence Cuvier and Brongniart concluded that the white chalk rests upon the gray. They were thus able to separate the chalk into two distinct mappable units which they called *formations:* the *Lower Chalk,* chiefly gray chalk and greensand; and the *Upper Chalk,* a massive white chalk with flint nodules. It is the Upper Chalk which appears across the English Channel in the "white cliffs of Dover."

Cuvier and Brongniart recorded more than fifty different kinds of fossil shells and other animal remains in the Upper Chalk, all of which resemble, although they are not identical with, those of animals now living in the sea.

If one climbs the steep slope of the hills on the Paris side of the chalk lowland, he finds scattered outcrops of a plastic clay in the bottoms of little ravines where the thin soil cover on the steep hillside has been washed away by the rains. This clay rests upon the chalk. Many pits show that near the base of the hills all the outcrops are of clay, but ravines farther up the slope show loosely consolidated sandstone resting on the clay. Still higher are exposures of limestone and marl (impure limestone). In all these beds the stratification is distinct, and each stratum slopes gently toward Paris.

The clay that overlies the chalk yielded no fossils, leading Cuvier and Brongniart to conclude that, although both chalk and clay were originally deposited in water, they were formed under different environmental condi-

FIGURE 8-1. Top, *map of the Paris Basin showing the rock formations as mapped by Cuvier and Brongniart. Bottom, relief diagram of the area east of Paris (outlined on the map). Note how the topography indicates the distribution of the inclined formations. (After W. M. Davis.)*

tions. Apparently, the animals that flourished while the chalk was accumulating did not live in the muddy water from which the clay was deposited. In some places, at the contact between chalk and clay, fragments of the chalk were found included in the lowermost bed of clay. Cuvier and Brongniart concluded that the chalk must have been a coherent rock when the deposit of clay began to form, and hence that considerable time had elapsed between the accumulation of the chalk and that of the clay.

Eventually, Cuvier and Brongniart systematically worked out the succession of the formations all the way to Paris. It included several limestones, sands, clays, and gypsum beds. Some beds contained marine shells; others bones of land mammals and birds, skeletons of fresh-water fish, and impressions formed by the leaves of land plants. Each formation thus showed certain characteristic physical features and also contained distinctive fossils by which it could be recognized. In its larger features the succession was constant throughout the Paris Basin, but not every bed, nor even a group of beds making up a formation, could be followed all the way around the periphery of the basin. In places, beds had been eroded away before the overlying beds were deposited; other beds had been deposited as lenses in discontinuous small basins, or were heaped up by currents in the water, as are present-day spits and sand bars.

Identifying Formations by Fossils.   Such variations made Cuvier and Brongniart's task of mapping difficult. For example, the distinction of one limestone from another, among so many, could not always be made on the basis of color, details of bedding, thickness, or other physical characteristics. Not all these difficulties were solved by Cuvier and Brongniart; modern geologic maps of the Paris Basin show many refinements in the subdivision and mapping of the formations that escaped their pioneer studies. But the basic principles they established have served well in all later studies. Perhaps the most important conclusion they reached was:

> **Each formation (closely related group of strata) contains its own characteristic assemblage of fossils.**

This generalization has proved to be of fundamental value in correlating isolated outcrops of strata, even across seas and oceans. It has been checked again and again by the Law of Superposition. For example, approximately the same changes in the fossils occur in the same order as one goes up through the strata of England, France, and North Africa.

William Smith's Geologic Map of England.   The first geologic map of England, published in 1815 by another of Cuvier's contemporaries, an English surveyor named William Smith, was also a great stimulus toward the wider use of fossils in geologic mapping and correlation. Smith's map represented a considerable advance over the maps of the Paris Basin prepared by Cuvier and Brongniart. As might be expected of a surveyor, Smith took pains to locate the contacts of the rock formations accurately and to plot them correctly on the map with reference to streams, roads, and other control points. Starting with the exposure of a contact on a canal or river bank, he would follow its approximate position across a soil-covered hill, using as guides the small rock fragments in the soil or at the mouths of fox or rabbit holes, until he could again locate the contact exactly in another valley where it was exposed in the bed of a stream.

Smith, like the French geologists, learned much about the succession of strata by studying man-made excavations. He was employed as a surveyor on many canals built just prior to 1800. On the Somersetshire Coal Canal, for example, Smith superintended the excavation work for more than six years, necessarily observing the details of the strata in several miles of cut. He noted that in many excavations the strata are not horizontal, but are inclined to the horizon at an angle. He also records that "each stratum contained organized fossils peculiar to itself, and might, in cases otherwise doubtful, be recognized and discriminated from others like it by examination of them."

Employed on many engineering projects throughout England, Smith often traveled as much as 10,000 miles in a year. On all these travels he made full notes of the succession and nature of the rocks in different parts of the country. Finally, after twenty-four years of observation, he published his colored geologic map of England which has taken its place among the great classics of geology.

Smith's map profoundly influenced the development of geology. A recital in words of the results of his twenty-four years of labor would be too bulky for use, but the principal relations that he found could be quickly grasped by any trained observer once they were plotted on a map.

Scientists saw that the answers to many problems could be had by carefully plotting the position of the rock masses that make up the earth's crust. Smith's skill in predicting the kind of rock that would be found in an excavation or tunnel, or the amount of overburden that lay above a concealed coal or clay bed, led men to think of the economic aspects of geology and of the practical ways in which a knowledge of the succession of rock strata may be used.

Perhaps Smith's most significant advance, however, was to determine the *stratigraphy,* the *order of succession,* of the different sedimentary formations for an entire country, and thus to prove the continuity of individual formations over great areas. He showed conclusively that, if a given bed or stratum occurs above another distinctive formation in one locality, it occurs nowhere else below it. Indeed, the succession of strata that he established was found to hold not only for England, but also for much of Europe. Eventually, the subdivisions that he and other pioneers established were extended and refined into the *Standard Geologic Column* to which we now refer sedimentary formations throughout the world.

Methods and Principles of Geologic Mapping

This brief account of the early geologic maps gives us glimpses and hints of the methods and principles used in geologic mapping. We should now summarize these principles, consider their validity, and show how they are applied in actual field work.

On William Smith's geologic map of England, lines representing the contacts between different rock formations are drawn for distances equivalent to hundreds of miles on the ground. Yet, in tracing an individual contact for 100 miles, Smith probably found, on the average, less than 50 exposures where the actual contact could be seen on a clean rock face. How, then, can his map record the real distribution of the rocks? Can it be anything but a guess? The eyes of a geologist are no more capable of seeing the bedrock through a cover of soil than those of any other observer. How, then, can the geologist make inferences about the position of the bedrock underground that will withstand objective tests, such as those provided by digging wells and sinking mine shafts?

The succession must be pieced together from scattered outcrops. Although in an area of a few square miles a geologist may find only one or two outcrops in which he can see the actual contact between two formations, he will doubtless find 100 or more outcrops composed entirely of rock belonging to one or the other formation. He thus has some clue as to the position of the contact with respect to each outcrop. The problem is roughly analogous to drawing a contour line to conform to the elevations determined at 100 or more control points (Appendix I). To use scattered rock outcrops, however, requires that the individual formations be correctly identified (correlated) from outcrop to outcrop, and this is not always easy.

Four Fundamental Postulates of Geologic Mapping.   There are four fundamental postulates that underlie the making of a geologic map. Two of these, the *Law of Superposition,* and the *Law of Original Horizontality,* have already been discussed (Chap. 5, p. 73). The third is merely a common-sense deduction from these two, and, like them, was first stated by Steno:

> **A water-laid stratum, at the time it was formed, must continue laterally in all directions until it thins out as a result of non-deposition, or until it abuts against the edge of the original basin of deposition.**

This is the *Law of Original Continuity.* An important corollary of this law, not fully appreciated by Steno in 1669, but well known to the geologists of France and England at the beginning of the Nineteenth Century, may be stated:

> **A stratum which ends abruptly at some point other than the edge of the basin in which it was deposited, must have had its original continuation removed by erosion, or else displaced by a fracture in the earth's crust.**

These four postulates—(1) superposition (the higher bed is the younger), (2) original horizontality (stratification planes are formed roughly parallel to the earth's surface), (3) original continuity, and (4) truncation by erosion or dislocation—are the basis for many of our interpretations of the relative relations of strata. They are not absolute rules that can be rigidly applied. For instance, some beds, once horizontal, have been highly tilted and even overturned by movements of the earth's crust (Chaps. 9 and 10), so that a stratum formerly beneath another may now lie inverted on it; other strata, as at the front of a steep delta, may have been deposited on appreciable slopes; the edges of many landslides end abruptly instead of thinning to a feather edge. Such exceptions, however, are not common and can generally be recognized easily by the geologist.

How do we correlate between scattered outcrops when we are in the field making a geologic map?

Correlation of Rock Outcrops. In a ravine on a grassy hillside we may see a bed of clay with well-marked horizontal stratification; we assume it continues horizontally into the hill at the same elevation, for how else can its horizontal stratification be extended? If we go 200 yards along the same contour, without seeing an outcrop, and then find a clay bed at the same elevation in another ravine, we may suspect it is the same bed. If both are gray, the probability is heightened. If both show lines of nodular lumps (concretions) along the stratification planes, and if the size and spacing of these lumps is about the same, we are still more confident of our correlation. If both rest on red limestone, both are overlain by fine-grained brown sandstone, and both contain the same kinds of fossils, we become practically certain of the correlation. We are now justified in assuming that the bed is continuous beneath the soil between the two exposures. If we go on a little farther and come to a deep ravine in the hill which shows a very large outcrop containing not a few feet of strata but several hundred, and if in this section there is only one clay bed with features identical to those we saw in the two small outcrops, we have still further evidence that our mapping is correct. We can now map the clay, for the contacts of the bed with those above and below will parallel the contours, as they are horizontal planes along the top and bottom of the clay bed. We have, furthermore, gained information on the thickness and relations of the red limestone below and the brown sandstone above the clay bed, because in this larger outcrop they are exposed through a much greater thickness. For example, we may find that the brown sandstone is 180 feet thick and is overlain by a thick bed of distinctive black limestone crowded with fossils.

If, now, we go on to still another outcrop and find here the clay bed is a little thinner than in the last outcrop, if instead of resting on red limestone it now rests on a pink limy sandstone, and if it has a pebbly green sandstone instead of the fine-grained brown sandstone above it, we would be less certain of our correlation, though we would not regard it as impossible. To check on the correlation, we might go to a lower elevation and study the relations of the sandstone in various outcrops. If we found it gradually varying from limestone to sandy limestone and then to pink sandstone, our correlation would be strengthened. It would be further confirmed if we walked up the hill and found that the pebbly green sandstone above the clay is overlain by a black limestone full of fossils exactly like those in the limestone that overlay the brown sandstone in the large exposure.

Thus in geologic mapping there is invariably an element of judgment. Some correlations are certain, others are reasonably sure, and for still others there may be a reasonable doubt. As to the doubtful ones, two geologists may disagree, just as two equally qualified physicians sometimes disagree in the diagnosis of identical, but not sufficiently definitive, pathological symptoms. But nearly all such differences in interpretation have to do with minor features in a stratigraphic succession. Thicker groups of beds commonly exhibit enough peculiarities to lead any two careful observers to identical conclusions.

Geologic Sections.   Geologic sections are commonly used along with geologic maps in all economic applications of geology. A geologic section represents the way the rocks would appear on the side of a trench cut vertically into the land surface. It resembles a geologic map in representing the projection of the rock boundaries to a surface, but this surface in a cross section is vertical and not the land surface of the earth.

As an example, let us make a geologic section of the horizontal clay bed just mentioned, taking our data from a geologic map showing the relations of the clay in a deep branching ravine where the beds are well exposed. The geologic map is shown in Figure 8-2. The completed section appears below the geologic map, together with the construction lines used in drawing it.

The geologic section portrays the strata as they would appear on the wall

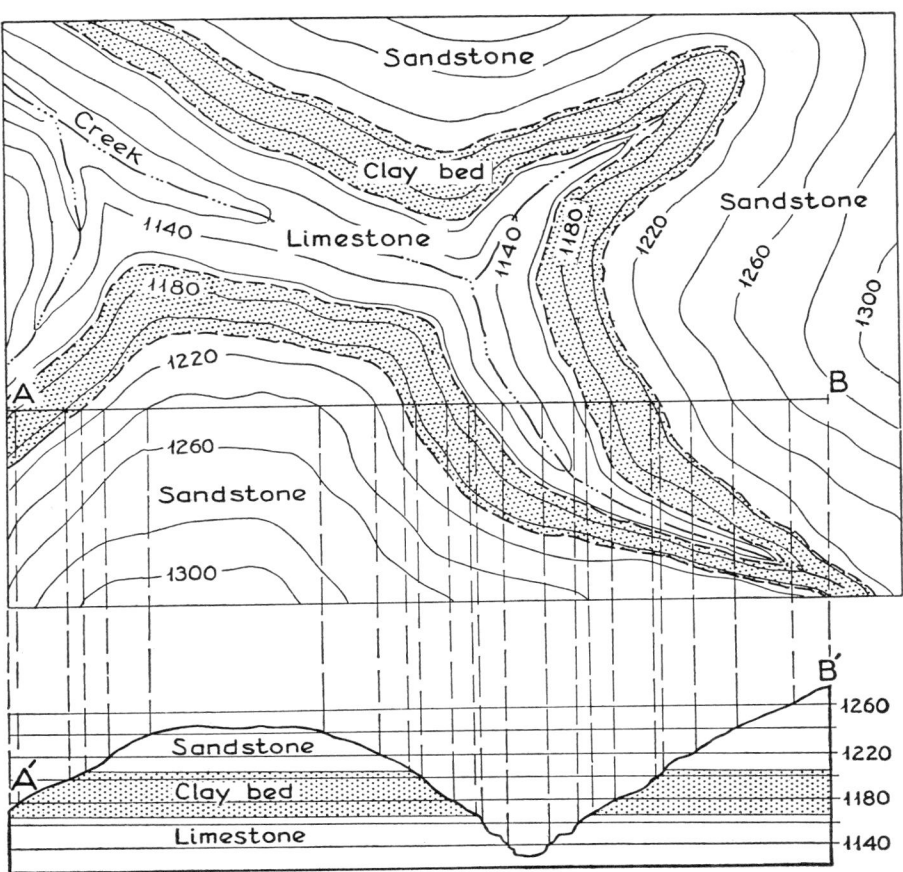

FIGURE 8-2. *Construction of a geologic section from a geologic map.*

of a trench dug along the line A–B of the geologic map. The section was readily drawn by making use of the contour lines on the map and relating the outcrops to them. First, it was decided to make both the horizontal and vertical scales of the section the same as that of the map, in order to avoid distortion. Then perpendicular lines were dropped from A and B (by means of a right-angle triangle) to the place below the map chosen for drawing the section. Then a vertical scale showing the range in elevation from 1120 to 1,260 feet above sea level was laid off on the line projected from B. Horizontal lines were drawn across the section, from the point representing the elevation of the respective contours from 1,140 to 1,260. Then a perpendicular was dropped from each point of intersection of line A–B with a contour line on the map, to the horizontal line on the section representing the appropriate contours. Thus, points were plotted on the section to guide the sketching of the *surface profile,* the irregular line A'–B' of the section. Now perpendiculars could be dropped to the profile from the intersection of the map line A–B with each contact of the clay stratum. There are five such intersections shown along the line of section—three on the top of the clay stratum, two on its base. Horizontal lines drawn through the projections of these points on the surface profile complete the cross section and show the relations of the clay bed to the underlying limestone and overlying sandstone.

Many practical uses can be made of geologic sections. If the clay in the bed is suitable for brickmaking, we can determine from the section how much overburden must be removed at any point along the section line so as to get down to the clay. If water is found in wells just above the clay, the section tells us how deeply we must drill at any point along the section in order to strike it. Mine shafts, tunnels, and drill holes all test the validity of our maps and sections and of the postulates underlying their construction.

Rock Structure and Geologic Mapping.   At many places the beds are no longer horizontal as at the time of deposition; they have been warped and folded (Chap. 9). We mentioned that the beds of the Paris Basin were tilted gently toward Paris. In projecting tilted beds on geologic maps and sections the same principles apply as with horizontal beds, but now the tilted bed will only rarely project parallel with a contour line. Instead, it will generally rise or fall in elevation when traced along a hillside.

Strike and Dip   To project a tilted stratification plane either along its surface outcrop or into the ground, the first step is to determine the *strike and dip* of the stratum. If a tilted stratum intersects the surface of a lake (Fig. 8-3) the trend of the intersection of the water's edge with the stratum —that is, the bearing, or compass direction, of the line of intersection of the lake surface (a horizontal plane) with the stratification (a tilted plane)— forms a definite line of reference. This trend, or compass direction, is called the *strike* of the stratum. We know that the stratum will project along it in the horizontal direction unless terminated by erosion or other causes. *All horizontal lines drawn on the stratum will have the same trend* as the line marking the edge of the lake against it; *each is a strike line.* Strike is meas-

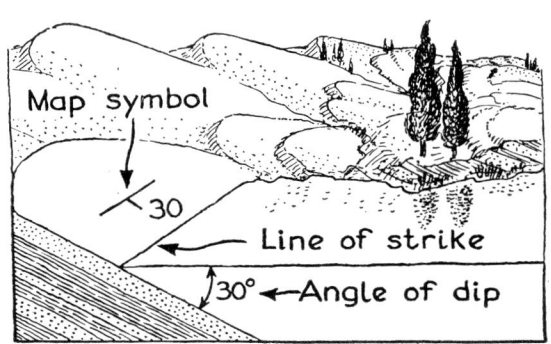

FIGURE 8-3. *The strike and dip of an inclined bed exposed along a lake shore.*

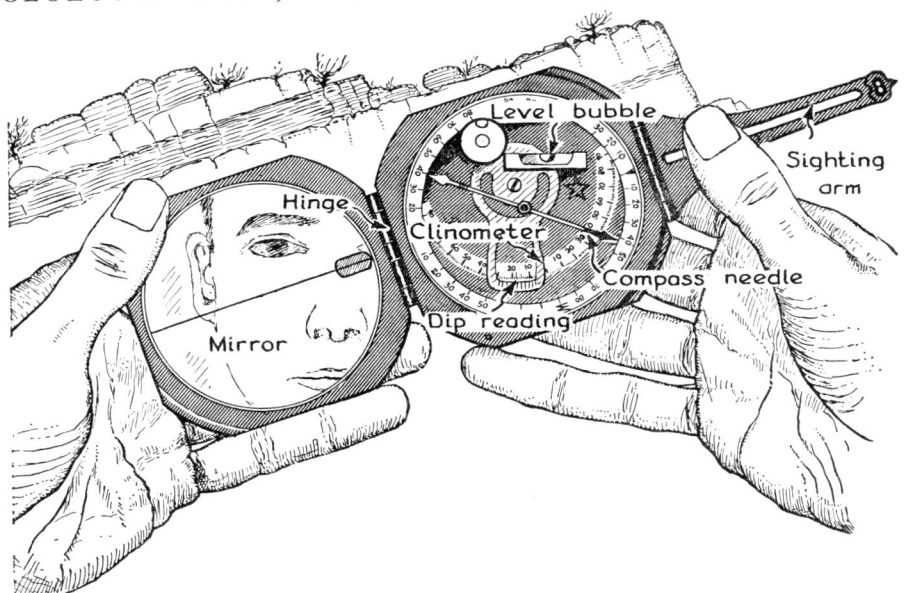

FIGURE 8-4. *Determining dip with the Brunton compass. The clinometer is rotated by a lever on the back of the compass until the bubble is centered, then the dip is read directly on the inner half circle.*

ured in the field by sighting along an inclined stratum with a compass (Fig. 8-4) equipped with a level bubble so that it can be held horizontally. The direction of strike is recorded from the north. Thus a bed which intersects a lake surface along a line running exactly northwest-southeast is recorded as having a strike of N45°W (which is, of course, the same as S45°E, or northwest-southeast). A bed on which a horizontal line trends 10° east of south would be recorded as having a strike of N10°W.

The dip of a bed is the maximum acute angle between a horizontal surface and the stratification plane. To go back to our example of the lake against an inclined stratum, the dip is the angle between the surface of the lake and the submerged continuation of the tilted bed, measured at right angles to the strike (Fig. 8-3). Dip angles are measured by a *clinometer,* which is simply a level bubble attached to an arm inside the compass that swings against a scale graduated in degrees (Fig. 8-4). The *direction of dip* must not be confused with the *angle of dip.* The direction of dip means the compass direction toward which the tilted bed is inclined down into the earth; in our example, it is the direction (not the inclination) taken by a drop of water lifted out of the lake and allowed to run down the surface of the inclined bed. *The direction of dip is always at right angles to the direction of strike.*

In recording strike and dip on a map a symbol consisting of two lines is used. The symbol ⟨symbol⟩ indicates that the stratum on which it is placed has a strike of north 45° east and a dip of 60° to the southeast. On most maps, the letters and figures indicating the strike are omitted; thus ⟨symbol⟩ indicates a bed striking north 45° west and dipping 55° to the northeast. The "N45°W" is omitted because the top of a map is always north (unless specifically marked otherwise), and therefore the direction of strike can be determined from the trend of the strike line, which is always accurately plotted with respect to the north line on the map. Similarly, the direction of dip need not be given, for the dip line points in that direction, but the angle of dip in degrees must always be given.

Topography and Geologic Mapping. A dipping bed can be projected along a contour only when the trend of the hillslope is parallel to the strike. If the strike does not parallel the contours, the trace of the bed along the hill will lie either higher or lower than the first outcrop, depending on whether the bed is followed in the up-dip or down-dip direction. This simple geometrical fact is the basis for most determinations of attitude of beds with low dips, for it is difficult to read a clinometer more closely than 1°. It is also the basic clue to reading a geologic map in order to glean from it the succession of the beds and their structure.

In the Paris Basin most of the streams drain to the Seine. East of Paris (Fig. 8-1) they flow westward, cutting approximately at right angles through the arcuate ridges. Applying the relation between the dip of a bed and its trace on the land surface, we find that, if we select some stratum such as one of the thin layers of greensand in the Lower Chalk, or one of the thin limestones that lies above the Plastic Clay, and trace the outcrops of this bed westward along the banks of a stream, the bed decreases gradually in elevation when followed to the west until it reaches the level of the stream; thereupon it crosses the stream bed, reverses its direction and climbs higher and higher when followed eastward on the opposite bank of the stream. In other words, the outcrop pattern is a "V" with the point of the "V" directed downstream. This can only be explained by the fact that the bed is inclined westward, and that its dip angle is greater than the fall (gradient) of the stream (Fig. 8-5).

The fact that the Upper Chalk crops out further west in the valley of the Marne and Aisne rivers (Fig. 8-1) than it does on the intervening divides proves that the contact of the chalk with the overlying Plastic Clay also dips to the west. By using such facts the east-west section forming the lower edge of Figure 8-1 (bottom) was drawn, although each formation shown on this section is projected to much greater depths within the earth than we can

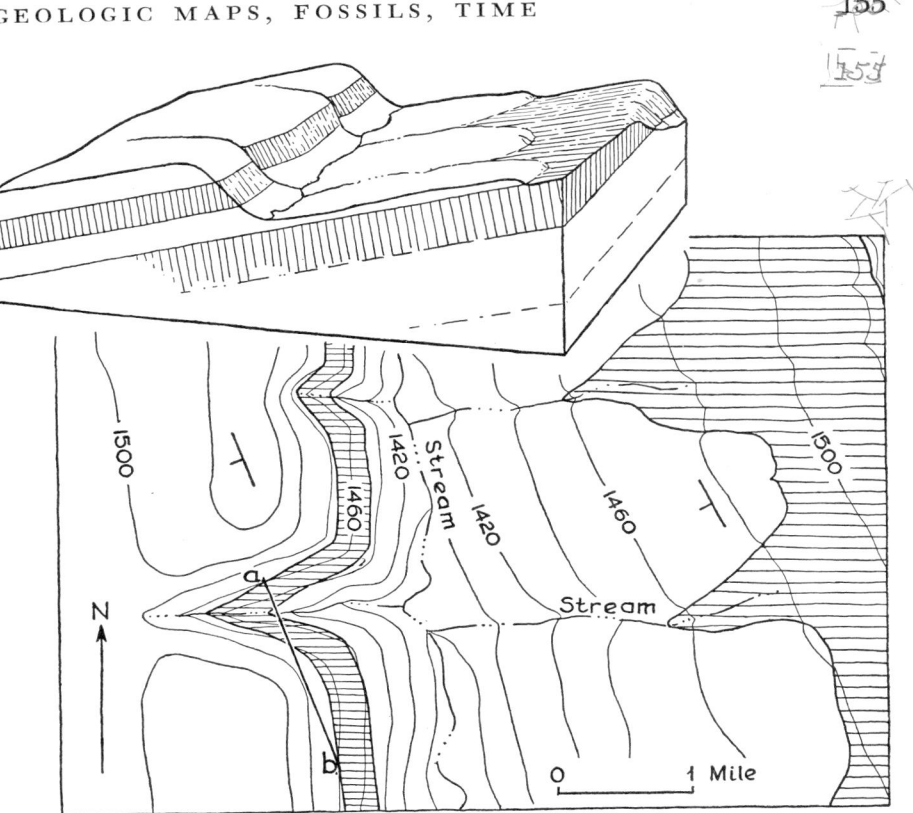

FIGURE 8-5. *Relief diagram and geologic map of the small area south of Epernay (see Fig. 8-1, bottom). Note the strong "V" in each contact where it crosses a stream valley, whether the stream flows east or west. The strike of the beds can be determined by joining points, as a and b, where a contact intersects a contour line. (Why?)*

actually observe in the field. The basic principle of original continuity is the same as for horizontal beds, but in projecting the bed to a geologic section, account must be taken of the dip of the beds.

Relation of Topography to Underlying Rocks. As shown in Figures 7-7 and 8-1, there is generally a relationship, faithful in areas of little soil but more generalized in areas of deeper soil, between the surface form of a country and the underlying bedrock. In the Paris Basin (Figs. 8-1 and 8-5) the curving ridges mark outcrops of resistant strata such as sandstones. Porous chalk layers and other easily eroded rocks are found in the lowlands. Weathering and erosion have etched the surface into relief, forming lowlands on the less resistant beds and leaving the more resistant ones standing as hills. The curving pattern of these hills and lowlands shows that the rocks of the Paris Basin have been warped into a shallow saucerlike basin. Indeed, the whole succession of strata resembles a pile of saucers of diminishing size, with Paris near the center of the smallest and uppermost saucer.

The correlation between topography and structure in the Paris Basin is by no means perfect—much closer adjustment is noted in deserts where the soil cover is thin or absent (see Fig. 7-7). However, the topography usually indicates the general trend of the underlying rocks. Generally, sandstone is more resistant to erosion than shale and hence forms a ridge. It is also more permeable to rain so that, if it rests on shale, there is likely to be a line of springs or a strip of flourishing plants along its lower contact. Thus, the minor topographic features of an area nearly always furnish clues to the characters and trend of the underlying bedrock. If we find conglomerate on top of a low ridge at one point, it suggests that conglomerate may underlie the ridge throughout its length. This is easily checked by seeking other outcrops along its summit. A road may expose chalk in a lowland. Perhaps a well has been dug in the same lowland a thousand feet away. Does it show chalk on its walls? By piecing together such information, the geologist can generally establish the stratigraphic succession, discover the variations in thickness of beds, and ultimately plot their positions on a map.

Even in country with a thick soil it is still possible to apply the principles of geologic mapping and correlation by using data gained from excavations, well borings, or geophysical methods (Chaps. 10 and 20).

Limitations of Scale.   As in all other maps, geologic maps require the rigorous selection of data; they emphasize some features at the expense of others. The smaller the scale of the map, the fewer the details that can be shown. On Figure 8-5, which shows only a few square miles of the Paris Basin near Epernay, details of thin layers above the chalk are shown that could not possibly be indicated on the small-scale map of the entire Paris Basin prepared by Cuvier and Brongniart (Fig. 8-1). Thus the geologist must select his *geologic map units,* or *formations,* to fit the scale of the map. The larger the scale and the better the exposures, the larger the number of formations that can be shown within a given area, although the number per square inch of map remains about the same (compare Fig. 8-1 with Fig. 8-5).

Formations.   The basic unit of the geologic map is the *formation. There are two criteria for selecting a formation: first, its contacts (i.e., the top and the bottom of a sedimentary formation) must be recognizable and capable of being traced in the field, and second, it must be large enough to be shown on the map.*

In their studies of the Paris Basin, Cuvier and Brongniart noticed faint stratification surfaces within the Lower Chalk. The chalk above and below these stratification surfaces, however, was so nearly identical in appearance and fossil content that such subdivisions of the strata were not regarded by the two geologists as either significant or capable of being successfully traced and mapped in the field. On the other hand, the contact between the Upper Chalk and the overlying Plastic Clay was mapped as a formation boundary because of the marked contrast in the rocks. A still more cogent reason for selecting this contact to map was the evidence that it represented a considerable interval of geologic time—enough for the underlying chalk to have become firmly coherent, so that it could be incorporated as pebbles and fragments in the lowermost bed of clay. The stratification within the clay was more prominent than the chalk, but again the similarity of each bed of clay to its neighbors made it difficult, if not impossible, to trace any particular stratification plane far.

The beds above the Plastic Clay posed a somewhat different problem to Cuvier and Brongniart. Here, there were many different kinds of rock in relatively thin layers—limestone, shales, sandstones, beds of gypsum, and clay. At some outcrops an individual bed, perhaps a clay only a foot thick, could be seen to thin and "lens out" within a few tens of feet. Or the overlying sandstone, perhaps 15 feet thick in this outcrop, could be seen to thicken when followed across country in successive outcrops, and then to thin again and perhaps ultimately disappear. Only on a very large scale could each thin layer be shown on a map, and it would take a prodigious amount of time to trace the contacts. Such a series of thin, variable beds, often including very diverse rock types, were grouped together by Cuvier and Brongniart as a single formation. Although the thin beds of such variable formations may be individually lenticular, and indistinguishable either from other beds of the same formation or adjacent ones, nevertheless the group as a whole is recognizably different from formations above and below it.

Depending on the scale of the map, the abundance of exposures, the character of the beds, and, not least, the discrimination of the geologist, very different map units (formations) might be selected in a certain section. Any differences are adequate to justify calling a particular bed, or any closely related group of beds, a formation, provided the differences allow the formation to be recognized in scattered outcrops, and provided the top and bottom of the formation can be traced in the field. The purpose of the map and its scale largely determine what units to select as formations.

Geologic formations in the United States are nearly always given a place name from some geographic locality near which the formation was first identified. This is followed by the name of the *dominant* rock variety composing it, or, if it is composed of a great variety of rocks, by the word "formation." Thus: the *Austin chalk*, the *Columbia River basalt*, the *Chattanooga shale,* the *Denver formation.* In Europe the practice is less formal; many formations are named from some characteristic fossil (the *Lingula flags,* a thin-bedded sandstone containing abundant fossils of the genus *Lingula*), some economic character (*Millstone grit*), or even a folk name (*Norwich Crag*).

Paleontology: The Use of Fossils in Geology

Paleontology (Greek: "science of ancient life") deals with the life of the geologic past. It includes the identification and classification of fossil animals and plants, and the study of their interrelationships and development.

Correlation and Faunal Succession

At first, Cuvier, Brongniart, and Smith used fossils to identify the beds in different outcrops very much as one might use unusual kinds of pebbles or chert nodules. But much more than this developed from their work. That many fossil shells found far inland are the remains of marine animals had been recognized for centuries. The great advance made by Cuvier and Brongniart was to recognize that there were systematic differences in the fossils in the strata of the Paris Basin. When they arranged the fossils collected from the various strata in the same order as that of the beds in which the fossils were found, they noticed that the fossil sequences differed from one group of beds to another, and that these differences were systematic: The fossils from the lower (older) beds were not so similar to animals now living as were those from the higher (younger) beds. No matter whether the particular fossil was a clam, a sea snail, or any other kind of organism, this was broadly true.

Geologic Chronology. This discovery was one of the most significant in the history of geology. Now fossils became recognized as diagnostic markers of a specific period of time during which the bed containing them was deposited. From studies of thousands of different species of fossils it has been discovered that some shells, even from very ancient rocks, differ little from shells of living organisms, but that such stable and long-continuing forms are very few. Although a rare long-ranging species may persist through a thick series of strata, its associates generally change gradually from bed to bed until none of its original companions in the lower part of the series are present in the higher. Conversely, an abundant species may become absent in higher strata, whereas its associates persist. Thus individual species range through different thicknesses of beds, which must mean different durations of time (Fig. 8-6).

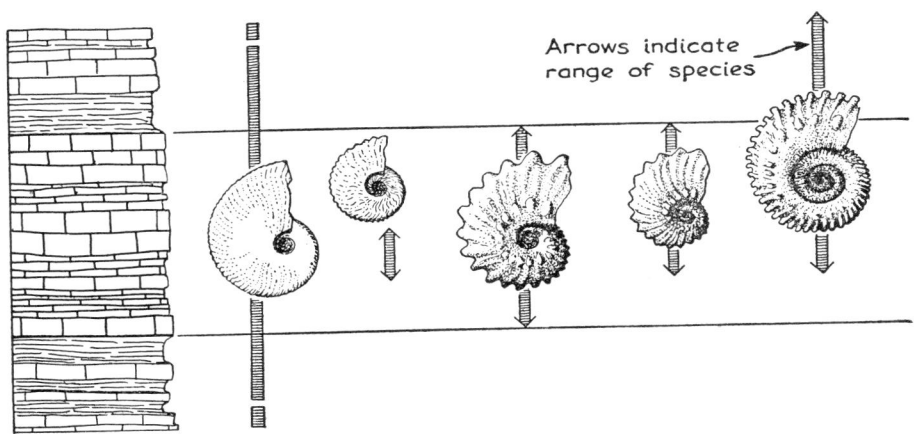

FIGURE 8-6. *Using fossils to date beds. The long-ranging species on the left is of no value in dating the limestone beds between the two horizontal lines, but the short ranges of the other species make them good time markers. The actual specimens were collected in England. (Data from S. W. Muller.)*

Here, the key was found to much of earth history. Once the systematic changes of fossils within the sequence of beds of the Paris and London areas were recognized, it became a challenge to workers to discover whether similar changes in the succession of fossils could be found elsewhere. Within a generation, parallel successions of fossils became known in the sequence of strata from many lands. Such successions, partly identical and partly supplementing one another, furnish the basis for geologic chronology. It is true that our geologic calendar built on fossil successions is not calibrated in years. Rough dates in years, however, are gradually being found from studies of radioactive minerals, as we shall see.

Few fossil forms have a truly cosmopolitan distribution, but there are some forms, especially among the free-swimming organisms, that are nearly world-wide in distribution. Wherever such wide-ranging forms have been found in the fossil record—whether in Europe, Africa, Asia, or America—the sequences of rock strata are consistent, one with another. Innumerable such examples justify the inference that:

> Like assemblages of fossil organisms indicate similar geologic ages for the rocks that contain them.

Sedimentary Facies and Facies Fossils. In correlating rock strata by means of fossils it is important to keep in mind the limitations to the spread of organisms imposed by their natural habitats. Many different depositional environments exist: floodplains of rivers, estuaries protected from the open sea, ocean beaches, coral reefs, and many others. Each has a characteristic group of animals and plants living contemporaneously with the somewhat different but equally characteristic groups of living organisms found in the others. For example, one does not expect to find antelopes in a coral reef, nor corals in a desert sand dune. Similarly, in the geologic past, we would not expect to find the same fossils entombed in all the different deposits formed during a particular span of geologic time. By analogy with similar modern organisms, we assume that some of the marine fossils represent free-swimming forms that could live in almost any part of the sea. Their shells, however, might be battered to pieces in the breakers and hence not be preserved in nearshore deposits. On the other hand, some fossil organisms were bottom dwellers, with habitat restricted to muddy bottoms; their remains would not normally be found in limestone or conglomerate, for they could not live in environments that permit such rocks to accumulate. Thus the fossils in ancient rocks of even the same age may differ with variations in the conditions of deposition of the rocks. Rocks of the same time span that were formed in different depositional environments are said to represent different "*sedimentary facies.*"

Most fossil organisms were restricted in their life habitats, and hence tend to occur typically in rocks of a definite kind of sedimentary facies. The term *facies fossils* is generally used for groups of fossil organisms limited to particular kinds of sediments. Thus we speak of the "limestone reef facies," which may give way along the strike of the beds to a mud-bottom-dwelling "shale facies."

One relationship that may occasionally be found between two facies fossils is shown in Figure 8-7. A long-range fossil (Fossil A) occurs above another shorter-ranging kind (Fossil B) in one region, but below it in another. Such reversals are generally the result of differences in environment of deposition of the two fossils, but they may occur simply because of the greater thickness of rocks during one fossil's time range, or even because of accidents of preservation. Likewise, it can be seen how a particular fossil may be absent from a certain bed containing a second kind, though it is found both below and above that bed. The environmental conditions (sedimentary facies) changed during the life-span of the longer-ranging form, excluding it temporarily from the area of deposition, but allowing it to return when the life habitat again became favorable.

Despite these complexities—and there are still others—Cuvier and Brongniart's generalization has been abundantly justified: The older the rocks, the less their fossils resemble living forms; the younger the rocks, the more their fossils resemble living forms, particularly those found in the same depositional environment.

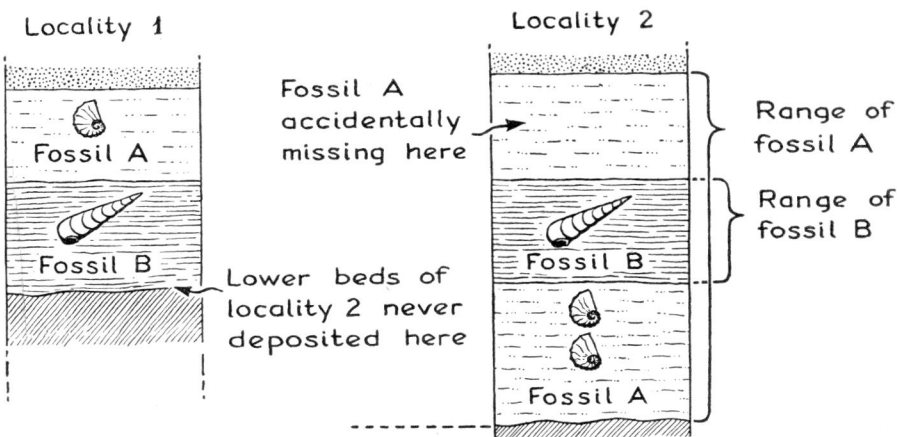

FIGURE 8-7. *The reversal of fossil sequence between two localities due to differences in deposition and accidents of preservation.*

The Geologic Column

The *standard geologic column* is the basis for the geologic time scale. The standard column is composed of a sequence of European formations because it was developed by expanding the stratigraphic sequences worked out by Smith, Cuvier, and Brongniart. Its fossil assemblages form the standard of comparison for sequences in all parts of the world. Correlations from one part of a continent to a distant part, or even to another continent, are based on assemblages of fossil plants and animals. Such correlations are sometimes handicapped by the local, rather than cosmopolitan, distribution of many fossils and by differing conditions of deposition. On the other hand, they are facilitated by the vast wealth of former living things; thousands upon thousands of varied forms are preserved in the fossil record.

The *standard time scale* is a scale representing the periods of deposition of the rock units in the standard stratigraphic column. This relative scale is slowly being calibrated in terms of the radioactive clock, described later in this chapter.

Simple Early Column. As early as the middle of the Eighteenth Century, Italian and German geologists had classed their local rocks into three groups: Primary (rocks like granite and gneiss, with neither bedding nor fossils), Secondary (cemented sedimentary rocks, generally exposed in the mountains), and Tertiary (weakly consolidated sedimentary rocks of the lowlands, which rested upon the Secondary rocks). Though fossils were known from both Secondary and Tertiary rocks, distinctions among them from bed to bed were either not made or not used in the Eighteenth Century classification.

Standard Column.   The present standard column was pieced together in the Nineteenth Century from strata exposed in various parts of Europe. The major subdivisions of the column are called *rock systems*. Most of the systems are represented, at least in part, in the area of Britain mapped by Smith and his early followers.

Smith grouped the oldest strata in his section as the *Old Red Sandstone*. These rocks contain few fossils and, in fact, are largely land-laid beds, but to the south they interfinger with fossiliferous marine beds widely represented in Devonshire. The Old Red Sandstone is, therefore, now considered a part of the *Devonian System*. Resting on the Old Red Sandstone is a group of strata called by Smith the *Mountain Limestone,* and this series is in turn overlain by a succession of sandstones, shales, iron ores, and coals which Smith called the *Coal Measures*. These rocks and the underlying Mountain Limestone are now grouped together as the *Carboniferous System*, named from the abundance of coal in the rocks. Above the Coal Measures Smith recognized a series of beds called the *Magnesian Limestone,* followed in turn by the *New Red Sandstone*. Both these formations are poorly fossiliferous and largely of continental rather than marine origin. The Magnesian Limestone is now considered part of the *Permian System,* named from a province in Russia just west of the Ural Mountains, where marine fossiliferous rocks occur in the same stratigraphic position. The New Red Sandstone, on similar grounds, is now regarded as part of the *Triassic System,* named from the sandwich of a marine limestone between two continental formations found just above the Permian System in Germany.

Overlying the New Red Sandstone, Smith recognized a whole series of richly fossiliferous beds which he divided into many formations, partly because the abundant fossils facilitated correlations and partly because the canals he constructed lay across these beds, and he thus had ample time to work out the details of their successions. Some years later many identical fossils were found in the rocks of the Jura Mountains of Switzerland and France. The name *Jurassic System,* to which these strata are now referred, recalls this fact. Jurassic rocks also underlie the Lower Chalk of the Paris Basin.

Toward London, the Jurassic strata are overlain by greensand, and this in turn by chalk. This series is grouped as the *Cretaceous System* (named from the Greek word for chalk). The fossils are like those of the chalk in the Paris Basin. Overlying the chalk, Smith found the *London Clay,* a group of beds rich in fossils that are like the fossils in the beds above the chalk of the Paris Basin. These beds, both in England and France, are now regarded

TABLE 8-1. *Geologic Column and Time Scale (as recognized by the U. S. Geological Survey)*

ERA	SYSTEM OR PERIOD	EPOCH	APPROX. AGE IN MILLIONS OF YEARS (FROM RADIOACTIVITY)[*]
CENOZOIC ("recent life")[†]	QUATERNARY (An addition to the old tripartite 18th Century classification)	RECENT PLEISTOCENE ("most recent")	
	TERTIARY (Third, from the 18th Century classification)	PLIOCENE ("very recent") MIOCENE ("moderately recent") OLIGOCENE ("slightly recent") EOCENE ("dawn of the recent") PALEOCENE ("early dawn of the recent")	17 (Miocene ?) 58 (Paleocene)
MESOZOIC ("middle life")	CRETACEOUS ("chalk") JURASSIC (Jura Mountains, Europe) TRIASSIC (from tripartite division in Germany)		
PALEOZOIC ("ancient life")	PERMIAN (from Perm, a province in Russia) CARBONIFEROUS (from abundance of coal in this system) DEVONIAN (from Devonshire, England) SILURIAN (from an ancient tribe, the Silures, on the Welsh border) ORDOVICIAN (from another ancient tribe, the Ordovices) CAMBRIAN (Latin name for Wales)		214 (Late Carboniferous) 255 (End of the Devonian) 350 (End of the Ordovician) 440 (Upper Cambrian, Sweden)
PRE-CAMBRIAN	Many local systems are recognized, but no well-established formal periods or smaller time units.		600 (Late pre-Cambrian, Belgian Congo) 1600 (Pre-Cambrian, Black Hills, U.S.) 1800 (Pre-Cambrian, Russian Karelia)

[*] (After Holmes, 1947, and Knopf, 1949.) [†] Definitions in quotations are from the Greek.

as part of the *Tertiary System*, a name that has survived from the primitive classification of the Eighteenth Century.

More than a generation after Smith, the stratigraphy of the highly disturbed rocks below his Old Red Sandstone was worked out. Three new systems were recognized: in descending order, *Silurian*, *Ordovician* (both names of old British tribes), and *Cambrian*, from the Latin name for Wales.

The Cambrian strata are the oldest rocks that contain fossils in any abundance, although sedimentary rocks thousands of feet thick are found beneath them. In this book such older rocks are grouped as the *pre-Cambrian*. They have great bulk and complexity and can be classified in various ways, in spite of the absence of fossils. Perhaps refinements in the dating of rocks by radioactivity may some day enable us to classify the pre-Cambrian rocks in an orderly chronological sequence, just as fossils have permitted us to do for the Cambrian and younger systems.

The systems of rocks are listed in Table 8-1, as the standard geologic column.

Geologic Time Scale

The standard geologic column is the basis for the *geologic time scale*. The names applied to the systems of rocks are also used for the *periods of time* during which the respective systems of rocks were deposited. Thus, we use the term *Carboniferous Period* for the time during which the *Carboniferous System*—the Mountain Limestone and Coal Measures of Smith—were laid down.

Table 8-1 presents the column and time scale as used by the U. S. Geological Survey. Appended to it are remarks as to the sources of the names and the ages of several minerals as found from studies of radioactivity. From these ages in years it is apparent that more subdivisions are recognized among younger rocks than among older. Just as in human history, the documents are more numerous and the gaps in knowledge less wide the more recent the period with which we deal.

Gaps in the Time Scale.   Divisions of the standard column are based on abrupt changes in the fossil assemblages of the strata in Europe. These points of division naturally were those marking the longer intervals of erosion or nondeposition in the European stratigraphic series. The longer the interval lost by erosion, the more prominent the fossil differences would be. As stratigraphic work was extended to other continents, however, fossil assemblages intermediate between those of adjacent periods as recognized in western Europe were discovered.

With more and more stratigraphic work the gaps in the type sections have continued to decrease. Beds containing fossil assemblages that pose "boundary problems"—that is, uncertainty as to their correlation with the upper part of one European system or the lower part of the next higher—are thus present at one place or another for nearly every systemic boundary. Even the boundary between the Paleozoic and Mesozoic Eras, long thought to represent an interval during which none of the present continents received deposits, is apparently bridged by an essentially complete succession of beds in the Himalayan Mountains. Even in Europe, where the divisions were first established, further research has served to narrow the lost gaps in the record. For example, a new division, the Paleocene, has been recently established for strata intermediate in age between Eocene and Cretaceous, where a long gap had been thought to exist. Now we find that different paleontologists refer the same strata (let us say, in Wyoming) to the Cretaceous or to the Paleocene, depending on their respective appraisals of the relative affinities of the fossils of these beds to that of one or another of the European sections.

Intercontinental Correlation. Thus, although the original divisions of the type section were based on stratigraphic relations of the beds, our correlations of distant strata are necessarily based on a *comparison of fossils,* not strata. The demonstration of an interruption in stratigraphic succession in the type area can give no help in correlating beds in a distant part of the world. A stratigraphic break in Nevada, for example, cannot be used as evidence that the beds above it are Permian and those below Carboniferous. This assignment of age to the Nevada strata can be made only on the basis of comparison of fossils with those of the type section of Europe.

The uncertainties posed by "boundary problems," however, are concerned largely with form instead of fact. Such disputed strata, it is generally agreed, must represent times between, or at least partially bridging, the gaps in the type areas of Europe. Further detailed stratigraphic work cannot but fill other parts of the gaps. This is indeed a tribute to the precision of correlation by fossils rather than a criticism of its uncertainties.

The Duration of Geologic Time

With the building of the geologic time scale there came a growing appreciation of the vast duration of geologic time. Contemplation of even the time required to lay down the 500-foot thickness of chalk in the Paris Basin is upsetting to the man who thinks of time only in terms of the span of human life. Some strata of the chalk are composed of the skeletons of minute animals and plants. Similar deposits are accumulating today at rates so low as to defy measurement—certainly no more than a few inches per century and probably much less. Yet the chalk represents only a small part of Cretaceous time, and the Cretaceous itself is a mere fraction of geologic time.

Not only the vast thickness of sediment impressed geologists with the immensity of geologic time, but also the vast parade of life recorded by the fossils—the development of thousands upon thousands of new species with advancing time and the gradual dying out of whole fossil assemblages and their succession by others—time after time—can only be conceived of as involving millions of years, unless greatly different rates of change prevailed in the past.

Time Represented by the Geologic Column

Spurred by their curiosity to translate geologic time into years, geologists tried many approaches. The simplest and most obvious one was to add the greatest thickness of strata of each period and thereby get the total thickness for all geologic time. This total, divided by the present annual rate of sedimentation, might be thought to give the length of geologic time in years.

However, so many assumptions are involved that such a calculation becomes meaningless. How can we arrive at an average annual rate of sedimentation when we know that it may take a century to lay down less than an inch of chalk, whereas a desert cloudburst may deposit 40 feet of gravel, sand, and boulders overnight? And how can we measure the time of scour and nondeposition in marine sediments when we know that a single storm may remove the accumulation of years from part of the sea bottom and perhaps pile up more sand on a nearby beach than was deposited during the whole preceding decade? In short, the rate of sedimentation varies so greatly and has been so little measured that the average annual rate can only be guessed at. This method can only yield an order of magnitude for geologic time. By such a calculation, the British geologist Sollas estimated (1899) that 34 to 75 million years have passed since the beginning of the Paleozoic —the larger figure including a guess at the length of the lost intervals in the record. These figures have a certain interest but surely are not reliable, as we shall see.

Age of the Ocean

The Irish scientist Joly made a more ingenious approach. He reasoned that the salt in the sea must have been brought in by the rivers, after being weathered from the rocks. Most of it has accumulated in the sea, as little is blown out by the wind or deposited by evaporation of sea water in desert lagoons or on tide flats. Hence the oceans must be growing saltier; and, if the amount of salt they now contain could be measured and divided by the annual increment, their age might be obtained. From the many elements in solution in the oceans, Joly selected sodium as the constituent best suited to this method.

The volume of the sea has been roughly computed from its average depth and area, and its composition is known from thousands of analyses. Thus the total sodium it contains can be roughly computed. Similarly, we have thousands of analyses of river waters, and from stream gages we know fairly well how much river water flows annually to the sea. From these figures Joly obtained a rough estimate of the annual increment of sodium.

Dividing the first amount by the second, he found:

$$\frac{15,627 \times 10^{12} \text{ tons of Na in the oceans}}{15,727 \times 10^{4} \text{ tons of Na added annually}} = \begin{array}{l} 99.4 \text{ million years,} \\ \text{the age of the ocean} \end{array}$$

Joly realized, of course, that this computation disregarded many factors. For one thing, river discharge was probably not constant throughout geologic time. Furthermore, not all the sodium in rivers is derived from weathering; much is from sewage and industrial waste, some is wind-blown from sea beaches, and much is leached from former marine sediments. Thus the present annual increment is higher than in the geologic past. How much higher cannot be ascertained.

Again, there are great deposits of rock salt among the stratified rocks, derived from evaporated sea water. If returned to the sea, they would increase the tonnage in the oceans. Again we have no way of getting an accurate correction. We have noticed how magnesium ion reacts with limy ooze to form dolomite. In a somewhat similar way, we now know that sodium ion reacts with clay minerals and is incorporated into marine muds and thus removed from solution. Finally, large amounts of sea water are mechanically enclosed in the pores of marine sedimentary rocks and the sodium so contained is then removed from the ocean.

When all these corrections are considered, it is obvious that Joly's 99 million years must be multiplied many fold, but, as no figure can be made quantitative (that is, subject to careful measurement for purposes of verification), the results yield nothing but the fact that the ocean has existed for a vast span of time.

Age Based on Classical Physics

Another estimate of the duration of geologic time was made by Lord Kelvin, the famous English physicist of the late Nineteenth Century. In his day, before the discovery of radioactivity, atom-splitting and reconstitution, and transformation of mass to energy, Kelvin could imagine no source of energy for the stars except the contraction of the gases composing them. Every astronomical body, including the earth, radiates heat into outer space, he knew, and, therefore, decided that each must be cooling and hence shrinking in size. He computed that, even if the earth had contracted from a size so great that it originally touched the sun, no reasonable assumptions as to original temperature could allow its age to exceed 400 million years. In fact, it was probably only about 20 million years old.

Lord Kelvin's dictum—despite his prestige as the leading physicist of his day—was rejected by geologists. Though the rates of sedimentation and of organic evolution cannot be measured, no geologist could bring himself to accept so short a time as Kelvin advocated. They regarded it as "unreasonably" short, despite his exact mathematical demonstration. We now know that his basic assumptions were the source of his error. The discovery of radioactivity and of other sources of energy not known to Kelvin demolished his basic assumption that all stellar energy arises from contraction.

The discovery of radioactivity also made it possible to establish some points on the geologic time scale in terms of years, and to show that all previous estimates of the duration of geologic time were far too short.

The Radioactive Clock

When Becquerel discovered radioactivity (1898), he opened new vistas in every science. One of the many byproducts of this discovery was the determination of some geologic ages in years.

A few elements, among them uranium and thorium, disintegrate spontaneously into other elements. The atomic nuclei of such elements are intrinsically unstable and emit *alpha* particles and *beta* particles (electrons). Each nuclear emission, whether of an alpha or beta particle, transforms the atom into a different element. Starting with uranium 238 (U_{238}) there are 15 steps in this natural process: 8 alpha particles and 7 beta particles are emitted, yielding ultimately a stable form of lead (Pb_{206}) which does not disintegrate further.

The rate of spontaneous disintegration varies tremendously with different elements. It is expressed in terms of the element's "half-life," which is the time required for half its atoms to disintegrate. The half-life of some members of the U_{238} series is only a fraction of a second; for U_{238} itself it is millions of years. The half-life period of the entire chain from U_{238} to Pb_{206} has been determined to be 7,600 million years. In other words, if we start with 1 gram of U_{238}, in 7,600 million years only 0.5 gram will be left, in another 7,600 million years there will be only 0.25 gram, and so on. The other half (or three-fourths) has changed into lead, helium ions, electrons, and very small amounts of intermediate elements in the series. In no experiment has the disintegration rate been changed by heat, pressure, state of chemical combination of the element, or time. The half-life of a radioactive element is thus considered a constant, that is, a fundamental property of the element.

Many minerals, most comparatively rare, contain appreciable amounts of uranium. Some of these are intimately associated with intrusive igneous rocks and have been segregated from the magma during the last stages of its consolidation, as shown by their occurrence in dikes and pods that grade into the intrusive rock. By analyzing such minerals and finding the ratio of uranium-derived lead to the uranium still remaining it is possible to find their age, and hence the time in years since the magma from which they crystallized was consolidated.

Such analyses are difficult. The proportions of the different lead isotopes must be determined so that uranium-derived lead (Pb_{206}) can be differentiated from any ordinary lead that may be present. Ordinary lead is a mixture of isotopes with an atomic weight of 207.21. Furthermore, all uranium-bearing minerals contain U_{235} as well as the more abundant U_{238}, and may contain thorium. Each of these also disintegrates to yield a lead having a different atomic weight. These additional radioactive substances make it difficult to find the ratios accurately. Absolutely fresh unweathered minerals must be used, for solutions circulating long after the mineral was formed might leach out the lead and uranium at different rates and thus produce great errors in the calculation of the age. Though many determinations have been made, only a few meet the rigid requirements with regard to analytical procedure and freshness of material. They have been entered in Table 8-1 in the right-hand column.

Thus we have a fairly reliable method for dating geologic time in years. Furthermore, for the rocks older than the Cambrian, it is the only method available for interregional correlation. The oldest mineral from the United States (1,600 million years) thus far determined comes from the Black Hills of South Dakota. It was taken from a segregation in a granite mass that cuts across schists. Therefore, it must be younger than the schists, which for many years had been considered late pre-Cambrian because they have not been severely metamorphosed. The Grenville gneiss of the Adirondacks and southern Canada, on the other hand, is highly contorted and severely metamorphosed, so that it has been regarded as extremely old pre-Cambrian. It turns out from the lead ratios that this rock is little more than one-third as old as the Black Hills rock. Such evidence shows the impossibility of using the degree of deformation and metamorphism of the rocks as a basis of correlation over long distances.

Facts, Concepts, Terms

GEOLOGIC MAPS
> Law of Original Continuity
> Correlation of outcrops
> Geologic sections
> > Strike and dip
> Effect of topography on outcrop pattern
> Topographic expression as an aid to mapping
> Formations

THE USE OF FOSSILS IN CORRELATION
> Stratigraphic facies
> Geologic dating
> The Standard Column
> Rocks and time

DURATION OF GEOLOGIC TIME
> Age determinations from radioactive minerals

Questions

1. When a geologic contact "V"s downstream, how is it dipping? (See Fig. 8-5.)
2. When a geologic contact "V"s upstream, how is it dipping? (See Fig. 8-2.)
3. Express a general rule regarding contours and outcrop patterns of horizontal beds.
4. If you traced a thick bed of lava for a mile and could not find its further continuation, what possible explanations of its termination can you suggest?
5. What map pattern would a bed dipping vertically and striking north make across a ridge trending due east?
6. If you follow the contact between a shale bed and a sandstone bed across nearly level country for several miles and it leads you in an oval path back to the starting point, what inferences can you make as to the dips in the area?
7. Some of the buttes of the New Mexico desert are capped by basalt flows; others are volcanic plugs. (See Chap. 5.) How would you expect these igneous rocks to differ in map pattern?

8. Draw a hypothetical geologic map showing a series of tilted beds that have been invaded by a sill and that also contain a buried lava flow. Indicate on the map a locality where you would expect to find fragments of the lava flow as inclusions in a sedimentary bed.

Suggested Readings

1. Mather, K. F., and Mason, S. L., *A Source Book in Geology*, McGraw-Hill Book Co., New York, 1939. (Especially pp. 181-191, 194-204.)
2. Adams, F. D., *The Birth and Development of the Geological Sciences*, Williams and Wilkins, Baltimore, 1938. (Especially Chaps. 7 and 8.)

9. *Movements of the Earth's Crust*

In chapter 7 we saw that many processes are working to wear away the earth's relief. In Chapter 8 we reviewed briefly the evidence for the very great antiquity of the earth. Clearly, if some other processes were not at work to counteract erosion, the lands would long since have been reduced to plains near the level of the sea. What is the evidence for the existence of such processes, and how do they raise new elevated tracts on which erosion can proceed?

The miner deep in an Illinois coal mine who unearths a tree stump with the roots spreading out in the position in which they grew concludes without question that this tree, now 1,000 feet underground and 600 feet below sea level, once grew on the surface of the earth. Similarly, the manager of a copper mine in northern Michigan whose ore comes from typical stream-deposited conglomerate or from the scoriaceous upper part of a basalt flow a mile underground reaches the same conclusion. The thought that land and sea are not always stationary but must locally have changed places has been entertained by nearly all peoples who have attempted to explain the occurrence of well-preserved marine fossils in the rock of plains, deserts, and mountains. How else can we explain the marine fossils in the rocks of the Paris Basin, and the deviation of the stratification in these rocks from the horizontal?

Measurable Displacements of the Earth's Crust

Displacements During Earthquakes

Spectacular but relatively small earth movements have accompanied many earthquakes. Earthquakes are dealt with more fully in Chapter 18. Therein

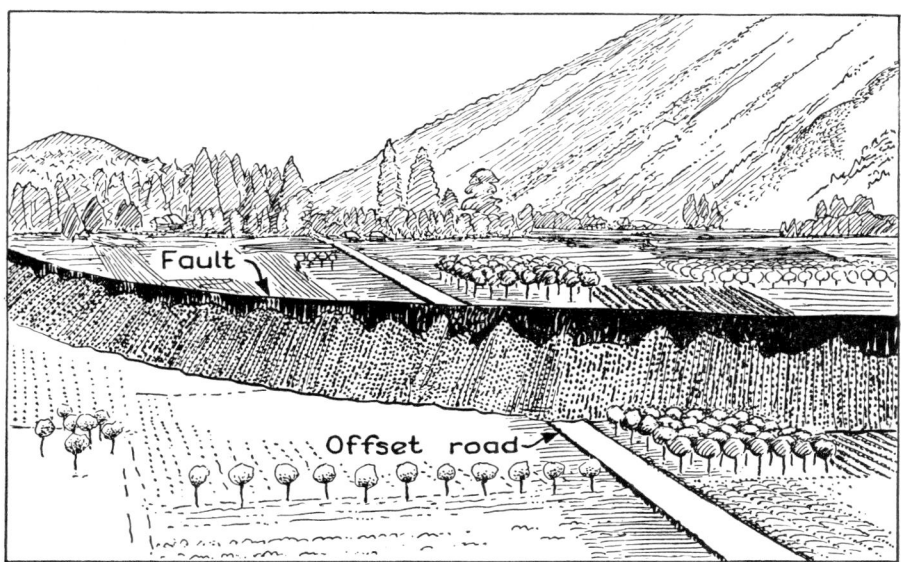

FIGURE 9-1. *Cliff formed during the Mino-Owari earthquake, Japan. The displacement, as measured on the offset road, was 18 feet vertically and 12 feet laterally. (From a photograph by Koto.)*

are described several destructive earthquakes that devastated densely populated areas. Most of these destructive earthquakes caused fissures to open in the ground and visible displacement of the earth's crust along the fissures. Such displacements are seen in the severed ends of roads, fences, strata, or other features that had extended uninterruptedly across the site of the break prior to the earthquake (Fig. 9-1).

Breaks in the earth's crust along which slipping has occurred are called *faults*. As will be shown later in this chapter, fault movement does not always produce an earthquake; slow, gradual movements along some faults have been observed without accompanying earthquake shocks. Sudden displacement along faults, however, appears to be the chief cause of earthquakes.

The fault displacements observed during severe historic earthquakes are small—seldom more than a few feet. In some, the movement was vertical, producing a small cliff along the fault; in others (for example, the destructive San Francisco earthquake of 1906), the fault walls slipped laterally, offsetting roads or fences that formerly crossed the break; and in still others (for example, the Mino-Owari earthquake in Japan), the slipping was oblique, with both vertical and horizontal components (Fig. 9-1). Table 9-1 lists some of the largest single displacements recorded along fissures during historic earthquakes.

TABLE 9-1. *Visible Fault Displacements Associated with Historic Earthquakes*

LOCATION	DATE	MAXIMUM VERTICAL DISPLACEMENT, FEET	MAXIMUM HORIZONTAL DISPLACEMENT, FEET
Assam, India	1897	35	0 (?)
Owens Valley, California	1872	23	12
Mino-Owari, Japan	1891	20	13
San Francisco, California	1906	3	21
Sonora, Mexico	1887	20	0
Pleasant Valley, Nevada	1915	16	0

Perhaps some historic earthquakes have produced even greater fault displacements on the sea floor. After an earthquake near Disenchantment Bay, Alaska, in 1899, the beaches stood 47 feet above the sea, and a wide expanse of sea floor was made into dry land. This is the greatest well-authenticated single displacement known to have accompanied an earthquake. The great Japanese earthquake of September 1, 1923, which largely destroyed Yokohama and Tokio with a loss of over 140,000 lives, was at first believed to have been accompanied by a much greater displacement, but the evidence is inconclusive. Comparison of soundings in Sagami Bay made before and after the earthquake showed local changes of more than 1,000 feet in depth of water. Detailed study suggests, however, that most and perhaps all of these great changes in the sea floor were not from fault displacement but more likely from large slides set up in the unconsolidated muds and silts on the floor of the Bay.

Thus, the historic record indicates that no single observed fault displacement accompanying an earthquake has been drastic enough to account for marine strata on mountain tops or rooted stumps in mines 1,000 feet or more below sea level.

Is it possible, however, that these small displacements are but minor episodes in the growth of the fault on which they occur? Examination of the structure of the rocks on either side of many earthquake fissures affirms this question. Although new faults with a few feet of displacement have formed during a few recorded earthquakes, most displacements have taken place by renewed movement along an old fault. The presence of such a fault is revealed by the impossibility of matching strata or other structures in the rocks on the two sides of the fault by merely restoring them to the positions they occupied prior to the earthquake.

The Pleasant Valley earthquake, which rocked the almost uninhabited desert of central Nevada in 1915, offers a good example. After the earthquake, the western base of the Sonoma Range was marked for 17 miles by a low cliff 1 to 16 feet high where none had existed before (Fig. 9-2). The cliff was formed by renewed movement on an old fault which bounds the low-lying, partially alluvium-covered Pleasant Valley on the west from the rugged Sonoma Range to the east. During the earthquake, the Sonoma Range, relative to Pleasant Valley, moved directly up the steep fault at its base, forming the new cliff and increasing the height of the range up to 16 feet.

In Figure 9-3, the 1915 displacement is clearly seen in the vertical cliff offsetting the surface of the alluvium. The small patch of alluvium clinging to the surface of the upthrown Sonoma Range block is 12 feet higher than the alluvial surface on the Pleasant Valley block. But the total displacement on the fault must be much greater; the dolomite of the upthrown block is against alluvium on the downthrown block for a vertical distance far greater than 12 feet. Furthermore, a short distance away, gullies cut in the Pleasant Valley alluvium show that it rests not on dolomite like that across the fault, but upon lava flows. Immediately across the fault, however, dolomite and other sedimentary rocks extend to the crest of the Sonoma Range over 2,000 feet higher. Still farther north, the higher peaks of the Range are capped by lava flows like those on the Valley floor. Here, however, the lavas rest on the dolomite and associated sedimentary rocks. Obviously, therefore, the total vertical displacement on the fault has been at least 2,000 feet, and perhaps it has been enough to account for the entire difference in level between

FIGURE 9-2. *Fault scarp* (white line) *at the base of the Sonoma Range, Nevada, formed during the 1915 earthquake. (Photo by B. M. Page.)*

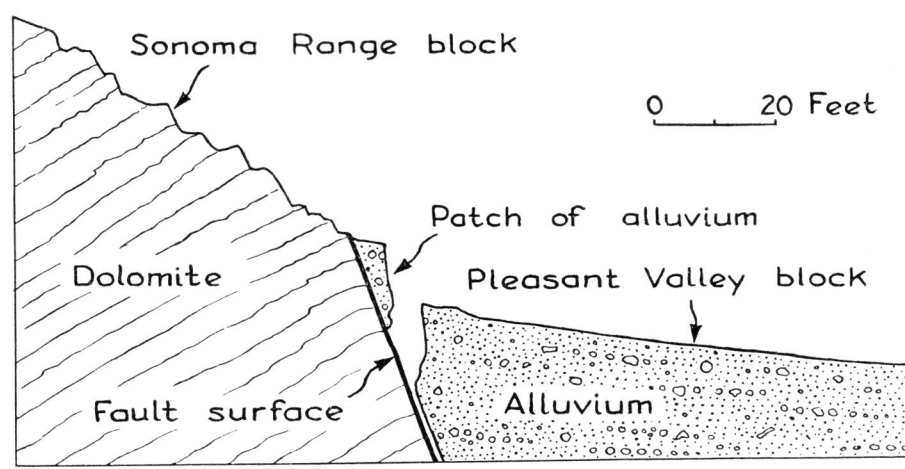

FIGURE 9-3.   *Cross section showing the relation of the new 1915 scarp to the old fault surface, Sonoma Range, Nevada. (After B. M. Page.)*

valley and range, which is locally more than 3,500 feet. Apparently, the 1915 displacement was only the latest of many similar movements.

Other historic ground displacements that have accompanied earthquakes show similar relations and warrant the conclusion that repeated small movements along a fault can ultimately displace the crust for thousands of feet.

Measurable Slow Movements Along Faults

The Buena Vista oil field is on a rounded hill rising above the flat San Joaquin Valley in California. Soon after the first oil wells were drilled, a road was built across the hill, and several pipelines were buried in shallow ditches along it. Within a few months, the roadbed cracked, and a vertical displacement of less than an inch appeared on its surface. It was patched, but repeatedly broke at the same place. Within a few years, eight of the buried pipelines had buckled up out of the ground, and they continued to arch upward slowly, a few inches each year (Fig. 9-4). The pipelines were thus shortened 9 to 19 inches in a period of 9 to 15 years. Cleaning tools lowered into the wells also revealed that 23 of the well casings were slowly being bent out of line. In some wells, bending continued until the casing col-

FIGURE 9-4.   *Buckling of heavy pipe line in the Buena Vista oil field, California. (Photo by T. W. Koch, courtesy of the American Assoc. of Petroleum Geologists.)*

lapsed; in others, the casing, when pulled from the ground, was found to have been displaced about 15 inches horizontally within the 10 to 15 years after the well had been drilled. The casing failures in different wells occurred at depths ranging from 76 feet to 794 feet. What caused these curious phenomena?

By plotting on maps and sections all of the points where such disruptions occurred, it was found that they lie along a smoothly curving surface. This surface slopes downward into the ground at a low angle, and the rocks it cuts are offset against it (Fig. 9-5). Both the disturbances of man-made structures and the offsets in the rocks prove that the surface is a fault. Apparently the rocks above the fault surface are moving slowly upward and southward with respect to those below, breaking the roadbed, bending the well casings, and buckling the pipelines out of the ground. Surveys show that the average rate of movement on the fault is 1 1/2 inches per year. No earthquake shocks arising from this fault have been detected, even by seismographs.

Other examples of slow continuous movement along faults are known. In some parts of Japan, differential movements between different blocks of the earth's crust separated by faults have been detected by resurveys of the position and elevation of a series of points established by precise triangulation and leveling (see Appendix I). For example, there were changes in level of a series of points along a 150-kilometer stretch of the northeast coast of the island of Honshu that occurred between leveling surveys made in 1900 and 1933. The changes indicate that throughout this area the earth's crust sank relative to sea level. The amount of displacement varies at different points, supposedly due to slow tilting and warping of fault-bounded blocks in the earth's crust.

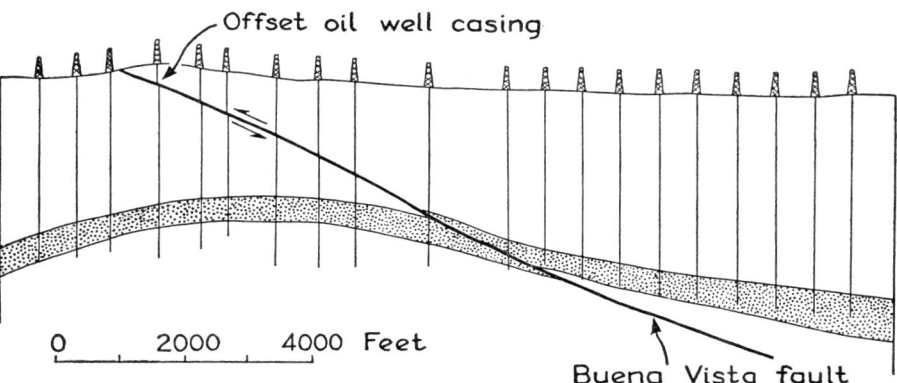

FIGURE 9-5. *Cross section showing the fault at the Buena Vista oil field. The half arrows indicate the relative movement. (From T. W. Koch, 1933.)*

Measurable Slow Movements Not Connected with Faults

The movements discussed so far have taken place along faults, but there are also widespread and important movements in no way related to faults. Although the rate of movement is slow, it has been measured at several localities. In a few places such movements have changed the landscape appreciably within historic time.

The classical and perhaps best known example of such movements is that of an ancient Roman ruin, the so-called "Temple of Jupiter Serapis" near Naples, Italy. Only three upright columns and a part of the floor are still in place (Fig. 9-6). On the columns about 18 feet above the floor is a line. Above this the columns are smooth; just below it they have been bored full of holes by a marine rock-boring clam. Shells of this clam can still be seen in some of the holes. It is reasoned that the temple was built on dry land, that slow downward movements then brought sea level to a point 18 feet above the floor, allowing the marine clams to bore the columns, and that later, the land rose again.

Along the shores of the Baltic Sea in Sweden and Finland a still more striking movement has been observed for many years. The farmland and fresh-water marshes fringing the Baltic are littered with marine shells identical with those in the Baltic today. Over 150 years ago these conditions were correctly interpreted to mean that the land had risen out of the sea. In order to find out whether the movements are still going on, monuments were set up at high tide line along the shore. Today many of these monuments are a few feet above sea level, and some as much as a mile inland. The maximum rate of uplift—3 to 4 feet per century—is in the northern Baltic. In some parts of Scandinavia there is little or no movement, and southern Denmark appears to be sinking about 2 feet per century. The rate of movement also appears to vary somewhat with time, as considerable differences from one decade to the next have been revealed by accurate tide gage records at many Baltic ports.

What causes this uplift? When the measured changes along the Scandinavian coast are compared with the obvious prehistoric evidence of uplift from raised marine beaches and terraces, an interesting relation appears. During the Pleistocene (Chap. 8) the Scandinavian area was covered by a huge sheet of ice such as mantles Greenland today. The Scandinavian uplift is greatest in the area where the Pleistocene glacier was thickest (as determined by several lines of evidence mentioned in Chap. 13). The ice sheet was thickest in the northern Baltic, and in the approximately 12,000 years since the ice melted (dated by varved clays in glacial lakes, see Chap.

FIGURE 9-6. *Columns of the so-called "Temple of Jupiter Serapis," Italy, with high-water marks about one-third of the way up. (Photo by E. F. Davis.)*

13) the northern Baltic has risen about 900 feet, an average rate of uplift of approximately 7.5 feet per century (Fig. 13-24, p. 324).

Because of the tendency of the earth's crust to reach isostatic equilibrium, uplift is what we would expect from unloading of a segment of the earth's crust by the melting of a large ice sheet, and the greatest uplift should take place where the greatest load was removed. Thus, this close coincidence between ice thickness and uplift is strong geologic evidence of the isostatic tendency. Since the uplift is still going on, we infer that Scandinavia has not yet risen to the elevation it must have had before the load of glacial ice began to accumulate and to depress the earth's crust under its weight.

Yet, there was no glacier or other large load upon the land at the Temple of Jupiter Serapis, which also sank and later rose. There must, then, be forces other than those tending to restore isostatic balance which are involved in crustal movements.

Examples of similar crustal movements could be multiplied. Tide gages show that harbors in Denmark, Japan, and elsewhere are slowly deepening. Although some Japanese harbors are sinking, others are shoaling, some at the comparatively rapid rate of 1 to 3 feet every 10 to 50 years. Ships can no longer enter some of them, and rocks formerly submerged are now parts of the land. Such records have been kept for so short a time, however, that some of the evidence is inconclusive. The few reliable measurements indicate that the movements are very, very slow. Nevertheless, the significant results that they have produced in the landscape indicate that in many places they must have been underway since long before human history began.

Geologic Evidence of Displacements of the Earth's Crust

Emergence and Submergence Along Coasts

Along most of the world's coastlines there is evidence of geologically recent (though prehistoric) crustal movement. The lobster fisherman who sets his traps in the water off the Maine coast now and then brings up fragments of the peat that floors the bottoms of some of the inlets and bays. In this part of the United States, almost every stream ends in a tidal estuary. Dredging from the bottom of these estuaries usually reveal river-deposited silts filled with decomposed grass roots, peat, or thinly stratified clay formed in fresh water lakes and marshes. All of these must have been deposited above sea level.

On a clear day, a person flying over southern San Francisco Bay can see the drainage channels of a former land on the shallow Bay floor. These channels, though interrupted by deltas or small wave-cut features at the shore, are clearly underwater continuations of the streams now entering the bay. They have apparently been drowned either by subsidence of the land on which they were flowing, or by a rise of the sea. Not far away, near Stockton, California, water wells drilled to as much as 1,000 feet below sea level penetrate buried soils, river silts with grass roots, peat accumulated in fresh-water marshes, and other materials that must have been deposited above sea level. Off Denmark and Japan, dredging in shallow marine water has revealed kitchen middens (refuse heaps left by primitive man) and accumulations of charcoal and ashes marking the position of ancient camp fires.

High above the reach of present-day waves, along the shores of Alaska, Labrador, Newfoundland, Chile, the East Indies, Scandinavia, Italy, and many other countries, are abundant relics of former shorelines. Such features include barnacle shells and coral still attached to the rocks in the position in which they grew; rocks bored by marine clams (many with the shells still in the holes); sea cliffs, sea caves, stacks (Chap. 16, p. 392), and other erosional features carved from solid rocks by former waves; and deposits of shell-strewn sand drifted along the ancient beaches by longshore currents. Conspicuous marine terraces border many sea coasts (Fig. 16-11, p. 396). Although locally masked by detritus washed down from the hills, or partly cut away by streams since their uplift, most of them preserve shell-filled sands or other clear proof that they are the uplifted floors of ancient seas.

Some of the most diverse examples of geologically recent changes of level are found in the East Indies, particularly in the region north of Sumatra, Java, and the southern Moluccas (Figs. 9-9 and 9-10). Fringing coral reefs abound in the warm clear seas near these tropical lands (Fig. 9-7). Growing

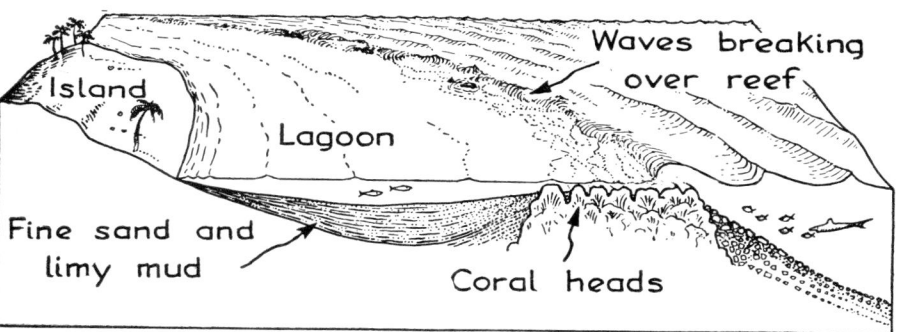

FIGURE 9-7. *Diagrammatic sketch of a typical coral reef.*

reefs generally have relatively flat tops with surfaces less than 150 feet below sea level. The reef-building coral animals cannot stand depths greater than 200 feet and are also killed by a few hours' emergence. Coral limestone, largely built of reef corals and associated organisms, is, however, by no means confined to the present zone of coral growth. Far above the reach of the highest storm waves are great terraces of white coralline rock exactly like that accumulating off the present shore. Some of these uplifted reefs are only a few feet above the sea, others fringe mountain tops 3,000 feet high (Fig. 9-8).

Some uplifted reefs form continuous level collars around the smaller islands (Fig. 9-9). Others have been tilted, for they may be 100 feet or more above sea level on one side of an island, but when traced around to the other side they gradually decline to sea level, or even pass below it. Still other uplifted reefs have obviously been faulted and warped.

Each of these raised reefs, of course, testifies to uplift of the land with

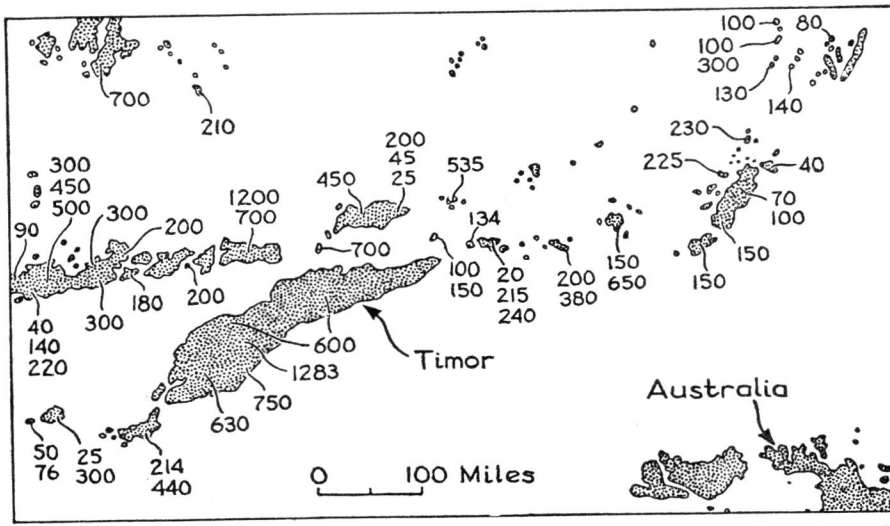

FIGURE 9-8. *Height, in meters, of uplifted coral reefs and stream channels in the southern Moluccas. (Redrawn from J. H. F. Umbgrove,* Pulse of the Earth *Martinus Nijhoff, 1947.)*

respect to the sea. But, as if to contradict this, we find equally clear evidence that some former land areas now lie depressed beneath the ocean. The chief evidence is from two sources: examination of samples dredged from the sea bottom, and fairly detailed contour maps of the ocean floor prepared from sonic soundings. Though coral reefs grow only in comparatively shallow water, at various points in the East Indies the dredge brings dead reef coral from depths of more than 1,000 feet. It also brings up land deposits such as muds that accumulated in fresh-water marshes and mangrove swamps or the poorly sorted silts and sands typical of river floodplains.

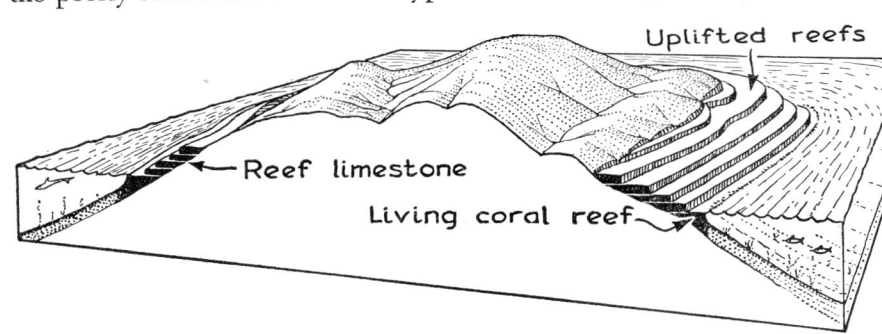

FIGURE 9-9. *Uplifted coral reefs on the island of Kissa, southern Moluccas (After Ph. H. Kuenen, redrawn from J. H. F. Umbgrove,* Pulse of the Earth *Martinus Nijhoff, 1947.)*

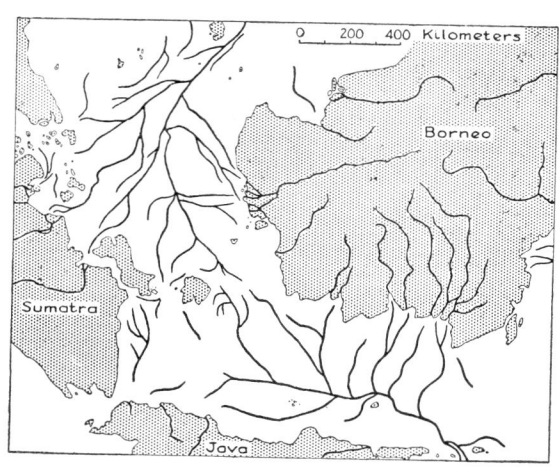

FIGURE 9-10. *The rivers of Borneo, Sumatra, and Java, with their submerged extensions. (Redrawn from Ph. H. Kuenen, Marine Geology, John Wiley and Sons, 1950.)*

Sonic soundings also show indubitable evidence that some of the shallow East Indian seas were once dry land. One example is the Sunda Shelf between Borneo and Sumatra, just west of the area of uplifted coral reefs shown in Figure 9-8. The Sunda Shelf is a shallow sea, mostly less than 250 feet deep. The soundings show clearly that the Shelf is a submerged plain with three trunk rivers. The headwaters of these rivers are the streams that still drain the northern parts of Sumatra and Java and the southern and western slopes of Borneo (Fig. 9-10). The underwater extensions of the streams have been located from detailed sonic soundings. Further proof of this submerged river system lies in the identity of the fresh-water fishes and other stream-dwelling animals of southwestern Borneo to those of eastern Sumatra, although a wide salt sea now separates the two areas.

Thus, a fairly small part of the East Indies yields clear geologic evidence of both submergence and emergence within comparatively recent time. Side by side, and interfingering with one another, lie the drowned topography of the Sunda Shelf and the uplifted, tilted, and faulted coral reefs of the southern Moluccas. Elevated reefs are also found in Java and Sumatra, pointing to early submergence and then uplift prior to the drowning of the river system on the Sunda Shelf. Obviously, the movements that have changed the levels here have not been simple. They involve not merely vertical uplift and subsidence, but concurrent bending and folding of the earth's crust. Indeed, many of the tilted reefs fit into a regional pattern of folds in the basement rocks beneath the surface. Study of the rock structure of the islands shows that most of the long curving island arcs are complex folds or arches in the rock, separated by basins and downfolds that still lie, in general, below the sea. The movements that produced these upfolds and basins are still going on, warping the fringing coral reefs above the sea at one point, drowning the reefs and the land-laid deposits behind them deeply at another, and forming a complex but slowly growing system of folds in the crust of the earth.

Contemporaneous Folding, Erosion, and Deposition

The folding of the earth's crust disclosed by the uplifts and submergences in the East Indies suggests that possibly such movements are related to the formation of mountains. Many mountain ranges show warped and bent strata that must have been deposited as nearly horizontal sheets. They are now distorted into patterns more complex than the East Indian warps, but suggestive of them. Can we find any geological links between the broad warping of the East Indies and the more intense localized folding of mountain ranges such as the Alps or the Appalachians? Let us consider first the relatively simple warps and folds in the lava flows that make up the eastern foothills of the Cascade Mountains in northwestern United States.

In central Washington and northern Oregon the walls of the canyons of Columbia River and its tributaries reveal flow upon flow of black basalt. The flows are 50 to 400 feet thick, and, unlike those at Stolpen near Freiberg, have relatively little interbedded sedimentary material. Here and there, however, thin sheets of river gravel separate the flows, showing that streams advanced onto the barren volcanic plain after some eruptions only to be overwhelmed by new lava flows. Elsewhere, narrow bands of red or black soil containing petrified (silicified) logs, or tree stumps with their roots still spreading out in the soil, appear between the flows. Obviously, time enough had elapsed for the scoriaceous flow tops to weather into soil and for forest to grow on it before a new sheet of lava devastated the area. Elsewhere, layers of a brilliant white rock with paper-thin stratification contrast strikingly with the basalt. Petrographic examination shows that these white rocks are made up almost entirely of the siliceous skeletons of minute one-celled plants called *diatoms*. Similar white diatomaceous deposits are being laid down today on the floors of shallow ponds and lakes nearby. In one part of central Washington, river gravels, floodplain silts, and diatomaceous deposits continued to accumulate on the surface of the lava plain long after volcanism had ceased, until locally the black basalt flows were buried under several hundred feet of these unconsolidated river and lake deposits.

The point to be emphasized is that at many places these basalt flows and associated sedimentary rocks are no longer horizontal (Fig. 9-11). They rise and fall in great arches and troughs. Here, on the flank of an arch, the scoriaceous top of a basalt flow and the diatomite strata above it dip at an angle of 20 degrees (Fig. 9-11). There, beds of gravel and sand between basalt flows dip at 70 to 85 degrees. Obviously, no stream tumbling down a 70-degree slope could deposit gravel and sand upon it, nor could a molten basalt flow drape itself over a steep-sided arch and maintain a uniform

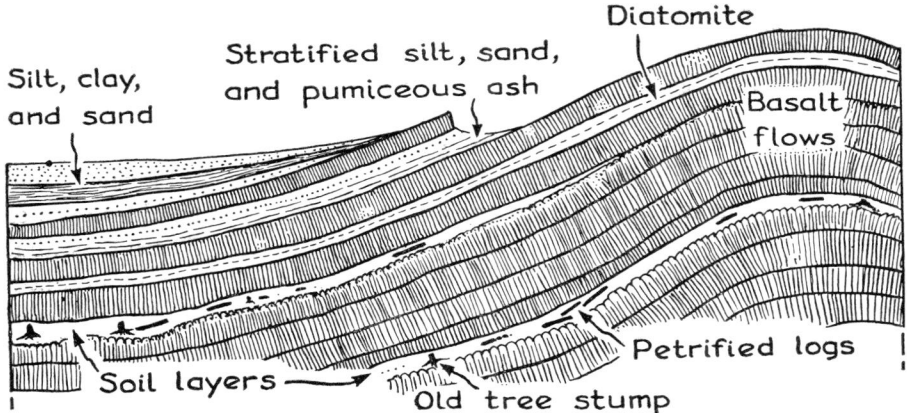

FIGURE 9-11. *Cross section of lavas and interbedded sediments near Yakima, Washington.*

thickness. Nor could the paper-thin sheets of diatomite remain on a 20-degree slope on a lake bottom. They must originally have been nearly horizontal. Clearly, the flows of basalt and sheets of gravel and diatomite were folded into the present arches and troughs after they had been deposited.

What is the relation of these arches and troughs to the present topography? Central Washington is rather exceptional among mountainous areas in that the major topographic ridges coincide with the arches in the lava. The flanks of the arches are scored by many ravines, nearly parallel, extending straight down the slopes. Turbulent floods course down these ravines in the wet season, cutting deep canyons into the sides of the arches. Generally, these streams, aided by rillwash and downslope movements, have removed most of the unconsolidated river and lake deposits from the tops and flanks of the arches, but the eroded edges of these unconsolidated beds can still be seen farther down the flanks of the folds. In some places, a few of the much more resistant basalt flows have also been eroded, and locally so many of them as to erase almost completely the fold as a topographic feature. In general, however, the smooth curving surfaces of the arch, intact except for small ravines, indicate that erosion is only getting started in demolishing the uplift. Many of these ravines do not have the typical concave-upward profiles of normal streams (see Fig. 12-21). The stream canyons and the folds have grown synchronously, and slow continuous arching of the lava sheets has modified the development of a normal stream regimen.

What is the relation of these topographic ridges to the troughs alongside? Conditions vary locally, but some of the partly dissected arches lie half buried in unconsolidated sediments deposited by streams and in lakes on the floors of the troughs. The uppermost beds of these trough deposits are nearly horizontal, abutting against the tilted rocks of the arches at high angles. In a few places, younger lava flows are interbedded with these young sediments, but they do not extend onto the arches. They were erupted after the folding had started, and poured down the troughs (Fig. 9-12).

In a few of the larger river canyons we can see, and elsewhere from records of wells drilled in the troughs we can infer, that the older beds of even the younger strata have also been somewhat folded. Where they abut against the neighboring basalt arch, they dip with the basalt, though at a lower angle (Fig. 9-11, upper left). Beds higher in the unconsolidated series dip similarly, but at still lower angles. In some troughs only the uppermost beds are nearly horizontal. These relations confirm our conclusion from studying the erosion of the arches: While the troughs were being downfolded, sedimentation was also going on. In a few places, sediments were deposited on the trough floors almost as fast as they grew, and have been progressively warped as the fold developed.

Some of the sediments filling these troughs were eroded from the adjoining growing arches, but much was brought in by rivers from the rugged Cascade Mountains to the west, as shown by the abundant quartz, pumice, and andesite fragments—constituents foreign to the adjacent arches.

Such evidence of contemporaneous erosion and sedimentation must mean that the folds grew very slowly. Individual flows can be traced from the crest of an arch down its flank until they disappear beneath the unconsolidated sediments of the adjacent trough; but across the basin they reappear and can be seen to rise in the next arch. Thus, originally horizontal flows have been folded until the surface of a particular flow on the crest of an arch is thousands of feet higher than it is at the bottom of an adjoining trough. But the folds did not grow overnight, for it is clear that folding, erosion, and

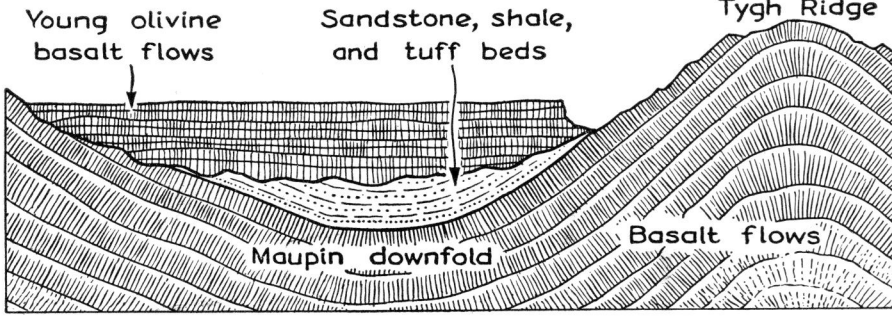

FIGURE 9-12. *Cross section showing the relation of folded lavas and sediments to younger horizontal lavas. Near Maupin, Oregon.*

sedimentation were all occurring simultaneously. The first stage of the folding blocked out the respective areas of erosion (arches) and sedimentation (troughs). The rising folds dammed streams and impounded lakes in the troughs, interfering with normal stream development. Sedimentation also was locally interrupted by tilting of already deposited beds. The concurrent erosion of the arches and deposition in the troughs as the folds grew shows that the movements were slow and gradual. There is meager but not wholly conclusive evidence from changes in the speed with which water is reported to flow in some of the irrigation canals built 10 to 50 years ago that the folding movements are still continuing, and thus changing the gradient of the canals.

We could cite many other examples to show that warping and folding go on slowly along with erosion and deposition. This has indeed been the rule in the young and growing mountain ranges fringing the Pacific Ocean, and in the Alpine-Himalayan belt across southern Eurasia.

Folds That Have Ceased to Grow

An older mountain range such as the Appalachian offers both contrasts and similarities to the structures of the East Indies and of the Cascade foothills.

Figure 9-13 shows a cross section of the folded rocks of a part of the Appalachians of central Pennsylvania. This section depicts in detail the position and attitude of the rocks deep underground as compiled from surveys of the extensive coal mines and many exploratory drill holes.

The independence between the folds of this section and the topography suggests immediately that the Pennsylvania folds are not growing today, nor have they in the recent past, as in the East Indies and the Cascade foothills. The folds have obviously long been dead, for the arches are no longer ridges nor are the troughs receiving sediments. In fact, throughout the folded Appalachians, it is common to find erosion-resistant rocks along the axis of a downfold forming a mountain summit, whereas adjacent upfolds with more easily erodible rock cores have been eroded to lowlands. Figure 9-14 shows a typical example. In the Appalachians, the topography is indeed closely adjusted to the rock structure, but the adjustment is of a kind quite different

FIGURE 9-13. *Folds and fault in the coal measures of Pennsylvania. The solid lines are mined coal beds; the dashed lines are their plotted continuations. The actual thicknesses of the individual coal beds is from 2 to 15 feet. (From N. H. Darton, 1940.)*

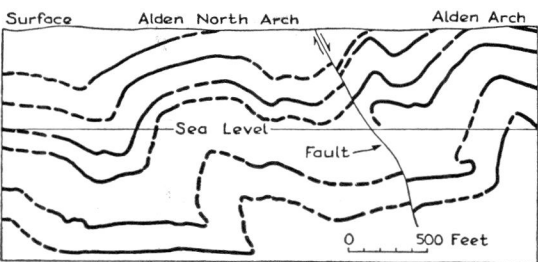

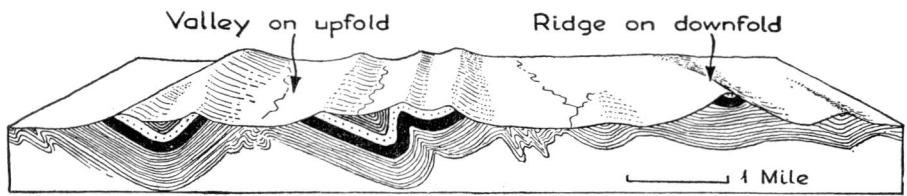

FIGURE 9-14.   *The relations between folds and ridges in the Appalachians near Rogersville, Tennessee.* (*From Arthur Keith, 1905.*)

from that of the East Indies and the Cascade foothills. In the Appalachians, all of the ridges are underlain by erosion-resistant rocks and the lowlands by easily eroded rocks. Stream erosion, uninterrupted by folding and warping, has had time to etch out the easily eroded rocks and leave the resistant ones standing in relief. This contrasts with the East Indies and central Washington, where soft sedimentary rocks often cap the very summit of a mountain. Clearly, the arches and troughs in Pennsylvania are not growing today, nor have they grown in recent geologic time.

Can slow crustal movements that took place in the far-distant geologic past, long before the present streams developed their courses, explain the folded beds of the Pennsylvania coal mines? The evidence is certainly less clear. Obviously, erosion has been much deeper in Pennsylvania than in central Washington or the East Indies. Gone are the sediments, if any, that were deposited in the troughs while the folds were growing. Any emerged shoreline features that may once have clung to the initially rising arches have long since been eroded away. We can easily measure the fault displacements shown by the offset of the individual coal beds in the section, but we cannot tell whether they required a few seconds or millions of years to form.

However, nothing in the Appalachian structure differs fundamentally in form from that of the East Indian arches and troughs. Although the underground form of the East Indian folds is concealed, their surface form and the canyon exposures suggest that the folds extend to great depths. In the higher and more rugged parts of the Cascades to the northwest of the foothills area, erosion has already torn down some of the basalt arches and stripped so much sediment from the intervening troughs that the best evidence for their slow growth has here disappeared. Thus, there is nothing about the Pennsylvania folds requiring different conditions of growth. Slow movements of the crust, like those now proceeding in the Dutch East Indies and Japan, are entirely adequate to explain the folds of Pennsylvania as well as those of the Cascade foothills. Again, we see another application of our fundamental postulate: "The present is the key to the past." Fossil marine shells at the tops of mountains and soils and fossil plants in deep mines are but what we should expect and predict from the evidence of slow but measurable changes of level that we can see taking place in different parts of the world today.

Furthermore, we can observe that, in some places and at certain times, the earth's crust has broken much like a brittle rock might be expected to break—along sharply defined faults. Elsewhere, and at different times, in the very areas of faulting the rocks have folded like plastic dough. The reasons for this differing behavior are obscure, for both kinds of deformation have taken place at the very surface of the ground. Nor is the ultimate cause of the deformation yet understood. These matters are further discussed in Chapters 10, 18, and 19. They are among the greatest problems of earth history.

Facts, Concepts, Terms

CRUSTAL MOVEMENT
Fault displacements associated with earthquakes
Fault displacements not associated with earthquakes

WARPING OF THE CRUST NOT ASSOCIATED WITH FAULTS
Slow vertical movements
Bending and folding of the crust

WARPING REASONABLY ATTRIBUTED TO ISOSTASY

WARPING INDEPENDENT OF ISOSTASY

WARPING AND FOLDING IN RELATION TO MOUNTAIN STRUCTURE
Characteristics of growing folds

TIME RELATIONS BETWEEN EROSION AND FOLDING

EROSIONAL PATTERNS ON GROWING FOLDS VERSUS THOSE ON DEAD FOLDS

Questions

1. In a certain mountain range all of the ridges and peaks are composed of erosion-resistant rocks that have been tilted, folded, and locally faulted. In another mountain range the rocks show about the same structure but soft, easily eroded rocks are locally exposed at the summits of the peaks, and many streams flow along the axes of downfolds. What can you say about the relative age of folding in the two regions? Explain.
2. If we assume a fold has risen many hundreds of feet above the sea, what would be the isostatic effect of erosion of the fold?

3. If a downfold receives a thick accumulation of sedimentary deposits, what effect may these deposits be expected to have on further development of the fold?

4. As many as 13 nearly level marine terraces have been recognized at heights ranging between 100 and 1,300 feet above the sea in the Palos Verde Hills near Los Angeles, California. What inferences can you draw as to the history of crustal disturbances here?

5. The rocks of the Swiss plain, just north of the intensely folded rocks of the Alps, show much gentler folds. The oldest rocks of the plain are marine, and about the same age as the youngest rocks of the Alps. The youngest rocks of the plain are nonmarine river gravels and sands which do not appear in the Alps. The Alps are like the Appalachians in showing no direct relation between folds and topography. What suggestions occur to you to account for these relations?

6. Marine Eocene rocks extend up the Mississippi Valley to Cairo, Illinois. They dip gently toward the Gulf. What do these facts mean as to crustal deformation?

7. The dolerite sill composing the Palisades of the Hudson across from New York City is several hundred feet thick where exposed along the west river bank. Sandstone beds above and below the sill dip to the west. How can you account for the absence of the sill east of the river?

8. Cincinnati is on Ordovician marine rocks. Both to the east and the west, Silurian, Devonian, and Carboniferous rocks are exposed, dipping away from Cincinnati. What is the simplest explanation of these relations that occurs to you?

9. Draw three sketch geologic maps to represent the evolution of one of the folds in the Cascade foothills. On the first map, show an almost uneroded arch in the basalt lavas and overlying sedimentary rocks, with a little new sediment formed in the adjoining troughs. In the second map, show the fold after the sedimentary rocks and some of the lava have been stripped from the crest of the arch, and the adjacent troughs are half filled with new sediment. In the third sketch, show the area after erosion has completely erased the arch as a topographic feature.

Suggested Readings

1. Umbgrove, J. H. F., *The Pulse of the Earth*, 2nd ed., M. Nijhoff, The Hague, 1947.

2. Daly, R. A., *Our Mobile Earth*, C. Scribner's Sons, New York, 1936.

10. *Records of Earth Movements*

EVIDENCE exists, then, that there are constructional processes to counteract the effects of erosion and so to prevent reduction of the lands to the level of the sea. One such constructional process is *volcanism*—the effusion of molten rock from the earth and its piling up on the surface. We have also seen (Chap. 9) that in places the surface of the earth is moving up or down with respect to sea level. A second major constructional process, then, is the slow gradual warping and uplift of segments of the earth's crust.

"Original Horizontality"—The Key to Structure

How can we analyze this interplay between destructional (erosional) processes and constructional processes such as volcanism and uplift? We have seen (Chap. 5, p. 73) that by the middle of the Seventeenth Century Nicolas Steno had worked out the "Law of Original Horizontality"—the proof that strata are deposited in nearly horizontal layers, parallel or nearly parallel to the surface on which they accumulated. But another century passed before geologists began using Steno's law to work out systematically the significance of the structural features found in the stratified rocks.

In lowland areas large-scale structural features such as folds and faults are generally masked by soil. It is to the mountains and deserts that the geologist must go to find large rock structures clearly displayed. Until late in the Eighteenth Century, however, the mountain areas of Europe were little studied because of the difficulties of transport and the fear of bandits.

Horace Benedict de Saussure (1740–1799), a Swiss geologist who was also an ardent mountain climber, did much to arouse the modern spirit of mountaineering and to develop interest in structural geology. While climb-

NW

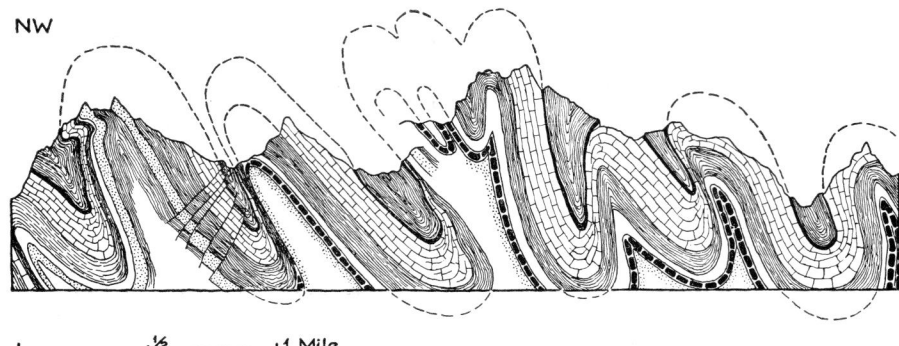

FIGURE 10-1. *Folded sedimentary beds, Saentisgebirge, Swiss Alps. (After A. Heim, 1922.)*

ing in the Alps, de Saussure noted that the strata appeared to be crumpled into folds like a rug pushed together in a heap on the floor (Figs. 10-1 and 10-19), but he did not immediately grasp the significance of the law advanced by Steno more than 100 years before. Instead, he at first assumed that the rocks had crystallized in the forms we now find them. It was not until he had carefully studied beds of conglomerate projecting vertically from the ground that he recognized it was absurd to believe that sheets of pebbles, interlayered with sand and all standing on end, could ever have been deposited in that position.

Since de Saussure's time generations of geologists have scaled the peaks of the Alps and those of most other mountain ranges and have plotted on maps and sections the geometry of the deformed strata they found. Aided by what has been learned from the structural features so well displayed in mountain cliffs and canyon walls, they have also been able to work out the underground structure of plains and lowlands, despite the extensive soil cover.

The key that unlocks the geometry of these structures, and through this the sequence in which they were formed, is the principle, proved by Steno more than 200 years ago, that stratified rocks were originally nearly horizontal.

Warps and Gently Tilted Strata

Over much of the earth the sedimentary strata appear to be still horizontal. If, however, we determine the elevations of several widely separated points on the upper surface of a particular bed, we generally find that the horizontality is not perfect. The bed either had an originally sloping surface or it has been slightly warped. Commonly, the gently inclined surface of the bed rises and falls so irregularly that its departures from horizontality cannot be explained entirely in terms of *original dip,* and so the surface must have been warped.

The rocks of the Mississippi Valley afford good examples of such broad warps. In small exposures the beds appear flat, but if we determine elevations on the upper surface of some widespread formation we find notable differences. For example, in wells drilled for petroleum in southern Illinois the Chattanooga black shale of Devonian age lies about 4,500 feet below sea level. When followed eastward it rises steadily until, finally, it appears at the surface about 500 feet above sea level near Louisville, Kentucky. It rises similarly, though less steeply, to the north, south, and west. Well records and mapping reveal that the top of the Dakota sandstone, a widespread formation of Cretaceous age, is about 3,000 feet below sea level near Williston, North Dakota, but is 2,000 feet or more above sea level near Shelby, Montana, and 1,000 feet above sea level near Sioux City, Iowa.

Such broad warps may pass gradually into folds—the Chattanooga shale is highly folded in the Appalachian Mountains, and the Dakota sandstone has been wrinkled into folds and broken by faults in the Rockies—but the characteristic feature of a broad warp is that the beds are still nearly horizontal, even though we find appreciable differences in elevation when large areas are considered. Such differences must have been caused by movements of the earth's crust.

Implications Regarding Isostasy

The crustal warping recorded by the Chattanooga shale and the Dakota sandstone raises many interesting problems. What forces have raised some parts of the Mississippi Valley and depressed other parts? Have these movements disturbed the isostatic condition of the area? Or have they resulted from isostatic readjustments of the crust to differences in load brought about by erosion in one area and deposition in another? Have the "unloading" of one area by erosion and the "loading" by newly deposited sediments in another led to the isostatic rise of the first and the sinking of the second, with the concurrent slow flowage of the dense subcrustal material beneath from the loaded to the unloaded area? Considerations of scale and size seem to demand that changes of level of the kind recorded in the warped surface of the Chattanooga shale must be compensated for in some way by transfer of material at depth to maintain or restore isostatic balance. Further discussion of these difficult problems is deferred to Chapters 18 and 19.

Showing Warps on Maps

How can we represent broad warps on maps and sections? If the topography is rough, ordinary geologic maps and cross sections such as are shown in Chapter 8, Figure 8-2, may suffice. However, in areas of gently dipping beds and low relief, few beds will crop out at the surface, and hence geologic maps are ill adapted to show the details.

If we are engaged in mining a coal bed by open-pit methods (stripping away the overburden above the coal) a dip of only a few feet per mile may be critical to our operation. Suppose at one place 25 feet of overburden must be stripped off to get to the coal bed. If the ground is flat, and the coal bed dips 100 feet per mile (a little more than 1°), it is obvious that half a mile down-dip our overburden will have increased to 75 feet. Up-dip, in a little more than one-fourth mile, the coal will rise to the surface and be eroded away. If the ground surface is hilly, as in the Wasatch Plateau coal field of central Utah (Fig. 10-2), or if the coal bed is irregularly warped instead of dipping uniformly, it may seem difficult to compute just what areas of coal can be mined economically by open pits and what portions could better be mined from underground tunnels and shafts. To help in the solution of such a problem, we may need to construct a structure contour map of the coal bed.

Structure Contour Maps.   A *structure contour* map, like a topographic

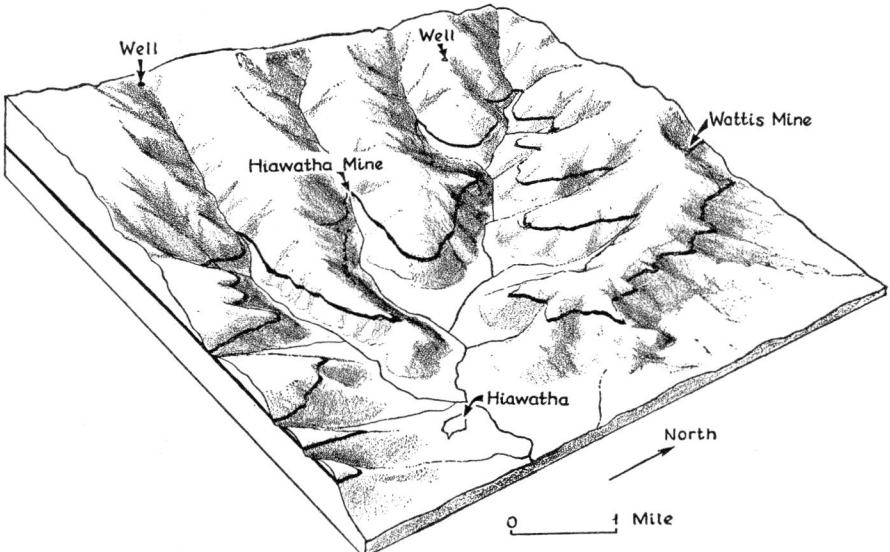

FIGURE 10-2.   *Relief diagram of a part of the Wasatch Plateau coal field, Utah. (From a map by E. M. Spieker, U. S. Geological Survey.)*

map, uses contours to represent lines of equal elevation. But, whereas a topographic map depicts the surface of the ground, a structure contour map shows the surface of a single bed as it would appear if all the beds above were stripped away. If we removed all the overburden from a coal bed and then made a topographic map of the newly exposed surface of the coal, the result would be a structure contour map of the coal.

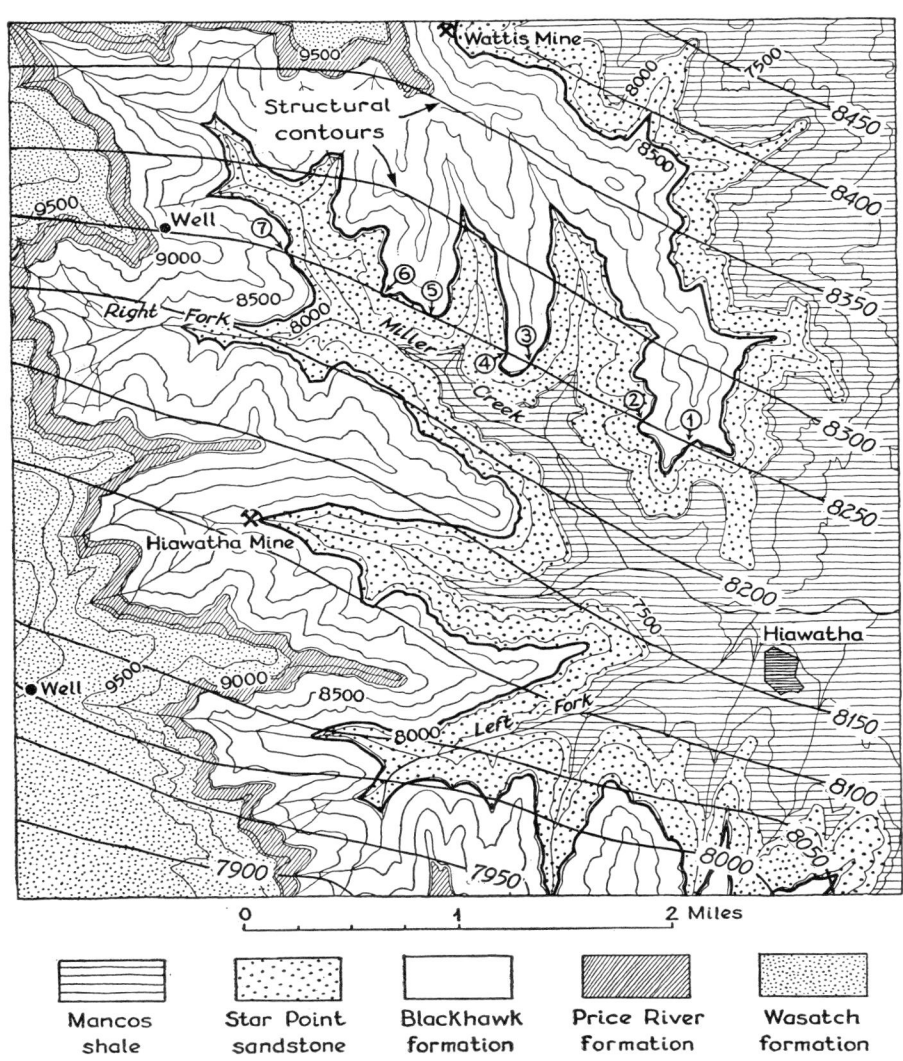

FIGURE 10-3. *Topographic, geologic, and structural contour map of the area shown in Figure 10-2. (Modified from E. M. Spieker, U. S. Geological Survey.)*

How is it possible to build up such a structure contour map without actually excavating down to the coal? Figure 10-3, which is a map of the area sketched in Figure 10-2, illustrates some of the steps by which we can get control points to draw the structure contours. The map in Figure 10-3 is a combined topographic and geologic map upon which the structural contours of the coal bed have been superposed. The outcrop of the Hiawatha coal bed is shown by a heavy black line. Notice that the contours showing the ground surface do not quite parallel the line marking the outcrop of the coal bed. For example, there are seven places (numbered 1 to 7 on Fig. 10-3) in the ravines north of Miller Creek where the 8,250-foot contour intersects the coal bed. At each of these points the top of the coal is 8,250 feet above sea level. Thus we have seven control points through which to draw the 8,250-foot structural contour. On the south wall of Left Fork Canyon there are ten points where the coal bed intersects the 8,000-foot topographic contour, giving us ten control points for drawing the 8,000-foot structural contour. Similarly, other control points are obtained by marking the intersections of other topographic contours with the coal. Still other points can be obtained from borings and wells. For example, a well drilled on the ridge north of the Right Fork of Miller Creek penetrates the top of the coal bed 750 feet below the surface. Since the curb (top) of this well is on the 9,000-foot topographic contour, the coal bed directly below must have an elevation of 8,250 feet. Thus we have another point through which to draw the 8,250-foot structural contour.

The coal bed is overlain by the Blackhawk formation. The thickness of this formation can be measured on the walls of several of the canyons. Five such measurements in different parts of the area give its thickness as 750, 775, 725, 740, and 760 feet. The variation is not systematic in one direction, so we can accept 750 feet as the average thickness. Therefore, at any point where a topographic contour intersects the top of the Blackhawk formation we can obtain the elevation of the coal bed directly below by subtracting 750 from the elevation of the topographic contour. The position of the 8,000-foot structural contour near the west edge of the map was determined in this way: A well located on the 9,750-foot topographic contour at this point does not go deeply enough to penetrate the coal, but at a depth of 200 feet it does cut the top of a thin bed of limestone which, from measurements on canyon walls, is known to lie 550 feet above the base of the Wasatch formation. The Price River formation lies between the Wasatch formation and the Blackhawk formation and is 250 feet thick, as determined by measurement at several points in the area. Thus at the site of the well the coal bed must lie 200 feet (depth to the limestone) plus 550 feet (distance from top of limestone to base of Wasatch formation), plus 250 feet (thickness of Price

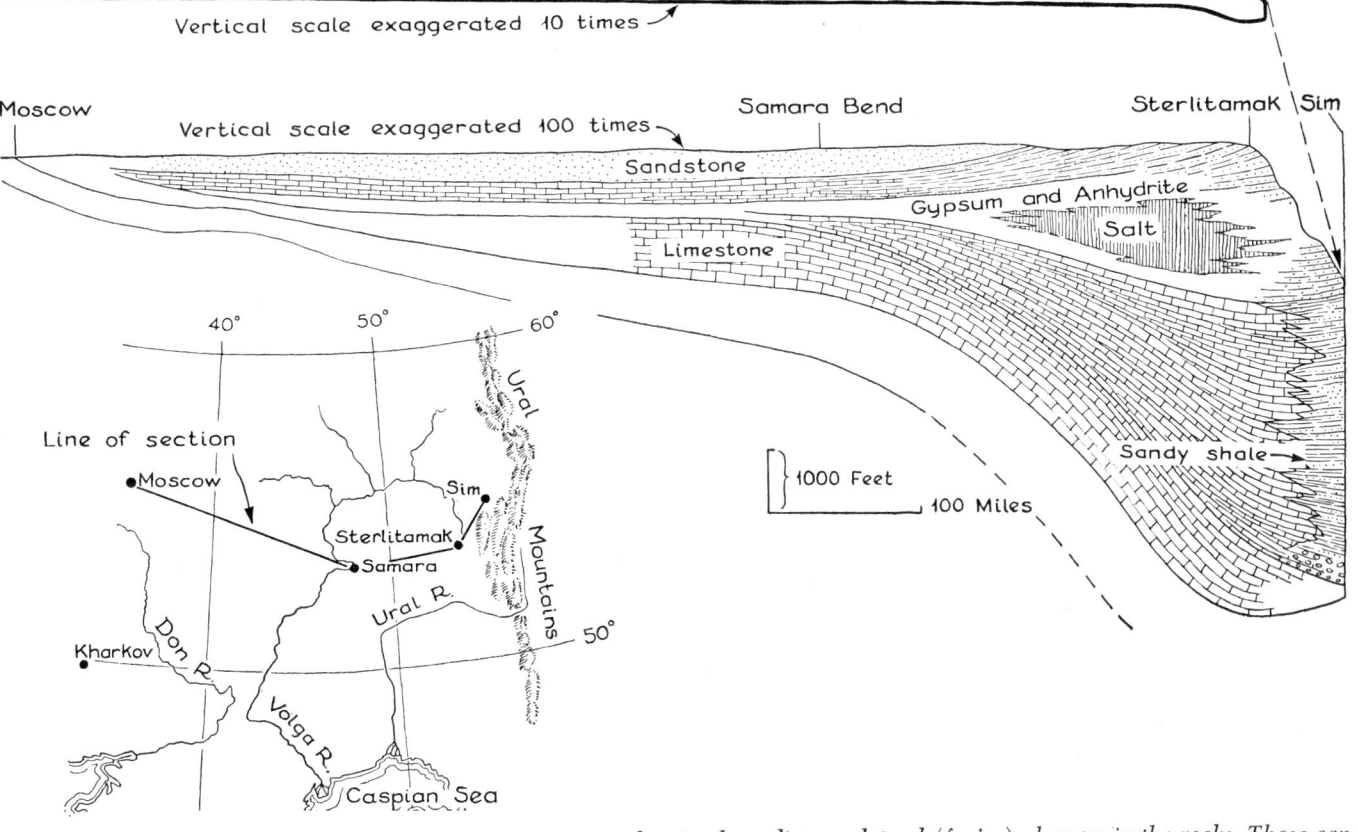

FIGURE 10-4. *Cross section of the Moscow Basin, Russia, showing long-distance lateral (facies) changes in the rocks. These can be plotted only because of the greatly exaggerated vertical scale. (After C. O. Dunbar.)*

River formation), plus 750 feet (thickness of the Blackhawk formation), equals 1,750 feet below the surface. Since the surface is at 9,750 feet, the coal is at 8,000 feet at this point; we thus have another point through which to draw the 8,000-foot structural contour. Many other elevations on the coal can be determined by these methods, and from them the structure contours on the Hiawatha coal shown in Figure 10-3 have been drawn.

Once such a structure contour map has been prepared, it is easy to tell the amount of overburden above the coal. Since the structure contour map represents the surface of the coal bed, and the topographic map that of the ground, the thickness of overburden at any point can be obtained by subtracting the elevation shown on the structure contour map from that shown on the topographic map for the point directly above. Thus, on Figure 10-3, wherever a structural contour and topographic contour cross we can find the thickness of overburden by subtracting the elevation of the structure contour from that of the topographic contour.

Cross Sections with Exaggerated Vertical Scale.   For certain purposes, *geologic cross sections* drawn with *greatly exaggerated vertical scale* are useful in depicting gently dipping rocks. Such exaggeration, of course, distorts the form of the features shown, and gives a wholly erroneous idea of the dip and the size of the structures, but it permits us to plot thin formations that could not otherwise be shown and to emphasize lateral changes in facies of the sedimentary rocks. Such sections also emphasize, even though they distort, minor changes in dip, small basins, or other minor structures. For example, Figure 10-4 shows the structure beneath a portion of the Russian plain. The large cross section is plotted with a vertical exaggeration of 100 to 1. On it the lateral variation of limestone to shale and of salt to various kinds of rock (sedimentary facies) is clearly shown, but such variations could not have been plotted on the section with 10 to 1 exaggeration shown in the same figure, and if the section had been drawn to true scale its entire thickness would have been encompassed within a single pencil line. But when the scale is exaggerated 100 times some of the beds in the section appear to dip 45 degrees or more, although their true dip is not more than 1 degree.

Continental Plates

Broad warps are the main records of earth movements in the relatively stable parts of the continents. Such stable areas of nearly horizontal sedimentary rocks are called *continental plates*. They differ from mountain belts in their gently warped strata as contrasted to the strongly folded rocks of most mountains. Also, drilling for petroleum shows that the sedimentary rocks of the continental plates are generally only a few hundred, or at most a few thousand, feet thick above a basement of granite and metamorphic rocks, whereas they are generally far thicker in mountain belts.

The great area from the Appalachians to the Rockies (except for the Ouachita, Arbuckle, and Wichita mountains) and extending northward through Canada in a broad belt east of the Canadian Rockies is characterized by this structure. This is the continental plate of North America. The great plain reaching from eastern Germany to the Urals, then extending from the Urals far into Siberia and southeastward toward Turkestan is another continental plate.

It should be said in passing that broad warps are also found in mountain belts, but there they are less readily recognized because more intense crustal movements largely obscure such relatively gentle deformation as these features record.

We pass now to a description of the records of more intense crustal movements. Although folds and faults are not confined to mountain belts, they are as characteristic of those regions as broad warps are of the continental plates.

Folds

Folds are the typical features of most mountain chains. They range in size from microscopic crinkles to great arches and troughs fifty miles or more across. Upfolds or arches in rocks are called *anticlines* (Figs. 10-5, 10-12, and 9-12); downfolds or troughs are *synclines* (Fig. 10-6). A *monocline* is a flexure in a series of beds which are nearly horizontal on either side of the flexure (Fig. 10-7).

Most folds have been greatly modified by erosion (Fig. 10-8). To visualize their forms we often reconstruct some of the eroded beds by drawing dashed lines beyond the truncated ends of the strata. In the monocline shown in Figure 10-7 and in the anticlines and synclines shown in Figure 10-1 the

FIGURE 10-5.   *Anticline in rocks of Cretaceous age near Livingston, Montana. (Photo by T. W. Stanton, U. S. Geological Survey.)*

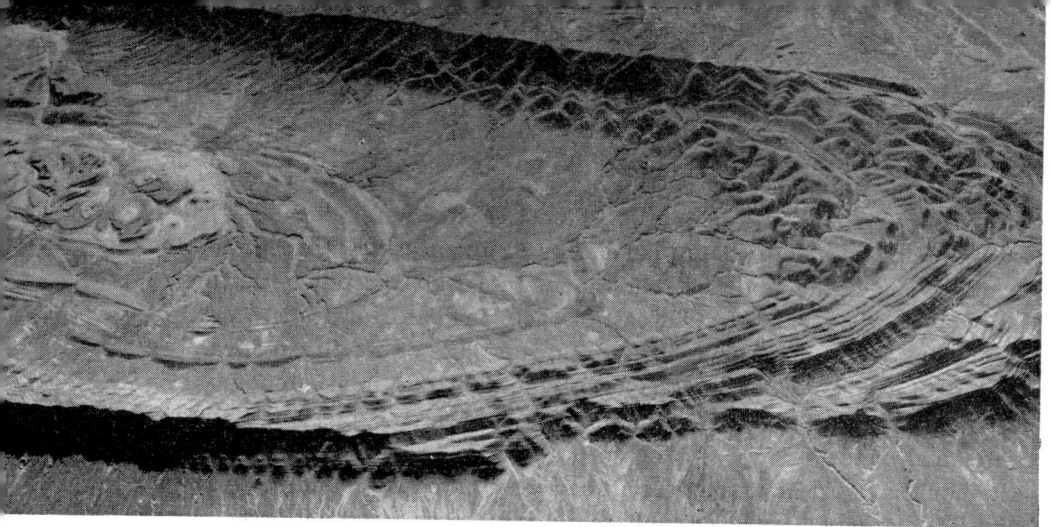

FIGURE 10-6. *Aerial view of an eroded plunging syncline in northwest Africa. The plunge is to the left. Note how the ridges formed on resistant beds clearly indicate dips toward the axis of the fold. (Photo by U. S. Air Force.)*

eroded upper part of the structures have been partially restored in this way.

Another kind of reconstruction often used to visualize the form of folds is shown in the lower part of Figure 10-9. The upper sketch shows a part of the Jura Mountains in Switzerland. The eroded surfaces of the folds are seen on the top of the upper figure, and a cross section forms the front. One bed of limestone, more resistant to erosion than most of the other beds, forms cliffs and bold ridges at the surface. The top of this limestone is shown as a heavy black line on the cross section. In the lower sketch the top surface of this resistant stratum is drawn as it would appear if suspended in space, entirely detached from the beds above and below it. From such a reconstruction we can readily visualize the gross form of the folds.

Structure Symbols. One of the best ways to portray a simple fold is by structure contours (Fig. 10-10). If information is too limited to get control for structure contours, special *structure symbols* on the geologic map may

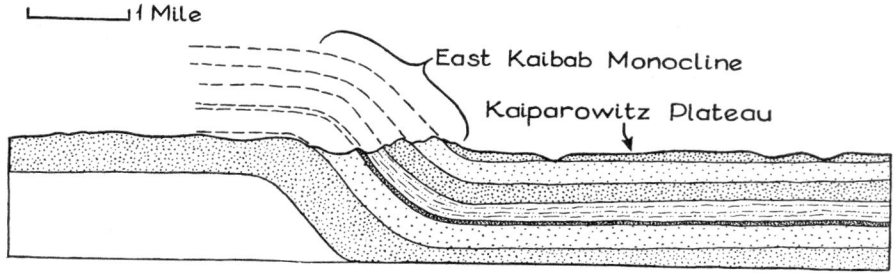

FIGURE 10-7. *Cross section of the Kaibab monocline, Utah. (After H. E. Gregory and R. C. Moore, U. S. Geological Survey.)*

help bring out the details of the fold. Figure 10-11 shows some of the more common symbols used to depict structure on geologic maps and sections.

The block diagram and geologic map (Fig. 10-12) of a small fold in the Jackfork formation of Arkansas shows how various kinds of structures may be represented by symbols. On the geologic map the strike and dip of the beds is shown at several places by the *strike and dip symbol* (see Chap. 8, p. 152, and Fig. 10-11). While the geologist is mapping, he plots these symbols on the topographic map wherever he determines strike and dip.

Note that two layers of shale interbedded with the prevailing sandstone of the Jackfork formation have been crinkled into a series of small, roughly parallel puckers that trend at an angle to the major fold. It is important in mapping to avoid mistaking the strike and dip of the strata in one of these minor puckers from the general strike of the beds in the major fold.

Another feature easily plotted on a geologic map is the position of the *axis* of the anticline. This is the line of opposed dips or, in other words, the line along which each bed exposed at the surface reaches its highest point as it arches over the top of the fold. On a geologic map an anticlinal axis is shown in symbol by a line crossed by opposed arrows (Figs. 10-11 and 10-12). The axis of a syncline is indicated by a line with converging arrows. Another measurable feature shown in Figure 10-12 is the inclination of the top of the fold, or its *plunge*. Note that at the crest of the anticline of Figure 10-12 each warped bed plunges eastward into the ground, following the trend of the fold. The measured inclination to the east, determined in de-

FIGURE 10-8. *Deeply eroded anticline, Green River, Colorado. (Photo by W. H. Jackson, U. S. Geological Survey.)*

grees from the horizontal, is 8 degrees; this is the *amount of the plunge*. Plunge is indicated on the geologic map by arrowheads and figures showing the direction and amount of the plunge (Figs. 10-11 and 10-12).

Kinds of Folds. A little study of Figure 10-12 will demonstrate why, after a *plunging fold* is eroded, the different strata that compose it show a curving or arcuate pattern at the surface of the ground. This relation is beautifully displayed in many airplane views of deeply eroded folds (Fig. 10-6). If the fold has no plunge (*upright fold*), and the topography is nearly flat, the beds will crop out as roughly parallel bands on either side of the axis. Resistant beds such as sandstone and limestone have been etched into ridges and reveal strikingly the details of the structure.

Many folds, such as the Paintrock anticline in Figure 10-10, are *asymmetrical*—the strata on one side of the axis dip more steeply than on the other side. Some are even overturned or recumbent. Superlative examples of overturned and recumbent folds are found in the Alps (Figs. 10-1, 10-19, and 19-10).

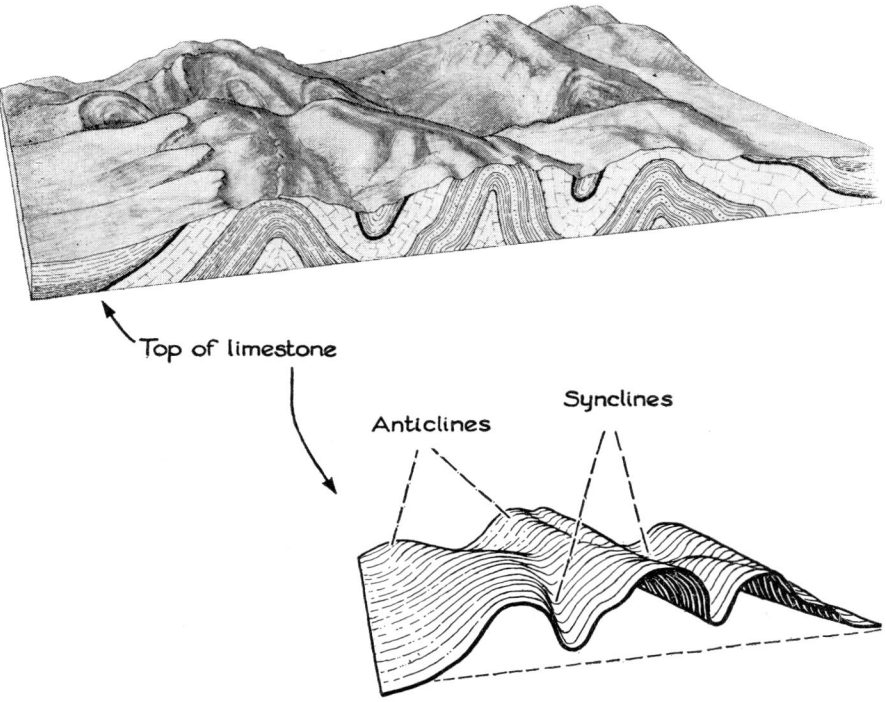

FIGURE 10-9. Top, *relief diagram of a part of the Jura Mountains, Switzerland. The top of a resistant limestone bed is shown by a heavy line.* Bottom, *sketch of the top of the limestone bed as it would appear if the beds above and below it were removed.* (*After A. Heim, 1922.*)

B
Paintrock Anticline
2000
Bonanza Anticline
Manderson Anticline
0
N
2000
-2000
A
1 Mile
Manderson Anticline
Bonanza Anticline
Paintrock Anticline
A
B
4000'
2000'
Sea Level
Greybull sandstone

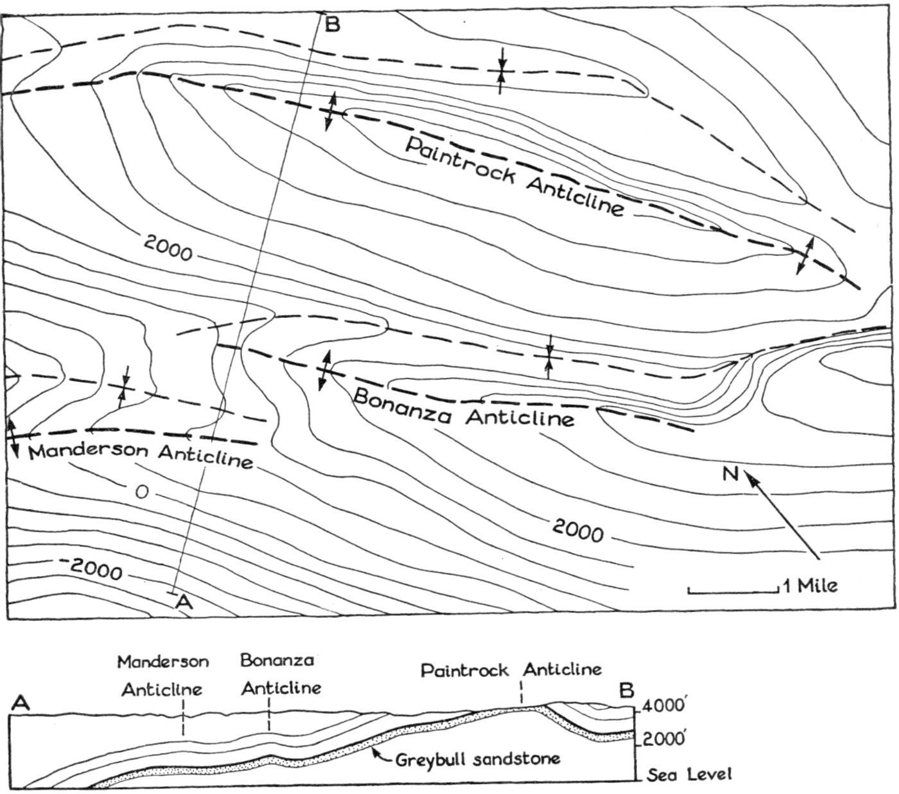

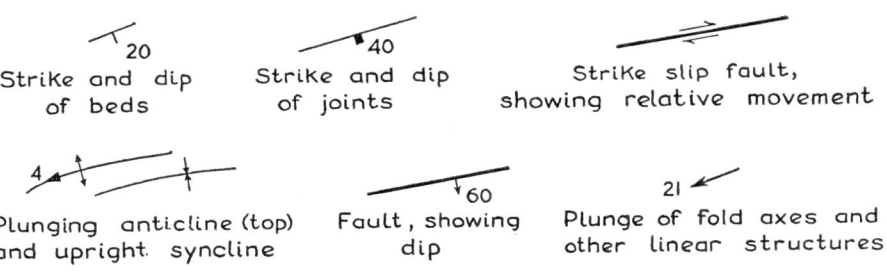

FIGURE 10-10. *Structure contour map and cross section of folds in the Big Horn Basin, Wyoming. The structure contours are on the top of the Greybull sandstone. (After D. F. Hewett and C. T. Lupton, U. S. Geological Survey.)*

20
Strike and dip of beds
40
Strike and dip of joints
Strike slip fault, showing relative movement
4
Plunging anticline (top) and upright syncline
60
Fault, showing dip
21
Plunge of fold axes and other linear structures

FIGURE 10-11. *Chart of common structure symbols.*

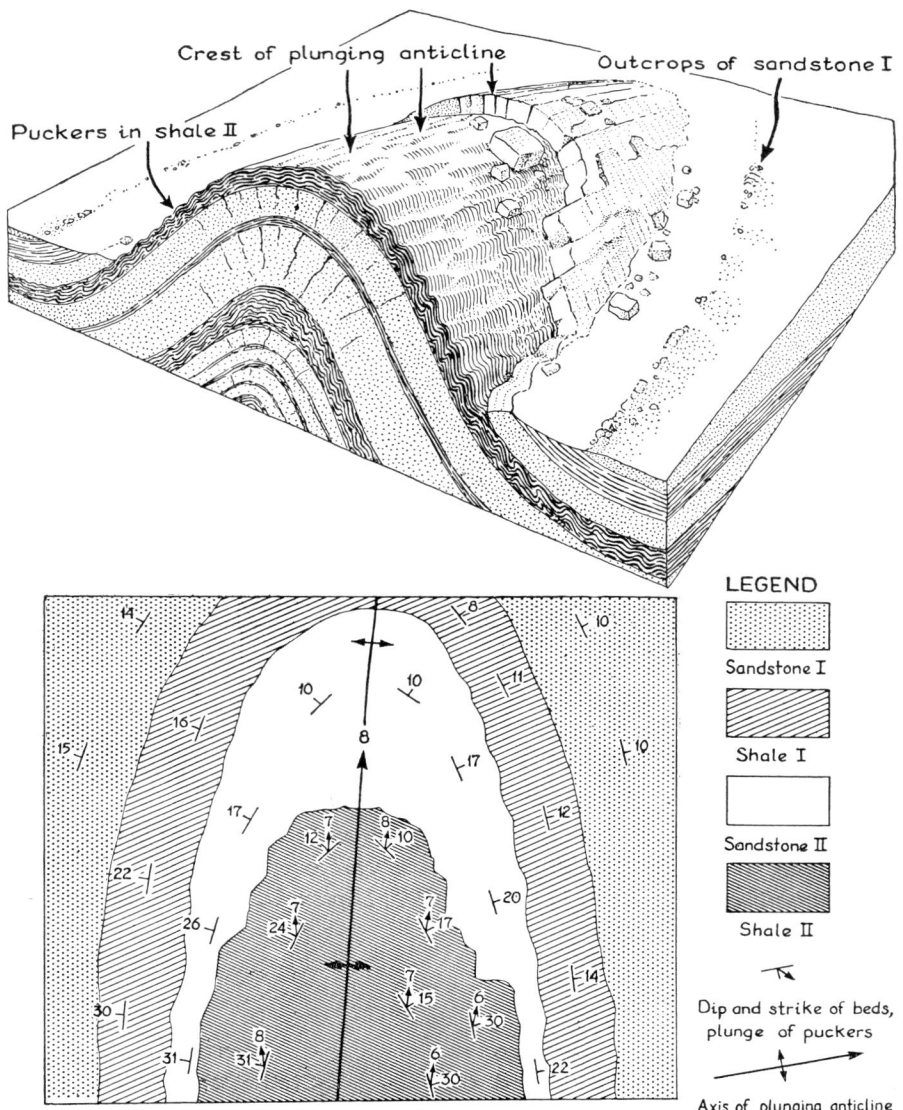

FIGURE 10-12. *Relief diagram and geologic map of a small fold in the Jackfork formation near Amity, Arkansas.*

Mountain Belts

Study of the figures shows that some of the folds illustrated could perhaps have been formed by uplift and depression much like the broad warps. But some, like the folds in Figures 10-1, 10-18, and 10-19, seem to demand that the rocks were squeezed together laterally.

Although some folds are isolated, the vast majority are arranged in long linear belts, forming the characteristic structural pattern of the mountain chains of the earth. Many of these folded belts comprise a whole series of roughly parallel anticlines and synclines so arranged as to imply horizontal, rather than vertical, forces as their cause. For example, many folds are overturned or even recumbent, forms that could hardly be produced without strong lateral movement. The evidence of such horizontal compression is further strengthened by the association of many such belts with great thrust faults (p. 210) along which the rocks above the fault plane have been shoved over those beneath for distances measured in miles. Detailed evidence of crustal shortening is given in Chapter 19; here, we merely point out that such structures characterize the Appalachians, the Andes, the Alps, and many other mountain chains.

This shortening implies the crowding together of a wide belt of rocks into a much narrower zone. The Appalachians or the Alps occupy such large areas that considerations of strength (Chap. 3) show that the folds cannot have grown merely by rising into the air above a solid basement like the folds in a rug pushed together over a smooth floor. The crust could not support such a thickened mass over so large an area. As the rocks are crowded together they must sink into the substratum enough to sustain themselves by flotation. This is proved, as we saw in Chapter 3, by the fact that the plumb bob is not deflected toward the Himalayas as much as it should be if the range were a pile of rock heaped onto an otherwise uniform crust. In other words, during the crustal shortening, the rocks of the folded belt must have been squeezed not only *upward* into the air, but also *downward* into the plastic substratum below, forming a *mountain root* large enough to buoy up the mountain chain. The existence of such a root beneath the Alps, Appalachians, and other ranges has been proved both by studies of gravity and by the records of earthquake waves that have passed through them, as will be described in Chapters 18 and 19.

Metamorphism in Mountain Roots

Where deep erosion has revealed the cores of large mountain chains it is common to find that the folded and faulted rocks along the edges of the mountain belt grade gradually into metamorphic rocks in the more deeply eroded interior. Furthermore, many mountain ranges show evidence of having been eroded to almost featureless plains, and then "rejuvenated" by the warping or faulting upward of the mountain stumps to form a new high mountain range (Chap. 19). In such "second cycle" mountains the proportion of metamorphic rocks increases. Also, large intrusions of granite or other plutonic rocks appear in increasing volume.

These relations indicate that in the formation of a large mountain belt by crustal shortening the brittle and cold rocks near the surface are broken by faults and thrown into large folds, but deeper within the mountain belt, where the load of overlying rock is greater and the temperature higher, deformation is accomplished mainly by granulation and recrystallization of the mineral grains to form new metamorphic rocks. Indeed, it seems reasonable that in the deeper parts of a mountain root the temperature may rise sufficiently to melt parts of the metamorphosed root into new masses of magma. It is not surprising, therefore, that the deeply eroded cores of mountain ranges are composed typically of metamorphic and plutonic rocks.

Joints and Faults

Fractures in rocks are classified by geologists as joints or faults. By definition, *joints* are fractures that have merely opened, without differential slip parallel to the walls; whereas *faults* are fractures along which there has been movement. In actual practice this distinction is often difficult; more commonly the separation is based on the scale of the mapping. A fracture cutting a plagioclase crystal, when seen under the microscope, may show a minute displacement of the plagioclase striae and so can correctly be described as a *microfault*. The same fracture probably would not be seen in a rock outcrop, but if it were observed it would surely be called a joint, for the small offset of the plagioclase striae would go unnoticed. In geologic mapping, fractures showing visible displacement of only a few feet are generally ignored or are mapped as joints; the mapped faults are larger features with considerable displacement.

Practically every rock outcrop shows numerous joints (Fig. 10-13). Because of their profusion and irregularity they are generally omitted from geologic maps. If there is some economic or other reason for showing them (for example, valuable ore veins may be developed along joints), the most

FIGURE 10-13. *Joints in granite, east of Hermosa Station, Wyoming.* (*Photo by Eliot Blackwelder.*)

FIGURE 10-14. *Aerial view of a fault near Great Bear Lake, Canada. The fault extends for 80 miles. In the photo, sandstone strata lie to the left, granite is on the right. Note the small faults and joints that intersect the fault at acute angles. (Data from A. W. Jolliffe, photo by Royal Canadian Air Force.)*

common method is to plot their strike and dip by the symbol shown in Figure 10-11.

Joints and small faults may be formed in several ways, but large faults that can be traced for miles (Fig. 10-14) and that displace their walls hundreds or thousands of feet have certainly been formed by crustal movements.

Kinds of Faults.   Faults are common in nearly all parts of the crust, but particularly so in the highly deformed rocks of mountain ranges. Geologists commonly distinguish three kinds of faults. The distinction is based on the direction of apparent movement along the fault fracture:

> **A normal fault is an inclined fracture along which the rocks above the fracture have apparently moved down with respect to those beneath (Figs. 10-15 and 10-17).**
>
> **A thrust fault is an inclined fracture along which the rocks above have apparently moved up with respect to those beneath (Fig. 19-2, bottom).**
>
> **A strike-slip fault is an inclined or vertical fracture along which movement has been predominantly horizontal (Figs. 10-16 and 10-17).**

Note the presence of the words "with respect to" and "apparently" in the definitions above. Actually, the movement is of course *relative,* and displacement with respect to a reference surface such as sea level is rarely known.

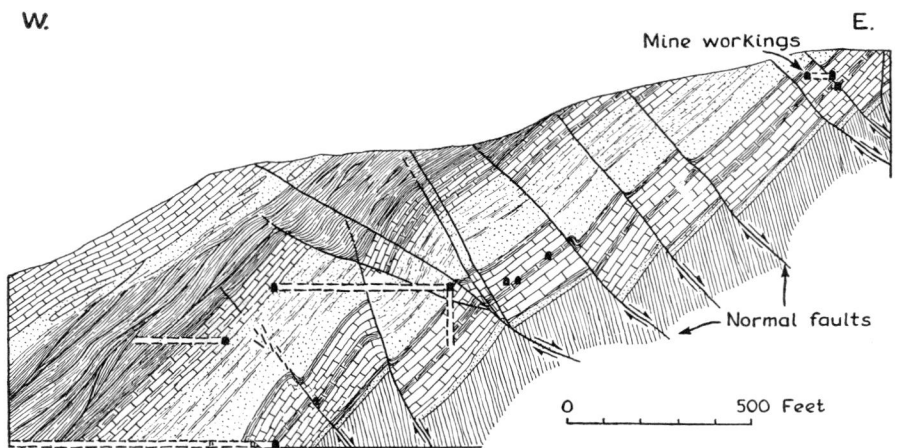

FIGURE 10-15. *Cross section showing drag and offset on normal faults, Magdalena mining district, New Mexico. Extensive underground workings aided the accurate plotting of these faults. (After G. F. Loughlin and A. H. Koschmann, U. S. Geological Survey.)*

A few times, after an earthquake, the actual direction of movement with respect to sea level has been measured (Chap. 9). More generally, the study of the geometric relations of the rocks on either side of the fault shows only the *apparent* movement. Abrupt bending of the strata as they approach the fault (called *drag*) may show us the direction of relative displacement, or we may be able to tell by the *offset* of the beds at the fault plane. Both drag and offset are well shown on several of the normal faults illustrated in Figure 10-15.

From offsets, measured either on the ground surface (horizontal plane) or on canyon walls and cross sections (vertical plane), we can at times determine the actual amount, as well as the direction, of relative movement along a fault (Figs. 10-16 and 10-18). By matching up the position of the Strontian and Foyers granite complex, the Moine thrust fault, and the areas of schists and other kinds of rocks on the two sides of the Great Glen fault of northern Scotland (Fig. 10-16), W. Q. Kennedy, a Scotch geologist, has inferred that a horizontal displacement of about 65 miles occurred on this great fault.

In areas of uniformly tilted strata it is not always possible to tell from the offset whether vertical or horizontal movement occurred along a fault. Either kind of movement can, after erosion, produce the same *apparent* displacement. Consider the examples sketched in Figure 10-17. Block A shows a series of tilted strata cut by a normal fault. B shows the same faulted terrain after erosion has removed the elevated block. Note the offset of the beds at the fault surface. C shows a similar series of beds displaced by a strike-slip fault. Compare B with C and note that the same offset can be produced either by normal faulting followed by erosion or by strike-slip movement. From mere examination of the offset of the beds in block B we would be unable to tell which kind of faulting, or whether a combination of the two, had produced the final result.

If two or more kinds of rock masses with different dips intersect, we can definitely tell the direction and amount of relative motion. A simple example is illustrated in blocks D, E, and F of Figure 10-17. Here there is a clear difference between blocks E and F. After elevation of part of block D by normal faulting, erosion has caused outcrops of both dike and strata to migrate down the dip, producing offsets in opposite directions, as shown in block E. With strike-slip movement (block F), all offsets are in the same direction, despite the difference in dip of beds and dike.

Examples like those shown in Figure 10-17 D, E, F are rare; in most cases a dike or other intersecting structure is absent, and we cannot tell the direction of relative motion. The word "apparently" in the definition calls attention to this ambiguity.

Folds and faults are related structures and may grade into one another. Some normal faults die out into monoclines along their strike. Some thrust

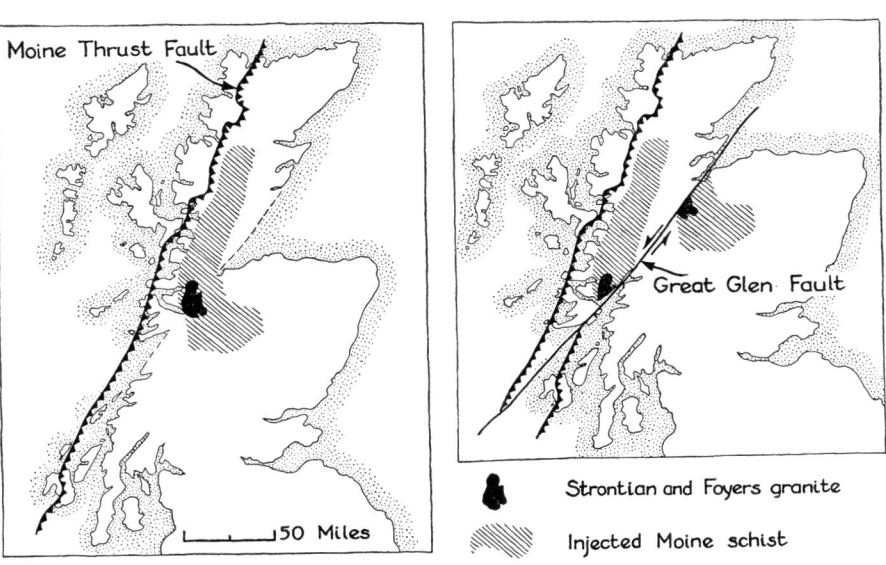

Moine Thrust Fault

Great Glen Fault

50 Miles

Strontian and Foyers granite

Injected Moine schist

FIGURE 10-16. *Maps of Scotland before* (left) *and after movement on the Great Glen Fault. Evidence for this great shift is found in the topography and in the offsets of the Moine thrust fault and the rock units shown on the maps. (After W. Q. Kennedy, 1946.)*

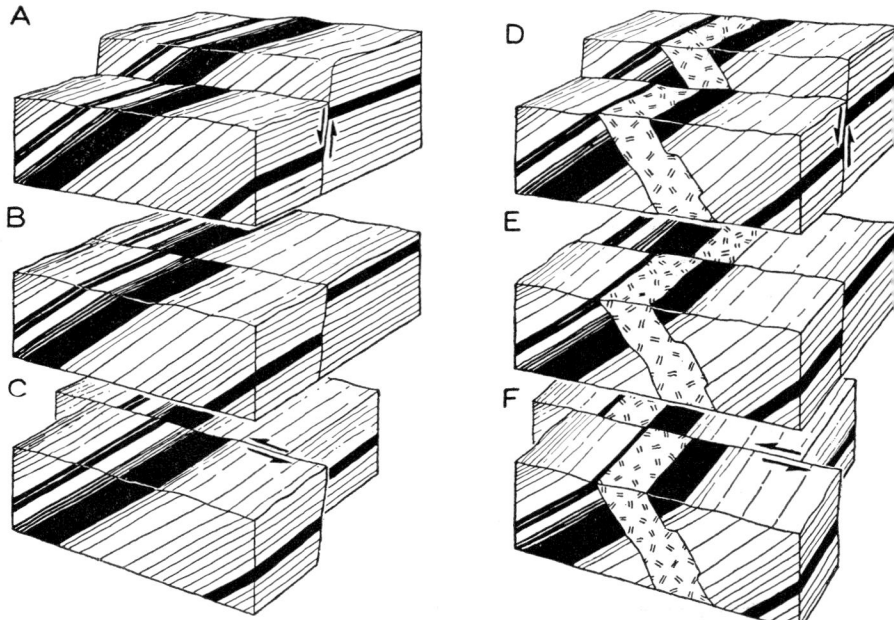

FIGURE 10-17. *Diagrams showing how normal and strike-slip faults can produce identical outcrop patterns (A, B, and C), and how these faults can be differentiated under ideal conditions (D, E, and F).*

faults are merely broken anticlines, although others show no evidence of having grown from folds. Even strike-slip faults may show interesting relations to folds. In Figure 10-18 the abrupt bending of the fold axes as they approach the faults, particularly near the points where the faults are dying out, indicates that both folds and faults were probably produced during the same period of deformation. The bending occurs over a distance of a few miles and so is not likely to be simply drag.

Although there are many exceptions, most normal faults dip steeply—commonly 65 degrees to vertical. All gradations occur, but most thrusts can be grouped into either *high-angle thrusts* or *low-angle thrusts*. The separation is based on whether the dip is greater or less than 45 degrees, but low-angle thrusts generally have dips of less than 30 degrees and high-angle thrusts greater than 60 degrees. Low-angle thrusts are conspicuous in many mountain zones (Chap. 19). Many have displacements measured in miles and attest to great horizontal shortening in the visible part of the crust. Some strike-slip faults also show displacements of many miles. Most strike-slip faults dip steeply and many change direction of dip when traced along the strike. Many normal faults and thrust faults show a little strike-slip

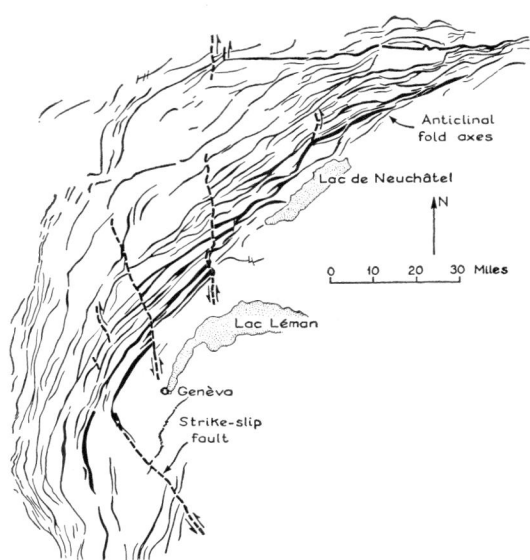

FIGURE 10-18. *Map of the anticlinal folds and strike-slip faults of the Jura Mountains, Switzerland. The weight of a line indicating a fold axis is proportional to the height of the fold. (After A. Heim, 1922.)*

movement; we use the term "strike-slip fault" only when this kind of movement becomes predominant.

The whole classification of faults, however, is artificial; not only are there all transitions between strike-slip faults and other kinds, but many high-angle thrusts, when traced along their strike, steepen in dip, eventually become vertical, and then dip in the opposite direction. Thus a thrust fault may change along the strike into a normal fault; a well-known example is the Uinta fault along the north side of the Uinta Mountains in Utah. Fault classification is also ambiguous if the fault planes have been later folded and warped. If we held strictly to the definitions given on page 207 we would have to call the irregular fault shown in Figure 10-19 a normal fault, for the rocks above the fracture (left side of section) have apparently moved down with respect to the rocks beneath (right side of the section). Detailed

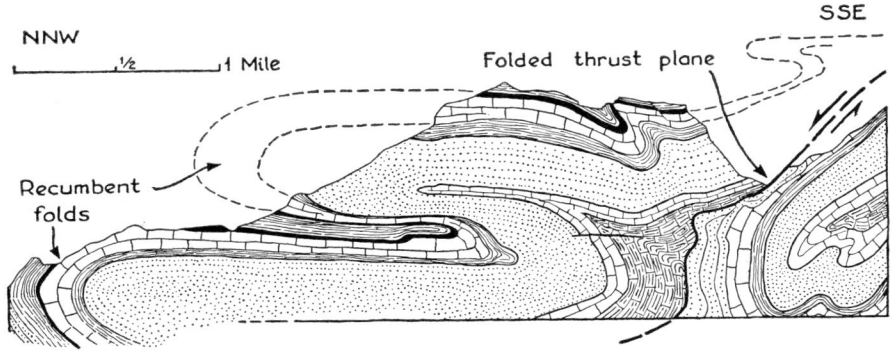

FIGURE 10-19. *Recumbent folds and a folded thrust surface, Swiss Alps. (After A. Heim, 1922.)*

study over a wide area beyond that of the section shows, however, that the fault formerly dipped southward at a low angle. It is really a thrust fault that has been warped and folded after the thrusting movement had ceased.

Unconformities

An *unconformity* is a buried erosion surface. To form an unconformity requires a reversal of the conditions of erosion and sedimentation; an area once being eroded has become one of sedimentation. The surface separating the newly deposited rocks from their basement of older rocks is the unconformity.

Typical examples of unconformities are shown in Figures 10-20 and 10-21. In Figure 10-20 well-stratified sandstones, forming the cliff, rest on granite. The surface of contact between the sandstone and granite is the unconformity. Examination of this contact shows clearly that the granite did not invade the sandstone, for there is no metamorphism of the sandstone by heat, nor does the granite penetrate the sandstone in dikes (Chap. 5, p. 82). On the contrary, the basal layers of the sandstone contain pebbles of the granite, and quartz and feldspar grains derived from the granite abound in it, proving that the sandstone is younger (Chap. 5, p. 81).

In Figure 10-21 an unconformity appears as a conspicuous dark line about

FIGURE 10-20. *Unconformity between granite and sandstone, El Paso County, Colorado. (Photo by N. H. Darton, U. S. Geological Survey.)*

FIGURE 10-21. *An angular unconformity, Wyoming. Tilted and eroded beds of sandstone are overlain by flat-lying clay and sandstone. (Photo by C. J. Hares, U. S. Geological Survey.)*

one-third of the way down from the top of the hill. The tilted beds of sandstone below it end abruptly against the surface of the unconformity. Above the unconformity and parallel with it lie beds of much-less-consolidated sandstone and clay. Their basal layers contain fragments of the underlying tilted rocks. Obviously, the younger poorly consolidated sediments rest on an old erosion surface that had been developed across the older sediments long after they had been laid down, consolidated, and tilted by movements of the earth's crust.

From these two examples it is clear that widespread unconformities record at least three important geologic events: (1) formation of the rocks below the unconformity, (2) erosion of these rocks, usually to a comparatively flat surface, (3) burial of this surface beneath younger strata. Most unconformities also record crustal movement between (1) and (2) giving the elevation necessary for erosion, but under some conditions—the erosion of a volcanic cone, for example—no crustal movement is required. Many unconformities also record a second movement of the crust between (2) and (3) during which the eroded surface was warped below sea level in order to receive sediments; but, again, a land area might be buried by river-borne sediment without movement of the immediate area of the unconformity.

The burial of an unconformity beneath new sediments is slow. Within the lifetime of one man no significant changes may be visible, yet, even today, slow implacable crustal movements whereby broad erosion surfaces are warped beneath the sea to become buried under sediments are in progress at many places. In Chapter 9 we learned of the drowned topography of the Sunda Shelf and of the subsidence in Denmark, Japan, the northeastern United States, and elsewhere. In each of these regions marine sediments are slowly burying the drowned surface, and ultimately, if not interrupted by a new uplift, they will cover it completely. Yet, though the erosion surface may disappear as a topographic feature, it remains as an unconformity and thus permanently records a definite series of events in the history of the earth. Unconformities, and the historical record they reveal, are not easily destroyed. Many are recognizable even after the rocks have been folded and metamorphosed. Only widespread uplift and subsequent deep erosion can obliterate them completely.

Many unconformities are covered by continental instead of marine deposits. Every river forms an unconformity when it sidecuts its banks and deposits a layer of gravel over the cut surface. Locally, river deposits bury extensive areas. In the Chaco region of Paraguay and in western Brazil the steep tributaries of the Parana and Amazon rivers are bringing vast quantities of detritus from the Andes and strewing it over low-lying plains at their foot. These sediments are slowly burying an erosion surface earlier developed across both the sedimentary rocks of the Andean foothills and the plutonic and metamorphic rocks of the Brazilian shield.

Basal Conglomerate.   In general, when a broad erosion surface of low relief is submerged it is at a rate so slow that the ocean shore migrates inland perhaps no more than a few yards per century. Hence the sediments that rest upon the unconformity at one point may differ in age from those which cover it a few miles away, even though the unconformity extends continuously between.

The slow advance of the sea upon an erosion surface produces several characteristic effects. Nearly every erosion surface has a mantle of soil and disintegrated rock and shows at least a few rounded hills and other irregularities. Waves are powerful agents of erosion (Chap. 16). As the sea slowly advances upon the land the waves have ample time to strip away the soil and perhaps some of the bedrock beneath. Irregularities on the erosion surface are planed down by wave attack, finally producing a nearly flat bedrock surface strewn with storm-swept gravel and sand. Finer debris is sorted out by the waves and carried into deeper water. The thin layer of conglomerate and coarse sandstone directly overlying the planed-off rock floor is called a *basal conglomerate*. Such conglomerates are the nearshore deposits of the slowly advancing sea.

In a few places submergence has been rapid enough to bury and preserve the soil and minor irregularities of the former land surface. Most such areas were probably inland arms of the sea, protected from the strong wave at ack of the open ocean. In Chapter 9 we mentioned the drowned drainage features on the floor of the southern end of San Francisco Bay. Twenty-five miles to the west are gravel and coarse sandy beaches of the open Pacific. The same submergence that lowered the floor of San Francisco Bay also affected the coastal section, but here strong wave attack along the open ocean has obliterated all evidence of subaerial topography. All the soil has been removed and a wave-cut platform covered with coarse gravel and sand is developing across the bedrock.

Unconformities covered by continental deposits may show quite different relations. Soils and weathered bedrock are not rare beneath stream deposits. Erosion surfaces buried under lava flows or tuffs may preserve every detail of the former landscape, including such ephemeral features as tree stumps still standing rooted in the position of growth (Fig. 9-11).

Kinds of Unconformities. In Figure 10-22 cross sections of six simple unconformities from different areas are sketched. Note in A and C that the beds both above and below the surface of unconformity are parallel. Such unconformities are called *disconformities.* Many disconformities are difficult to recognize; the series of beds may appear to be *conformable,* that is, to have been deposited continuously without erosional breaks. In C, however, disconformity is certain because Silurian strata are absent, as proved by fossils collected from the beds immediately above and below the unconformity. Obviously, the area could not have been under the sea throughout all Silurian time without receiving some deposits. In A the break recorded by the unconformity is vastly greater, embracing nearly all of geologic time since the Cambrian.

Unconformities in which the beds above the unconformity transgress the eroded edges of folded and tilted beds, like those shown in Figure 10-21 and in B and D of Figure 10-22, are called *angular unconformities.*

In another common kind of unconformity—Figures 7-6, 10-20, and E and F of Figure 10-22 show examples—bedded rocks rest on an eroded surface of plutonic or metamorphic rocks such as granite, gneiss, or gabbro. The term *nonconformity* has been used for this kind of unconformity, but, unfortunately, this term is also used in some geologic literature to describe angular unconformities.

Dating Geologic Events by Unconformities. The examples sketched in Figure 10-22 also illustrate how unconformities can be used to date various geologic events. For example, at Siccar Point, Scotland (B of Fig. 10-22), the Ordovician and Silurian beds were obviously folded before the unconformity was formed, because the overlying Devonian sandstone is unaffected by the folds. On the other hand, the beds in central Alabama (C of Fig. 10-22) were folded after the development of the unconformity, as both sets

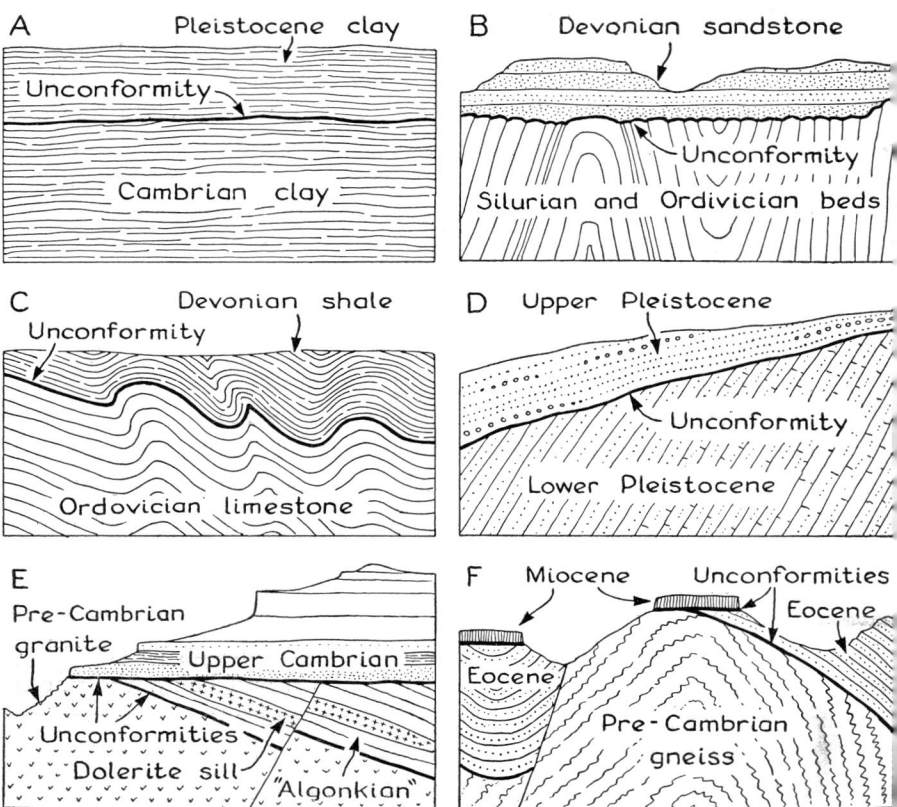

FIGURE 10-22.   *Cross sections showing different kinds of unconformities. A, Baltic region of Russia; B, Siccar Point, Scotland; C, central Alabama; D, Palos Verdes, California; E, Grand Canyon, Arizona; F, central Washington.*

of beds as well as the unconformity itself have been arched by the folds. The relations show merely that the folding was later than the deposition of the Devonian shale, and may have occurred at any time between Devonian and the present. At Siccar Point, however, the folding must have occurred after the deposition of the Ordovician and Silurian strata but before the Devonian sandstone had accumulated. The unconformity shown in D definitely proves that the tilted lower Pleistocene beds of southern California were folded during the Pleistocene because undeformed Pleistocene beds lie above the unconformity.

In the unconformity shown as E we see that the "Algonkian" of the Grand Canyon was invaded by a dolerite sill and cut by a normal fault before Upper Cambrian time, because the basal nonconformity of the Upper

Cambrian truncates these features as well as the tilted "Algonkian" beds and the granite upon which the "Algonkian" itself rests noncomformably.

Similarly, the relations of the two different unconformities shown in F—one beneath the Eocene beds, the other at the base of the Miocene lava flow—tell us that the fault cutting these unconformities has undergone two periods of movement—one during post-Eocene but pre-Miocene time, the other after the faulted area had been eroded to a plain and the Miocene lava spread out over it.

Time Significance of Unconformities. Every unconformity marks a time interval not represented by sedimentary deposits at that particular spot. Either there were no strata deposited here during this time interval because the land was being eroded, or, if any deposits were laid down early in the time interval, they have been eroded away during the later part. In Chapter 8 it was noted that the breaks in the geologic time scale, as built up in Western Europe, represent times of local nondeposition, although many are recorded by "transitional series" elsewhere. These breaks correspond with the major unconformities in the Western Europe section.

It is important to try to estimate the time value of an unconformity. A common fallacy is to assume that an angular unconformity necessarily implies a greater time lapse than a disconformity. True, in some areas the unconformities that are most widespread and that show the greatest time lapse happen to be angular unconformities, but this is not true in general. Contrast, for example, the unconformities shown in A and D of Figure 10-22. At several places in the Baltic region clays deposited in Pleistocene lakes lie directly on marine clays that contain Cambrian fossils. The interval of time represented by the disconformity includes all of the Mesozoic and nearly all of the Paleozoic and Cenozoic—a gap of about 400 million years. Yet, in some places the unconformity can hardly be located, so similar are the Cambrian clays to those of the Pleistocene. In Figure 10-22D, however, the marine sediments both above and below the angular unconformity are of Pleistocene age, as shown by their abundant fossils. Thus the events during Pleistocene time in southern California included deposition of the lower Pleistocene strata; folding, upturning, and erosion of the newly deposited beds to a comparatively flat surface; sinking of the erosion surface below the sea; and deposition of the upper Pleistocene strata. All of these local events have taken place in perhaps not more than one million years.

Such examples make it clear that the angle between the beds above and below an unconformity is independent of the time represented, and depends only on the nature of the deformation of the older beds. The time interval can only be established by the fossil content of the adjacent rocks. One reasonably safe generalization, however, is that a long lapse of time—perhaps equivalent to at least a geologic epoch—is recorded by a nonconformity where sedimentary rocks lie above coarse-grained plutonic and metamorphic rocks. Since such rocks are formed at considerable depths within the crust, to expose them at the surface must have required a long period of erosion (Chap. 5).

Conclusions

The picture of the earth that has been developed in this chapter is a lively one. The earth's crust is not static and dead. Clearly recorded in its structure is the evidence that many rocks do not long remain in the horizontal position in which they were deposited. They have been warped upward and downward, thrown into folds, broken and displaced by faults, invaded by igneous bodies, and crushed and recrystallized into metamorphic rocks. Unconformities allow us to date many of these events and also show that the earth movements responsible for the complex structure of the crust did not occur all in one great paroxysm, but have been going on throughout geologic time.

In Chapter 9 evidence was presented which proves that such movements take place very slowly, and in Chapter 8 we learned of the great span of geologic time. When we take into account the vast amount of time that must be involved in these movements, it is not surprising to find (as indicated in Chap. 3) that the earth's crust is still essentially in isostatic balance, despite the great changes in position of rock masses that have been brought about by the constructional processes of folding, faulting, and warping on the one hand, and by the destructional agencies of erosion on the other. By slow plastic flow of subcrustal material the earth tends to restore the disturbances in its isostatic balance generated by the forces that elevate mountain tracts and by the agents that erode them away. But perhaps it is significant that the greatest gravity anomalies that have been measured, and hence the segments of the earth's crust that depart farthest from isostatic equilibrium, are found in the Dutch East Indies, a region (Chap. 9) where folds are still rising and where the mysterious but implacable forces that fold and elevate mountain ranges are even now actively at work.

Facts, Concepts, Terms

WARPS

    Structure contour maps

Exaggerated scale cross-section
Continental plates

FOLDS

Plunging, upright, asymmetrical, overturned, and recumbent folds
Folded belts
Mountain roots

JOINTS AND FAULTS

Normal, thrust, and strike-slip faults
Apparent versus real displacements on faults

UNCONFORMITIES

Disconformities, angular unconformities, nonconformities
Basal conglomerate
Dating events by unconformities
Time involved in an unconformity

STRUCTURE SYMBOLS

IMPLICATIONS REGARDING ISOSTASY FROM WARPED AND FOLDED ROCKS

Questions

1. How do you know that the 5,000-foot difference in elevation of the top of the Dakota sandstone between Shelby, Montana, and Williston, North Dakota, is the result of warping since deposition of the sandstone, rather than uniform deposition in an ocean which was shallow near Shelby and 5,000 feet or more deep near Williston?

2. On pages 194-198 the method of getting control for drawing structure contours on a deeply buried bed is described. Describe how it would also be possible to get control to project above the ground surface the former position of a bed that has been eroded completely from the top of an anticline.

3. Draw a geologic map showing two anticlines and an intervening syncline that have been eroded almost to a flat surface. All three folds plunge to the north, and all are asymmetrical. Four sedimentary formations are exposed on the map.

4. Draw a geologic map showing a syncline plunging to the south which has been cut by an east-west normal fault that dips north. Assume that erosion has reduced the area to an almost flat surface.

5. Why is it more common to find overturned folds grading into thrust faults along their strike than into normal faults?

6. In the field, what criteria would you use to tell an angular unconformity from a thrust fault?

7. In the field, what criteria would you use to distinguish between a noncon-

formity developed across a granite mass and an intrusive contact formed by the invasion of the sediments by molten granite magma?

8. An extensive but thin basal conglomerate contains Lower Jurassic fossils in one locality but Middle Jurassic fossils when traced westward 230 miles. How is this possible?

9. Why are broad warps difficult to detect in mountain belts?

10. Draw *one* cross-section showing *all* of the following features:

    a) A series of folded marine sediments that lie nonconformably on granite and metamorphic rocks.

    b) A series of nearly flat lava flows that lie with angular unconformity on the folded sediments.

    c) Two thrust faults that are older than the lava flows but younger than the sedimentary rocks.

    d) A normal fault that is younger than the lava flows.

    e) A dike that is younger than the thrust faults, but older than the normal fault.

Suggested Readings

1. Bucher, W. H., *The Deformation of the Earth's Crust*, Princeton University Press, Princeton, N. J., 1933.

2. Umbgrove, J. H. F., *The Pulse of the Earth*, 2nd ed., Martinus Nijhoff, The Hague, 1947.

3. Cloos, Hans, *Einfuehrung in die Geologie*, Borntraeger, Berlin, 1936.

11. *Downslope Movements of Soil and Rock*

In Chapter 7 varied examples of erosion have been described, and in Chapters 12-16 more searching analyses will be made of the erosive effects of geological agencies such as streams, glaciers, waves, wind, and ground water. These agents are ultimately powered by gravity. Gravity also moves rock more directly in ways to be considered in the present chapter.

Varieties of Downslope Movements

The masses of rock and soil moved directly downslope by the force of gravity range in size from tiny bits of rock rolling down fissured cliffs, or small masses of water-soaked soil gliding down a hillside, to great landslides thousands of feet across crashing down the slopes of mountain ranges.

Some downslope movements are slow distortions (*creep*) of soil, rock, or other incompletely consolidated materials. Typically, they take place on grass-covered slopes where there may be no obvious surface indications of movement.

A second group of *slow slides* and *slow debris flows* differs from creep by showing at the surface distinct boundaries of the individual slides or flows. Such a slide moves as a unit bounded by a well-marked slip surface. The fragments of a debris flow are jumbled together during movement. Slides and flows rarely cover whole hill slopes, but are localized into tongues or broad rounded masses. They leave gaping cracks or rough scars at their sources. Slow slides and debris flows range from a few feet to several miles in length.

A third group, characterized by rapid movement, comprises *rapid debris flows, rapid landslides,* and *rock falls.* Rock falls include small blocks loosened from steep rock surfaces and rattling down in brief free falls to join the accumulations of similar blocks called *talus piles.* Most of the rapidly moving masses are small, but some great rock falls and landslides have destroyed villages and dammed rivers.

Rainwash

Rainwash on hill slopes marks the initial stage in the movement of the surface waters that are gathered later into well-defined streams and rivers. It is considered here because its effects cannot always be distinguished from movements of water-soaked soil, or even from the downslope creep under gravity of relatively dry masses of soil and rock.

During a heavy rainstorm, hill slopes may be covered by a thin sheet of moving water heavily charged with mud, silt, and humus, churned up by the pelting rain drops. Some of the water soaks into the ground; some moves down the slope. Generally, such sheets of flowing water quickly separate into numerous small, shallow rills, coursing almost straight downslope, but often dividing, reuniting, and finally losing their individuality at the edge of a stream or on the nearly flat valley floor (Fig. 11-1). Most of these shallow rill-courses are short-lived furrows that may be filled and obliterated in a single season, but some grow into larger, steep-walled permanent trenches called *gullies.*

FIGURE 11-1. *Rill furrows in fresh volcanic ash from Paricutin volcano, Mexico.* (*Photo by Konrad Krauskopf.*)

FIGURE 11-2. *Effect of creep on decayed boulders. (Photo by S. Capps, U. S. Geological Survey.)*

Creep

Creep is the most widespread of all downslope movements. Typically, it is a matter of slow downhill distortion in soil and mantle rock, or in thin-bedded or poorly consolidated sediments. Ordinarily, the movement dies out gradually at shallow depth without developing a slip surface at its lower boundary. Distortion caused by creep is shown by many rock bodies whose previous shape or structure is known. For example, decayed boulders (Fig. 11-2) that once must have been nearly round are now streaked out into thin ribbons of varicolored soil. Creep is also well displayed in the bending of thin-bedded, steeply inclined strata as they come to the surface on a steep hillside (Fig. 11-3). On many slopes, creep is perceptible through its effects

FIGURE 11-3. *Bending of thin vertical strata by creep, Washington County, Maryland. (Photo by Stose, U. S. Geological Survey.)*

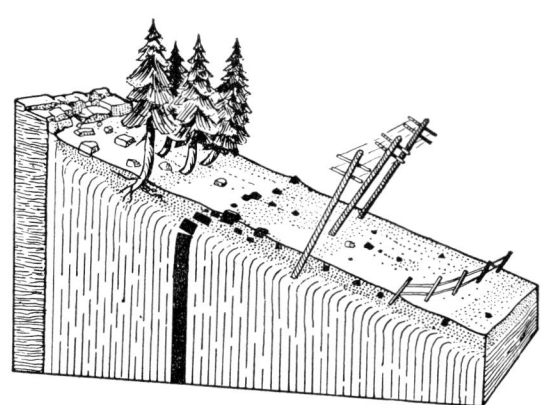

FIGURE 11-4. *Common effects of creep.* (*After C. F. S. Sharpe,* Landslides and Related Phenomena, *Columbia University Press, 1938.*)

on rock structures, trees, posts, buildings, etc. (Fig. 11-4). Observations extending over several years in climates with marked wet and dry seasons show that soil creeps most actively during the wet season and at the beginning of the dry season.

The *mechanisms of creep* are somewhat conjectural. Wetting of clay-rich soil—or freezing of water contained in pores in the soil—causes the soil to swell upward, as shown in Figure 11-5. A particle of soil at *a* is pushed upward at right angles to the surface by the swelling, and comes to rest at point *b*. On drying of the clay, or melting of the ice, the soil contracts again, but the particle will not return to point *a,* but will move vertically to *c,* or, if the soil is wet enough to flow slightly under the action of gravity, it may glide still farther down the hillside. By repeated wetting, or freezing, the particle is moved, step by step, farther and farther downslope. Because ice expands 9 per cent on freezing, the upward swelling may be very marked in regions with cold winters. It is called *frost heaving.* The thickening of ice layers in cracks after repeated freezing adds to the heaving. Soil crusts may be raised several inches and paved roads may be heaved upward and broken into rubble by the expansive force. Upon thawing, the shrinkage does not follow exactly the original direction of expansion because the pull of gravity causes a net movement downslope.

Other processes commonly involved, varying with the climatic conditions, include wedging by plant roots; the moving of soil by earthworms, rodents, and other burrowing animals; and the downslope movement of the soil beneath the feet of animals such as sheep or cattle. Any random movement of loose material gives gravity a chance to make a net displacement downslope.

Creep passes imperceptibly into downslope movement of more definitely bounded masses. In the California Coast Ranges a black sandy clay soil develops on several types of sedimentary rock. On many moderate slopes (e.g., 15°–20°) this soil and broken rock are intimately mixed but are sharply separated from the fresh rock beneath by a smooth surface, suggesting that the surficial layer in slow downslope transit across the underlying rock has

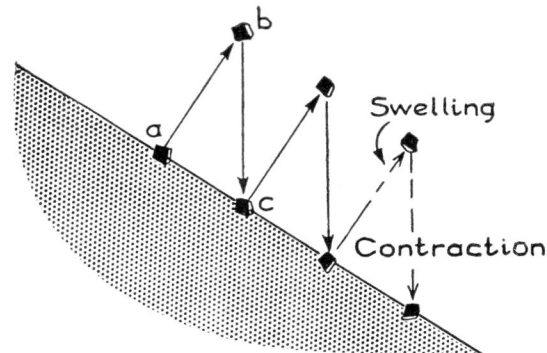

FIGURE 11-5. *Downslope creep. The arrows show the motion of a particle with alternate swelling and contraction.*

scraped it clean. Marked movement in the last 40 or 50 years is shown by the tilting of fence posts and telephone poles, but the lack of scars or irregularities on the surface indicates that the movement is general rather than concentrated in landslide tongues.

Solifluction

In arctic and subarctic regions, mass movements called *solifluction* (literally: "soil flow") are intermediate between creep and debris flows. Here, frost action and other forms of weathering produce abundant rock fragments of all sizes, including some material fine enough to be called soil. When saturated with water, these accumulations of debris may move slowly down

FIGURE 11-6. *Solifluction lobe on Victoria Island, Canada. The rows of stakes were used to measure the rate of flow of the lobe. (Photo by A. L. Washburn.)*

moderate slopes (5°–20°) and along the floors of steep valleys in the form of sheets, lobes, and tongues. A. Lincoln Washburn, a student of the Arctic, has suggested, from studies on Victoria Island northwest of Hudson Bay, that most of the movement takes place immediately after the thawing of the frost-heaved surface in spring. Rows of stakes Washburn had driven in the lobe shown in Figure 11-6 indicated slight local tilting and forward movements during the summer of 1940. The maximum forward movement of 1¾ inches occurred almost entirely in the 33 days beginning June 14. Apparently in arctic and subarctic regions, solifluction proceeds by very slow flow of water-saturated soil, year after year.

Solifluction is also prominent above timber line on mountains in temperate regions.

Rock Glaciers

In southwestern Colorado and elsewhere, the flat floors of large amphitheaters are strewn with fine debris marked by surficial ridges (Fig. 11-7) somewhat like those on solifluction lobes, but also resembling those that form on rock falls (Fig. 11-15). These ridged debris piles have been called *rock glaciers,* perhaps because the ridges resemble the debris ridges on glaciers. Rock glaciers probably do not result from rock falls or rapid slides, because the debris is relatively fine grained and the ridges are sinuous and irregular, but the mechanisms of their motion are somewhat uncertain.

FIGURE 11-7.   *Rock Glacier, Engineer Mountain, Colorado. (Photo by W. Cross, U. S. Geological Survey.)*

FIGURE 11-8. *Slide near Orinda, San Francisco Bay region, California. Four-lane highway and large dump trucks give scale. (Photo by Bill Young, courtesy of the* San Francisco Chronicle.)

Slow Slides and Flows

If part of the soil and loose rock on a hillside moves downslope a little more rapidly than the general hillside creep, it is set off from its surroundings by a gaping crack at its upper edge, and by cracks or ridges at its sides (Fig.

11-8). If such a mass glides more or less as a unit along a sharply defined basal slip surface, it is a slide; if its constituent parts are broken up and jumbled together during movement, it is a *debris flow*. Many such masses are debris flows throughout, but others start as slides and become broken and mixed accumulations farther downslope. Some slow slides and flows are very large. An example from Wyoming will serve to show the distinctive features of the group.

Gros Ventre Debris Flow

The Gros Ventre River (pronounced Grow Vont) is a minor westward-flowing tributary of the Snake River south of Yellowstone Park, in Wyoming. In

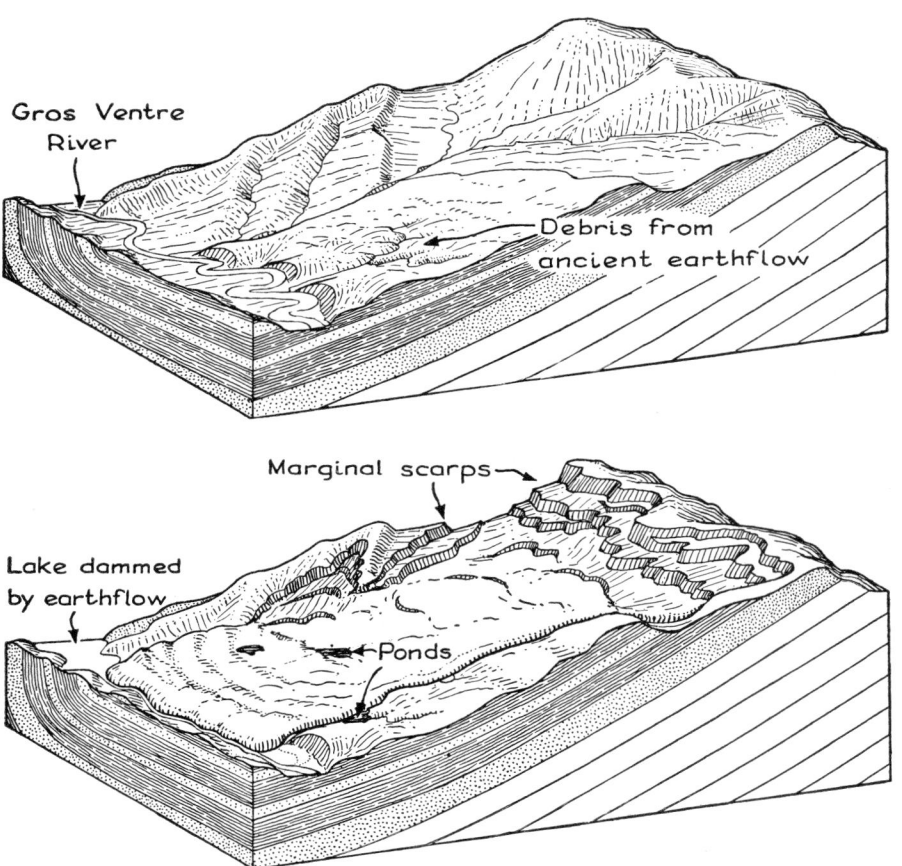

FIGURE 11-9.   *The south slope of the Gros Ventre River valley before* (top) *and after* (bottom) *the 1909 debris flow. (After Eliot Blackwelder, 1912.)*

the spring of 1909, the Gros Ventre River was dammed in midcourse by rock debris that had begun to flow as early as May, 1908, and to move slowly down the moderate (10°–20°) slope of the Gros Ventre Mountains on the south side of the river (Fig. 11-9). The rocks involved in the sliding were soft shales, with some interbedded thin sandstone and limestone layers. The shales had been thoroughly soaked by heavy rains. Some of the strata probably slipped along the bedding planes, which almost parallel the land surface. The sliding was imperceptible to an observer but wrought notable changes in the landscape within a few weeks. Telephone poles tilted slowly downhill, snapping the wires. A wagon road paralleling the river soon became so hopelessly twisted and broken that repairs were futile. Eventually, it was so obliterated by movement that traces of it were hard to find.

The flow did not move in one mass, but in sections, beginning at the east side and spreading week by week. The sliding and crumbling mass moved fastest in the wet spring months of 1909 and slowed noticeably by autumn. Nevertheless, incessant movement toward and into the river bed continued through 1910. It almost entirely ceased in 1911; the river was then able to cut 10 feet down into the debris dam and partially drain the lake that had formed on the upstream side.

Numerous gaping fissures formed at and near the places of initial movement on the east and south sides of the Gros Ventre debris flow. Further down, where the material of an earlier flow, faintly indicated in Figure 11-9, was set in renewed motion, earthen domes swelled up and broke open in broad cracks. The mass as a whole thickened toward the river, developed a very irregular surface, and was churned into a jumbled mixture of clay and coarser material.

Slow Slides

Many slides break up far less completely than did the Gros Ventre flow. A mass of rock along a cliffed shore may move on a flat, water-lubricated slip surface for many feet without breaking up at all, leaving only a gaping fissure on the inland side as evidence of movement. In weaker materials, small slide blocks commonly rotate on cylindrical slip surfaces (*soles*), as shown in Figure 11-10, forming a depression at the head of each slide unit.

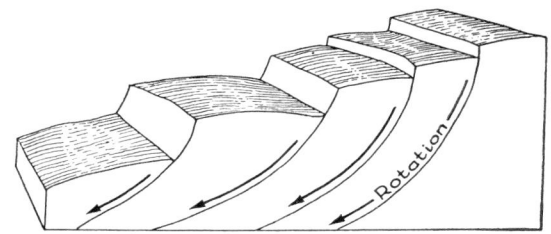

FIGURE 11-10. *Rotation of slide blocks on curved soles.*

Rapid Slides, Flows, and Falls

Small Rapid Slides

The great majority of rapid gravity movements involve only small volumes of soil or rock, but the aggregate effects are large. In humid, temperate regions, small rapid slides with barely recognizable slip surfaces are extremely common in soil or weak sedimentary rock, especially on steep grassy slopes and after unusually heavy rains. The depression, or scar, left by a typical slip may be 10 feet wide and 20 feet long in the direction of slope. At the base of the scar, the soil piles up in more-or-less crumpled and disordered masses (Fig. 11-8). The distinctive features of such slides are present in the miniature examples in dry volcanic ash shown in Figure 11-11. One of these little slides is very fresh; the others have lost their original sharpness.

Talus Piles

Quantitatively, the most important falls of rock are countless multitudes of small fragments, from a fraction of an inch to a few feet in diameter, that drop from cliffs or steep rock surfaces and accumulate in talus piles at the

FIGURE 11-11. *Slides in a 3-feet high stream bank cut in fresh Paricutin volcanic ash. (Photo by Carl Fries, Jr.)*

bases of the steep slope or cliff (Fig. 11-12). A talus pile maintains a fairly constant surface slope as it grows. The slope angle, commonly about 30°, is called the *angle of repose* because it is the maximum angle at which the material will remain stable.

In their quantitative importance, small rapid slides and talus piles are perhaps second only to creep among gravity movements. The widening of steep-walled valleys in arid regions is largely by the fall and continued streamward movement of large and small talus fragments.

Rate of Talus Formation.   Those who climb steep mountain slopes of closely jointed rocks know well the sound of, and also the danger from, falling rock fragments that come from high aloft and bound erratically down long talus slopes. Some talus piles grow rapidly, others slowly. The mere presence of talus is no proof of rapid change in the configuration of the cliffs above. Great talus piles at the foot of granite cliffs in southern Arizona are composed of huge blocks so thoroughly weathered that they could not have withstood the impact of fall. They must have weathered in place, on the talus

FIGURE 11-12.   *Talus slope at the base of a basalt butte, Grand Coulee, Washington. (Photo by courtesy of the Washington Department of Conservation and Development.)*

FIGURE 11-13. *Changes (lettered points) on Castle Rock, Kansas, between 1894 and 1941. (Courtesy of H. T. U. Smith. Upper photo by Williston; lower photo by H. T. U. Smith.)*

pile. Weathering is so slow in this climate that a long time is implied for the talus growth.

In a few places, the rate of spalling has been roughly determined. Figure 11-13 shows Castle Rock, Kansas, made of soft sedimentary strata. In the semi-arid Kansas climate, it has suffered the loss of perhaps 3 or 4 per cent of its volume between 1894 and 1941. At this rate, Castle Rock might be consumed in 1,200 to 1,600 years.

Large Rock Falls and Mudflows

Large slides of rock occasionally avalanche down steep slopes, or even break loose and hurtle through the air. Large and rapidly moving mudflows also occur, and, if well lubricated, may move swiftly over very gentle slopes. Several such rapid slides, flows, and falls have destroyed villages or portions of cities. From the many such catastrophes, we select for description a rock slide and fall in the Swiss Alps, and mudflows in Norway and Switzerland.

Elm Rock Slide and Fall. Perhaps the best known rock fall occurred at Elm, a village in the northeastern Swiss Alps, in 1881. A steep slope 2,000 feet high, the prow of a ridge between two valleys, had been undercut half way up in quarrying for slate. In the course of one and one-half years, a curving fissure 30 feet deep had been formed, extending 1,100 feet above the quarry and outlining the future slip mass (Fig. 11-14). This fissure was approximately perpendicular to the stratification and foliation of the rocks. In the late summer of 1881, the local runoff from heavy rains poured into the fissure, saturating the shattered rocks. Late one September afternoon, two small slides occurred, starting just above and on either side of the quarry. A few minutes later, the whole mass outlined by the fracture crashed down, filled the quarry, and shot part way across the narrow valley as a free-falling

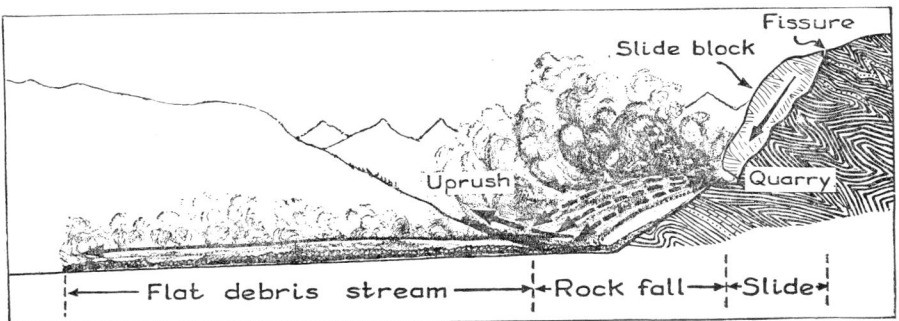

FIGURE 11-14. *Cross section of the rock slide and fall at Elm, Switzerland, showing the original position of the slide block. (After A. Heim, 1882.)*

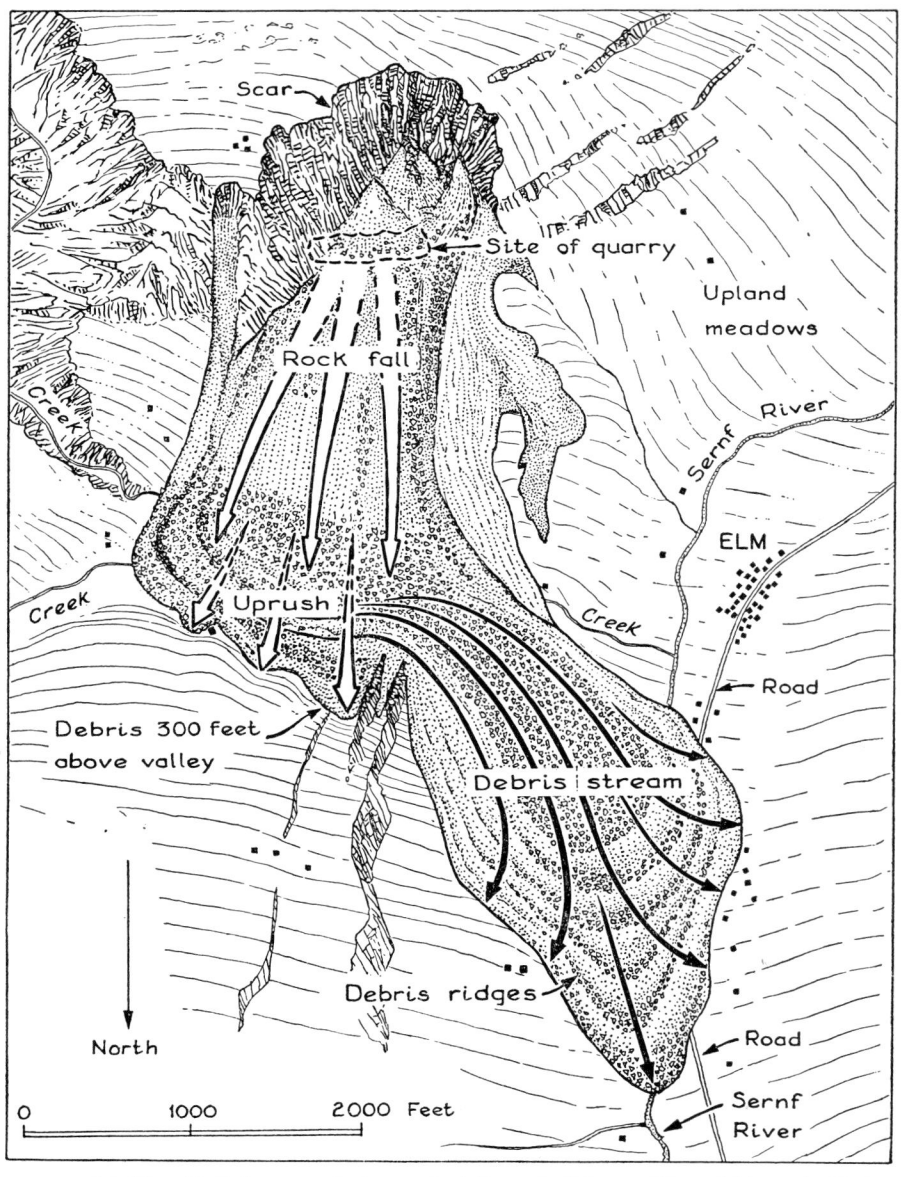

FIGURE 11-15. *Relief map of the rock fall and slide at Elm. (After A. Heim, 1882.)*

rock avalanche (Figs. 11-14 and 11-15). On striking the valley floor, it rushed obliquely up the opposite slope to a height of 300 feet, turned, and shot down the valley in a debris stream, destroying houses and everything in its path, and killing 115 people. Thirteen million cubic yards of rock fell an average of about 1,450 feet, and spread as rubble over a third of a square mile to a depth of 30 to 60 feet. A dense cloud of dust covered the scene for several minutes.

All observers agreed that, after the sliding mass had fallen steeply to the quarry floor, it shot forward into free fall. The villagers could see across the valley beneath the cascading torrent of shattered stone to the hillside beyond.

The Elm debris was composed largely of fragments a few inches in diameter, including much intermingled soil and rock dust. Many larger blocks, 5 to 20 feet across, were scattered through it, especially in the central, higher, and later portions of the debris tongue. The margins were abrupt. The top of the debris stream was hummocky and irregular, and low ridges festooned its surface. By 1928, 47 years after the event, almost all the devastated area had been restored to pasture and potatoes.

From eyewitnesses' reports, it was estimated that the debris at the front of the tongue traveled one and one-half miles in 45 or 50 seconds. Physical considerations make this appear reasonable. The physical calculations are based on the acceleration of gravity. In free fall the acceleration or increase of velocity is 32 feet per second in each second. The short preliminary sliding and the free fall at Elm were calculated to have taken 17.2 seconds, and the velocity when it hit the valley floor about 186 miles per hour. Including the time of braking to a stop down valley, the whole duration was calculated at 53.4 seconds—in good agreement with estimates by eyewitnesses. The average velocity of the farthest traveled blocks was calculated to have been 93 miles per hour.

A rock fall such as that at Elm illustrates the physical concepts developed in Chapter 7 concerning the principle of the *conservation of energy*. Before the fall, the poised rock had energy of position, or *potential energy,* measured by its mass and its elevation above a convenient reference plane. During fall, it lost potential energy and gained *energy of motion,* measured by mass and velocity. When brought to rest by friction, energy of motion was changed to *heat energy*. The law of conservation of energy requires that the sum of the potential energy, energy of motion, and heat energy resident in the rock mass or developed from it remains constant. Hence, the heat energy developed by braking must have been equal to the potential energy lost during the fall. Geologists are necessarily concerned with the source of energy in all geological processes. In dealing with erosion and downslope movements, the principle source is that of the earth's gravitational field, as we have just seen.

Rapid Mudflows. The Elm fall affected hard rocks, and was made possible by a steep slope. Masses of mud, on the other hand, may break through weak barriers and flow fairly rapidly for long distances even on gentle slopes.

In northwestern Europe and eastern Canada, many valleys are floored with unconsolidated clays and sands, extending many miles inland and rising to elevations of several hundred feet. These soft sediments become saturated with water after the spring thaw. A mass of mud as much as a square mile in area may suddenly break through a slightly more solid barrier, and roll down valley as a destructive mudflow, the front moving as much as 3 or 4 miles in a single hour, and then piling up in a temporary dam of sandy clay, behind which a lake forms that in turn overflows and breaches the dam. In the source area, such a mudflow leaves a clean-cut scar, with rough floor and steep walls a few feet or a few tens of feet high. One night in 1893, a mudflow of this kind moved down a Norwegian valley so fast that 111 persons were caught by it and lost their lives. Others escaped after being carried long distances in their wooden houses.

In arid regions, still another kind of mudflow takes place. A thick mass of silt, sand, and coarser debris scoured by a flood from the canyon floor may roll forward for miles, damming, piling up, breaking through, and finally coming to a stop by gradual thickening through loss of water, or by spreading over a plain.

Some of the largest and most destructive mudflows have developed in the pyroclastic debris ejected from active volcanoes. They are described in Chapter 17.

Underwater Flows. Of all downslope movements of solid or semi-liquid material the most obscure and difficult to study are those which glide along the floors of lakes or seas. One thoroughly studied example started at the shore of the Swiss lake of Zug and squirted out under water along the lake floor.

Zug Underwater Flow The source area in the city of Zug extended from a lake front retaining wall for about 200 feet inland to a street lined with buildings. Most of the area was filled land (Fig. 11-16). In the two years after the retaining wall was built, water began to appear in previously dry cellars behind it, and, in the spring of 1887, the ground had settled and the pavements had cracked slightly. The first disastrous movement, in mid-afternoon of July 5, dropped a small section of the retaining wall and three houses beneath lake level and caused several deaths. A half hour later, wooden piles from the broken retaining wall suddenly rose to the lake surface a hundred yards offshore, showing that lateral movement as well as sinking had occurred. The main flow came three hours later, with streets and houses suddenly settling beneath the lake level. The average drop was about 25 feet, and some buildings moved 30 to 60 feet lakeward.

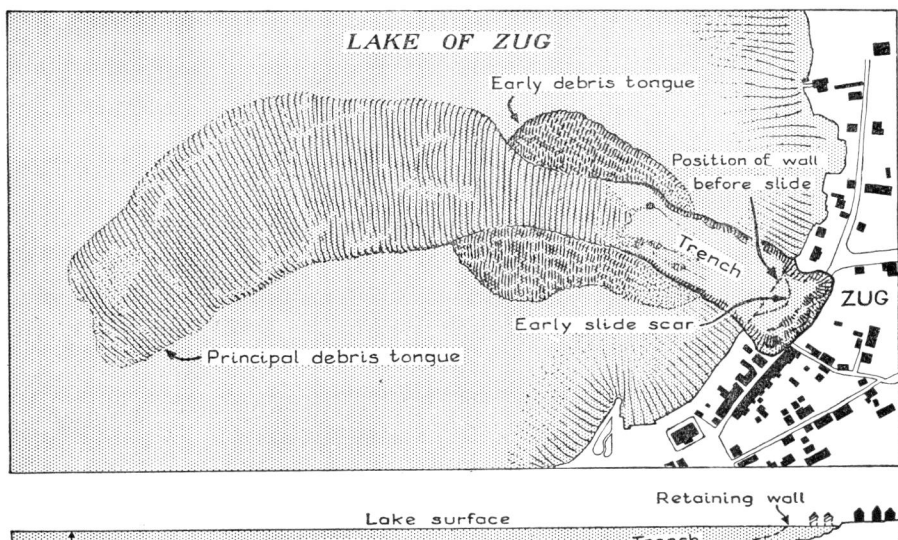

FIGURE 11-16. *Map and cross section of the underwater flow at Zug, Switzerland. (After A. Heim, 1888.)*

The material that flowed out into the lake from beneath the source area was water-saturated silt and fine sand from the submerged part of a small stream delta. The silt flow excavated a trench 200 feet wide, up to 20 feet deep and extending almost 1,000 feet along the lake floor in the similar silt and sand on the bottom of the lake. From the end of the underwater trench a new deposit of sandy silt, from the source area and trench, extended out over the lake floor as a thin debris tongue 2,500 feet long, with an uneven hummocky surface (cross section, Fig. 11-16). The end of the tongue was 140 to 150 feet below the land surface in the source area, so the slope on which it flowed was less than 3°. The deposit was more than 15 feet thick in places, but averaged less. About 200,000 cubic yards of material were moved, somewhat less than half from the land; the remainder eroded from the lake floor by the flow. The Zug flow was a gravity current of silt and sand, heavier than the lake water. It moved down the slope of the lake floor somewhat as a stream of water on land moves downslope beneath a cover of air.

Several other similar debris flows into lakes have occurred in Switzerland in historic time, including a disastrous earlier flow at Zug, March 4, 1435. All appear to have been caused by the loading of weak sediments by artificial fill and buildings. Because started by man's modification of existing conditions, their geological significance is somewhat doubtful. Nevertheless, there are abundant evidences of similar flows in marine sediments, so we shall refer to them later in connection with submarine topography.

Distinction Between Products of Slow and Rapid Movement

The division of slides and debris flows into two groups, one whose movement is imperceptible to the casual observer; the other, more rapid, is satisfactory if the movements are actually observed. The great majority of slides and debris flows, however, are recognized only by their effects. In these, the distinction may be difficult or impossible. Rapid movement usually leaves a well-marked scar or amphitheater in the source area, and commonly a symmetrical pile of rather finely divided debris downslope, often with transverse or concentric corrugations (cf. Fig. 11-15). Slow *slides,* because of their less severe impacts with the valley floors, are less likely to break up into fine fragments, except near their lower margins. Most are composed of units that can be mapped, like fault blocks.

Prehistoric Slides and Other Downslope Movements; Reconstruction of Events

Ancient Slides and Falls

Thousands of ancient slides have been recognized in all parts of the world. Perhaps the largest is just over the hill from Elm, the enormous Flims landslide (at least 15,000 million cubic yards—1,000 times as big as that at Elm). This mass probably started movement as a unit, but has now been thoroughly broken up, especially some portions which had traveled the farthest. It started to move downslope along the slip plane of an Early Tertiary thrust fault. Long ago, probably in Late Pleistocene time, it blocked the valley of the upper Rhine, forming a lake. Now the Rhine has eroded the landslide dam, cutting a gorge 2,000 feet deep and 9 miles long. All these things can be seen clearly: the hummocky surface of the great slide, somewhat modified by erosion; the well-marked margins of the landslide dam; and the sudden change in the Rhine from a broad, open upper valley to the steep-walled gorge through the slide, and then to the open valley below. One thing remains uncertain: Did the slide move rapidly or slowly? A hint is perhaps furnished by the hummocky surface, much like that of the slow Gros Ventre slide.

A more complex assemblage is found where the San Bernardino Mountains rise above the southern margin of the Mojave Desert in southern California. A partly eroded landslide mass is spread out below the outcrops of two thrust faults. On the lower of these faults, granite overrides Late Tertiary sediments; on the upper, crystalline limestone overrides the granite. The major slide block, composed of limestone, also lies on Tertiary sediments; it may have reached its present position in Late Tertiary or Pleistocene time. The most puzzling element in the assemblage is a broad lobe of poorly cemented breccia that extends 4 or 5 miles out over the flat desert floor (Fig. 11-17) and is made up in most places exclusively of pebble-sized limestone blocks. It is somewhat eroded—one stream from the mountains has cut entirely through it—but its margins are still 25 to 100 feet high and sharply marked. Can this be the debris from a giant rock fall, perhaps 100 times as large as that at Elm? Several features suggest this, particularly the blocks being exclusively of limestone with no admixture of any of the other rocks of the region; the piling up of the breccia against low hills far out on the desert, with a slight deflection around the hills somewhat like that at Elm; and, finally, at the edge of the low hills, the exposure by erosion of layers of two underlying breccias, the upper composed of sheared and crushed granite, the lower of the Tertiary sediments. A somewhat similar but less sharply defined segregation by rock types was present in the Elm debris.

FIGURE 11-17. *Dissected breccia lobe on the Mojave Desert. Note how the breccia laps up against the range of hills at the left. The lobe is 2½ miles across at its near end. (Photo by Robert C. Frampton, Claremont, California.)*

The corrugations of the breccia surface, unlike those at Elm, are almost straight. They nearly parallel the trend of the partially overridden low bedrock hills to be seen at the left of Figure 11-17, and may represent waves set up by these barriers. The differences from the Elm flow, however, raise doubt as to the origin of the limestone breccia, and stimulate study both of this and of any active slides or falls that may in the future be observed.

Slides or Flows Preserved in Sedimentary Rocks

The form and textures of certain breccias, or brecciated strata, now parts of Paleozoic, Mesozoic, and Tertiary formations, suggest that these are ancient slides or debris flows. Most convincing are distortions in thin-bedded marine sediments and lenses of heterogeneous breccia apparently in troughs like that developed in 1887 on the lake floor near Zug. These have been recognized in Wales, Peru, Quebec, and California, to mention only a few of many localities.

Significance of Downslope Movements

Grand Canyon Example

The Grand Canyon of the Colorado River not only gives evidence of extensive erosion (see Chapter 7, p. 138, and Fig. 7-6), but makes possible at least a rough estimate of the parts played by different erosional processes in forming it. Figure 11-18 is a structure section across the Grand Canyon approximately from south (left) to north (right), showing the Granite Gorge in which the Colorado River flows, and the stepped profiles in the slightly inclined strata overlying the gneiss of the Granite Gorge. The total depth of the canyon is about 1 mile, and the width from rim to rim about 7½ miles. The Granite Gorge is about 1,300 feet deep at the line of section,

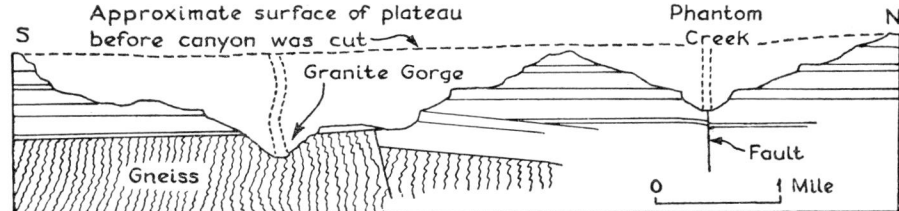

FIGURE 11-18. *Section across Grand Canyon. The parallel dashed lines above Granite Gorge and Phantom Creek indicate the narrow trenches that would have been made by stream cutting alone. (After Bright Angel quadrangle, U. S. Geological Survey.)*

and the inner gorge of Phantom Creek, about 900 feet deep. The Granite Gorge has relatively straight side slopes, the inner gorge of Phantom Creek has side slopes that are concave upward. Phantom Creek furnishes an especially favorable case for analysis of the relative roles of stream transport and downslope movements in denudation. The stream follows a fault of 100 or 150 feet vertical displacement. The position of the stream was apparently determined by the fault; presumably the stream has eroded downward along the fault, fixed in position by the ready erosion of the shattered rock along the fault zone. The dashed lines in Figure 11-18 indicate a trench as wide as the creek bed to represent the erosion ascribed to the stream. But if the stream directly eroded only this narrow vertical slot, what processes account for the much wider actual valley? Judging the past in terms of the present, these other processes include rainwash concentrated into minor rills and along better defined gullies, and also several kinds of downslope movements of rock. In this arid climate the initial fall and continual streamward movement of talus fragments is an important cause of the widening of Phantom Creek valley, but many other kinds of downslope movement also contribute.

The Grand Canyon itself has presumably been cut by the Colorado River and widened by rainwash and downslope movements in the same manner as Phantom Creek. Since the course of the Colorado was not determined by a fault line, however, the river may have wandered somewhat to reach its present position (Fig. 11-18) instead of cutting straight down, as did Phantom Creek. A similar development may be taken as probable for streams and stream systems in other regions, especially those in canyons or deep valleys. Going one step further, we may assume tentatively that in a stream-cut valley whose floor is little or no wider than the stream, the depth of the valley is a measure of downcutting by the stream and the valley's width is an indication of the amount of material contributed by rainwash and downslope movements of rock and soil. Steep side slopes indicate the predominance of downcutting by the stream, gentle slopes the effectiveness of the widening process.

Downslope movements of rock and soil are, of course, not limited to the matched sides of valleys. They are equally effective in reducing the steepness of single slopes or cliffs, as along shores. Gravity movements extend to the very summits of ridges, reducing their height. Rocky summits are lowered by the fall of fragments loosened by weathering (Fig. 11-13). This process has contributed to the reduction of the summit ridge between Phantom Creek and the Grand Canyon, so that it now lies well below the level of the plateaus to north and south (Fig. 11-18). Soil-covered divides, on the other hand, must be lowered chiefly by creep and rainwash.

Effect of Plant Cover

Downslope movement of material toward drainage lines varies with the plant cover, which, in turn, is dependent primarily upon climate. With regard to plant abundance, the most important distinction among nonarctic climates is that between humid and arid climates.

Even in humid regions, the density of the plant cover varies widely, being thin on cold moors and very thick in tropical areas with year-round rainfall. Typically, as in eastern United States or west-central Europe, trees, shrubs, and grass cover almost the whole surface, and topographic forms are smoothly rounded. Here, creep is the principal variety of downslope movement. Minor differences in rock resistance, as on a single slope, are masked by the almost continuous cover of soil and vegetation, but major differences become apparent in the steepness of the side slopes and in the general pattern of ridges and valleys. The etching out of weaker rocks, as shown in Figures 7-7 and 9-14, is called *differential erosion.*

In arid regions, trees and shrubs are few, and grass may be absent. As a result, no soil covers the resistant rocks, which stand out in knobs, ridges, or cliffs. Weak rocks are reduced more rapidly by downslope movements and rainwash, and their areas of outcrop are obscured by talus accumulations derived from the resistant rocks (Fig. 11-12). Such differences in rock resistance through differential erosion become strikingly apparent even when slight, as in the set of horizontal strata shown in Figure 11-13. On a larger scale, compound cross profiles of hill slopes resulting from differential erosion are more sharply defined in arid than in humid regions.

Conclusions

Considered in terms of the areas affected, downslope movements and rainwash are actually the most important of all processes of erosion. For every square foot that is directly subject to erosion by a permanent flowing stream there are hundreds of square yards over which soil is slowly creeping—flowing a few inches downhill after frost heaving or clay hydration; being pushed downslope by burrowing rodents and under the feet of grazing animals; or gliding and falling in debris flows, slides, and talus.

The critical factor for most gravity movements is the presence of enough water to lubricate slide surfaces or to produce a semi-liquid flow mass. Many rocks that are relatively strong when dry are weakened and become plastic when wet. If great volumes (Chap. 3) become so weakened, gliding may take place on a great scale, even though individual pieces of the mass involved are still strong by our human standards.

Downslope movement is a very important and widespread process of erosion, but it rarely goes on alone. Streams cut valleys, gravity movements widen them, and the streams carry away the debris. Glaciers gouge valleys deeper, and gravity movements load the glacier margins with rock fragments that the ice transports away. Waves driving on shore undermine cliffs, whose upper portions then fall into the sea, to be broken up and carried away by wave-generated currents. In brief, the role of gravity movements is to supply material to the agents of long-distance transportation.

Downslope movements differ from most other erosional processes in that they can operate below the sea as well as above. Though the buoyant effect of the water lessens the effective pull of gravity on material immersed in it (Archimedes' principle), the soaking of the material is complete. As a result, even though weathering is thought to be slight below the sea, downslope movements should be common there. As we saw in Chapter 9, and other examples will be cited in Chapter 16, such movements are known to have taken place within historic time. The record of their existence in the geologic past is found in disturbed beds and brecciated areas in the marine sedimentary rocks. It is probable that such movements play a large part in molding the submarine topography at depths far below sea level. Thus, downslope movements may contribute to some extent to both of the two broad levels of the earth—one near sea level, the other corresponding to the floor of the ocean (Chap. 2, Fig. 2-9). Downslope movements on land bring debris to the streams, and they in turn build the shorelines seaward. Downslope movements in the oceans tend to reduce submarine heights to the level of the ocean floor.

Engineering Applications

Gravity movements of soil and rock may affect man-made structures such as roads, bridges, dams, or houses. In the United States alone, clearing highways of the debris that slides or rolls from the slopes above, and the repair of sections of roadbed that have settled and slumped downslope, costs millions of dollars annually. Many newly built railroads and highways have had to be rerouted within a few years in order to detour slide areas or cliffs which shed many talus fragments.

Expensive mistakes of this kind can be avoided by careful geologic inspection of the ground along the right-of-way. Areas of active sliding show readily recognized features: The hill slope is generally hummocky and contains undrained depressions, the rock may show slip surfaces roughly parallel to the surface of the hill, landslide scars and curving debris ridges are likely to be present, fences and telephone poles are tilted downhill, and tree trunks bend uniformly as they enter the ground.

Obviously, the cut made for a highway or railroad increases the danger of sliding because the excavation removes support on the downslope side of the slide. Many an ancient landslide or debris flow that had not moved for tens or even thousands of years has been reactivated by the removal of debris from its toe during the construction of a roadbed.

Although highways and railroads can generally be relocated to detour a dangerous slide area, it is sometimes impossible to move other kinds of structures that may be endangered by gravity movements. Geologists and engineers must then take steps to stop, or at least to slow down and minimize, the danger. An oil field near Ventura, California, affords a typical example. Oil wells drilled through a slide were slowly bent out of line as it moved. Some of the well casings were sheared completely off at a slip surface along the base of the slide. Movement was most rapid in the winter when the ground was saturated by seasonal rains. Unless the slide could be controlled, a valuable oil field would have to be abandoned. Geologists and engineers solved the problem by digging and boring galleries along the base of the slide and installing drain tile to carry off the seepage from the rains. They also paved the entire hillside with asphalt. Most of the rain now runs rapidly off the pavement, and the little that does penetrate the slide is drained off. By thus preventing access of water, which transformed the clays into a lubricant, they stopped the motion of the slide.

During the building of the huge Grand Coulee Dam on the Columbia River, construction work was threatened when a tremendous mass of water-soaked silt and sand started to creep into the excavation dug for the north abutment of the dam. To stop this threatened slide, engineers hit on the ingenious idea of penetrating the silts with numerous pipes in which a refrigerant was circulated. The refrigerant froze the water in the pores of the silt, thus cementing the particles and increasing the strength of the silt. This effectively stopped the movement, and the area was kept refrigerated until the concrete for the dam was poured, and the excavation thus filled and stabilized.

Similar problems must be met when heavy structures such as large bridges and dams are built on soft clay or on creeping ground. Many soils behave plastically when loaded, and will flow radially outward from beneath the load. One of the piers of the San Francisco Bay Bridge was made purposely large at its base so that the weight of the heavy structure could be distributed over a wider area of the hard clay on which the pier was set. Many kinds of laboratory tests have been devised to determine the load that various kinds of sand, clay, and other loose foundation materials will bear. These tests are the basis of a relatively new branch of engineering science called *soil mechanics*. Field conditions, including attitude of stratification or other slip planes, amount of contained water, slope, and other factors must be taken into account, as well as the laboratory characteristics of the materials. By these methods it is often possible to forestall or control gravity movements, even in places where weak materials are covered by massive structures.

Facts, Concepts, Terms

GRAVITY MOVEMENT OF ROCK

CREEP, SLIDES, DEBRIS FLOWS, ROCK FALLS, TALUS

RAINWASH

CAUSES AND EFFECTS OF CREEP

SOLIFLUCTION; ROCK GLACIERS

SLOW DEBRIS FLOWS

RAPID SLIDES AND ROCK FALLS

UNDERWATER FLOWS

RECOGNITION OF ANCIENT SLIDES AND ROCK FALLS
    On land
    In the sea

RELATIVE ROLES OF DOWNSLOPE MOVEMENTS AND STREAM EROSION

ENGINEERING APPLICATIONS

Questions

1. What is the lowest level to which downslope movements on land can deliver material? Downslope movements in the sea?
2. On steep mountain slopes that receive a heavy snowfall even trees that are rooted in rock crevices have trunks that tilt downhill as they emerge from the ground, then bend to a vertical position a few feet above the surface. Why?

3. In warm humid regions compact clay-rich soils are more subject to creep than sandy or gravelly open soils, but the latter move readily under arctic conditions. Why?

4. Describe the structures you might find in a marine sedimentary rock that would indicate the area had been the site of an ancient underwater flow.

5. A symmetrical, almost perfectly round soil-covered hill is underlain by vertical soft shales containing two thin beds of hard sandstone. One sandstone is red in color, and it exactly bisects the center of the hill. The other sandstone is white, and it extends through the hill about halfway down from its summit. Draw a map or sketch showing the two sandstone beds and indicate on the sketch the areas of the hill over which you would expect to find (a) numerous loose fragments of red sandstone, (b) numerous loose fragments of white sandstone.

6. Building sites on a hill with a fine view are restricted to two locations: Both are underlain by soft shale, but at one site the stratification dips steeply into the hill, at the other it dips roughly parallel to the hill slope. Which site would you choose, and why?

7. Basalt cliffs on the Columbia Plateau have large piles of coarse talus at their base. On the Colorado Plateau, under similar climatic conditions, equally large cliffs of crumbly sandstone have little or no talus. Can you suggest an explanation?

Suggested Readings

1. Sharpe, C. F. S., *Landslides and Related Phenomena*, Columbia University Press, New York, 1938.

2. Howe, Ernest, "Landslides in the San Juan Mountains, Colorado, Including Consideration of Their Causes and Their Classification," U. S. Geol. Survey Prof. Paper 67, 1909. (Many photographic illustrations.)

3. Heim, Albert, "Bergsturz und Menschenleben," Vierteljahrschriften Naturforsch. Gesellschaft zu Zürich, 1932. (Data concerning Swiss and other landslide rock falls and debris flows, including Elm and Flims.)

12. *Stream Erosion and Deposition*

As SHOWN in Chapter 7, the sediment load of rivers and smaller streams provides undeniable evidence of the erosive power of running water. Surface runoff picks up and transports small amounts of mud and sand from its first confluence in rills and tiny brooks, and eventually enlarges these miniature valleys into gullies. Rills join to form small streams, and these, in turn, unite into rivers. The whole system erodes the surface of the land and transports the debris to the ocean. Knowledge of how streams erode, transport, and deposit rock waste is essential to an understanding of the development of our present landscapes and to the interpretation of the origin of ancient sedimentary deposits.

Furthermore, the study of stream flow contributes to the efficient use of rivers by man. Without a knowledge of stream mechanics we cannot hope to solve the problems of recurring flood disasters, erosion of rich valley soils, water storage, and river navigation.

River Systems

Geologists and engineers often use the term *stream* in a general sense, that is, to designate bodies of running water of any size. In this chapter, for instance, the term *stream erosion* refers to the action of rivulets as well as rivers. A *river system* includes the various streams that join within a drainage area. A river or *trunk stream* is formed by the confluence of *tributary streams*, themselves formed by smaller tributaries, and so on to the individual contributions of springs, rills, and rainwash.

Several characteristics of river systems change fairly regularly from the smallest headwater tributary to the mouth of the main trunk river. The most obvious variables are: (1) the *gradient* of the stream, commonly measured by the *number of feet of fall per mile of course;* (2) the form of the stream valley; and (3) the amount of erosion or deposition along the course.

Most upper tributary streams flow with steep gradients (commonly over 50 feet per mile) through topography marked by valleys and ridges. They actively erode the bedrock along their courses. On the other hand, the gradient of a trunk stream, especially in its lower part, is ordinarily less than 2 feet per mile. Here deposition equals or may even exceed erosion and the river flows in a sinuous course upon a smooth plain of its own deposits.

It is along the lower, gently sloping parts of river systems that the problems of stream control are most keenly felt. It is here, also, that man uses rivers for navigation, builds great cities, and tills the richest of farmlands.

The relations of rivers to the valleys in which they flow have long been recognized. In 1802 the Scotch mathematician and naturalist John Playfair observed the nice adjustments within a river system and concluded that rivers must surely have cut their valleys:

> Every river appears to consist of a main trunk, fed from a variety of branches, each running in a valley proportional to its size, and all of them together forming a system of vallies, communicating with one another, and having such a nice adjustment of their declivities, that none of them joins the principal valley, either on too high or too low a level, a circumstance that would be infinitely improbable, if each of these vallies were not the work of the stream that flows in it.

Today, with a thousandfold more data than were available to Playfair, it is found that stream flow, erosion, and deposition are in extremely sensitive balance. In evaluating this balance quantitatively, geologists and hydraulic engineers have recognized several variable factors. Among these are the gradient and shape of the channel; the velocity of the stream, including variations of velocity within the channel; and the nature of the material that is being eroded and transported by the stream. Before these variables can be discussed even qualitatively, however, it is necessary to understand the nature of fluid motion itself.

Laminar Flow and Turbulent Flow

If needle-thin streams of colored fluid are admitted to a stream of water flowing very slowly through a glass tube, the colored threads advance in parallel lines downstream from the points of contamination. Where an obstruction is met, as in Figure 12-1, *left,* the flow lines move around the object in a "streamlined" way. This phenomenon, called *laminar flow,* is apparently produced by the sliding of thin layers of fluid over one another. It is

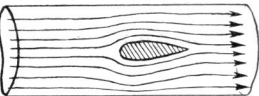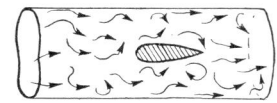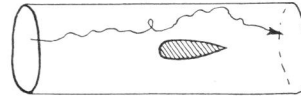

FIGURE 12-1.   Left, *laminar flow;* center *and* right, *turbulent flow.*

characteristic of viscous liquids, such as cold lubricating oils or sticky molten lava. It can be produced in water only at velocities less than an inch or so per second (1/16 mile per hour).

If the flow is speeded up, the pattern of colored threads abruptly changes. Instead of parallel lines they begin to trace irregular, tortuous paths, then swirl in all directions to the extent of losing their identity by mixing in the water stream. This is called *turbulent flow.* Figure 12-1, *center,* shows the traces of many such filaments during a short time interval, and Figure 12-1, *right,* shows the trace of a single thread during a longer interval. As the velocity of the stream increases, or if the walls of the tube (Fig. 12-1) are roughened, the turbulence is increased.

Velocities and Turbulence of Natural Streams

Many measurements show that nearly all natural streams, even the smoothest flowing, move rapidly enough to produce turbulent flow. The power of a stream to erode and transport apparently depends on this turbulence, and also on the distribution of velocities and turbulence within its channel.

Range of Velocities

Stream velocities range widely, depending mainly on the *gradient,* the *shape of the channel,* and the *discharge,* that is, the quantity of water passing a given point, usually measured in cubic feet per second.

The effect of channel shape on velocity is well shown by the Rhine, which flows at a rate of 2 to 3 feet per second in the broad, flat channel between Basel and Bingen, whereas in the narrow, deep gorge guarded by the Lorelei, the velocity increases to 6 to 10 feet per second.

The effect of discharge on velocity is strikingly evident during floods. Mississippi River velocities measured near Vicksburg in 1937 and 1938 were mostly between 2 and 7 feet per second, but increased to 10 and 14 feet per second during the February, 1937, flood. The Columbia River, about 170 miles from its mouth, had a mean velocity of 1.9 feet per second on March 12, 1945, when the discharge was only 78,000 cubic feet per second, but a mean velocity of 11 feet per second at flood flow of 1,018,000 cubic feet per second on May 3, 1948. Where stream gradients are very steep, as in mountain torrents, velocities in excess of 30 feet per second (more than 20 miles per hour) have been measured.

Distribution of Velocity and Turbulence

The velocity of a stream varies greatly in different parts of its cross section. Near the bed or banks a thin film of water, adhering closely to the solid surface, moves exceedingly slowly and probably entirely by laminar flow. Away from the banks the velocity increases rapidly through a foot or so of the stream. Several feet away from the banks or bed the current is nearly

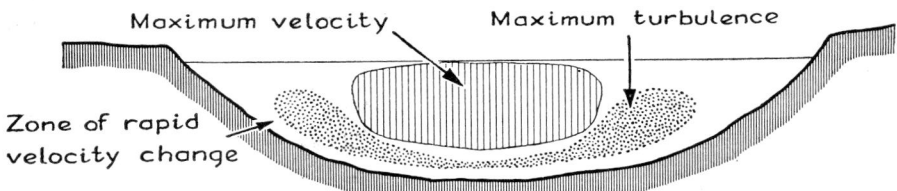

FIGURE 12-2. *Distribution of velocity and turbulence in a symmetrical channel. (After John Leighly, 1934.)*

uniform (Fig. 12-2), though there may be a slight increase to the maximum in the stream center.

Turbulent flow is most marked in the zone of sharp velocity change, bordering the stream's channel. Spiraling and rolling eddy currents are thrown from this layer of unbalanced flow into the main body of the stream. In Figure 12-2 the regions of most prominent turbulence are shown in their relation to the velocity distribution.

How a Stream Carries Its Load

The Suspended Load

The turbulence in streams accounts largely for their ability to carry a load of silt and clay particles. Whether the particles are brought by rainwash and rills or are eroded directly from the stream's bed or banks, it is the *upward* currents of turbulence that allow their transport. Once a grain has been lifted by an upsweeping eddy, it is kept from falling out again only by the force of other upward moving filaments. True, there are as many downward currents as upward, but both are randomly distributed, so a particle may long be sustained before striking a downward swirl. Even if the grain is eventually dropped, or carried down in an eddy, other grains will be picked up by currents acting in the opposite direction. The important result is that, in the time interval during which the grain is picked up, jostled around, and deposited again, it has moved downstream a distance that depends on the average forward velocity of the water. To an underwater observer moving with the current, the grains will appear to rise, gyrate, bob, and fall; but to an observer on the bank the stream appears a homogeneous mixture of water and sediment, all flowing downstream. Because turbulent currents seem actually to suspend sediment in the stream, this part of the stream's load is called the *suspended load.*

The Bed Load

Even-flowing rivers seldom develop enough turbulence to lift particles larger than medium-grained sand from their beds. Coarser material may, however, be pushed or rolled along the bottom by the swirling currents of the boundary zone, thus becoming part of what is called the *bed load.* Ex-

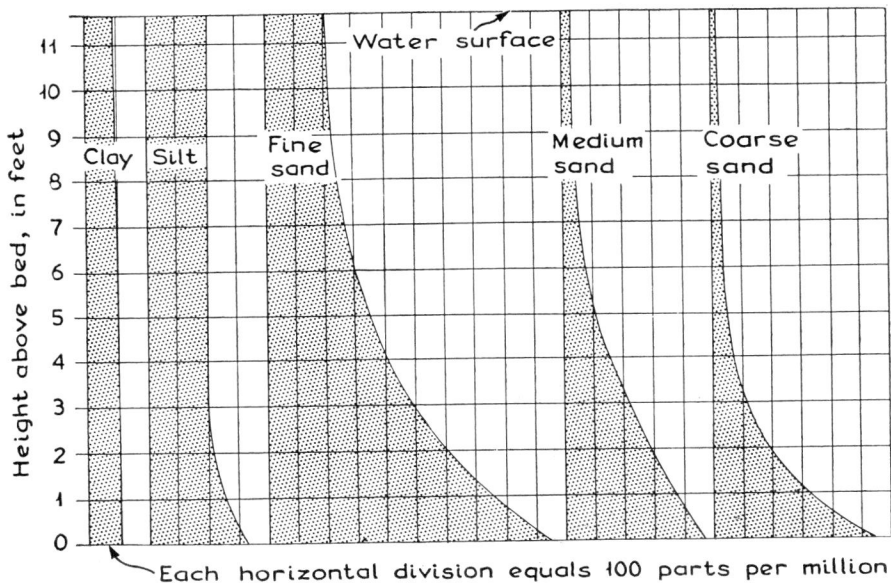

FIGURE 12-3. *Graph showing the variation in sediment content with depth, Missouri River, near Kansas City, Missouri. From samples taken on January 3, 1930. Note the great concentration of the coarser grains near the bed and the nearly equal distribution of the suspended clay and silt. (After L. G Straub, in* Hydrology, *courtesy of Dover Publications.)*

cept when stream velocities are high, the grains of the bed load do not move continuously but progress downstream by many stops and starts. Consequently, the bed load usually moves much more slowly than the suspended load. The motion of the bed load has been observed through windows in the walls of wooden flumes. Most of the grains roll, though some slide along the bottom, and others bounce along or vault into suspension.

As the velocity of a stream flowing on a bed of sand is gradually increased, the motion of particles on the bed progresses through: (1) short transport of a few individual sand grains, (2) spasmodic movement and deposition of groups of grains, and (3) smooth, general transport of many grains. Under some conditions the scour and deposition of grains produces ripples on the bed. At high velocities large numbers of grains may go into temporary suspension, proceeding as such dense clouds that a distinction cannot be made between the immobile part of the bed and the moving, partially suspended load. A comparable situation is found in natural rivers, where, except at low velocities, the bed and suspended load apparently grade into one another. This gradational situation is indicated graphically in Figure 12-3.

The Dissolved Load

A third way that streams carry material is in solution. The dissolved particles are evenly distributed in the water by diffusion. Here transport does not depend in any way on the nature of stream flow. The chief source of the dissolved load is from inflow of ground water that has percolated slowly through weathering mantle. Probably very little is dissolved from the channel walls except where streams flow on limestone.

Thousands of chemical analyses show that few rivers carry more than 1,000 parts per million (0.1%) of dissolved materials. A general average for many American rivers is about 200 parts per million. It should be noted, though, that this may sometimes form a fourth or a third of a stream's total load.

Load Competence

A stream moving at a given velocity exerts a corresponding force upon the particles of its bed. For particles of the suspended load, the upward-moving currents of turbulence exert forces that depend on their velocities. The total force a current exerts against a grain depends on the grain's surface area, whereas the grain's weight is a measure of its resistance to movement (inertia). Since the ratio of the surface area to weight increases with decreasing size, small grains are more easily moved than large ones. Thus for any stream, the current will suffice to move particles only up to a certain limiting size. This maximum grain size gives us a simple measure of the stream's transporting power or *competence.*

Considerable experimental work indicates that the diameters of sand grains and coarser debris that can be moved by a stream increase approximately as the square of the velocity. That is, if the velocity is doubled, the diameter of grains that the stream can move is increased four-fold; if the velocity is tripled, the diameters increase nine-fold, and so forth. The high velocities of steep streams, especially when in flood, make their currents almost irresistible. Some move boulders over 10 feet in diameter. When the St. Francis concrete dam in southern California broke in 1928 the great mass of water suddenly released carried blocks of concrete that weighed as much as 10,000 tons ($63 \times 54 \times 30$ feet) for half a mile downstream.

The Nature of Flow in Steep Streams

The flow of steep, tributary streams differs markedly from the even or tranquil flow of broad trunk rivers. Not only are their average gradients and hence their average velocities greater, but both gradient and velocity fluctuate more. Changes in gradient of the stream bed create what hydraulic engineers call *nonuniform* flow.

A typical case of nonuniform flow is illustrated by Figure 12-4. Water from A passes over the uniform, gentle gradient AB with a moderate velocity. At B the steeper gradient sharply increases the velocity, and hence the stream shallows from B to C. In the stretch from C to D the gradient and, therefore, the velocity are again moderate, and the stream deepens. The alternating rapids, pools, riffles, and waterfalls of most mountain streams illustrate such variations in depth and velocity.

The flow in the A to B stretch of the figure may be considered typical of all streams with low-to-moderate gradients. Turbulence disturbs the surface with mild boiling, rolling, or whirlpool motions. Surface waves produced by

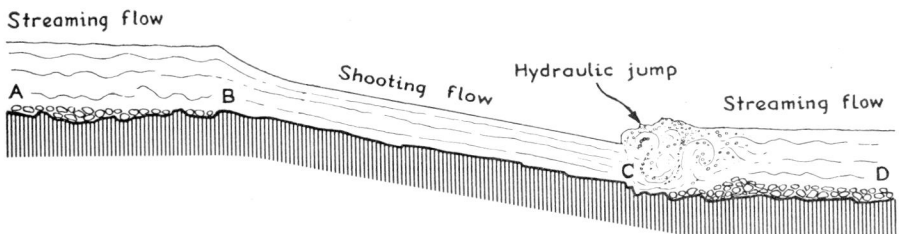

FIGURE 12-4. *Nonuniform flow in a natural stream.*

impact of the current against the bank or bed—or by dropping a stone into the stream—travel along the surface in all directions, disturbing the general downstream flow of the water. This kind of stream flow, because of its unhurried, though somewhat disturbed, nature, is called *tranquil* or *streaming* flow.

Where such a stream suddenly quickens on a steeper slope, as at *B* in Figure 12-4, its appearance changes abruptly. It breaks away from the surface of the tranquil stream in a curving lip, and shoots swiftly down the steeper slope. If the bed is reasonably smooth, the stream surface is smooth and shimmering; where the stream flows over obstacles, there is a smooth roll or a boiling wave, but the wave's disturbance does not move upstream. Turbulent eddies are apparently flattened and attenuated, for disturbances of the surface are greatly reduced. This kind of flow is called *rapid* or *shooting* flow.

The critical velocity separating shooting from streaming flow cannot be simply defined, for it depends on many hydraulic variables. In general, shooting flow will be produced in a stream one foot deep and several feet wide if the slope is greater than about 3°; in larger streams shooting flow develops on lesser gradients.

At *C* in Figure 12-4 there is a particularly interesting feature of nonuniform flow. Here the surface of the swift shooting stream must suddenly rise to the level required by the slower velocity below *C*. This abrupt rise, produced wherever the channel gradient sharply decreases, is appropriately called the *hydraulic jump*. The impact of the shooting water against the slower produces a strong rolling vertical eddy or a seething, air-charged mass of very high turbulence. Except for the waterfall, which is, in a sense, an extreme case of the hydraulic jump, the jump is the greatest dissipator of energy in a stream.

Erosion on Steep Gradients

What mechanisms, inherent in the nonuniform flow of steep streams, allow them to erode their beds and banks? Whether a stream moves by shooting or streaming flow, most of its eroding power depends on the energy of large fast-whirling eddies. These eddies are produced mainly by impact against, or deflection around, obstacles in the channel, such as boulders on the bed, bedrock projections, gravel bars, shallow depressions, and deep holes.

During floods, swirling currents may be able to free and move jointbounded blocks of bedrock, a process called *hydraulic plucking*. Large nonuniform-flowing streams produce eddies of great competence. Below Great

Falls, Maryland, where the Potomac River moves by streaming flow through a gorge, 15-inch boulders have been lifted as high as 60 feet above the channel bottom during floods. The impact of such missiles must effectively erode both bed and banks. In slower streams, swirling currents armed with coarse sand and small pebbles strongly abrade both moving rock fragments and exposed bedrock. Even fine sand may be an effective abrading agent if the currents are swift. Effects of eddy erosion can be seen along many streams during low-water stages. Among them are rounding of the projecting rock, fresh angular scars where blocks have been torn from channel walls, and cylindrical cavities (potholes) drilled in solid rock by stones caught in holes on the surface but rolled about by the swirling water (Fig. 12-5).

The strong turbulence of the hydraulic jump causes relatively rapid erosion. Erosion at waterfalls illustrates an extreme in the competence of nonuniform flow. Water falling freely is accelerated by gravity, so that at the end of a second its velocity is about 32 feet per second (20 miles per hour).

FIGURE 12-5. *The bed of the Susquehanna River, during the drought of 1947, showing potholes and pitted, abraded surfaces. The location is Conewago Falls, Pennsylvania, where a sill of resistant dolerite has produced a rapid in the river's course. (Photo by* Lancaster Intelligence Journal; *courtesy of Herbert H. Beck, Franklin and Marshall College.)*

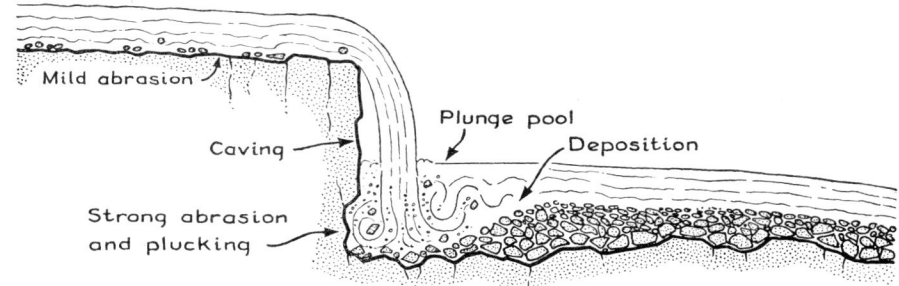

FIGURE 12-6. *Erosion at a small waterfall.*

The Niagara River falls at about 50 miles per hour at the base of the 150-foot high Niagara Falls. Thus many waterfalls develop deep *plunge pools* at their bases which are cut back under the falls (Fig. 12-6). Migration of waterfalls upstream by undercutting at the plunge pool and caving of the walls above may contribute greatly to the erosion of a stream valley. On the Niagara River, for instance, the Niagara Gorge between Queenstown and the Falls, a distance of about 7 miles, was cut principally by headward erosion of the waterfall, not by slow downward erosion along the entire channel.

The Nature of Sinuous Rivers

The lower courses of most river systems show features that contrast strongly with the pools, rapids, and waterfalls of steep streams. Trunk rivers commonly flow in broad, smooth valleys sloping so gently seaward as to appear flat. Apparently disdaining the shortest route to the sea, the rivers wind in intricate serpentine courses. The Mississippi River, from Cairo, Illinois, to the mouth of Red River, has a typically sinuous course (Fig. 12-7). Such streams are often called *meandering rivers*, though, strictly speaking, bends called *meanders* should turn through 180° or more. Alternating with sinuous stretches are relatively straight *reaches*, such as that just below Cairo in Figure 12-7.

The broad flat on which such rivers flow is called the *river floodplain*, for it is more or less flooded when the river overflows its banks. Some floodplains are no wider than the meandering course of the stream—their bends impinge against the valley walls. Many floodplains, however, are far wider than the meanders, and may broaden greatly where joined by sinuous tributaries, as does the Mississippi floodplain, shown by the dotted area of Figure 12-7, *left*.

Low indistinct ridges on floodplains, called *natural levees*, commonly

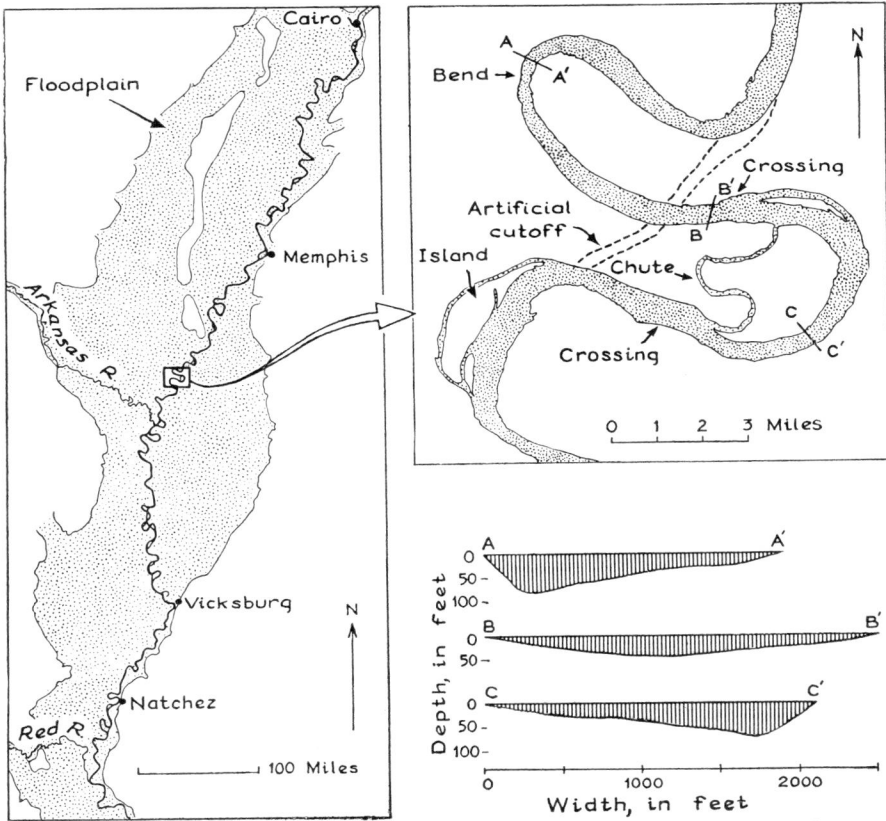

FIGURE 12-7. Left, *the sinuous course of the Lower Mississippi River;* right, *map and cross sections of three bends. The artificial cutoffs are man-made channels, dug in 1941 and 1942. (After H. N. Fisk, Mississippi River Commission, 1947.)*

border sinuous river channels. When streams overflow their floodplain, the great decrease in velocity and turbulence of the overbank flow results in deposition of even the finest suspended load. The coarsest part of the suspended load, dropping out just as the floodwaters overtop the banks, builds up the low natural levees. The fine silt and clay are carried farther and deposited on the lowlands of the floodplain, behind the natural levees. These muddy floodplain accumulations are individually thin, but may build up to considerable total thicknesses over long periods of time. In the lower Nile Valley, silt has risen as high as 15 feet about the ancient Egyptian structures on the floodplain, indicating a depositional rate of 4½ inches per century. The floodplain behind the natural levees may be poorly drained, and lakes, swamps, or indistinct sinuous channels are common.

The Channels of Sinuous Streams

Sounding of the Mississippi's bed reveals marked changes of the channel shape from bend to bend. As shown in Figure 12-7, *right,* the channel is deepest near the outer bank of each bend. The inner slope is gentle or even convex, and often consists of shifting sand bars. In the short straight stretch between bends the river shallows considerably, and the channel is more or less symmetrical. These shallow channels between bends, called *crossings,* troubled the old-time river pilots. Some are less than 10 feet deep at low water, as compared to "bendway" depths of 45 or more feet.

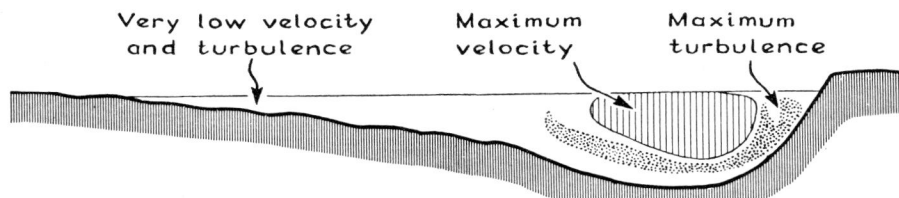

FIGURE 12-8.   *Cross section showing distribution of velocity and turbulence at a bend. (After John Leighly, 1934.)*

This variance in channel shape is directly related to the distribution of velocity and turbulence. In straight reaches or crossings the zone of maximum velocity, bordered by two zones of high turbulence, is midway in the channel. At a bend the conditions are different. The water moves forward into the bend with the direction given by the straight stretch above, so that the zones of maximum velocity and turbulence impinge against the outer bank of the bend. These competent currents contrast strongly with the sluggish flow over the convex slope on the inside of the bend (Fig. 12-8).

Erosion and Deposition in Sinuous Channels

The bank materials are generally unconsolidated sands, silts, and clays; therefore the outer bank of a bend will be easily eroded by the strong currents directed against it. During a flood, when velocities and turbulence are greatest, the channel deepens, causing the outer banks to cave off rapidly. The fine-grained parts of the load thus acquired, the clays and silts, are readily carried away in suspension. Ordinarily, however, the coarser material is not moved far. Experiments show that almost all the sand and gravel is deposited on the next crossing, and over the inside slope of the next bend (Fig. 12-9).

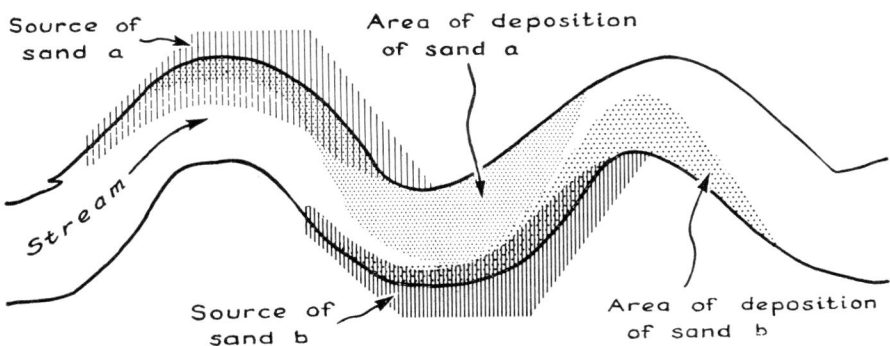

FIGURE 12-9. *Localization of erosion and deposition of marked sands in an experimental sinuous stream. (After J. F. Friedkin, U. S. Waterways Experiment Station, 1945.)*

Erosion of the outer bank and deposition on the inner slope of bends cause them to shift in position. This *migration* of the channel, or shifting of the meanders, allows a stream to wander widely over its floodplain. If the floodplain materials are homogeneous, the meanders tend to sweep evenly downstream without lateral shifting. Over a long time the stream reworks the deposits left by previous meander courses and deposits material that is later cut away again as its channel continues to migrate. Thus the coarser sediments of floodplains move only slowly toward the sea.

Erosional and depositional features on many floodplains clearly indicate former positions of sinuous channels. In Figure 12-10, *top,* the low stripelike ridges that almost parallel the inside of the bends are beaches or bars left behind as the bends migrated. The crescent-shaped lakes are abandoned river bends. Such *oxbow lakes* develop either by *cutoff* or by *chute* formation. Cutoffs form when one meander is slowed in its downstream migration by a resistant bank, allowing the next bend upstream to catch up with it and cut through the narrowing neck between the two bends. Figure 12-10, *bottom,* illustrates how this might have taken place on a bend in the river of Figure 12-10, *top.* A chute, on the other hand, forms when a river in flood simply flows over the bar and beach deposits at a bend, and then continues to use this shortcut channel. Chutes sometimes serve as alternate channels around bends, as illustrated in the middle distance of Figure 12-10, *top,* or on the enlarged map of a part of the Mississippi (Fig. 12-7). Generally, the cutoff or chute channel is more efficient than the channel of the old bend, for, since the distance is shorter, the gradient is steeper, and therefore the velocities and competence are higher. Velocity and turbulence in the cutoff bends become low or almost nil; consequently, sand is immediately deposited at

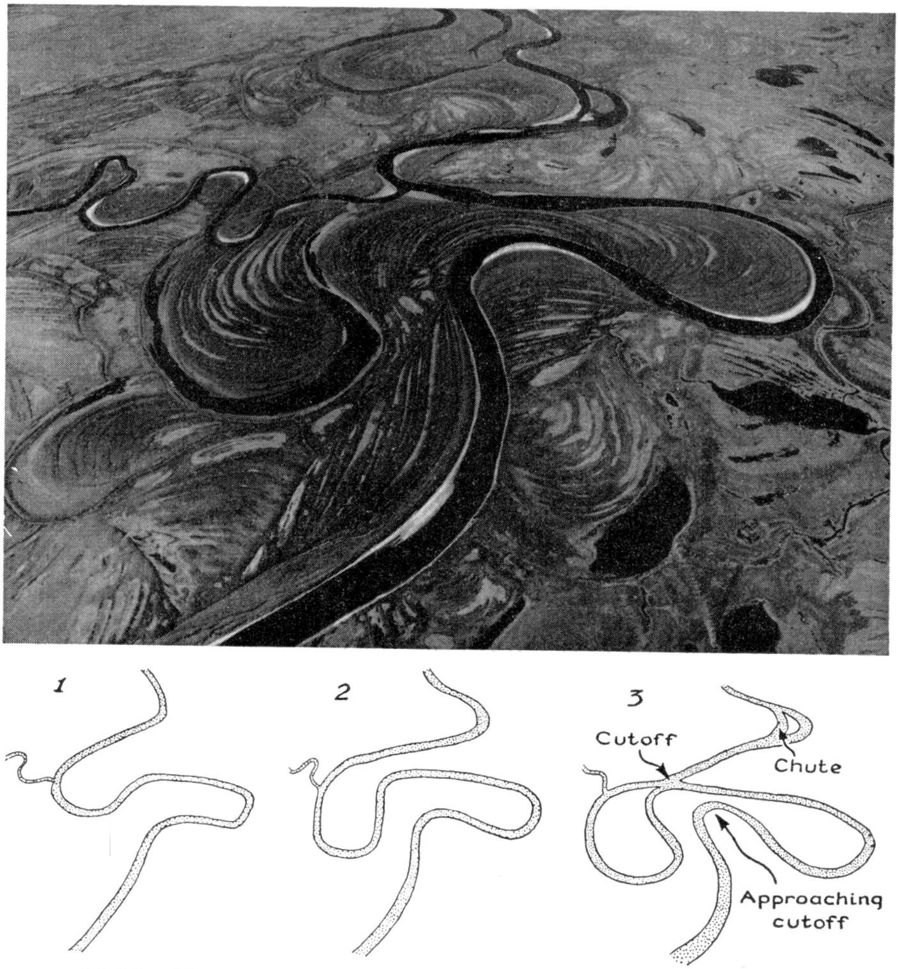

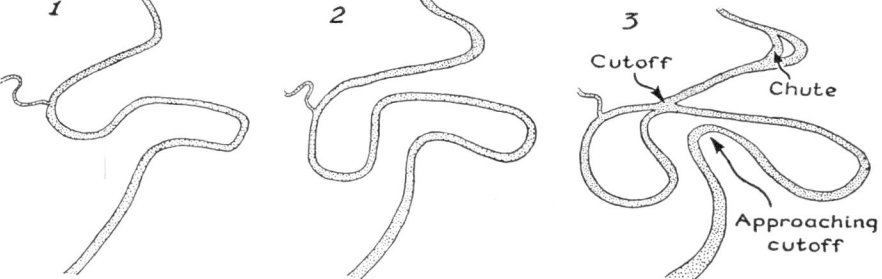

FIGURE 12-10. Top, *a river meandering on a broad floodplain. Note the low-ridged deposits on insides of bends and the abandoned meander channels, including oxbow lakes.* Bottom, *possible stages in the evolution of the meander cutoff that has recently taken place in the center of the view.* (*Photo by U. S. Air Force.*)

both ends of the old bend, eventually damming it off from the new channel, and converting it into an oxbow lake.

Significance of Sinuous Channels

Why should streams meander down their floodplains rather than flow in straight channels directly to the sea? In seeking an answer to this question geologists and hydraulic engineers not only have uncovered clues to the cause of meandering, but have come to understand many variables affecting river flow. Observations of the discharge, load, and velocity of natural rivers have contributed much to our knowledge of flow in sinuous channels, but many variables are too intricately interrelated to examine one at a time, and hence their effects on sinuosity are difficult to evaluate. Because of the variability of natural rivers, investigators are turning more and more to stream models for help in solving both the theoretical and practical problems of stream flow. Simple wooden troughs, used by early experimenters, have been replaced by carefully constructed scale models that not only exactly reproduce channel and valley shapes, but also have measuring and control devices for nearly all the variables of flow. Such models are used considerably in planning major projects of river regulation and construction, and are thus the very intricate "test tubes" of the hydraulic engineer.

Many river model studies of meandering have been made at Vicksburg by the Mississippi River Commission. The models were not mere table-top arrangements, but were of considerable size (Fig. 12-11, *left*). In one of the most revealing experiments a straight channel was carefully molded in uniform Mississippi River sand, and water was allowed to run in it for three days. The stream quickly developed a sinuous course, whose bends (Fig. 12-11, *right*) swept evenly downstream in the homogeneous, easily eroded floodplain material. The meandering began when the stream, cutting one of its banks more rapidly than it could transport all the debris, deposited a sand bar along that bank.

FIGURE 12-11.   Left, *molding a straight channel in Mississippi River sand at the U. S. Waterways Experiment Station, Vicksburg, Mississippi;* right, *sinuous course produced after 72 hours of flow in an initially straight channel. (Photos from J. F. Friedkin; courtesy of the U. S. Waterways Experiment Station.)*

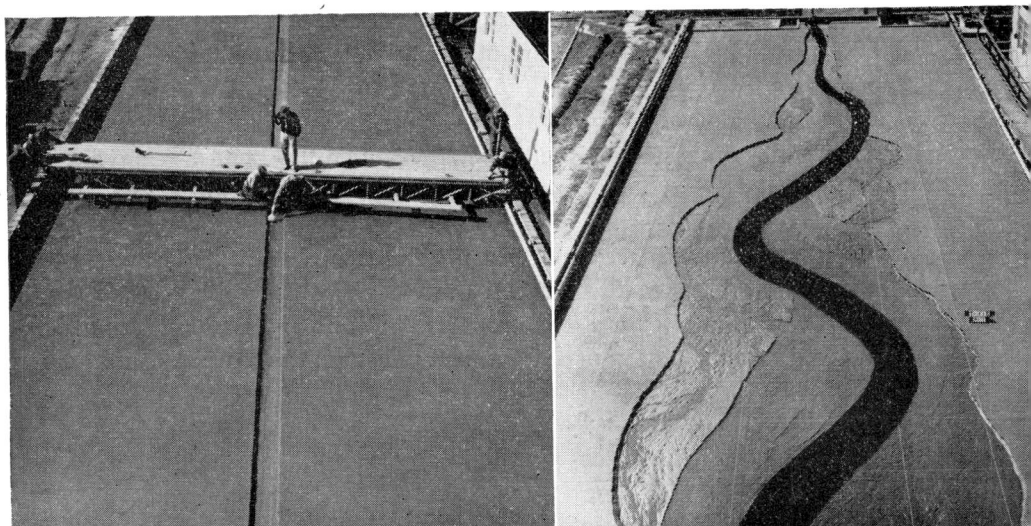

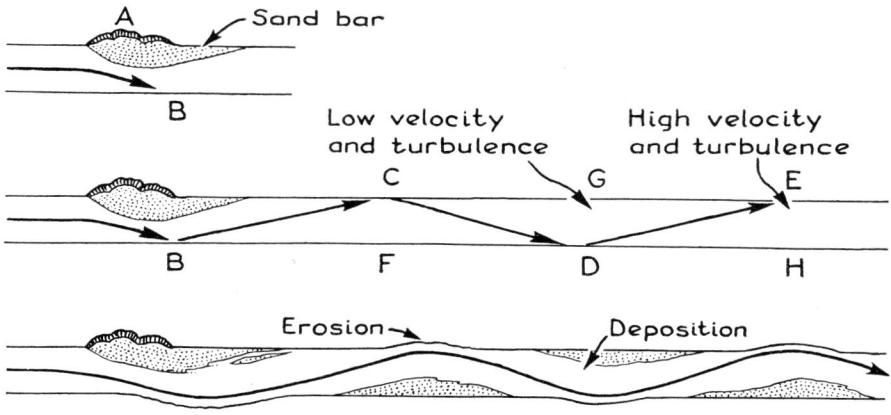

FIGURE 12-12. *Vertical views showing how a cave-in, at A, started the development of a sinuous course in the experiment of Figure 12-11.*

Figure 12-12 illustrates the sequel to this initial cave-in. The sand bar at A forced the stream against the opposite bank, B, from which the current was deflected through the points C, D, E, and so forth, like a ball oscillating against the sides of an inclined trough. These deflections made the currents at B, C, D, and E more competent than at F, G, and H. The banks under lateral attack were eroded, and the load thus acquired was deposited in the first downstream area of low turbulence. Deposits on one bank crowded the stream against the opposite bank, so that the angle of attack on this bank became greater and greater, gradually forming a series of migrating bends.

During the experiments discharge, valley slope, load, and erodibility of bed material were individually varied, in an attempt to ascertain their effects on sinuosity. With increased discharge, the meanders widened in a regular and predictable way, verifying the general rule in nature that big rivers have big bends, little rivers little ones. For each set of flow conditions there was a certain maximum meander width. The chutes and cutoffs are apparently the mechanism—a sort of safety valve—that keeps this neat balance.

Likewise, the valley gradient directly and predictably affected the meander width: the steeper the slope, the wider the bends. A stream with steep gradient thus lengthens its course by forming wide bends, thereby decreasing its gradient.

When the streams were not given any load of sand at the head of the model channel (as in Fig. 12-11), the meanders developed only *after* the stream was able to gain a load from its banks downstream. In experiments with channels molded in partially cemented sand, the streams could gain no load from their banks and did not meander at all. Thus, as suggested by the mechanism of meander formation shown in Figure 12-12, in order for streams to develop sinuous courses, they must have easily erodible banks. This generalization is perhaps the most significant result of the Vicksburg experiments. It is strongly supported in nature by the common association of sinuous streams with wide valleys underlain by loose alluvial deposits.

Braided Streams

A limiting case was reached in the experiments when streams were fed more sand than their gradients allowed them to carry away. Here the channel quickly choked with sand bars. The stream was then deflected in all directions into such a spread-out and intricately subdivided cross section that no continued attack could be made on a bank and no meanders developed. Such streams are called *braided streams*. They are common below glaciers where sediment chokes the stream (Fig. 13-13); they also abound in sandy regions where more material is washed to the streams than can be transported in low-water stages (Fig. 7-7). The Platte River in Nebraska is an excellent example.

The Concepts of Base Level and Grade

All streams are limited in their ability to deepen their valleys by the level of the sea—or lake—into which they flow. If this limiting level, called the *base level* of the stream, remains unchanged, the stream tends to stabilize itself on that gradient which is just necessary to transport its load.

Sinuous streams like the lower Mississippi sidecut slowly through the alluvium of their valleys, but alter their gradients very little over long periods of time. Sand and gravel are moved from the outer banks of bends to crossings, and thence to the growing bars on the convex slope of the next bend. Thus over a given stretch of the river, erosion and deposition practically cancel one another's effects, although load is being moved continuously downstream. Similarly, the average gradient remains the same over considerable stretches of the river, though the bed gradient varies from deep bendways to shallow crossings. If the gradient and average channel shape remain the same, the hydraulic characteristics of the stream—the distribution of velocity and turbulence—do not change, and thus the pattern of erosion and deposition are stable. Stream flow is just adequate to deliver at the lower end of a given stretch a load equivalent to that introduced at the upper end of the stretch. Streams, or parts of streams, that are thus balanced between erosion and deposition are said to be *graded* or to be flowing at *grade*. If the slope of a graded stretch is reduced—as by slight movements of the earth's crust—the velocity and turbulence of the stream is reduced, and the stream deposits sediment until its channel is built up to the slope required by grade. If the gradient is increased, the velocity and turbulence increase, and the stream erodes its bed to the limiting slope of the graded condition. Changes in load or discharge similarly force streams out of the graded condition, until a new balance is attained. The steep tributaries of river systems are said to be *above grade* because they are obviously eroding downward into their beds. Rivers flowing *below grade*, characterized by braided courses, tend to deposit their load or *aggrade* their channels.

Stream Deposits

Besides the alluvial deposits of floodplains, stream deposition may produce deltas, alluvial fans, and extensive alluvial plains.

Some floodplains apparently result from long-continued stream erosion. A large number have been formed by the accumulation of stream deposits in pre-existing valleys. The aggradation of most of these valleys seems ultimately to have been caused by crustal warping, changes in sea level, or widespread glaciation.

Erosional Floodplains.   Under ideal conditions, floodplains may develop by protracted valley cutting. This ideal process has been summarized by some geologists under the heading "Stages in the Evolution of a Valley." This theory is probably best explained by applying the principles of stream erosion and deposition thus far discussed to a simple hypothetical example. Consider the stream of Figure 12-13. As long as it flows on a slope steeper than is required by the graded condition, it will cut into its bed, thereby tending to steepen its valley walls. This steepening initiates downslope

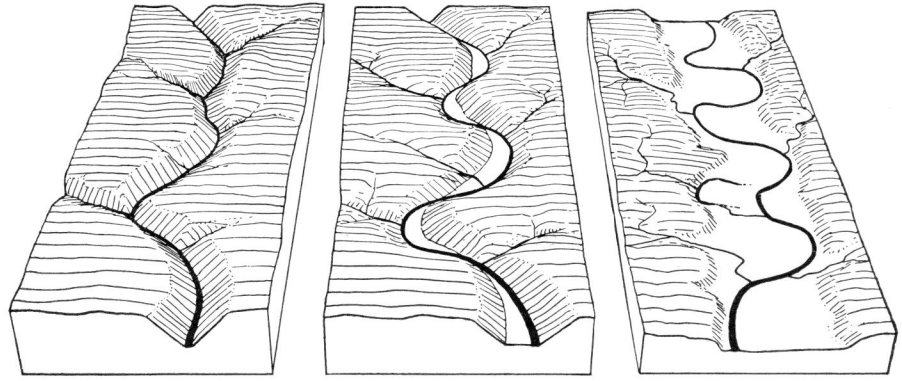

FIGURE 12-13.   *The evolution of an erosional floodplain.*

movements on the sides of the valley. As long as the stream is competent to remove this debris and, at least intermittently, to attack its bed, its valley will have a V-shaped cross section (stage of youth) (see Fig. 12-13, *left*).

Since the stream cannot cut below its base level, bed erosion gradually reduces the average gradient, thus diminishing its ability to cut down. Eventually the stream will become graded, and, except during the strongest flood flows, its bed will be protected by a veneer of sand or gravel.

Though such a stream can no longer cut downward, it may erode its banks by lateral attack against the valley walls. This will happen if the river is crooked at all. As in meandering streams, the mechanism of lateral erosion is simply the impingement of vigorous, turbulent currents against the outer banks of bends. The weak currents along the inner banks of these bends will simultaneously drop their load as beach or bar deposits. As the stream slowly cuts into its valley walls, it leaves behind a planed-off bedrock surface covered by a thin veneer of sand and gravel (stage of maturity) (Fig. 12-13, *middle,* and *right*). Such a process can eventually produce a broad continuous floodplain. The attainment of grade separates the stage of youth from that of maturity.

This sort of lateral erosion must be much slower than that of streams meandering on alluvium, primarily because of the greater erosional resistance of most bedrock. Another factor is the enormously greater load acquired by lateral cutting and downslope movements. When a stream cuts even a foot into the wall of a deep valley, it makes it possible for downslope movement to add tons of debris to its load.

Because of the great amount of time required to cut erosional floodplains, the probability that some change of conditions in or on the earth will upset the ideal progression of events is correspondingly great. It seems, indeed, that good examples of simple, laterally eroded floodplains are rare. Most present-day streams have been sufficiently affected by crustal movements, sea level changes, or climatic changes to produce more complex valleys. Widespread floodplains seem to be chiefly of aggradational origin.

Aggradational Floodplains.   Careful studies of the valleys of many sinuous streams show that their floodplains, instead of being alluvial veneers on a truncated bedrock surface, are underlain by deep deposits of sand, gravel, and mud. Among these streams are: the Lower Mississippi River; many stretches of the Missouri, Miami, and Ohio rivers; the short sinuous stretches of the lower Colorado, Columbia, and Snake rivers; and the Sacramento-San Joaquin River system of California. It is a notable fact that through the sinuous channels of these streams flows no less than two-thirds of the total runoff of the United States.

The evidence of aggradation is from drill holes and water wells in the floodplains. The Mississippi River Commission used several hundred drill-hole and water-well records in its study of the deposits underlying the Lower Mississippi floodplain. These showed that the plain is underlain by alluvial debris ranging in depth from about 100 to over 400 feet. Near Natchez this debris is 260 feet thick, with its base 215 feet below sea level. Drill holes across the present floodplain indicate that the buried surface, the valley of a prehistoric Mississippi, is not a smooth plain but a steep-sided, though shallow, valley with many tributaries.

A detailed study of the Santa Ana River in southern California demonstrates similar aggradation. Here an ancient channel is buried beneath stream deposits 140 feet thick at the present mouth. This filled channel, traced upstream in drill holes, has about the same gradient as that of the modern stream.

What has caused the marked filling in of such valleys as these? The deep filling of the Mississippi River Valley was probably induced in part by crustal downwarping (perhaps from isostatic sinking due to the load of the Mississippi Delta), and in part from a greatly increased stream load in the northern tributaries. The aggradation of the Santa Ana River was probably caused by a rise in sea level since the ancient channel was cut. The ultimate cause of this change in sea level and in the load carried by the Santa Ana and other ancient rivers was apparently the growth and melting of very large glaciers during the recent geologic past. These glaciers robbed the atmosphere, and ultimately the ocean, of enough water so that sea level was distinctly lowered. As the glaciers melted, not only was this water returned to the ocean, so that the sea level rose, but also large quantities of rock waste released by the glaciers were dumped into the existing river systems. Glaciers (described more fully in the next chapter), affected the streams of North America in many ways, upsetting the ideal evolution of stream valleys toward erosional floodplains.

Sinuous Streams in Glacial Valleys. Some of the best examples of meandering streams occupy valleys that once contained tongues of moving ice. As noted at the Nisqually glacier of Rainier National Park (Chap. 7), the load carried by glaciers is large; streams that flow from the snouts of glaciers commonly have very large loads supplied to them. As the glacier melts away, the load diminishes and the streams stabilize themselves on the flat, debris-filled valley floors. In the process they take on a sinuous course, and slowly rework the fluvio-glacial deposits which they have inherited. Literally thousands of streams in northern America, Europe, and Asia owe their meandering courses directly to glacial erosion and deposition.

Deltas. Any laden stream must drop its load when it enters quiet water such as a lake or sea. Where the Nile emerges from its valley near Cairo, it splits into channels. These further subdivide and flow to the sea on a broad plain of river deposits. Because of its triangular shape, the name *delta* was applied by Herodotus to this flat depositional plain.

Deltas may be triangular, often with a convexly curved border against the sea, or irregular with lobelike extensions like that of the Mississippi (Fig. 12-15). The shapes and sizes of deltas are considerably affected by the wave and tidal characteristics of the water bodies in which they are formed. Thus, the Mississippi and Colorado rivers, which empty into relatively calm and shallow gulfs, have prominent deltas, but the Columbia and Congo rivers have no subaerial deltas at all. The Columbia's load is distributed by ocean waves and currents for hundreds of miles along the sea coast; the Congo's is somehow washed down a long, deep submarine canyon into the depths of the Atlantic.

Where sand-laden streams flow into a deep, still body of water, the layers of deltaic sediment are not simply parallel to the lake or sea bottom, but

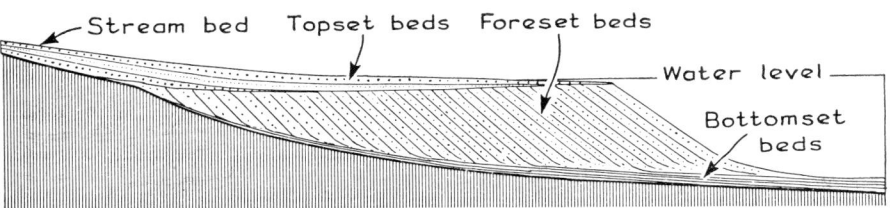

FIGURE 12-14. *Diagrammatic cross section of a simple delta.*

show a characteristically discordant arrangement (Fig. 12-14). By either bar deflection or overflow on the delta surface the stream tends to seek new channels, often subdividing into several *distributaries* over the delta surface. The stream deposits formed on the top surface of the delta are called *topset beds* and are commonly thin, because the distributaries must maintain certain minimum gradients across the delta. Thicker deposits, called the *foreset beds*, are laid down on the frontal slope of the delta. The finest part of the load, kept in suspension for a long time by weak currents moving down the frontal slope, is deposited on the lake or sea bottom, in front of the advancing delta. These *bottomset beds* are eventually covered by later foreset layers, often with notable angular discordance.

Most large deltas are much more complex than this ideal example. The Mississippi delta, like that of most other large rivers, shows little or no discordance between topset, foreset, and bottomset beds, probably because the load is fine grained and the sea bottom on which the delta is growing is only slightly inclined.

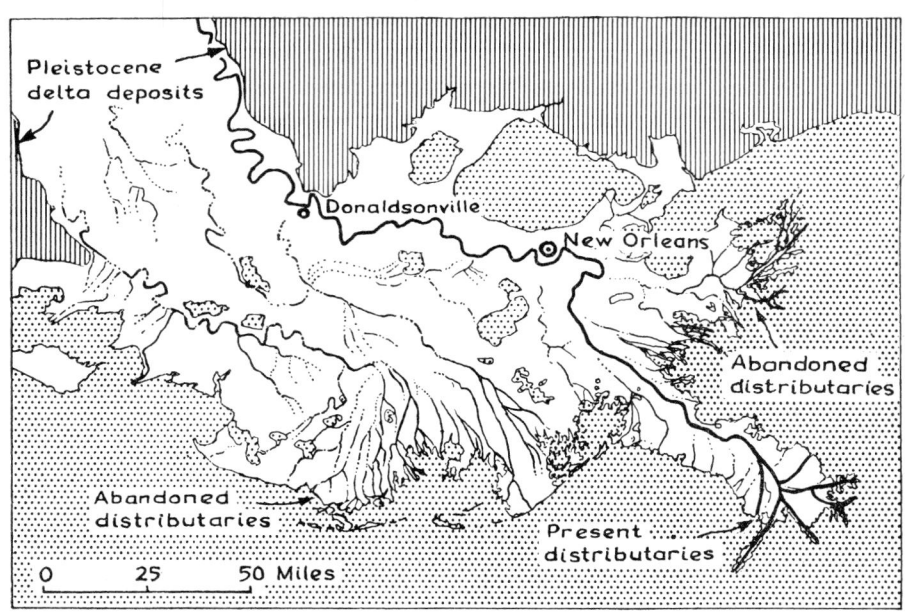

FIGURE 12-15. *The delta (white) of the Mississippi River. Note the old meander courses, the numerous lakes, the positions of abandoned distributaries, and the "bird's-foot" pattern of the present distributaries. (After H. N. Fisk, Mississippi River Commission, 1945.)*

As can be seen in Figure 12-15, although the delta extends far upstream, the main river divides into distributaries only within about 20 miles of its mouth. On the ancient upstream part of the delta the river is as sinuous as elsewhere on its floodplain; below Donaldsonville the gradient drops sharply and the channel becomes increasingly straight and deep. Here the natural levees are exceedingly broad and flat, showing a relief of only 5 or 10 feet in a breadth of one or two miles. Below New Orleans the delta surface outside the natural levees consists of sea-level marshes.

That the river has not always held its present course is proven by natural levees and other abandoned channel features on the marshy plain. The most recent of these abandoned courses is east of New Orleans. Others are shown on the map of Figure 12-15.

Alluvial Fans.   A stream emerging from a steep, narrow valley onto a broader lowland may build up a gently sloping conical deposit with an apex at the mouth of the narrow valley. Such a deposit is called an *alluvial fan* (Figs. 12-16 and 12-26). These depositional land forms are especially common in two locales: where steep tributaries debouch onto floodplains, and at the borders of steep-sided structural depressions such as downfaulted or sharply downfolded basins. They are particularly abundant in arid and semi-arid regions.

Deposition on an alluvial fan is like that on a small delta. The gradient lowers abruptly and the stream competence falls. The stream can no longer carry its load but must deposit it and thus build up a gradient that would permit it again to carry the load. Two factors delay this: (1) the diversion of the stream into many small channels and (2) the readily erodible walls of its channels in the fan deposits.

The slopes and structural characteristics of different fans vary with the size of the stream and the sort of load it is carrying. Small streams transporting coarse loads may construct fans with slopes as steep as 10° or 15° (as Fig. 12-16). Bedding is commonly indistinct or absent in such accumulations. The slopes of large fans (Figs. 15-3 and 15-4) commonly decrease from 3° to 5° at their apices to less than 1° near their bases. A decrease in the average grain size of the fan deposits goes hand in hand with this decrease in slope.

Basin-Filling by Streams. Streams do not everywhere move their loads to the ocean. Many relatively low areas form catchment basins for sediments. Such basins, *intermontane basins,* may be produced by natural dams of lava or sediment, by downfaulting or warping of the earth's crust, by glacial scour, or by combinations of these.

The runoff from more than 80 per cent of the area covered by the state of Nevada is confined to such basins. Here alluvial fans are abundant, often coalescing and overlapping at the foot of straight, steep mountain fronts to form continuous *bahadas* or *piedmont alluvial slopes.* The sediments in the central part of large intermontane basins may be so slightly inclined as to appear flat; the name *alluvial plain* is given to such a broad aggradational feature.

FIGURE 12-16. *Small alluvial fans formed when leaks in a high-level canal caused rapid gullying, near Leadville, Colorado. (Photo by M. R. Campbell, U. S. Geological Survey.)*

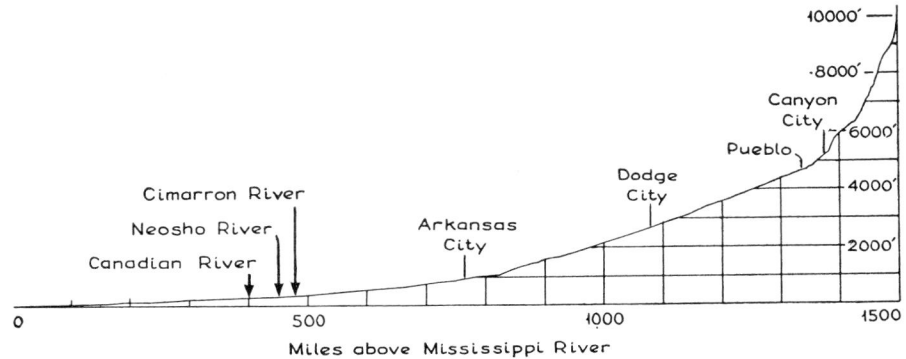

FIGURE 12-17. *Long profile of the Arkansas River, from Tennessee Pass, Colorado, to the Mississippi River. Note the great vertical exaggeration. (Modified from Henry Gannett, U. S. Geological Survey.)*

The Long Profiles of Streams

A fruitful way of finding the degree of balance between erosion and deposition in river systems is to make a graphic study of the stream gradient over long stretches of the stream's course. This is done by plotting the elevations of points on the stream's course against their distances from the mouth, and connecting these points with a line. Such a line is called the longitudinal or *long profile* of the stream. In order to show variations in gradient it is necessary to exaggerate greatly the vertical scale of such a plot, for the length of streams is commonly very great compared to their vertical fall.

The long profile of the Arkansas River, which rises in the central Rocky Mountains and flows across the Great Plains, joining the Mississippi 440 miles above its mouth (Fig. 12-17), shows slopes consistent with the general description of river systems given in the introduction to this chapter. In the upper course the gradient is steep and irregular; here the river flows in high mountain valleys and, just above Canon City, through the chasmlike Royal Gorge. In contrast, the gradient of the lower 200 miles of the river is even and very low, nowhere exceeding 1 foot per mile. Here the river meanders in a tightly serpentine pattern on a nearly flat, broad floodplain bordered by low bluffs. The notable thing shown by the profile, however, is that between these two extremes the gradient varies regularly and so permits the profile to approximate a smooth concave curve, steepening markedly near the river's origin. Probably at least half of the thousands of long profiles that have been drawn show this general form, though many others show departures from it. Before discussing the exceptional cases, we must try to discover the significance of the concave form.

The Profile of Equilibrium

Examination of many streams shows that it is especially their sinuous portions on floodplains that have nearly smooth concave profiles. Since these sinuous courses are commonly the effect of a graded condition, it may be said that graded stretches of streams show nearly smooth profiles. The Arkansas River, for instance, shows a fairly smooth concave profile between Pueblo, Colorado, and its confluence with the Mississippi; over nearly all this 1,350-mile course it flows on a floodplain, neither eroding nor aggrading its bed to any marked degree. The profiles of such graded rivers as the Mississippi, the Missouri, the Amazon, the Po, and many others display nearly smooth concave curves everywhere except in their uppermost courses. Nearly smooth concave profiles, then, are the ultimate goal toward which all streams work as they approach grade. They are the commonest type of the *graded profile* or the *profile of equilibrium*.

The general concavity, itself, of graded profiles raises an important question. If such rivers as the Arkansas are, indeed, graded over certain long stretches, how is it that their gradient changes almost continuously (Fig. 12-17)? That is, if a river is at grade on one slope, how can it also be at grade on a steeper one upstream? The factors that apparently allow this are: (1) the increase in discharge downstream and (2) the change in the grain size and quantity of load downstream.

Increase in Discharge Downstream. The profile of Figure 12-17 does not represent a stream of constant size throughout; like most rivers, the Arkansas grows in size downstream by the confluence of tributaries and the seepage of underground waters.

There is a decrease in the ratio of friction surface to channel cross section as a stream grows. Consider the junction of tributaries in Figure 12-18. If the channels are all one foot deep with nearly vertical banks, the result of the union would be the removal of four cross-sectional feet of friction sur-

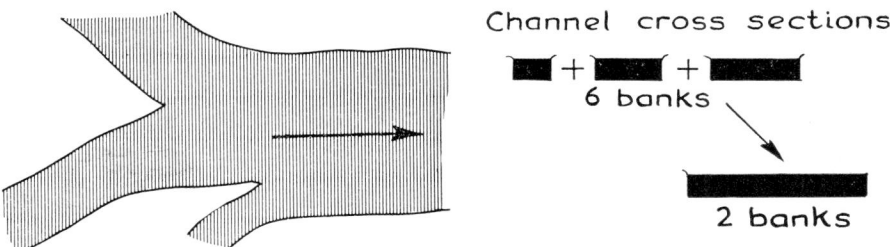

FIGURE 12-18. *Decrease in channel frictional area due to confluence of tributaries.*

face, in the diagram a 30 per cent reduction. The energy previously dissipated in friction is available for erosion of the channel, resulting in a lowering of the gradient below the junction.

Can the actual increases in discharge be correlated with changes in the slope? Though data on the discharge of all the tributaries, particularly the amount of addition from ground water, are not available, a general assessment of the relation between slope and discharge is possible. The Arkansas River is initially formed by the confluence of dozens of mountain streams which are, generally, flowing above grade. From Pueblo, where the river enters the plains, to Dodge City, contributions are restricted to small intermittent streams and seepage of underground water. In summer much of this stretch is braided. Between Dodge City and Arkansas City the discharge is increased by many small tributaries, and below Arkansas City by three major tributaries: the Cimarron, Neosho, and Canadian rivers. As shown by the profile, the gradient flattens below the junction of these tributaries. A similar relation has been noted in the Lower Mississippi River, whose gradient and discharge have been carefully studied.

Some statistical studies have been made to determine the cause of the concavity of a stream's profile. For example, the discharges of rivers in the French Alps have been measured regularly, day after day, at 73 places. When the mean maximum discharge at each station was graphically plotted against its local gradient, the points approximated a smooth curve—strongly suggesting a causative relation between slope and discharge.

Effect of the Load.   All graded streams do not show such a simple relation, probably because of the varying loads brought to the trunk stream by different tributaries. If the competence or capacity of a tributary is con-

FIGURE 12-19. *Rapids over a debris dam at the junction of Tapeats Creek with Colorado River, Arizona. (Photo by Freeman, U. S. Geological Survey.)*

siderably greater or less than that of the trunk stream, the confluence will force a steepening or flattening of the trunk stream's gradient. Such a situation may be illustrated by an extreme case on the Colorado River of Arizona. Here tributaries are dry through most of the year, but during the seasonal floods their swift currents move great quantities of coarse debris into the main river. Much of this material is too coarse to be moved by the more sluggish Colorado, and so it piles up as an apron of boulders, called a *debris dam* (Fig. 12-19). Since these boulders can be transported only after they

TABLE 12-1 *Variation in Different Grain Size Fractions of Lower Mississippi River Sediments, 100 to 1,000 Miles Below Cairo. (After Charles Nevin; from data of the U. S. Waterways Experiment Station, Vicksburg, Mississippi.)*

Miles Below Cairo →	100	300	500	700	900	1,000
Gravel, %	29	8	14	5	trace	none
Coarse sand, %	30	22	9	8	1	none
Medium sand, %	32	50	46	44	26	9
Fine sand, %	8	19	28	41	70	69
Silt, %	trace	trace	2	1	2	10
Clay, %	trace	trace	1	trace	1	10

have been slowly ground down by abrasion, from the standpoint of human history, debris dams cause practically permanent convexities on the river's profile.

Decrease in Grain Size Downstream. Though the debris dams of the Colorado provide striking examples of the effect of the grain size of the load on a river profile, more important, though more subtle, examples may be found in graded, sinuous rivers. A careful study of the bed materials of the Mississippi River shows a marked decrease in the size of the particles downstream. The change in grain size of 600 samples that were collected between Cairo and the Gulf of Mexico is indicated by the data of Table 12-1. The average composition of the delta sediments, which are 70 per cent silt and clay, is further evidence of this general change. On the Rhine River, between Basel and Bingen, there is a marked decrease in pebble size downstream that is closely correlated with a flattening of the river profile. It is noteworthy that the discharge increases only slightly over this stretch. Thus in some rivers the slow attrition of the bed load downstream appears to be a factor in allowing the stream to flow on an ever-decreasing slope.

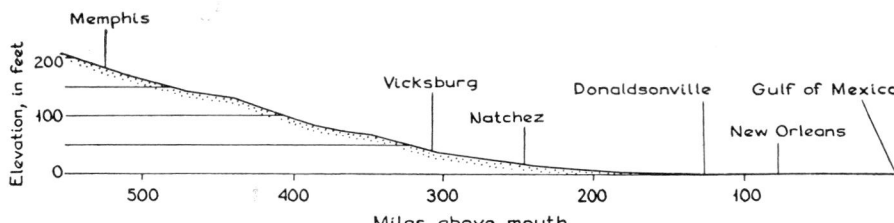

FIGURE 12-20. *Long profile of a part of the Lower Mississippi River at low water. (After H. N. Fisk, Mississippi River Commission, 1947.)*

Departures from the Ideal Profile

The smoothly concave curve of the ideal profile is closely approached but probably never attained by natural streams. Even in such rivers as the Lower Mississippi, probably as well-graded a stream as could be found, the profile shows slight but abrupt changes of slope between many individual stretches (Fig. 12-20). This is not at all abnormal. For one thing, the inflow of most tributaries demands such changes in gradient. Note that the vertical scale of this profile has been exaggerated 2,100 times in order to show these slight changes. Other streams show much more marked changes in slope that must be considered to be actual departures from the ideal profile.

One of the principal values of the concept of the ideal profile is that marked departures from it call attention to exceptional circumstances. The debris dams on the Colorado are an example. Lava flows, landslides, or drifting sand dunes may upset the graded condition of a stream and impose an irregularity on its profile. Glaciation in the recent geologic past has greatly affected the slopes of many streams. Most, if not all, of the major streams of the northern United States have irregular profiles inherited from glacial erosion and deposition.

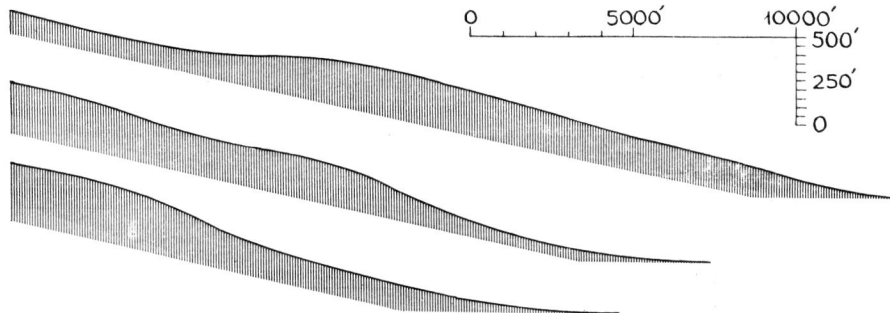

FIGURE 12-21. *Profiles of streams flowing down the flanks of anticlinal ridges in eastern Washington. (From maps of the U. S. Geological Survey.)*

Earth movements are also a factor, if they do not allow the stream quickly to adjust by degradation or aggradation. Some of the anticlines in the south central part of the state of Washington (Chap. 9) have apparently been up-lifted rapidly and recently enough to outstrip the streams flowing down their flanks. The profiles of Figure 12-21 illustrate the local broad convexity of stream gradients on some of these folds.

Even in streams that have long been cutting toward grade in a relatively stable region, there may appear distinct changes in slope—called *nick points* —on the profile caused by local outcrops of especially resistant rocks. The profile both above and below such a nick point may show a smoothly con-cave shape.

Considerable time is required for the ideal profile to form in regions of resistant rocks. Therefore all kinds of structural or climatic "accidents" may intervene before the graded profile can be formed. It is a tribute to the erod-ing power of running water that many large streams flow on grades that ap-proximate the profile of equilibrium.

Sculpture of the Land by Stream Erosion and Downslope Movements

Land forms result from the interplay between two opposing forces. On the one hand, earth movements and depositional processes raise certain parts of the crust; on the other, erosional forces constantly work to level them. Streams and downslope movements, attacking uplifted areas, produce land forms whose shapes depend chiefly on the structure of the underlying rocks,

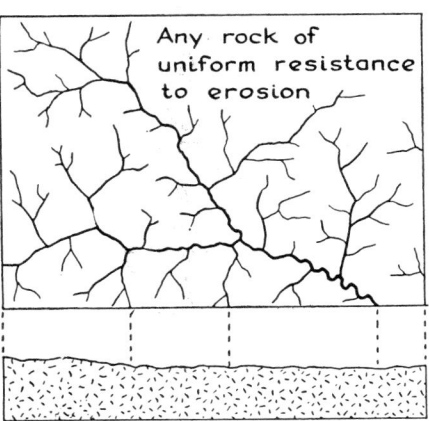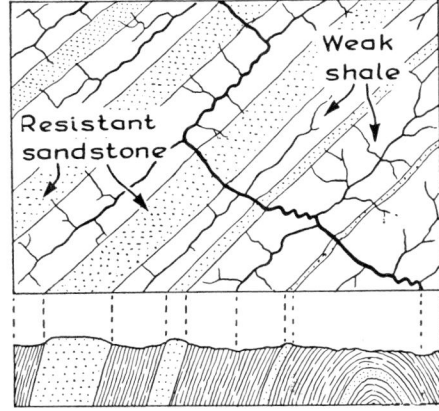

FIGURE 12-22. *Structural control of stream patterns;* left, *dendritic pattern;* right, *trellis pattern.*

FIGURE 12-23. *Structural control of land forms, subdued by soil creep, San Jose Hills, Los Angeles County, California. (Photo by Robert C. Frampton, Claremont, California.)*

the climate, and the nature of the uplift that initiated erosion. The first two of these factors can ordinarily be determined, and, if the processes of stream erosion are well understood, it is often possible to interpret the nature of the constructional force. The remainder of this chapter will serve to enlarge briefly on this possibility.

Headward Erosion and Landscape Patterns

Erosion in the steep headwater tributaries of a stream system not only deepens its valleys but also extends them by *headward erosion*. Stream systems thus tend to expand as they reduce the land, and, coupled with downslope movements, create new ridge and valley systems or modify old ones. Headward erosion is guided by the arrangement of slopes and the structure of the rocks being dissected. These two factors thus control the resulting pattern of stream courses and ridge lines. If the rock and its mantle offer uniform resistance to erosion, and if the slopes of the land surface are more or less randomly arranged, tributaries will branch and erode headward in a random fashion, producing a *dendritic pattern*, resembling the branching limbs of a tree (Fig. 12-22, *left.*) If the rocks being dissected are of unequal resistance, the extension and downcutting of tributaries will be most rapid on the weaker rocks, which will thus come to underlie valleys or lowlands between ridges or uplands of resistant rocks. Thus in a stream system flowing across a region of steeply dipping parallel beds, tributaries tend to concentrate on the belts of weak rocks, forming a rectangular or *trellis pattern* (Figs. 12-22, *right*, and 9-14). Other examples of the *structural control* of drainage patterns and ridge lines are the concentric patterns on eroded domes and basins (Fig. 8-1) and the strikingly linear stream courses along some faults (Fig. 10-14).

In temperate regions the structural control of land forms is usually more subtle than in the examples just cited. In Figure 12-23, for example, the major ridges are underlain by resistant rocks and the larger valleys by soft rocks. Yet neither the drainage pattern nor the arrangement of slopes gives much clue to the underlying structure. Even when on the ground, it is difficult to tell what resistant rock underlies a given ridge, because downslope creep of a layer of grass and soil has masked the rock contacts, forming smoothly rounded hills.

Structural Terraces and Plains. The structural control of stream erosion and downslope movements can go further than simply orienting individual

FIGURE 12-24. *A broad structural plain along the Colorado River, Arizona, formed where soft shale overlies a more resistant rock. The shale is overlain by a massive sandstone bed, remnants of which form cliffed buttes in the center of the view. (Photo by Robert C. Frampton, Claremont, California.)*

FIGURE 12-25. *New terraces formed after relative uplift of the Sonoma Mountain block, Nevada, at the time of an earthquake in 1915. (Photo by Ben Page.*

valleys and ridges. In thick horizontal strata of unequal resistance to erosion a stream will cut quickly through weak beds but will be arrested by a resist ant layer. Thus, a general, though temporary, base level is formed, and head ward erosion by tributaries may strip the weak layer over a large area, form ing a *structural plain* on the resistant bed. In an arid or semi-arid climate the dissected upland border is eroded back as a steep embayed slope, and temporary upland remnants called *mesas* and *buttes* are commonly left be hind to attest to the stripping of the plain (Fig. 12-24). If a stream cuts its valley in a sequence of alternating weak and resistant beds, each resistant bed will form a step or *structural terrace* in the side slope of the valley. Such terraces are common in the Grand Canyon of the Colorado (Fig. 7-6) and where the Columbia and Snake rivers cut into the horizontal lavas of the Columbia Plateau.

Stream Terraces

More common than structural terraces are valley terraces that cannot be explained by differences in the resistance of flat-lying strata. Some such terraces are paired across a valley, others are single. Some are made up en tirely of river deposits, others are cut in bedrock but have veneers, channel fillings, or residual patches of river gravels on their flat surfaces. From such facts as these, it seems certain that these terraces are the remnants of old floodplains, now incised by the streams that once made them. Why a stream

FIGURE 12-26. *Faulted and entrenched fans at the mouth of Tuber Canyon, Panamint Range, California. (Photo by John Shelton.)*

thus dissects its floodplain is in many cases difficult to understand. In general, terraces are initiated by earth movements, changes in sea level, or any factor that puts the stream above grade.

Terraces obviously caused by earth movement were formed shortly after the Pleasant Valley earthquake of 1915 (Chap. 9). The 12-foot offset along the front of Sonoma Range (Fig. 9-2) produced a sharp fall in the profile of streams crossing the fault. Erosion at this step was rapid, so that by 1930 it had retreated far upstream, leaving a 12-foot channel that broke the old floodplain into a pair of matched terraces (Fig. 12-25). Because this stream now sidecuts at its new graded level, the terraces will be reduced in width, though probably parts of them, particularly downstream from projecting ridge spurs, will remain for a very long time. It is of interest to note that even if the 1915 fault offset had not been observed, a clue to the movement could be found by examining the stream terraces, for they parallel the present stream profile in the Sonoma block, but end abruptly at the fault line.

Another example of terraces caused by faulting occurs on the west side of the Panamint Mountains, California (Fig. 12-26). After faulting formed a series of small scarps across the large alluvial fans bordering the range, streams cut trenches in the uplifted portions of the fans and formed new fans below the scarps. Since stream valleys within the mountains had no floodplains, the terraces extend only from the new fault line to the mountain front.

Careful mapping of terraces in the Mississippi River Valley has suggested earth movements of a broader and more complex nature. From Cairo to Natchez, Mississippi, several terraces are approximately parallel to each other and to the present river profile, a few tens to a few hundreds of feet above the latter. South of the Mississippi-Louisiana line, the terraces converge downward, and lie on Pleistocene alluvium instead of on older rock. Near Baton Rouge they disappear beneath the present floodplain. This pat-

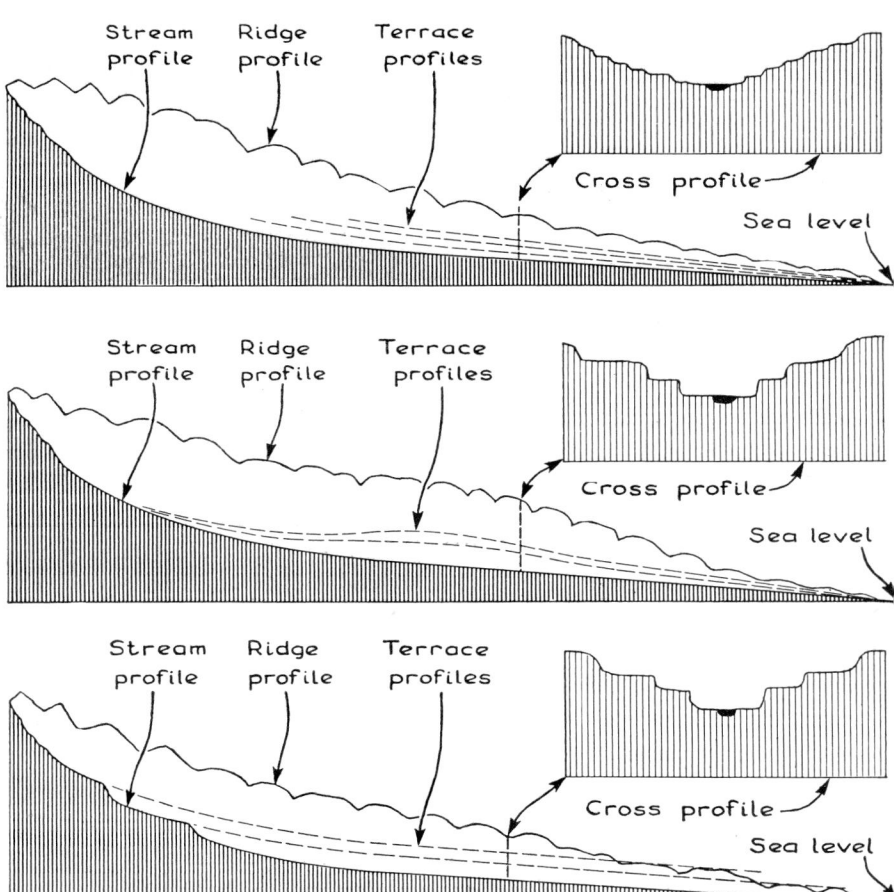

FIGURE 12-27. *Ideal profile patterns along terraced stream valleys. Top, many low unmatched terraces formed during normal valley sculpture. Note that they gradually converge to base level. Middle, matched terraces initiated by intermittent upwarping in the midcourse of the stream. Bottom, matched terraces formed after two drops in sea level. Note that they parallel the stream profile except upstream where they can be correlated with two nicks in the stream profile.*

tern of warped terraces indicates a striking example of slow earth movements, for apparently the delta region has subsided while the area crossed by the river valley has progressively risen.

Terraces may be produced by climatic changes that upset the graded condition of a stream. They may also form even during the normal progress of valley sculpture because a trunk stream flowing through hilly areas will have its load reduced as its tributaries slowly lower the surrounding uplands. The trunk stream thus deprived of load is able to flow on a lesser gradient and to incise its floodplain, producing terraces that are commonly numerous, unmatched, and of small areal extent. These terraces record only the lateral sweeps of the stream as it slowly lowered its floodplain. Though, in general, terraces formed after earth movements are matched and more extensive than these, slow movements or changes of sea level can produce similar forms. The only way terraces can be used to establish, with a fair degree of certainty, recent events of earth history is to trace them out mile by mile and plot their profiles in relation to the profile of the present stream. The simplified diagrams of Figure 12-27 suggest ideal cases for interpretation.

Extensive Erosion Surfaces of Low Relief

An example of a well-marked erosion surface of fairly low relief, now being destroyed by erosion, but too extensive to be called a terrace, occurs in the

FIGURE 12-28. *The erosion surface at Harrisburg, Pennsylvania (foreground and middle distance). The level summits of the distant ridges are remnants of an older, more thoroughly dissected erosion surface called the Schooley surface. (Photo by George H. Ashley; courtesy of Department of Internal Affairs of Pennsylvania.)*

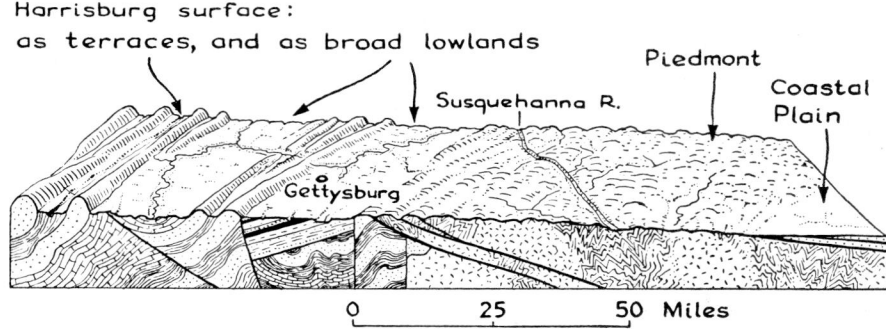

FIGURE 12-29. *Relief diagram showing the distribution of the Harrisburg surface in a strip across southern Pennsylvania, just south of Harrisburg. The underlying structure is generalized. (After maps and folios of the U. S. Geological Survey.)*

central and southern Appalachian Mountains. This old surface is well developed in the vicinity of Harrisburg, Pennsylvania, and so is called the "Harrisburg surface" (Fig. 12-28). In some areas at or near the east edge of the range this surface is many miles wide, undulating, and mostly underlaid by deep residual soils. The surface has been formed on all but the most resistant rocks of the region, cutting across flat and tilted beds alike. The Susquehanna and other rivers have cut 200- to 300-foot gorges into it, and lesser streams have partially dissected it. The Harrisburg surface exists only as valley terraces in the parts of the Appalachians that are dominated by ridges of resistant sandstone (Fig. 12-29). The terraces are 200 to 300 feet above the streams in their lower courses, but the terrace gradients are commonly less steep than those of the present streams, so that upstream the terraces decrease in height. Most river valleys in the southern Appalachians have similar terraces that can be approximately correlated with the Harrisburg surface. In some places broad passes between drainage systems, followed by roads and railways, are on the Harrisburg surface, though it is of interest to note that locally the surface is not at the same elevation in different valleys on the two sides of such passes.

Widespread erosion surfaces, commonly called *peneplains,* similar to the Harrisburg surface but generally with still lower ridges, occur in many parts of the world. Some, as in the central part of the United States, from Missouri and Kansas south, lie near the graded level of through-going streams. Others, as will be seen in Chapter 19, have been uplifted and so deeply eroded that only scattered remnants are left as flat mountain summits. How are such extensive, nearly flat surfaces formed? Careful study of the better preserved surfaces reveals that many are the result of erosion by streams flowing very near their base level. Evidence for such an origin lies in the presence of a deep zone of thorough weathering, the fact that all but the hardest rocks—in all attitudes—are truncated, the presence of residual stream channels and stream gravels, and the complete absence of surficial marine or glacial deposits. Though these surfaces are similar to terraces in some respects, they differ from them in other ways besides their size. Floodplains may be formed by the sidecutting of a single stream, but surfaces such as that at Harrisburg are produced by downslope movements combined with the work of many streams. For this reason these surfaces are not flat, but undulating, their low crests and gentle slopes being the last remnants of hills and ridges.

Though it is tempting to suggest that such extensive plains are to be expected as the common end product of subaerial erosion, the present status of the earth's landscapes does not bear this out. All such surfaces of low relief that have been found by careful study to be the result of stream erosion are, like that at Harrisburg, variously uplifted, warped, dissected, or otherwise imperfect. However, the existent remnants of these surfaces are evidence of more stable periods preceding the present, and, like stream terraces, locally enable us to estimate the nature of crustal and climatic changes during the relatively recent geologic past.

Facts, Concepts, Terms

LAMINAR FLOW; TURBULENT FLOW

RELATION OF VELOCITY TO TURBULENCE

STREAM LOAD
 Suspended load; bed load; dissolved load

COMPETENCE OF STREAMS

SHOOTING FLOW AND STREAMING FLOW

ABRASION BY STREAMS

NATURE OF SINUOUS RIVERS
 Meanders; floodplains; natural levees
 Cutoffs; chutes; oxbow lakes

SCALE MODELS OF SINUOUS STREAMS
 Factors in the formation of artificial meanders

BRAIDED STREAMS

BASE LEVEL; GRADED STREAMS

EROSIONAL FLOODPLAINS; AGGRADATIONAL FLOODPLAINS

DELTAS; ALLUVIAL FANS; BAHADAS; ALLUVIAL PLAINS

LONG PROFILES OF STREAMS; PROFILE OF EQUILIBRIUM

SCULPTURE OF THE LAND BY STREAMS

   Headward erosion; dendritic and trellis patterns
   Stream terraces
   Peneplains

Questions

1. Would you expect the dissolved load per cubic foot of water to be higher in the Columbia (high rainfall) or in the Colorado River (low rainfall)? Why?
2. The Shenandoah River has a meandering course in the Great Valley. It then flows into the Potomac, which is cutting through a rock-walled canyon. In light of what was said of the erosion of the Appalachian Mountains in Chapter 9, how can this anomaly be explained?
3. Can you draw simple cross sections, like those of Figure 12-27, to show the arrangements of terrace profiles, stream profiles, and ridge profiles for the Sonoma and Panamint Range examples given on page 279?
4. Engineers have made many *artificial cutoffs* (Fig. 12-7) in the Lower Mississippi and other sinuous rivers. Considering the nature of sinuous channels, can you suggest reasons for these projects?
5. List several criteria for distinguishing between floodplain, delta, and alluvial fan deposits in ancient sedimentary rocks.
6. Suggest how climatic changes might produce stream terraces in areas with which you are familiar.
7. Experiments show that streams can transport clay-sized particles more easily than silt, but that clay is much more difficult to *erode* than silt. In view of these facts, can you suggest a reason for the absence of meanders on the Mississippi River delta, below New Orleans?
8. The St. Lawrence, one of the great rivers of the continent, has no delta, even though it runs into a landlocked estuary. Can you suggest why?
9. The profile of equilibrium resembles the land portion of the graph in Figure 2-9. Can you offer any explanation of this?
10. Where is the sediment coarsest on an alluvial fan? Why?

Suggested Readings

1. Cotton, C. A., *Landscape*, Cambridge University Press, Cambridge, 1941. (A systematic and well-illustrated book on the development of land forms by stream erosion.)

2. Fisk, H. N., *Fine-Grained Alluvial Deposits and Their Effects on Mississippi River Activity*, Waterways Experiment Station, Vicksburg, Miss., 1947. (A clear, well-illustrated account of the Mississippi floodplain, with an interesting theory for its origin.)
3. Gilbert, G. K., *Geology of the Henry Mountains* (pp. 99-150, "Land Sculpture"), U. S. Geographical and Geological Survey of the Rocky Mountains Region, 1877. (The classical American statement of the principles of stream erosion.)

13. *Glaciers and Glaciation*

GLACIERS are slow-flowing, thick masses of ice. *Snowfields* are thinner and essentially motionless masses of permanent snow (Fig. 13-1).

Snowfields and glaciers are savings banks in the water economy of the earth. Each year, some of the water evaporated from the seas and lands falls as snow. Most of the snow melts in summer, but in polar regions and at high altitudes part of it remains throughout the year in glaciers and snowfields. If withdrawals by melting and evaporation exceed deposits as snowfall, the balance in the bank decreases and the glacier recedes, but if precipitation exceeds withdrawals, the glacier grows and spreads outward.

The Snowline

The lowest altitude at which permanent snow rests is called the *snowline*. Obviously, it does not persist always in one position but is affected by variations in the annual snowfall, temperature, wind direction (which controls drifting), and topography (which controls both avalanching—snowsliding—and shading from the sun).

Mean annual temperature decreases at higher altitudes and at higher latitudes, hence the snowline is higher near the equator than in polar regions. The amount of snowfall also affects its position. Thus, the snowline stands over 5,000 feet higher (8,000 to 9,000 feet) on the dry eastern side of the St. Elias Mountains along the Alaska-Yukon border than on the wet western side (2,500 to 3,000 feet), and it is lower in well-watered Norway than on the colder but dry Taimyr peninsula. In much of Siberia, northern Alaska, and Canada, the mean annual temperature is low enough, but the

FIGURE 13-1. *Snowfields above the head of a valley glacier, Alaska. The crevasses mark the start of the flowing glacier. (Photo by U. S. Air Force.)*

precipitation is too scanty to support permanent snowfields and glaciers. In such areas, the pore moisture in the soil and rock is frozen to considerable depths, forming a great sheet of *permanently frozen ground.*

Snowfields

Above the snowline, on all except the steepest and windiest slopes, lie permanent snowfields (Fig. 13-1). Excavations show that the beautiful geometric patterns of new-fallen snowflakes (Fig. 13-2) do not persist at depth. Instead, the snowfield consists largely of small rounded granules of ice about the size of birdshot. This material, called *firn,* has been formed by compaction of the feathery snowflakes and by alternate thawing and refreezing of their edges. This melting and refreezing is not due entirely to variations in air temperature above the snowbank. Water expands 9 per cent when it freezes; hence, pressure tends to lower the melting point of ice. Thus, even at temperatures slightly below the normal freezing point (0° Centigrade or

FIGURE 13-2.   *Forms of fresh snowflakes. (After A. E. H. Tutton, 1927.)*

32° Farenheit), the snowflakes deep in a snowbank may be so tightly packed by the weight of later snowfalls that a thin film of water forms at the points of contact between them. This water film may then move to a nearby pore space, where the pressure is lower, and refreeze into ice, even though there has been no change of temperature.

Rains falling on the snowbank and meltwater formed on a warm day are absorbed as in a blotter by the granular firn. Perhaps the following night this water freezes, assisting in the compaction of the firn by excluding air spaces. Thus, deep in the snowfield the firn gradually changes to interlocking granules of solid ice. Crevasses in glaciers commonly show all steps in the transformation: Snowflakes → Firn → Solid Ice.

Glaciers

If snow and firn merely continued to accumulate each winter, even the highest mountains in arctic regions would eventually be buried. Ice, however, is not a strong mineral, and its strength limits the height to which it can accumulate. Before any great thickness is reached, the ice at the bottom begins to creep and flow outward under the overlying weight. Thus, from the snow and firn a *flowing ice mass*, or *glacier*, is formed.

Glacier Motion

Glaciers move so slowly that the motion cannot be seen, but the fact that the ice actually is moving can be readily proved by driving a straight row of stakes across the surface of the glacier. In a few days or months the straight line becomes a curve, bent downstream in the center of the glacier where the speed is greatest. The rate varies in different glaciers and also with the season of the year. In a few of the Alaska and Greenland glaciers, velocities of more than 150 feet per day have been recorded, but such rates are exceptionally high. Figure 13-3 shows the rate of movement characteristic of glaciers in the Alps. The lower parts of some glaciers on the coast of Alaska appear to be practically stagnant.

Indirect proof of glacier flow is given by the rocks frozen into the glacier.

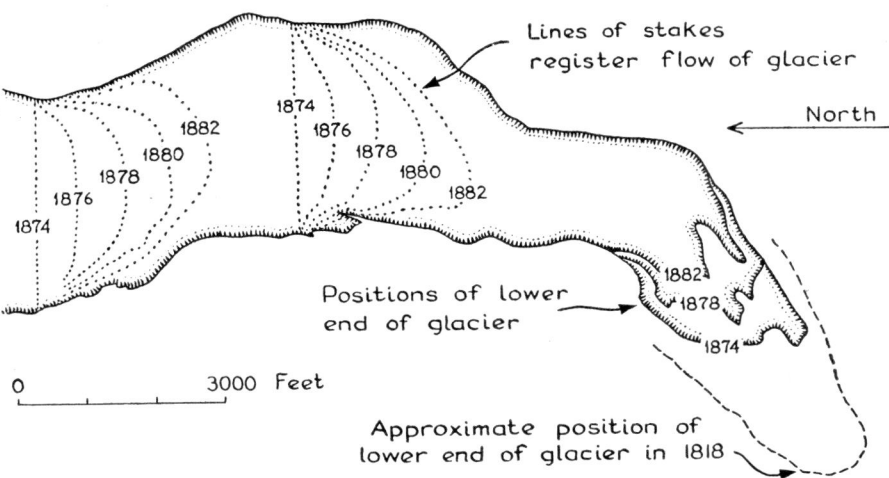

FIGURE 13-3. *Records of flow and frontal shrinkage of the Rhone glacier, Switzerland. (After A. Heim.)*

Most large glaciers contain abundant boulders which must have been carried from far upstream by the moving ice, because no similar rock is found in the valley walls nearby. The bodies of climbers who crashed to their death in the treacherous crevasses of the Bossons glacier in the Alps were released 41 years later at the terminus of the glacier several miles below.

Nature of Glacier Flow.   The upper surface of glaciers is riven by crevasses, and the ice appears to be hard and brittle. It shatters like glass under a sharp blow, just as does an ice cube from the refrigerator. It is difficult to imagine such a substance flowing. Actually, there is little or no evidence that the brittle, upper part of a glacier does flow. The abundant crevasses prove that fracture is the chief form of yielding. Deep within the glacier, however, deformation is almost entirely by flowage. Glaciers are perfect examples of the effect of scale on strength (Chap. 3). Ice in small masses is rigid and brittle, but in glacial masses it flows under its own weight. The brittle, crevassed surface ice is rafted along on the flowing ice below.

The mechanism of glacier flow is complex. Microscopic studies of ice obtained from glaciers show that some of the ice crystals have been bent, others have glided along the sheets of atoms parallel to the base of the hexagonal crystals, and still others have been granulated and sheared. Furthermore, the crystals tend to be arranged in parallel planes, as in foliated metamorphic rocks, suggesting that they recrystallized during or after movement. Such recrystallization is demonstrated by the coarser crystals of glaciers as compared to those in snow and firn. Crystals in the firn of Alpine snowfields average less than ¼ inch in diameter, but at the glacial snout the grains average about an inch in diameter. Crystals 6 to 8 inches in diameter may be collected from the edge of the Malaspina glacier in Alaska. The large crystals must have grown by recrystallization of smaller ones.

The ice at the front of a glacier, or in deep glacial crevasses, is generally layered. The layering superficially resembles stratification. It lies roughly parallel with the floor of the glacier and curves upward along its walls. Measurements at the glacier front have shown that the ice in adjacent layers moves at different speeds, proving that the layers are separated by shear surfaces. This layering is one manifestation of shear flow—the layers are due to the development of many closely spaced surfaces of shear within the glacier along which the ice and contained rock detritus are dragged forward and streaked out by movement. The shear-banding of glaciers closely resembles the foliation of many gneisses and schists. Clearly, ice under a relatively small load is no longer brittle, but reacts to differential pressure by flowage and recrystallization. We have, in glacier ice, an example of a metamorphic rock, completely transformed from the original snow and firn into a wholly different streaky crystalline mass by movement and recrystallization under pressure.

Kinds of Glaciers

The form of a glacier is largely controlled by the topography over which it flows.

Valley glaciers are streams of ice that flow down the steep-walled valleys of mountain ranges. Fed by snowfields on slopes and in catchment basins above the snowline, the glacier may extend far below the limit of perpetual snow. All glaciers end at the point where the rate of melting at the front equals the rate of ice replenishment by flowage.

Valley glaciers are typical of lofty mountain ranges the world over. Most of those in the United States, except for the extensive glacier system on Mount Rainier, Washington, are short blunt ice streams only a few hundred feet thick. It is difficult to distinguish many of them from snowfields; indeed there are all gradations.

The Rocky Mountains, the Cascade Range, and the higher parts of the Sierra Nevada contain hundreds of small, rounded or irregular ice masses that lie in steep-walled clefts or else occupy amphitheater-like depressions opening out over steep slopes or cliffs (Fig. 13-4). Such masses are called *cliff glaciers* or *hanging glaciers*.

In contrast to these puny ice streams, many of the valley glaciers in the Himalayas and Alaska are 30 to 70 miles long and over 3,000 feet thick. Each

FIGURE 13-4. *Small cliff glaciers, Sierra Nevada, California. Note how the rock in the foreground has been shattered by frost action. (Photo by François Matthes, U. S. Geological Survey.)*

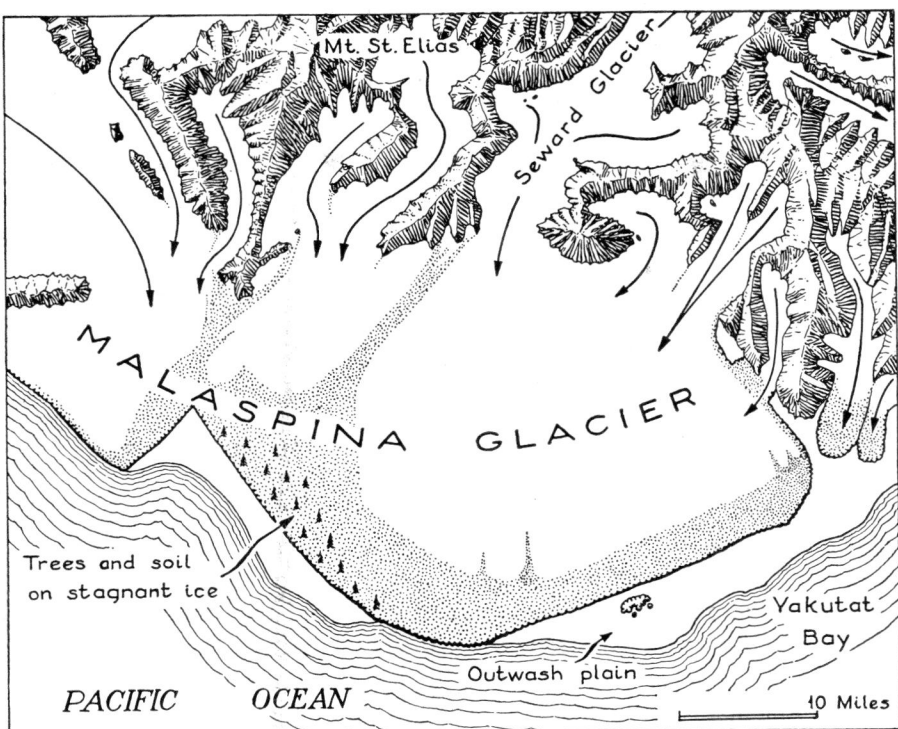

FIGURE 13-6. *Map of the Malaspina glacier, Alaska. The arrows indicate the flow of valley glaciers that feed the Malaspina. (After R. S. Tarr and L. Martin, 1914.)*

glacier has numerous tributary ice streams that merge to form an integrated system (Fig. 13-5) that may drain an area of several hundred square miles.

At the foot of the St. Elias Mountains in Alaska several of these long systems of valley glaciers emerge from the mountain front and spread out on the plain below to form the Malaspina glacier, a lobate mass of ice covering 800 square miles (Fig. 13-6). Such glaciers are called *piedmont glaciers* (compare with piedmont slopes, p. 269).

The largest glaciers are huge ice sheets, called *continental glaciers*. They are found today only in the polar regions. All of interior Greenland (about 637,000 square miles) is covered by ice, leaving only a narrow fringe of land along the coast. The Greenland glacier spreads outward in all directions from two high points in the interior. Locally, the coast is bordered by lofty

FIGURE 13-5. *Valley glacier with numerous tributary glaciers, Alaska. (Photo by U. S. Air Force.)*

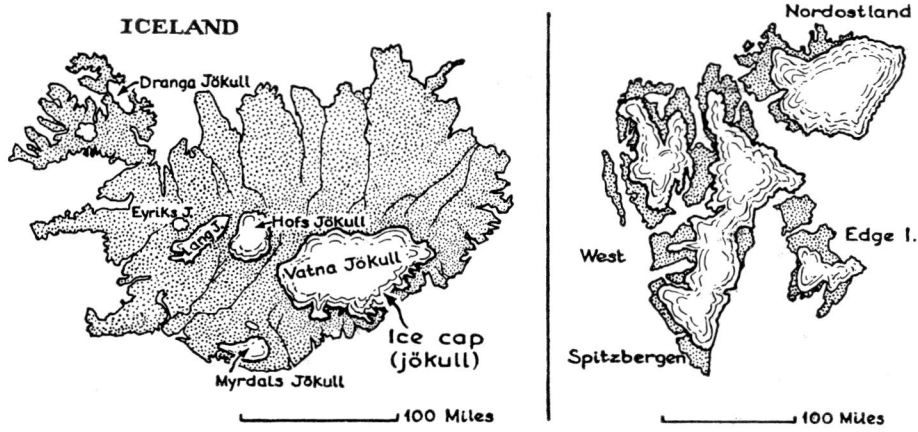

FIGURE 13-7.   *Ice caps of Iceland and Spitzbergen.* (*After* Stieler's Atlas.)

mountains through which the glacier spills, splitting up between the peaks into valley glaciers which then descend the coastal valleys to the sea.

Antarctica supports a much larger continental glacier, estimated to cover 5,000,000 square miles—an area larger than that of the United States and Mexico. The Antarctic glacier overrides the coast and projects into the "shelf ice" formed by the freezing of the sea. In places, the Antarctic glacier is held back by mountains through which the ice escapes in a series of large valley glaciers. The famous Beardmore glacier, ascended by several of the early explorers in their quest for the South Pole, is one of these outlet rivers of ice from the high polar ice mass. The Beardmore glacier is 300 miles long, 12 miles wide, and extends far out into the shelf ice even after it reaches the sea.

Small masses of radially spreading ice are found on Iceland, Spitzbergen, in parts of Scandinavia, and on the islands north of Canada (Fig. 13-7). They are called *ice caps* to distinguish them from the huge continental glaciers.

Erosion and Deposition by Glaciers

Glaciers are much more powerful agents of erosion than rivers, but, because ice is restricted to high altitudes and to polar regions, running water moves far more rock debris each year than does ice.

Acquiring of Load

Frost Weathering and Avalanching.   Glaciers acquire rock debris in several ways. Valley glaciers commonly cover only a small part of the mountains in which they lie. On the gentle slopes between and above them lie extensive snowfields, but on steep windswept slopes even snow is unable to cling throughout the year. Hence, great expanses of craggy ridges, peaks, and cliffs rise above the glaciers (Figs. 13-4 and 13-5). These bare rock surfaces are strongly shattered by frost action (Fig. 13-4). During the daytime, meltwater from snowbanks seeps into the rock crevices, and then, perhaps that same night, freezes and breaks blocks loose from their parent ledges. Small grains, loose chips, and even great blocks of rock thus freed tumble down the steep slopes and accumulate in talus piles on the edge of the glacier. Much debris is also carried down by avalanches (snowslides) during the spring thaw. Freshets from melting snow or summer rains also wash much debris onto valley glaciers.

Gradually, by being buried under new accumulations of snow or avalanche deposits, and by falling into crevasses opened by the motion of the glacier, much of this rock debris becomes incorporated within the ice, especially along its edges. The edge of a valley glacier may contain so much debris that it shows up in photographs as a conspicuous dark streak (Figs. 13-5 and 13-13). These stripes of dirty ice and loose rock along the edges of the glacier are called *lateral moraines*. Where two valley glaciers unite, the adjacent lateral moraines merge and form a *medial moraine*. If tributary glaciers are numerous, many medial moraines may streak the surface of the main glaciers (Figs. 13-5 and 13-13). The moraines are not just surface features, they are wedgelike masses that extend deep into the glacier.

Meltwater Shattering. The exposed crags above a glacier are subjected to strong frost action but a glacier blankets its floor from extreme temperature changes. However, the rock floor of a thin glacier may be exposed to meltwater at the bottom of crevasses, and here a change in temperature will cause some frost shattering. At the head of most valley glaciers is a deep arcuate crevasse, or series of closely spaced crevasses, called the *bergschrund* (Figs. 13-1 and 13-8). The bergschrund gapes open in summer, but is filled or bridged over by snow in winter. It is inferred that the bergschrund is formed by downstream movement of the plastic ice in the lower part of the glacier which breaks the brittle ice above and pulls it away from the rock wall.

Adventurous observers have descended into bergschrunds on ropes and found the lower parts to have a rocky wall on the upstream side and an ice wall on the other (Fig. 13-8). The rock wall is shattered and riven by joints, some of which gape open and disclose slight displacements of the angular blocks of rock that they outline. Some of the blocks are free but have moved only slightly; others lean out against the ice in precarious unbalance; and

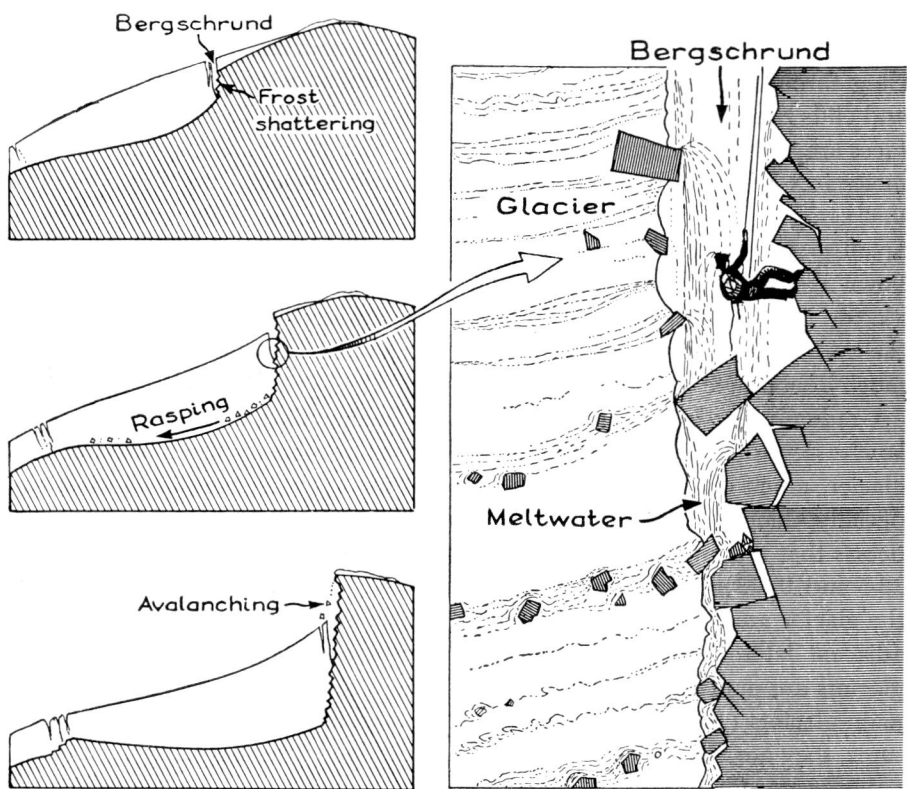

FIGURE 13-8. *Progressive stages of erosion at the head of a valley glacier* (left) *with a detail of the bergschrund.* (*In part, after W. V. Lewis, 1938.*)

the rest, entirely free from their parent ledges, are now incorporated in the glacier (Fig. 13-8).

During the day, water from the melting snow pours over the face of the bergschrund, filling the joints. The almost nightly frost loosens new blocks. Obviously, here is a zone of active glacial erosion, sapping the headwall by the plucking of frost-rifted blocks, steepening the cliffs behind, and gradually extending the glacial channel headward.

An open crevasse, however, cannot extend to depths of more than about 200 feet. Experimental tests on ice under pressure show that at loads corresponding to greater depths ice will flow plastically and slowly close the deeper openings. Nevertheless, certain typical features of glacier heads indicate that shattering somehow does go on to greater depths.

One factor is that the meltwater itself opens channels in the ice at the bottom of the bergschrund. On hot summer days much meltwater cascades into the bergschrund. A "chinook" wind or a warm summer rain may greatly increase the volume. Despite this great volume of meltwater, the bergschrund practically never fills with water to the point where streams emerge and flow over the ice. Apparently, the inpouring water melts its way down between the rock face and the glacier far below the bottom of the bergschrund, possibly, indeed, to the very base of the glacier. Actually, local melting of the deep ice should be expected from physical principles; the compressed ice at depth melts at a slightly lower temperature than the ice and snow at the surface. Meltwater pouring into the crevasse also carries heat downward to melt more ice deep within the glacier. Plastic flow of the ice tends to close the meltwater tubes, but as long as heat is being continuously transferred downward, melting of the walls may offset the constriction by flowage. Furthermore, water in the tubes may be under high hydrostatic pressure, and the pressure opposes plastic closing.

Thus, we have a mechanism by which the rock wall at the head of a glacier may be wetted periodically to great depths. But, just as periodically, freezing cuts off the supply of water and the accumulated meltwater within the glacier and in the cracks in the rock wall may ultimately freeze. As a result, the rock wall is shattered and disintegrated to depths much greater than the bottoms of the crevasses. Frozen meltwater also adds volume to the glacier. Such a glacier, now frozen tight to the headwall, moves slowly forward, drawing out the shattered blocks of rock with it, and tending to open new cracks and crevasses to be filled by a further flow of meltwater.

It must be remembered, however, that meltwater in the deeper part of the glacier is protected from day and night changes in temperature. Deep within a glacier the temperature remains almost constant at the freezing point, and ice and water are in physical equilibrium with one another. Change from liquid to solid may occur here chiefly in response to variations in pressure. A crack opened by the pull of the downsliding glacier will almost immediately fill with ice—the lowered pressure within the crack causes any moisture gaining access to it to freeze. But, in freezing, a gram of water gives up 80 calories of heat, and, if this heat is not dissipated, freezing will stop. For these reasons, meltwater shattering must be less effective at great depths than on the wall of the bergschrund.

Cirques Downstream flowage of the glacier removes the shattered debris and exposes new rock surfaces to attack. Where valley glaciers have melted away completely, exposing their headwalls, the glaciated valley generally ends in a semicircle of high cliffs that bound an arcuate rock basin generally occupied by a beautiful mountain lake. Such cliffed valley heads are called *cirques* (Figs. 13-4 and 13-17). The curving cliffs at the head of the cirque are shattered clear to their bases and they generally meet the smoothed and polished floor of the valley at a sharp angle. The jagged and shattered cliff face—the product of frost action and meltwater shattering—contrasts strikingly with the smooth valley floor—the effect of rasping by glacial ice.

The Glacial Rasp.   We have seen that the head of a valley glacier is a zone of rapid accumulation of snow and ice heavily charged with rock debris. If accumulation has been rapid enough to produce a steep downstream gradient of the top of the glacier, outward movement in a thin ice mass may occur along curving shear planes similar to the movement in a landslide (Fig. 11-10). In a thicker and longer glacier, extrusion flow of plastic material may be less concentrated along shears, but it will follow similar paths. As the ice moves across the bedrock floor, rock debris frozen in its base is dragged across the glacier bed, abrading the bedrock (Fig. 13-8). The glacier thus becomes an extremely effective cutting and polishing tool; it acts like a gigantic rasp, gouging into and scraping off irregularities on its bed, and smoothing and grooving the rock beneath it.

Debris in Continental Glaciers   Practically the entire landmass occupied by a continental glacier is covered with ice; thus frost shattering and talus accumulation can take place only around relatively rare islands of rock that project through it. Also, little debris can be won by meltwater shattering of the rock floor at the bottom of crevasses. Yet, the ice at the margins of the Greenland glacier and other ice caps is heavily charged with rock debris, just as in valley glaciers.

Some of this debris is the eroded soil that formerly covered the land over which the glacier flowed. In many localities, where continental glaciers have

FIGURE 13-9.   *Rock surface showing glacial grooving, polishing, and, at the left, plucking, near Mount Baker, Washington. The ice flowed diagonally from the upper left to the lower right. (Photo by H. A. Coombs.)*

FIGURE 13-10. *Till deposited by a valley glacier, West Walker River, Nevada. The largest boulers are 1½ feet in diameter. (Photo by Eliot Blackwelder.)*

melted back along their edges, a striking feature of the rock floor formerly covered by ice is the almost total absence of residual soils and of preglacial unconsolidated mantle such as stream gravels and floodplain deposits. The soil and loose rock have been frozen into the lower part of the glacier and dragged off bodily by the moving ice.

As in large valley glaciers, rock floors under continental glaciers are smoothed and polished. On these polished floors are numerous scratches and even deep grooves (Fig. 13-9). Obviously, the glacier has gouged and abraded material from its bed.

Rate of Erosion by Glaciers The rapidity with which a glacier abrades its bed depends on four factors: the resistance to abrasion and plucking of the particular bedrock, the abundance of cutting tools (rock fragments) frozen into the base of the ice, the speed of flow, and the weight (thickness) of the ice. Thus, thick continental glaciers flowing over weak or shattered rock are powerful agents of erosion. Similarly, on the floor of a cirque, glacial abrasion is generally great because here the ice is likely to be thickest and it is abundantly supplied with cutting tools.

Debris Released by Glaciers

Till. That glaciers are effective eroding agents is well shown by the vast amount of debris released at the glacier front by melting. Piled in hummocky ridges (*moraines*) along the front of the ice are heaps of boulders, sand, and silt, heterogeneously mixed without appreciable sorting or stratification. Such unsorted debris (Fig. 13-10), deposited directly by the ice, is called *till*. The proportion of boulders to fine material in till ranges widely. Some till is largely coarse boulders, but that from thin ice caps eroding shale and limestone may be chiefly clay and silt, with only scattered boulders.

Rock fragments found in till are not shaped like those in stream or marine deposits. Stream and beach pebbles are commonly rounded, but most of the fragments in till are subrounded or sharply angular. Some have been cracked and broken by the crushing weight of the overriding glacier; many, especially in valley glaciers, are joint blocks loosened by frost action and little modified during transport. Some fragments show one or more nearly flat, grooved, and polished surfaces. Obviously, these flat surfaces were faceted and smoothed as the boulder was dragged against the bedrock floor, or against other rock debris within the ice.

Rock Flour. The effectiveness of the glacial rasp is especially obvious from the large amount of fine material (silt and fine sand) released at the glacier front. Turbulent streams of milky water generally cascade from one or more tunnels in the front of a glacier (Fig. 13-11). If we collect a beaker of the roily water and let it stand, most of the suspended material promptly settles to form a thick layer of silt and fine sand on the bottom of the beaker. A very small amount of fine clay particles may remain suspended for hours or days. The silt, examined with a microscope, differs strikingly from the mud brought down by freshets in an unglaciated region. The minerals composing the silt of the glacial streams are not the ordinary soil minerals characteristic of chemical weathering. Instead, small sparkling cleavage frag-

FIGURE 13-11. *Silt-laden stream emerging from an ice tunnel in the front of a glacier, Tanana district, Alaska. (Photo by S. Capps, U. S. Geological Survey.)*

ments of unweathered feldspar and other undecomposed minerals are abundant, and there is an almost complete absence of the yellow iron stain, the black and gray colors from humus, and the slimy clays so characteristic of chemical weathering. Clearly, most of this material is not an ancient soil removed by the glacier, but is *rock flour* made by the grinding and crushing of rock fragments against one another and against the bedrock.

Glaciers release an amazing quantity of rock flour. Glacial streams are almost invariably overloaded. They quickly deposit the coarsest material (rubble and coarse sand), building up a sloping alluvial plain (Fig. 13-13) over which they spread in braided courses. Most of the rock flour, however, is carried far beyond the front of the glacier and builds floodplain deposits farther downstream, or is spread into lakes as extensive delta and bottom deposits. Such rock flour forms great *terraces of white silt* along the rivers of British Columbia and Alaska.

The powdered rock is also widely scattered by the winds. Extensive un-

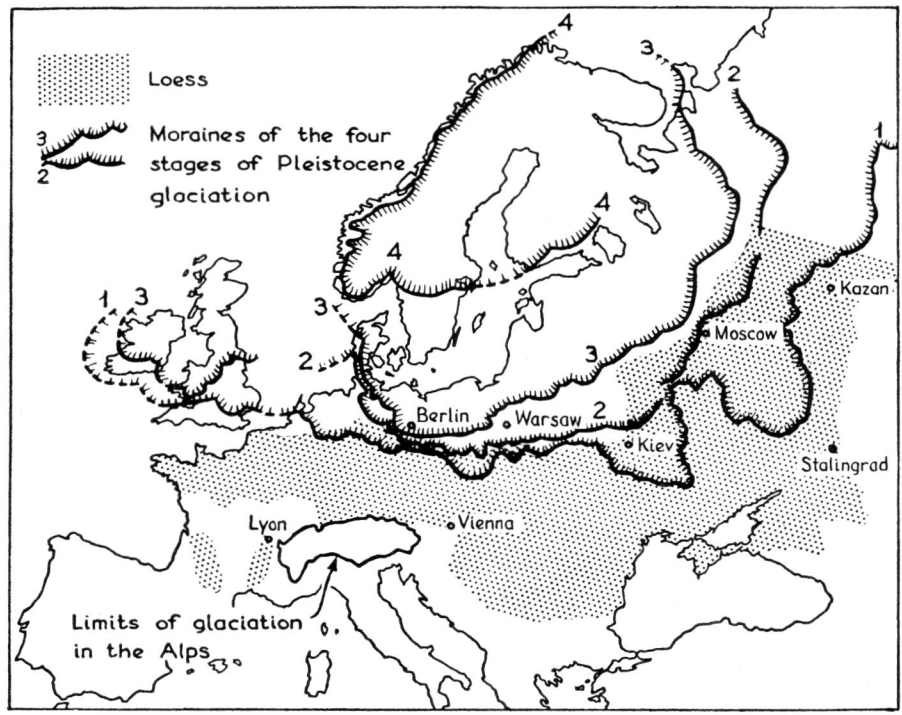

FIGURE 13-12. *Relation of the areas covered by loess to the terminal moraines of the four glacial advances in Europe. (Adapted from R. F. Flint,* Glacial Geology and the Pleistocene Epoch, John Wiley and Sons, 1947; *and R. A. Daly,* The Changing World of the Ice Age, Yale University Press, 1934.)

stratified deposits of *loess* (a wind-deposited soil composed chiefly of undecomposed silt particles) are found just beyond the limit of glaciation in
many areas (Fig. 13-12). Loess characteristically contains abundant small
tubelike holes where grass roots have grown.

Forms of Glacial Deposits

The debris dumped by glaciers and by the streams flowing from them is
called *glacial drift*. The drift assumes characteristic and easily recognized
topographic forms.

Moraines.   Unstratified glacial drift (*till*) is deposited directly by the

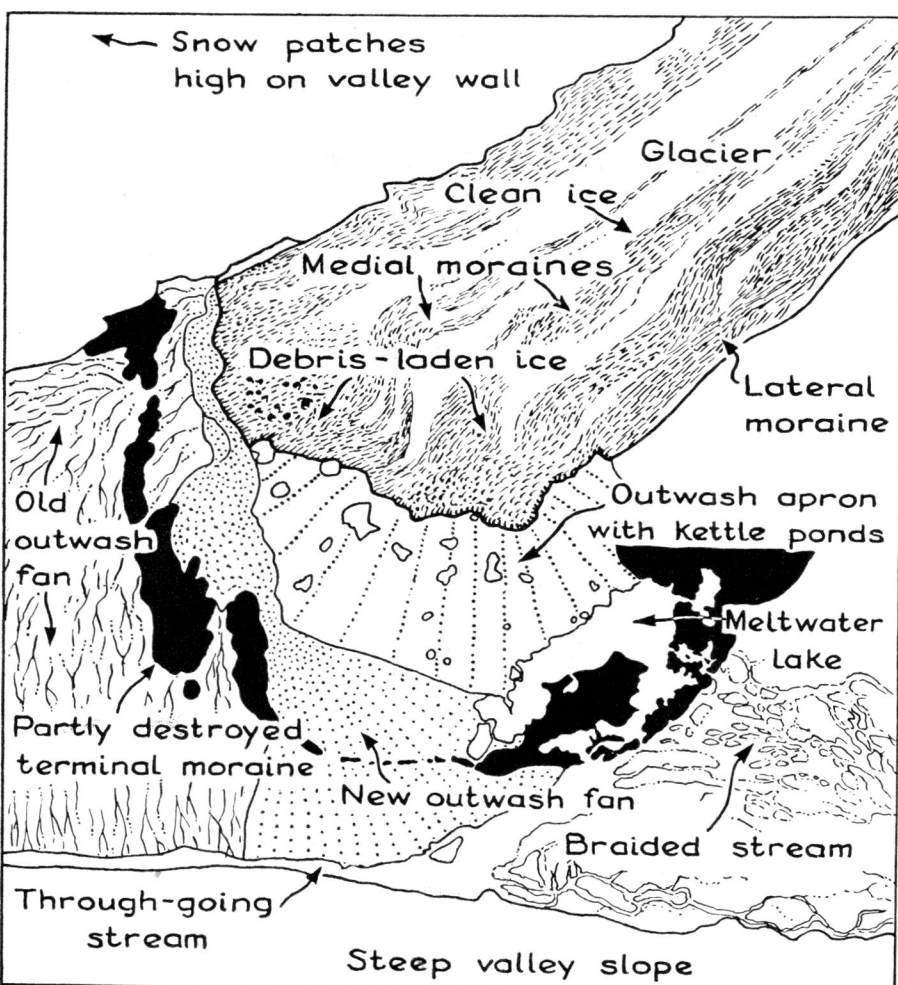

FIGURE 13-13. *Vertical aerial photo and explanatory sketch of deposits at the foot of a valley glacier, Alaska. Note how the recessional moraine (black in sketch) has been partly buried by later outwash fans. Note also how the course of the through-going stream becomes braided where it acquires the load of present glacial outwash. (Photo by U. S. Air Force.)*

ice and occurs chiefly in the form of moraines. The term *moraine* is used both for material deposited by the ice and also for the material upon or within the moving ice. The largest and best developed moraines are generally at the front of the glacier. If the ice front is in equilibrium, so that the rate of flow equals the rate of melting, it will remain essentially fixed in position for a long time. Here, the debris carried by the ice is released and accumulates in great hummocky morainal ridges. Because the ice front fluctuates with mild climatic changes, however, more than one morainal ridge is likely to form. The moraine that marks the farthest advance of the glacier is called the *terminal moraine*. Moraines that mark stages of halt as the glacier recedes are called *recessional moraines*.

Most terminal and recessional moraines of continental glaciers are broadly lobate in plan and can be followed, with minor interruptions, for miles (Fig. 13-12). The moraine at the front of a valley glacier is generally a high crescentic ridge curving around the front of the valley glacier and extending up its sides as lateral moraines. Small recessional moraines of valley glaciers may be almost buried in outwash debris, as in the case illustrated in Figure 13-13.

During the advance and withdrawal of a continental glacier some till is deposited beneath the ice and overridden by it. Till may also be scattered over a glaciated area by melting out of the englacial (ice-enclosed) debris as the ice recedes (Fig. 13-18). This irregularly scattered till, called *ground moraine*, is not concentrated into definite ridges like the terminal moraine. Some is found packed into hollows such as preglacial stream canyons; some is plastered around and upon low hills of the bedrock floor. Actually, the ground moraine is the most widespread deposit of a continental ice sheet. In most places it is spotty and only a few feet thick, but near the margin of the ice sheet it may cover hundreds of square miles in an almost continuous veneer.

In some areas, the ground moraine has been moulded into clusters of closely spaced hills, each shaped like the bowl of an inverted spoon, with the steeper slope facing the source of the ice. These streamlined hills are called *drumlins*. They vary in size, but are commonly about a mile long, ¼ mile wide, and 50 to 150 feet high. Drumlins occur in clusters of scores or hundreds. Each drumlin in the cluster has its long axis roughly parallel to the direction of flow of the former glacier (Fig. 13-14).

Although we cannot see drumlins forming in modern glaciers, they are doubtless masses of subglacial debris overridden and moulded into streamlined forms by an actively flowing ice sheet. Excavations show they are composed largely of till, commonly a sticky variety containing much clay. Some drumlins have cores of bedrock; others are composed entirely of glacial debris.

Stratified Glacial Drift.   Meltwater from the glacier carries off much of the finer debris and deposits it either on the beds of the overloaded streams (*glacio-fluvial deposits*) or in lakes (*glacio-lacustrine deposits*) and in seas.

A widespread glacio-fluvial feature, especially at the margin of continental glaciers, is the *outwash plain* (Fig. 13-13). This is an apron of coalescing

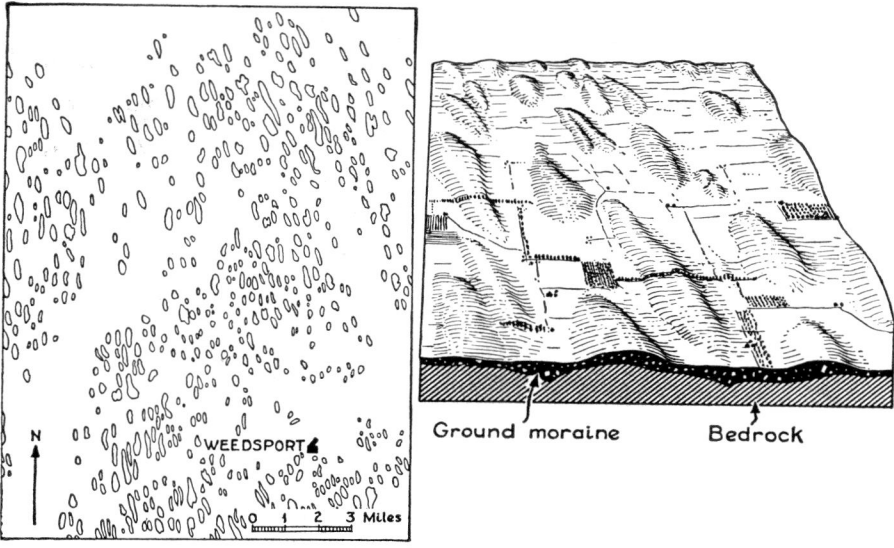

FIGURE 13-14. *Map of drumlins near Weedsport, New York, and sketch showing their low, elongate forms. (After Weedsport quadrangle, U. S. Geological Survey.)*

alluvial fans deposited by the heavily overloaded meltwater streams emerging from the ice. Braided streams flow across this apron, dumping their gravel and sand in rudely stratified layers. The outwash apron generally spreads out just in front of a terminal or recessional moraine.

In places, the deposit may consist of *pitted outwash*, characterized by numerous undrained depressions ranging in size from mere dimples a few feet wide up to great holes a mile long and more than 100 feet deep. These holes are called *kettles*. They are formed by the melting of ice blocks that were detached from the main glacier as it wasted away, and were then partially or wholly buried in stream-deposited glacial debris (Fig. 13-15).

The streams emerging from valley glaciers generally have somewhat steeper gradients than those from continental ice sheets. Nevertheless, they are so greatly overloaded that they immediately aggrade the valley floor to form a long narrow floodplain of coarse gravel and sand which grades downstream into finer sands and eventually into wide floodplains of silt. These aggraded valley bottoms, floored by glacio-fluvial debris, are called *valley trains*. When the glacier melts completely, the stream, being no longer overloaded, generally erodes the valley train into a series of terraces.

Stratified sand and gravel, containing here and there an irregular patch of till, forms long winding ridges in some areas from which continental glaciers have melted. Most of them are less than 100 feet high and a few hun-

FIGURE 13-15.   *Small kettle lake in outwash gravels of the Baird glacier, Alaska.*
(Photo by A. F. Buddington, U. S. Geological Survey.)

dred feet wide, but they may be several miles or even tens of miles long.
They are called *eskers*, a term derived from the Gaelic meaning "ridge."

Eskers appear to be the deposits of aggrading streams that flowed in
tunnels beneath the ice, or along the bottoms of crevasses, for some of
them merge downstream into fans of outwash or into deltas. Most eskers
must have formed when the ice sheet had wasted to a thin, almost stagnant
mass, for, obviously, the strong thrust of an active glacier would close the
tunnels and crevasses and scatter the esker deposit into ground moraine.

Where a glacier-fed stream enters a lake, a *delta* forms which may grow
until it converts the lake to a swampy plain. The delta material is typically
cross bedded and may range in grain size from coarse gravel to fine silt.

On the floors of glacial lakes, mud and silt accumulate as thinly stratified
layers. Each layer is commonly only a fraction of an inch thick, giving the
deposit a conspicuous paperlike stratification (Fig. 13-19).

Many glacial lakes are dammed on one side by the ice itself, and the
water may extend into crevasses and irregular-shaped holes in the wasting
glacier. Such lakes are generally temporary, and fluctuate in level as movement or melting of the ice diverts their outlets. Sediments accumulate in
them. Because of fluctuations in lake level and movements of the glacier,
these lake deposits are generally interlayered with stream deposits and
with till. Upon melting of the glacier, patches of these sediments that filled
former crevasses and holes in the ice are left in terraces and in isolated buttes
called *kames* and *kame terraces*.

Modification of Topography by Glaciers

It is apparent that glaciers differ greatly from rivers in their mechanism of flow, and in their erosional and depositional features. The topography shaped by them contrasts sharply with the topography shaped by running water. Such differences are conspicuous in areas recently uncovered by glacial recession.

The polished and grooved bedrock floor is a feature not found in normal stream channels. Glaciated areas also contain swarms of lakes, ponds, and marshes that fill either shallow rock basins scooped out by the ice, or depressions dammed by moraines and outwash. The streams that connect these abundant lakes are totally ungraded. An immature stream pattern with numerous waterfalls and lakes is typical of recently glaciated areas.

The upstream sides of peaks, hills, or small rock knobs overridden by ice are commonly rounded, polished, and grooved; the downstream sides are irregularly jagged (Fig. 13-9). It is inferred that the glacial rasp has sanded off the upstream side, but rock has been removed from the lee side, chiefly by quarrying of joint blocks caught by angular boulders dragged over them. Pressure relations may also be important. Pressure on the upstream face of the knob causes melting, and the water freezes in the low-pressure area, causing plucking on the lee side of the knob.

The usual mountain stream valley has a V-shaped cross profile (Chap. 12), but in valleys from which glaciers have melted the lower part of the

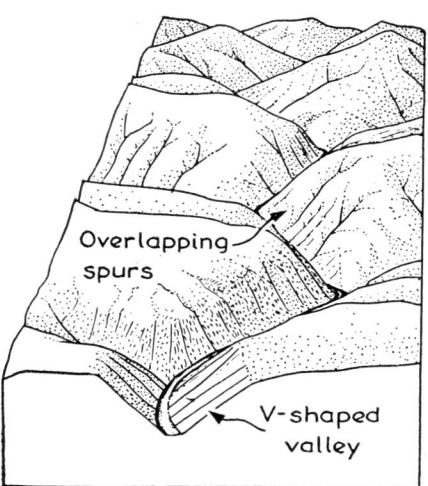
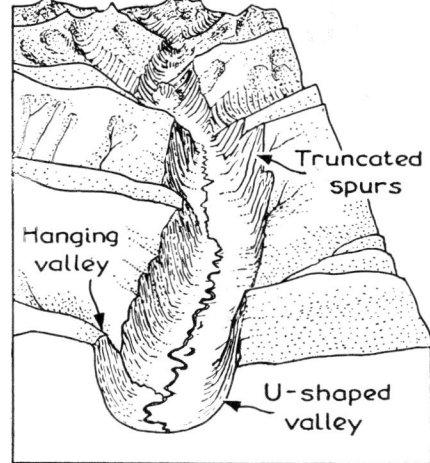

FIGURE 13-16. *A hypothetical stream valley and its associated land forms before and after being modified by a valley glacier.*

profile is characteristically U-shaped, though remnants of a former V-shape may be preserved on surfaces higher than the top of the vanished glacier. It is inferred that the spurs projecting between tributaries along the walls of a river canyon have been quarried away by the ice (Fig. 13-16). A glacier cannot negotiate the sharp curves around spurs as easily as water; because of its bulk and brittleness it overrides the spurs and planes them into triangular facets forming a straight line of cliffs. At the same time, the floor of the valley is deepened by the glacial rasp. The widening, deepening, and straightening of former stream canyons by ice transforms the original V-profile to a U.

The long profile of a valley formerly occupied by a glacier is irregular and is commonly interrupted by abrupt steps above and below which the smooth U-shaped valley may continue with flat gradients. Several such "cyclopean steps" may appear in a single glaciated valley, alternating with polished rock floors and lake-filled basins. Valley deepening by glaciers is not controlled by a smooth grade profile related to sea level, as in streams,

FIGURE 13-17.   *Mount Assiniboine, a glacial horn, near Banff, Alberta. Note the many cirques, some with hanging glaciers. (Photo by Department of Mines and Resources.)*

but by the thickness of the ice, the speed and duration of its flow, and the structure of the rock in its bed. In restricted parts of its course, it is not unusual for a valley glacier to deepen its trough far below sea level.

Unlike stream valleys, glaciated valleys commonly head in cirques. Abandoned cirques (Fig. 13-17) are conspicuous features in recently deglaciated mountains. These gigantic semicircular basins may be so closely spaced that the divides between them are reduced to knife-edged comblike ridges, or to triangular mountain "horns" (Fig. 13-17).

Stream valleys and glaciated valleys also differ in the relation of the tributaries to the main stream. The surface of a tributary glacier may accord with that of the main glacier—though, because of the rigidity of a thin ice mass, this is not always so—but the floor beneath the glacier and tributary is cut to depths depending on the volume and speed of the respective ice streams and the kind of rocks over which they move. The thin tributary glacier, unless favored by less-resistant bedrock, cannot keep pace with the scouring and plucking by the larger glacier. As a result, when the ice retreats, the tributary valleys are commonly left "hanging" high above the floor of the main valley, and their streams descend into the main valley in high waterfalls or steep rapids.

Now that it is obvious glaciers leave clear marks of their former presence upon a landscape, let us see if there is evidence of widespread former glaciation in areas that today have a temperate or even a tropical climate.

Former Periods of Glaciation

The Iowa farmer, sweating under an August harvest sun, may doubt that "the present is the key to the past" if he is told that the soil he tills has come from deposits left by an ice sheet that once covered most of northern North America in the Pleistocene epoch. Similarly, the dark-skinned native of the Talchirs in India, who pauses to rest from the steaming tropical heat on a polished and striated ledge of rock, will doubtless consider the idea fantastic, though alluring, if told that his perch is the floor of an ancient glacier that spread widely over India in Late Paleozoic time.

Conclusions such as these so tax the imagination that, in spite of the clear-cut evidence, they were not accepted even by geologists until every possible alternative—from the Biblical flood to some mysterious kind of volcanic activity—had been tested and found inadequate to explain the facts recorded in the rocks and in the drainage patterns upon them. Yet, evidence of the presence of former ice sheets can be readily gathered by any interested observer.

The Pleistocene Glaciations

In most of Scandinavia, southern Canada, Labrador, and parts of northern United States, the normal soil profile (see Chap. 6) formed by weathering is either poorly developed or missing. Instead, rounded hilltops expose smoothly polished rock ledges like those beneath present glaciers. In places, the polish and smaller striations have been destroyed by weathering, but where protected by even a thin veneer of till, outwash, or peat, they may be as fresh and clear as beneath a modern ice sheet.

Boulders—many of huge size—are scattered erratically over the polished surface (Fig. 13-18). Many boulders are of rocks entirely different from the local bedrock. In the fields of the Iowa farmer, boulders of gneiss and granite rest on flat-lying limestones and shales. The headwaters of streams in the area expose no such bedrock, so these boulders could not have been brought in by floods. In parts of Iowa, chunks of copper are occasionally found like the copper ore mined from rock ledges on the Keeweenaw peninsula in Michigan or Isle Royale in Lake Superior, but not found elsewhere. Boulders of an unusual variety of granite called "rapikivi," known in outcrops only north of the Gulf of Finland, are scattered widely across Estonia and even far into Poland. The basalt plateau of eastern Washington is strewn with huge granitic boulders of kinds found only many miles to the north. To reach their present position they must have been transported directly across the Columbia River canyon, which is here 1,500 to 2,000 feet deep (Fig. 13-23). A

FIGURE 13-18.   *Glacial boulders resting on a surface polished by a Pleistocene valley glacier, Sierra Nevada, California. (Photo by Eliot Blackwelder.)*

large nickel deposit in northern Finland was actually discovered by tracing its strewn ore fragments northward to their source near Petsamo.

Development of the Glacial Theory. The implications of such relations seem obvious to us now. Nevertheless, in the period from 1821–1835, when two European geologists, Venetz and Charpentier, showed that boulders of rocks characteristic of the central Alps were widely scattered across the broad Swiss plain to the north, and correctly inferred a former extension of the present Swiss glaciers, their ideas met little but skepticism.

In 1836, however, Charpentier induced one of these skeptics, a young Swiss geologist named J. L. R. Agassiz (1807–1873), to accompany him on a visit to the active glaciers and, lower in the valley, the huge abandoned moraines of the Rhone glacier. Convinced on this expedition that the evidence for former more widespread glaciation was even stronger than Charpentier had claimed, Agassiz set to work on the problem. He promptly saw and correctly interpreted the relation of the transported blocks to the polishing and grooving of the bedrock, and showed that these smooth striated forms could not possibly have been produced by water. Agassiz's work, too, met with general disbelief. As his biographer relates:

> Men shut their eyes to the meaning of the unquestionable fact that . . . the former track of the glaciers could be followed, mile after mile, by the rocks they had scored and the blocks they had dropped.

Agassiz, however, was not easily discouraged. In 1840, he visited much of Scotland, the north of England, and parts of Ireland. There, he demonstrated glacial phenomena identical with those on the Swiss plain, and he announced that not only had glaciers once existed in Britain, but that they formerly covered most of the country. At first this conclusion raised a furore of objections, but it also set geologists to observing and more carefully analyzing the evidence. Faced with the facts so easily gathered over much of the British countryside, it was not long until Agassiz's views prevailed.

Agassiz emigrated to America and began the studies of glaciation in New England since followed so fruitfully by many investigators. Studies since Agassiz's time have supplied a wealth of information. Modern maps portray in detail the erosional forms and deposits of the former ice sheets. Many abandoned moraines have been traced in the field, lobe by lobe, and plotted on maps until their distribution in North America and Europe is well known (Fig. 13-12). It is not difficult to map the moraines. Although locally absent or poorly developed, most are conspicuous, easily identified, hummocky ridges of till that are readily traced for miles. Behind them, like signposts pointing the way, lie striated rock floors, locally strewn with till and foreign boulders. Here, too, are innumerable lakes and marshes where the vanished glacier gouged its floor or blocked older drainage by debris it left on melting. In front of the moraines are coalescing fans of outwash and, still farther beyond, sheets of loess, or terraces of silt. All assist in locating the position of the edge of the vanished glacier, and all testify to its former existence. The landscapes of New York, New England, the Great Lakes states, and the Northwest are nearly everywhere stamped with these glacial imprints.

In mountainous areas, such as Yosemite National Park, the Cascade Mountains, or the Rockies, former glaciers are recorded by the numerous U-shaped valleys; the fresh moraines; the abundant cirques; the superb waterfalls pouring from "hanging valleys"; the clear mountain lakes nestled in rock-scoured basins; and the terraces of outwash gravel, sand, and silt extending downstream from the moraines.

Advance and Recession of Pleistocene Ice Sheets.   Careful study of Pleistocene glacial deposits shows that they are not the result of a single advance and retreat of the ice. Instead, they record at least four stages of ice advance separated by long periods during which the glaciers withdrew to greater altitudes or latitudes than they occupy today, or even disappeared.

At many places in North America and Europe, roadcuts or natural exposures reveal two tills, one on top of the other. The upper layer of till may contain boulders of almost fresh granite and gneiss, some with polished facets and striae. In the lower till, however, beneath a layer of thoroughly weathered clayey soil, the outlines of similar boulders can also be made out, but the rocks have weathered to the consistency of cheese—the feldspars are thoroughly rotted to clay, and the ferromagnesian minerals have decomposed to limonite and other materials. Only chemically resistant rocks like quartzites are found intact in this lower till.

It is apparent that the older till had undergone long weathering and that typical A and B soil horizons (see Chap. 6) had developed on it before the younger till covered it. The B-horizon of the weathered till is usually rich in clay that cements it into a tough sticky mass. Such tough clayey subsoils, whether of glacial origin or not, are popularly called *gumbo. Gumbotil* is the technical term used to designate deeply weathered clayey tills.

Moraines with mantles of gumbotil, unlike younger unweathered moraines, have generally been considerably eroded since they were formed.

Careful study of the degree of weathering of different glacial deposits and mapping of their mutual relations have permitted the discrimination of four main stages of ice advance. Between the ice advances the climate appears to have been mild and warm, as shown by the fossils found in stream and marsh deposits between the tills.

Dating Pleistocene Deposits by Varved Lake Sediments.   Lake deposits are abundant in glaciated regions. Deltas grow rapidly into glacial lakes

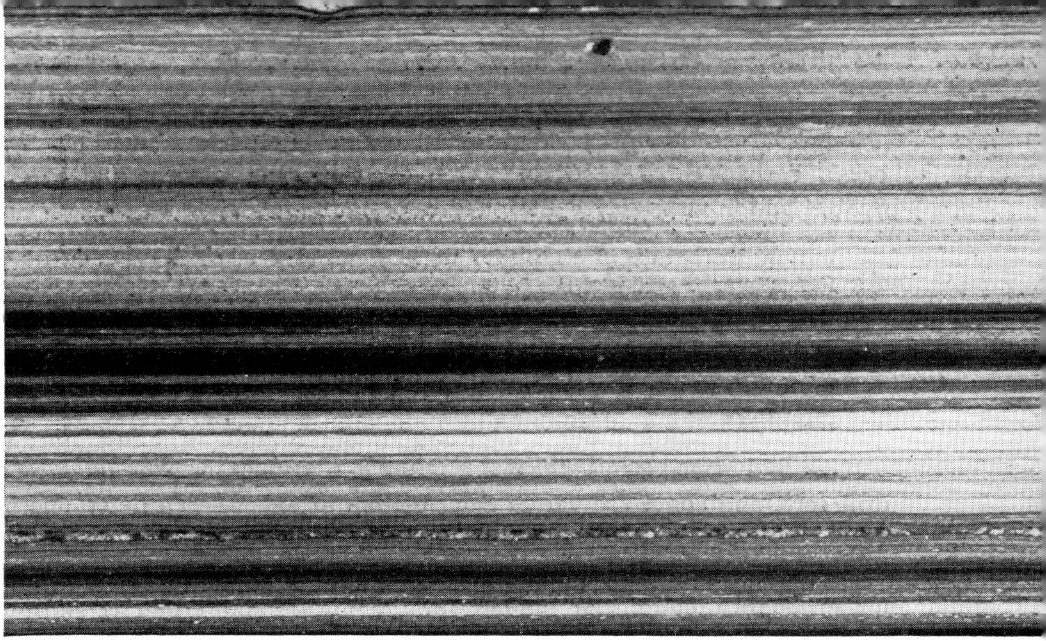

<figure>
FIGURE 13-19. *Cross section showing bedding in varved clay. (Photo by W. Bradley, U. S. Geological Survey.)*
</figure>

where torrents of turbid meltwater pour into them. The finest silt and mud settles slowly and is spread throughout the lake. Many glacial lakes are deep green, owing to the dispersion of light by the abundant particles of suspended clay.

In the winter, the lakes freeze over, and the small tributary streams may freeze solid. During this quiet period, the suspended clay particles beneath the ice settle slowly to the bottom. By spring, most glacial lakes have nearly cleansed themselves. In summer, however, the surface ice melts and the streams become gushing torrents of debris-charged water which sweep coarse as well as fine material out into deep water. The coarse material settles promptly to the bottom, covering the fine winter deposit with a thin layer of coarse silt. Then, with the progress of the season, the silt is buried by successively finer deposits that gradually grade upward into the fine dark-colored mud of the next winter. In the summer, microscopic organisms flourish in the warm surface layers. They extract calcareous matter from the water. The carbonaceous bodies of the organisms sink only slowly on death. Thus, the summer layer is richer in light-colored silt, the winter layer in black organic matter. Such cycles have been observed to form in many existing lakes.

These thin laminae of alternating fine and coarse material, each pair representing the deposit of one year, are called *varves*, from a Swedish word meaning "seasonal deposit." In most varves the double band representing a full year is, on the average, only a fraction of an inch thick (Fig. 13-19).

Some varved deposits contain fossils that definitely prove their seasonal origin. Perhaps the most striking example is a series of Miocene shales from Switzerland in which layer after layer repeats the following sequence:

> . . . at the bottom of each layer are blossoms of poplar and camphor trees, indicating the spring of the year; above this is a thin zone containing winged ants and fruits of the elm and the poplar, indicating summer, and the summer zone in turn is overlaid by a zone containing fruits of the camphor tree, wild grape, and date plum, indicating autumn. The three zones together constitute a varve in which the progression of the seasons is thus marvelously recorded.[*]

If we find glacial lake sediments exposed in a roadcut, or penetrated by a drill core, and count the varves from bottom to top, we can determine in years how long a time the deposit represents. Furthermore, since the varves record unusual seasons (for example, an exceptionally warm year yields an unusually thick and coarse summer layer), it is often possible, by plotting the thickness of varves on graphs, to correlate the upper layers in a southern lake with those near the bottom of a more northerly younger lake that lay along the line of ice recession.

By such methods it has been determined that the last ice sheet retreated from the site of Stockholm, Sweden, about 9,000 years ago, that southern Ontario lay under ice 13,500 years ago, and that about 4,300 years elapsed while the ice retreated from a front near Hartford, Connecticut, to a front at St. Johnsbury, Vermont—a distance of 190 miles.

Drainage Changes Beyond the Limits of Glaciation.   The great climatic changes that brought glaciers down over northern Europe and North America could not fail to effect major changes in landscapes farther south

Lake Bonneville   In Pleistocene time, Nevada and western Utah were not barren, semidesert areas as they are today. The nature of the sediment on the desert floors and the fossils contained in them shows that fresh-water lakes, their shores clothed with pine trees and luxuriant grass, occupied most of the intermontane valleys. The wave-cut shoreline of the largest of these vanished lakes, called Lake Bonneville, makes conspicuous horizontal notches on the cliffs and steep slopes fronting the Wasatch Mountains over 1,000 feet above Great Salt Lake. At this stage, Lake Bonneville had an outlet into the Snake River and thence by way of the Columbia to the sea. As it overspilled the divide at the outlet, the water quickly eroded a canyon through unconsolidated alluvium until it reached bedrock 350 feet below At this resistant bedrock sill the level remained nearly constant for a long time, and the greatest deltas and shore terraces were built. Moraines from

---

[*] Knopf, A., "Time in Earth History," from *Genetics, Paleontology, and Evolution* Princeton University Press, Princeton, N. J., 1949.

valley glaciers in the Wasatch extend to the old shoreline. Some rest on lake sediments and are themselves cut by beaches. The glaciers were, therefore, about contemporaneous with the expansion of the lake—perhaps a little older than the maximum expansion.

As the glaciers waned, the area became drier, streams dwindled, and evaporation from the lake began to exceed inflow. Gradually, the outlet dried up and the lake shrank to successively lower levels, as shown by a series of shoreline features cut on the delta fronts and beach deposits formed during the higher stand of the lake. Today, Great Salt Lake and the Bonneville salt flats west of it remain as the last desiccation pool of this once-vast body of water (Fig. 13-20).

Great Lakes and Missouri Valley Area   Abandoned Pleistocene lake beds and stream courses are not confined to deserts. The advance and retreat of the continental glaciers produced great drainage changes, not only in the areas overriden by the ice sheets, but also far south of the ice margin. Some of the most striking changes affected the Great Lakes region and the Missouri Valley. Indeed, the glacial blocking of the former northward-flowing courses of the Yellowstone and Milk rivers, and the detouring of their waters along the margin of the glacier into the lower Missouri River, established about 350 miles of the present course of the Missouri. Similar ice-marginal drainage largely determined the present position of the Ohio River.

A long and complex series of Pleistocene drainage changes, involving the development of several large glacial lakes, preceded the establishment of the present Great Lakes and interconnected river systems. Two stages in the evolution of the Great Lakes are shown in Figures 13-21 and 13-22, but these maps represent merely two episodes in a long history worked out by the glacial geologists. The evidence for the drainage changes and abandoned lakes is extensive and convincing. It consists of abandoned shorelines high above present water bodies, marked by beach ridges, wave-cut cliffs, deltas, spits, and bars. These features can be traced for miles, but when followed northward, generally end abruptly against a moraine or apron of outwash that marked the glacial front when the lake was in existence. Vast areas

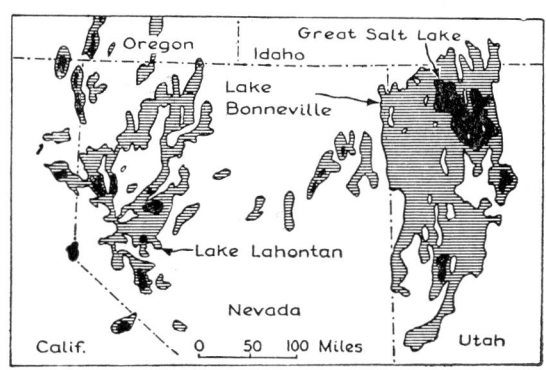

FIGURE 13-20.   *Map showing the extent of the great Pleistocene lakes in the western United States. Present lakes (in part, ephemeral) are in black. (After G. K. Gilbert.)*

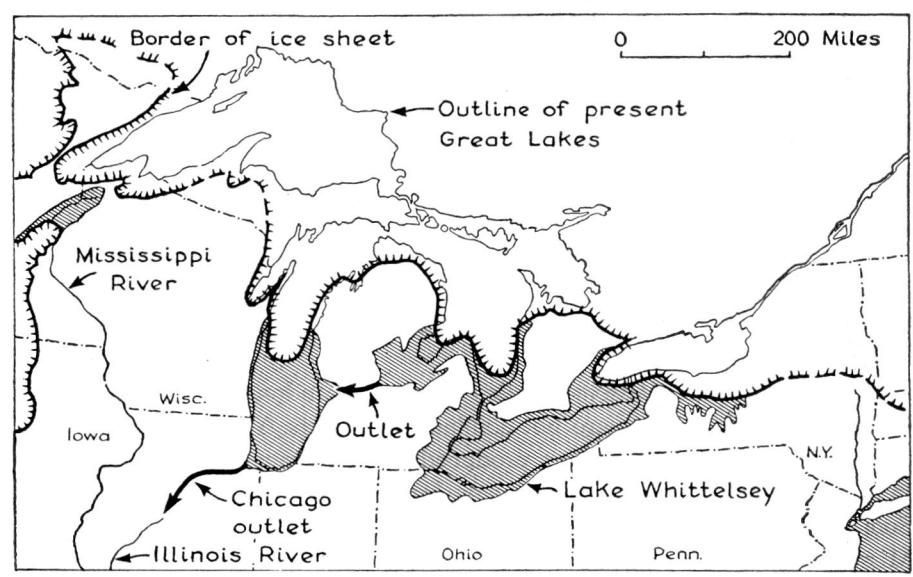

FIGURE 13-21. *Meltwater lakes formed during a recessional stage of the Pleisto-cene ice sheet in the Great Lakes region. The outline of the present lakes and state lines are shown for comparison. (After F. Leverett and F. B. Taylor, 1915; F. Leverett and F. W. Sardeson, 1932; and W. S. Cooper, 1935.)*

enclosed by the shorelines are covered with varved silts and clays—the deep-water deposits of former glacial lakes. The evidence of abandoned river courses is equally convincing. Some abandoned valleys have been choked and partially obliterated by deposits of glacial drift; others cross present divides but have been left high and dry by the uncovering of a lower outlet when the ice melted; and still others are river segments buried beneath the sediments of expanding ice-marginal lakes.

As shown in Figures 13-21 and 13-22, a series of temporary glacial lakes occupied the southern end of what is now Lake Michigan. Their outlet was a short stream from the site of Chicago via the Illinois River to the Missis-sippi. Simultaneously, a larger glacial lake, called Lake Whittelsey, occupied the expanded basins of Lake Erie and southern Lake Huron. Lake Whit-telsey drained westward by a river across central Michigan into an ancestor of the present Lake Michigan. Slightly later, the ice shrank northward and uncovered a lower outlet across New York via the Mohawk and Hudson rivers (Fig. 13-22). Immediately, Lake Whittelsey shrank greatly, the cross-Michigan outlet dried up, and huge volumes of meltwater that formerly flowed to the Gulf of Mexico now entered the Atlantic at the mouth of the

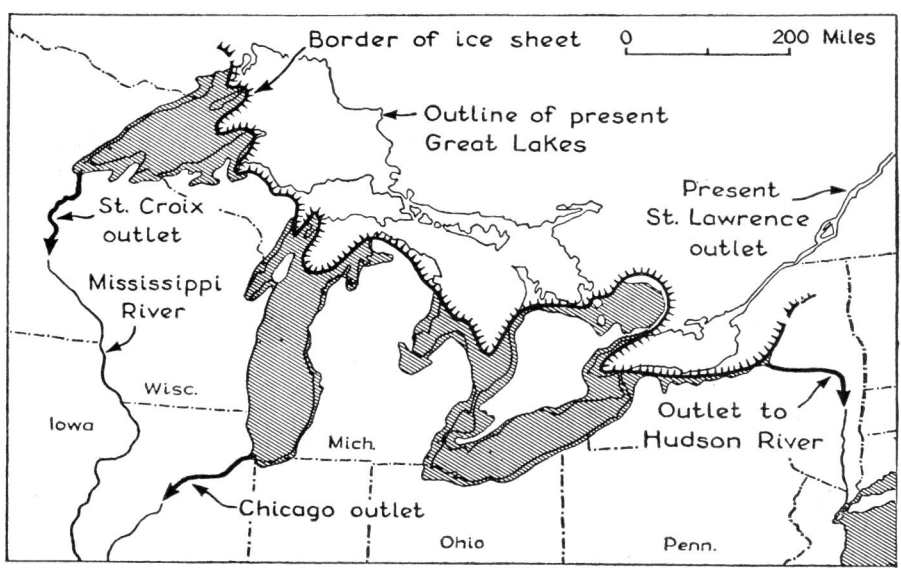

FIGURE 13-22. *Same as Figure 13-21 at a somewhat later stage. (After F. Leverett and F. B. Taylor, 1915.)*

Hudson. Still later, the ice melted away from the St. Lawrence outlet. Both the Chicago and the Mohawk-Hudson outlets were then abandoned, and the present outlines of the Great Lakes were established.

As the ice withdrew still farther, an enormous glacial lake—larger than all the Great Lakes combined—developed in the Red River Valley of Manitoba, Minnesota, and North Dakota. This lake has been named Lake Agassiz, after the famous Swiss glaciologist. Its water spilled southward into the Minnesota River, ultimately entering the Mississippi at the site of St. Paul. With further glacial retreat, Lake Agassiz was drained. The lake sediments now form the fertile wheat lands of Manitoba, northern Minnesota, and eastern North Dakota.

Grand Coulee The Columbia River was diverted by ice in eastern Washington (Fig. 13-23) with striking effects. A great lobe of the ice sheet encroached at right angles upon the westward-flowing Columbia River, filled its 2,000-foot canyon, and spread southward upon the Columbia Plateau. It dammed the river, causing it to spill out across the plateau. The surface of the plateau in this area slopes southward with an average tilt of several feet per mile. The debris-laden river, cascading on this steep gradient, quickly eroded a network of canyons on the plateau. With changes in the ice front, the streams shifted; one channel would no sooner be occupied than a slight advance would block its head and open a new channel elsewhere. For a

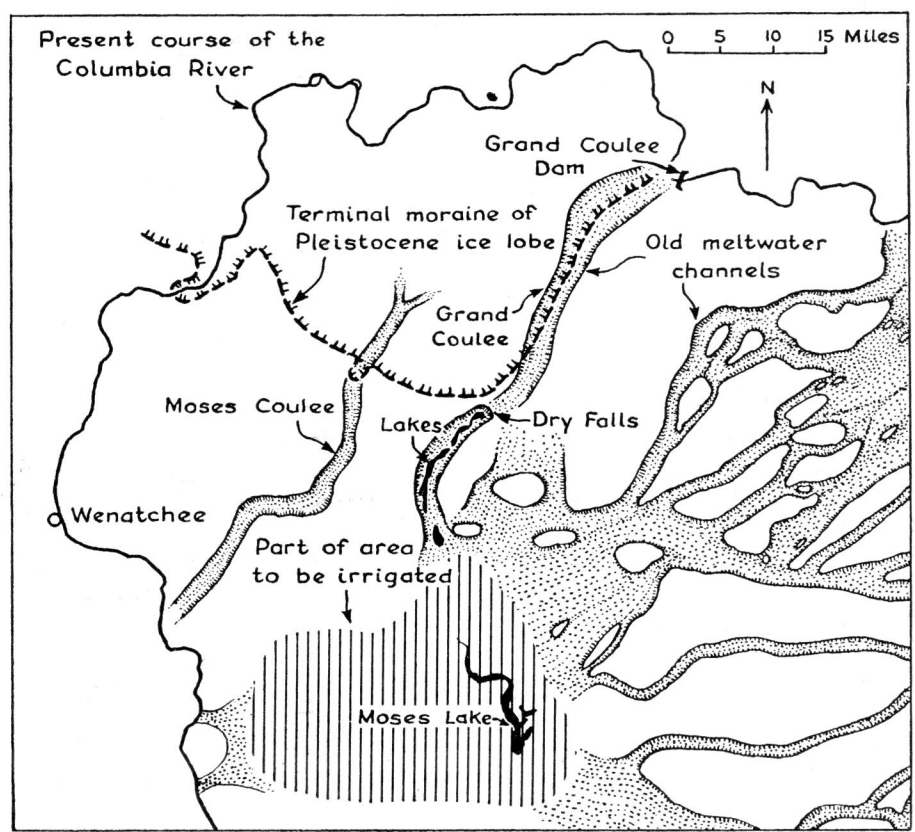

FIGURE 13-23. *Map of central Washington showing Grand Coulee and other features formed by the glacial diversion of Columbia River. (In part, after J. H. Bretz.)*

considerable time, however, much of the drainage was concentrated into one main channel now called the Grand Coulee. Here, the river excavated a great canyon 500 to 1,000 feet deep and 1 to 15 miles wide in the basalt flows of the Columbia Plateau. Midway in its course is a gigantic waterfall 400 feet high and nearly 3 miles wide, which during the ice age must have thundered with the roar of a thousand Niagaras. But, today, the Grand Coulee is dry and the voice of its waterfall is stilled. Waning of the glacier lobe uncovered the preglacial canyon and allowed the river to return to its former course.

Today, men are busy restoring at least a part of the waterflow through Grand Coulee. Across the Columbia River at the head of the coulee, they have built Grand Coulee Dam, and will use a part of the hydroelectric power generated to pump water from the lake behind the dam into the Grand Coulee, from where it will flow southward by gravity to irrigate millions of acres of rich but arid land. Constructed at great cost, Grand Coulee Dam is the largest engineering project built by man, yet how puny it appears compared to the ice dam thrust across the river in the same position some 20,000 years ago!

Pre-Pleistocene Periods of Glaciation

Many ancient sedimentary formations are composed of unsorted debris showing all the characteristics of till except that they are tightly cemented. They contain striated and faceted stones and are associated with varved shales and slates or with sandstones and conglomerates showing the typical features of outwash deposits. These features, and their local occurrence above polished and grooved rock floors, leave no doubt that they were formed during an ancient glacial epoch. Such cemented glacial tills have been called *tillites*.

Although small bodies of tillite have been found in rocks of many different ages, two periods in the earth's history before the Pleistocene are characterized by particularly widespread glacial deposits. During the Late Paleozoic, ice sheets spread widely over India, South Africa, Argentina and southern Brazil, southern Australia, the Boston Basin in the eastern United States, and many other localities. In India, Australia, and South Africa, grooved rock floors covered by the hard tillite may be seen at hundreds of localities. Their preservation today is due to the fact that they were buried under a succession of younger sediments. Deep burial protected them from weathering and erosion, and the unconsolidated glacial deposits were slowly cemented into rock. Recent uplift and erosion of the overlying deposits has now exposed them to view.

In the Late pre-Cambrian also, glacial climates appear to have been widespread. Ancient tillites of this age have been found on every continent except South America.

Causes of Glacial Climates

Geologists and climatologists have tried for more than a century to explain the recurrence of glaciation on a continental scale. Theory after theory has been suggested, but all explain too little or too much. None can be considered satisfactory, at least in its present form; yet they have an interest that justifies at least brief mention.

Facts to Be Explained

1. Continental glaciers (in Greenland and Antarctica) occupy about 10 per cent of the land surface today. They formerly covered about three times as much area during four different epochs of Pleistocene time.

2. Climatic zones during the Pleistocene were about parallel to their present positions but were displaced to the south during the glacial maxima and to the north during interglacial epochs. In low latitudes high rainfall was contemporaneous with glacial climates at higher latitudes.

3. Estimates of the duration of the several glacial and interglacial stages of Pleistocene time do not suggest periodic recurrence; the climatic fluctuations seem to have been irregular.

4. Evidence of continental glaciation on scales approaching that of the Pleistocene is also found in Late Paleozoic rocks (about 200 million years old) and in pre-Cambrian rocks (at least 500 million years old), but not in strata of intermediate age.

Suggested Explanations

The many theories that have been offered may be grouped under the following general categories: geographic, atmospheric, and oceanic; geophysical; astronomical; and cosmic.

The geographic theories attempt to account for glaciation by climatic changes that would be expected from changes in the areas of the continents with respect to the oceans, changes in oceanic and atmospheric circulation due to changes in the shape of ocean basins, mountain uplifts, and so forth. These causes are inadequate because there is no evidence whatever of any notable shift in the distribution of landmasses or of violent fluctuations in their altitude since Pliocene time, yet we have conclusive evidence of four great glacial advances and of milder-than-present climates in some, at least, of the interglacial times.

It has been suggested that changes in the amount of carbon dioxide and of volcanic dust in the atmosphere would bring about climatic variation. Carbon dioxide absorbs some of the heat radiated to space from the earth's surface. If its quantity were greater, more heat might be "blanketed in" and the temperature would rise, whereas, if it were less, the climate would become colder. However, this mechanism is quantitatively inadequate, especially because the effect of simultaneous change in amount of water vapor would practically compensate for the variations in carbon dioxide. Volcanic dust undoubtedly screens out some of the sun's radiation, but there is no sign that volcanoes were more active during the stages of ice advance than during interglacial times. Changes in salinity of sea water (and consequently modifications of the oceanic currents and their climatic influences) are likely results from glaciation but they can hardly have brought it about.

Geophysical theories that attribute glaciation to shifts in the positions of the continents are obviously not applicable to the Pleistocene climatic fluctuations, nor is there any strong evidence that these could ever have sufficed for earlier glaciations. Although it is conceivable that the amount of heat given off from the interior of the earth may have fluctuated from time to time, measured fluctuations are so small compared to the heat received from the sun (less than 1/10,000) that they seem negligible as a climatic factor.

The astronomical theories take account of the periodic variations of the earth's motion with respect to the sun. There are several such variations and there can be no doubt that they would have some climatic effect. Each affects the distance of a particular point on the earth's surface from the sun and hence the amount of solar radiation it receives. They could not have alone produced the glacial periods, however, because their effects on northern and southern hemispheres, though not always exactly opposed, would not be identical, either. Yet, we now have recession of glaciers taking place simultaneously in both hemispheres and it is reasonable to think that earlier fluctuations were also simultaneous. Furthermore, since all these astronomic motions have been going on through geologic time, we should have had continental glaciation every few hundred thousand years. Instead, we must go back to the Permian to find anything comparable to the Pleistocene glaciation in the geologic record. These geologic arguments seem conclusive, but several meteorologists contend, in addition, that the temperature changes produced by this mechanism would be entirely too small to produce glaciation even in one hemisphere.

The cosmic theories are of two kinds: (1) that the solar system from time to time encounters cosmic dust clouds or (2) that the sun varies in the amount of heat it radiates. The cosmic dust clouds may either screen out the sun's radiation (if the component particles are fine enough) or blanket the earth's radiation to outer space (if the particles are somewhat larger). This is, of course, a very flexible hypothesis, as there is no way of testing it; with one assumption we can get a colder climate, with another a warmer. Accordingly, this idea is not seriously considered by most geologists, even though it may be the true explanation, for the scientific approach is to consider seriously only those hypotheses that are susceptible of testing in some fashion, however indirectly. The second suggestion, that the sun's radiation fluctuates somewhat over long periods of time, is on a better footing.

The sun's radiation does, in fact, fluctuate, and with it the amount of heat received by the earth. Variable cloudiness and other atmospheric factors make accurate measurements difficult, but short-term variations of as much as 3 per cent in the heat output seem well established. The American astronomer, Abbot, a leader in this research, has declared that larger solar changes may well have taken place in the past, even though it is not yet possible to prove it. Several climatologists have tried to deduce the consequences of such moderately large changes, if they do, in fact, occur. We have not space to review all the hypotheses that have been suggested; perhaps the one that has won most support from meteorologists is due to the British meteorologist, Sir George Simpson.

Simpson (1934) reasoned that if solar radiation increased, the air temperature would rise all over the earth, winds would increase and so would cloudiness and precipitation. The increased precipitation would lead to greater accumulation of snow and the increased cloudiness would lessen summer melting. The ice caps in both hemispheres would expand. However, the rising temperature would eventually prevent further advance of ice and ultimately cause melting. At the high point on the radiation curve, the world climate would be milder and wetter (more "oceanic") than it is today. When radiation again decreased, the reverse sequence would be followed: appearance and advance of the ice; then retreat because of lower precipitation; and, finally, when radiation fell to its present value we would be back to present conditions, with polar accumulations of relatively small size. Thus, the somewhat paradoxical condition would arise, that a *rise* in the general temperature of the air would, by increasing precipitation and cloudiness, lead to continental glaciation of much of the present temperate zone. On this theory, one rise and fall of radiation would produce two glaciations in high latitudes, separated by a warm and wet interglacial time. The climate following the second glacial would be cold and dry like the present. In low latitudes there would be a single epoch of increased rains that would endure through both high-latitude glacials and the intervening warm, wet interglacial.

The four Pleistocene glaciations recognized in Europe and North America would, on this hypothesis, require two cycles of increased solar radiation, separated by a time during which the climate may have been much as it is today. Simpson computed that, during the maximum of solar radiation, the mean temperature was 5° to 10° C. warmer than at present and that the average cloudiness was increased from the present 54 per cent to between 70 and 80 per cent. Simpson computed that these changes are reasonably to be expected if the sun's radiation fluctuated about 20 per cent above its present value.

This hypothesis is attractive because it would account for the increased precipitation that would be necessary to feed ice sheets. The present line of mean annual temperature at the freezing point lies far closer to the equator than the limit of present ice sheets, and this clearly indicates that low temperature alone is not enough to produce glaciation but that increased precipitation is needed. With the present configuration of land and sea, which is not greatly different from that during the Pleistocene, increased precipitation can only be brought about by increased evaporation from the oceans. The theory also accounts for the warm, moist climate suggested by plant fossils from some interglacial deposits. On the other hand, there is no general evidence that the last interglacial period was generally warmer and moister than the second, as it should have been on this hypothesis. Nor is there general agreement that the increased rainfall in lower latitudes lasted throughout both the last two glacial epochs and the intervening interglacial epoch. Finally, inasmuch as most astronomical processes are cyclic, if the variation of the sun's radiation is also cyclic, we should have had numerous other glacial episodes throughout geologic history. We must conclude that, even if this theory is satisfactory in some respects, an auxiliary hypothesis is needed to account for the long nonglacial periods of the geologic record.

We thus see that no satisfactory hypothesis has yet been advanced to explain the widespread glaciation of the continents during the Pleistocene. It is impossible to say whether or not the present is simply another interglacial epoch. We cannot tell whether glaciers will again encroach on the sites of Chicago, Copenhagen, and Warsaw within a few thousand years, or whether the climate will remain about the same as it is today. Much more work will have to be done to win an answer to this fascinating problem.

The Effect of Glacial Loads on the Earth's Crust

As we saw in Chapter 3, there is strong evidence from deflection of the plumb line and from gravity measurements that large segments of the earth's crust are essentially in isostatic balance. Glaciers of continental dimensions also afford clear geologic evidence in support of this conclusion.

The fact that former glaciers flowed over and eroded the ground on which they rested is proof that their surfaces sloped down from their feeding grounds to their termini. Wherever the glaciers spilled around mountainous topography, it is possible to get some measure of their thickness. In New England, for example, the highest peaks were unglaciated, and a measure of more than 4,000 feet has been made for the Pleistocene ice in that region. In southern Canada the ice would certainly have been considerably thicker.

Although the density of ice is only a little more than one-third that of ordinary rock, a load of 3,000 feet of ice over an area as large as the Great

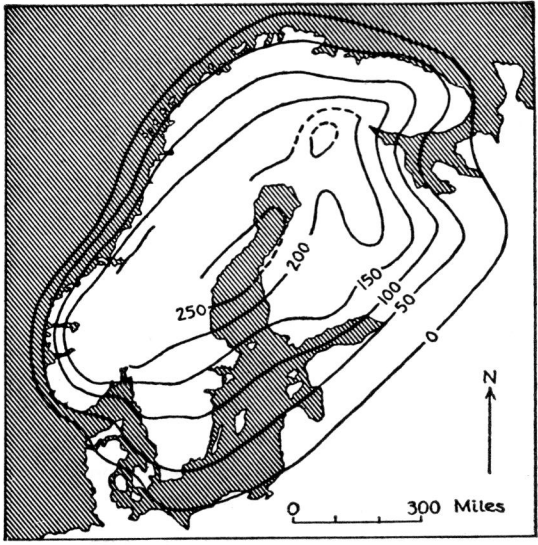

FIGURE 13-24. *Postglacial uplift in Scandinavia. The heavy lines connect points of equal uplift, in meters, of the highest strandline of the sea which flooded the area just after the melting of the glacier. (After R. A. Daly,* The Changing World of the Ice Age, Yale *University Press, 1934.)*

Lakes and southern Canada would be the equivalent of a load of more than 1,000 feet of rock placed on the area, and should bend the crust down. Though the isostatic response by subcrustal flow might take considerable time, owing to the high viscosity of the subcrustal material, the lowering should be considerable and measurable, if isostasy really is general.

Fortunately, the shorelines of ancient glacial lakes give us a means of testing the idea. They were level surfaces when formed. After the ice load melted away the rise of the crust in response to unloading should tilt the shorelines, causing them to rise to the north. This is exactly what is found, not only in the Great Lakes region, but in the Baltic (Fig. 13-24), in Labrador, and even in the basin of Lake Bonneville, where the load was only 1,000 feet of water covering a much smaller area than that of the great continental glaciers.

We will return to this geologic evidence of isostasy when we discuss the origin of mountains in Chapter 19. Suffice it to say here that the testing of the isostatic principle by glaciers is in full agreement with the geodetic evidence from which the theory was first announced.

Facts, Concepts, Terms

SNOWFIELDS AND GLACIERS
Transformation: snow → firn → ice

Nature of glacier flow
Combined brittleness and plasticity in glaciers

VALLEY GLACIERS; PIEDMONT GLACIERS; CONTINENTAL GLACIERS

ACQUIRING OF ROCK DEBRIS BY GLACIERS
Frost weathering; avalanching; meltwater shattering; plucking;
rasping
Till; rock flour

DEPOSITS ASSOCIATED WITH GLACIERS
Moraines: lateral; medial; terminal; recessional; ground
Stratified drift: outwash aprons; kames and eskers; valley trains and silt
terraces; deltas; varved clays
Loess

TOPOGRAPHIC FORMS ASSOCIATED WITH GLACIERS
Cirques; horns; U-shaped valleys; hanging valleys
Smoothed and grooved rock surfaces
Lakes and swamps; immature drainage patterns
Pitted outwash; terraces of silt
Morainal ridges; drumlin clusters

THE PLEISTOCENE GLACIATIONS
Development of the glacial theory
Evidence of advance and recession
Weathered tills; superposed tills; interbedded interglacial deposits
Drainage changes
Evidence from fossils of climatic changes

PRE-PLEISTOCENE PERIODS OF GLACIATION

CAUSES OF GLACIAL CLIMATES

ISOSTATIC RESPONSE TO GLACIAL LOADING AND UNLOADING

Questions

1. What is the evidence that recrystallization is an important process in the transformation of snow to firn and ice?
2. Why are the crevasses on a glacier less than 200 feet in depth?
3. Explain the processes by which the head of a valley glacier grows in volume as a result of (a) new accessions of snow and ice, (b) inclusion of rock debris.
4. Draw a longitudinal profile through a valley glacier and label the following features: cirque, terminal moraine, meltwater tubes, bergschrund, shear banding in ice, snowfield, rasped bedrock, plucked and shattered bedrock.
5. How does rock flour released from a glacier differ from the fine grained materials formed by weathering?

6. List, compare, and contrast the different kinds of glacio-fluvial and glacio-lacustrine deposits.
7. How can varved clays be used to tell the duration of a lake in years?
8. Draw a hypothetical sketch map showing the location of all the following features: (a) a lobate terminal moraine, (b) a recessional moraine, (c) pitted outwash, (d) ground moraine, (e) a drumlin cluster, (f) a plain underlain by varved clay, (g) an esker, (h) kame terraces, (i) an abandoned stream course, (j) spits and bars formed in a former lake that was dammed by the ice.
9. Outline the evidence that establishes the fact that a large lake, called Lake Bonneville, formerly covered much of northwestern Utah.
10. Outline briefly, and illustrate by sketches, the kind of evidence found in the field that indicates that there was more than one period of ice advance during the Pleistocene.
11. How would you tell a tillite from a conglomerate? From an alluvial fan deposit? From cemented talus or landslide debris?
12. Name the three chief periods of geologic time during which extensive glaciation occurred.
13. On what continents is there evidence of extensive glaciation within the tropics?
14. Ice floats in water. How, then, is it possible for a glacier entering the sea to erode its bed below sea level? How deep below sea level is it theoretically possible for a glacier 1 mile thick to erode its bed?
15. What is known about the causes of glacial climates?

Suggested Readings

1. Ahlmann, H. W., *Glaciological Research on the North Atlantic Coasts*, Royal Geog. Soc., Research Series No. 1, 1948.
2. Coleman, A. P., *Ice Ages, Recent and Ancient*, Macmillan Co., New York, 1926.
3. Daly, R. A., *The Changing World of the Ice Age*, Yale University Press, New Haven, 1934.
4. Flint, R. F., *Glacial Geology and the Pleistocene Epoch*, John Wiley & Sons, New York, 1947.
5. Gilbert, G. K., "Lake Bonneville," U. S. Geol. Survey, Monograph I, 1890.
6. Matthes, F. E., "Geologic History of Yosemite Valley," U. S. Geol. Survey, Prof. Paper 160, 1930.
7. Zeuner, F. E., *The Pleistocene Period, Its Climate, Chronology, and Faunal Successions*, Ray Society, London, 1945.

14. Ground Water

GROUND WATER is the water that saturates the pores and cracks in soil and rock beneath the land surface. It comes to the surface in springs, and also swells the volume of many streams by seeping into them from their beds and banks. Ground water is also obtained for human use by sinking wells.

Source of Ground Water

Water at the earth's surface has one obvious source—rain and snow—and, as Perrault showed long ago (Chap. 7, p. 126), there are good reasons for considering these sources as practically the only ones. They are also the ultimate source of practically all ground water. Most soils and rocks are porous. Small voids and openings occur between grains of soil, larger openings are formed by burrowing animals, and cracks are formed by the shrinkage of drying soil clays. Even well-consolidated rocks may be riven by joint cracks or other openings into which water can penetrate. After a rain part of the water seeps into these openings and percolates beneath the land surface. Some of the rain water that enters the soil remains near the surface, absorbed by the soil colloids, or held against gravity in the smaller voids by surface tension (*capillarity*, the force that pulls water up a very slender tube and holds it there against the action of gravity). Deeper water may even be drawn back toward the surface of the ground by capillarity, thus helping to replace the losses from evaporation and transpiration by plants. At least a part of the rain water, however, percolates deeper and deeper through the larger soil openings until it ultimately reaches a zone beneath the surface where all the pores in the rock are completely filled with water.

The Water Table

The name *water table* is given to the upper surface of this zone below which all pores are filled with water (Fig. 14-1). Except for the discontinuous films

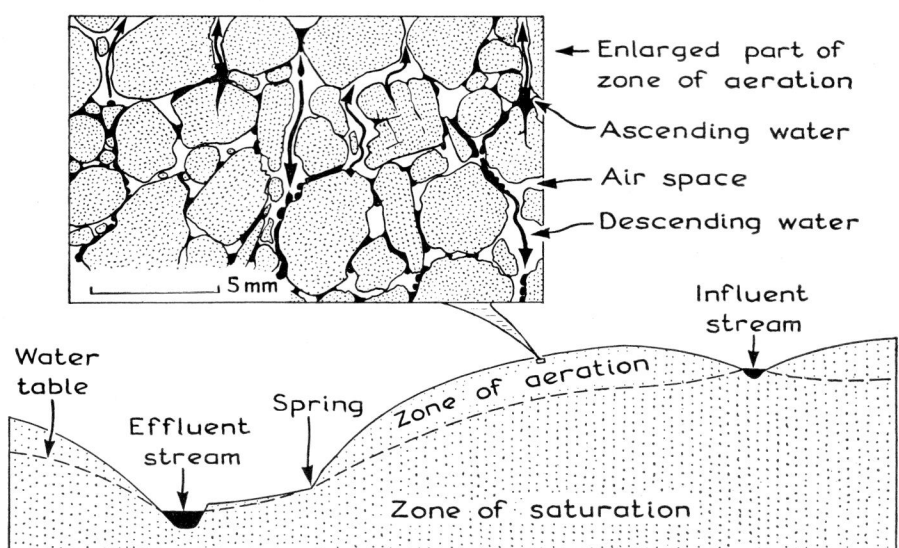

FIGURE 14-1. *Cross section showing the water table and its relationship to streams and a spring. The greatly enlarged inset shows the movement of water* (black) *in the zone of aeration.*

and irregular masses of water held by capillarity, or in process of transit downward, the pores above the water table are filled with air, and so the zone between the water table and the surface is called the *zone of aeration.* Here weathering and other chemical changes go on under oxidizing conditions. Below the water table is the *zone of saturation.*

In most places the water table is a few feet, or a few tens of feet, below the ground surface, but in arid regions it may be hundreds of feet below. In marshes, on the other hand, the water table practically coincides with the ground surface, as it does at the edge of surface-water bodies such as lakes and streams. If many water wells have been dug, the form of the water table below the ground surface can be readily determined and accurate contour maps of it drawn from the elevation of the water surface in the various wells. Such maps show that, in general, the water table is a subdued replica of the surface topography. It rises under the hills and sinks beneath the valleys, but is usually much smoother than the land surface. The level of the water table also fluctuates slightly with the seasons.

Subsurface Pores, Cracks, and Channels

If subsurface pores are abundant and large below the water table, water in considerable volume can be pumped from wells. When wells are pumped, water from pores in the nearby rock flows rapidly into the well and replenishes the water being extracted. On the other hand, if the pores are small and widely spaced, only a little water can be recovered by pumping because replenishment is slow. The shapes, sizes, and aggregate volume of the open spaces—pore spaces—in each kind of soil and rock are thus primary factors in determining the occurrence, amount, and rate of movement of ground water.

Porosity and Permeability

Porosity is the ratio of pore volume to total volume, expressed as percentage. It is generally determined by testing samples of rock in the laboratory. Porosities of most uncemented clastic sediments range between 12 and 45 per cent. Variations in the *shapes* of the grains, their *sorting,* their *packing,* and the *degree of cementation* cause these differences (Fig. 14-2).

Mineral and rock grains vary in shape from thin plates and irregular

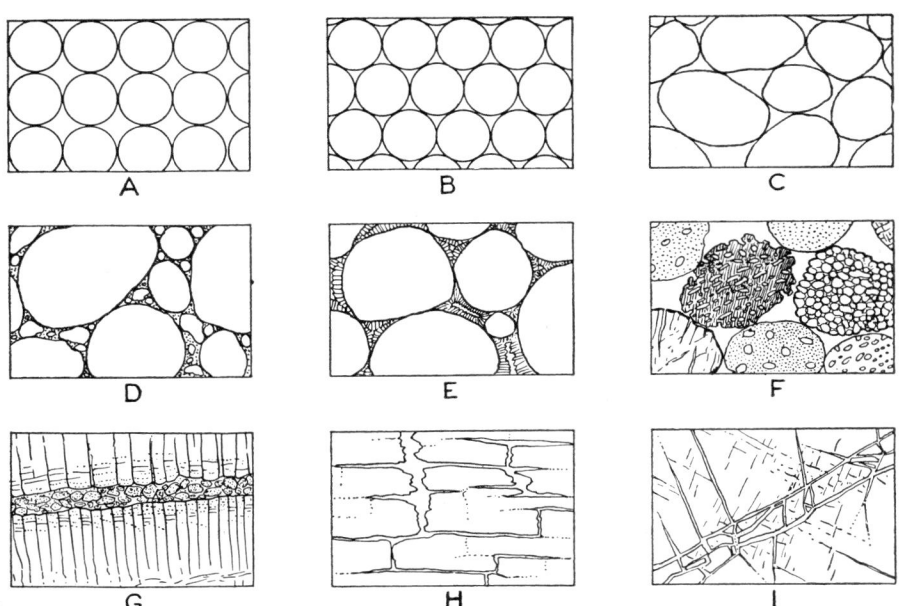

FIGURE 14-2.   *Porosity in rocks. A and B, the decrease in porosity due to compaction of spheres; C, a natural sand with high porosity due to good sorting; D, same, with low porosity due to poor sorting; E, low porosity due to cementation; F, very high porosity produced by loose, well-sorted grains that are themselves porous; G, porous zones between lava flows; H, limestone rendered porous by solution along joints; I, massive rock rendered porous by fracturing.*

angular bodies to nearly perfect spheres. The arrangement of the grains with respect to one another (the kind of *packing*) greatly affects the amount of pore space. Commonly, the packing is loose when a sediment is first deposited, and the pore space is relatively high. This space is then progressively reduced by *compaction* from the pressure of new sediments subsequently deposited above. *Cementation* (deposition of mineral matter in the pores) further reduces the pore space.

Closely packed spheres of uniform size have 26 per cent pore space. This is true whether the spheres are 1 millimeter in diameter, 5 feet in diameter, or any other size. Porosities above 26 per cent usually indicate nonuniform packing, or that some grains are themselves porous. Shape is also a factor, but the presence of nonspherical grains may either raise or lower the porosity. Porosities below 26 per cent usually result from poor sorting, or from compaction and cementation of the voids between grains.

The total capacity of rock or soil to *hold* water is determined by the porosity, but the capacity to *yield* water to the pump depends on the size of the pores as well as on the total amount of pore space. Not all the water in the pores will flow toward a pumping well. Much is retained as water films stuck to the walls of the pores. In rocks with very small pores practically all the water may thus be retained, even though the porosity is high. *Permeability, the capacity of a porous medium to transmit a liquid,* and not porosity, is, therefore, the most important physical property to be determined in estimating the reservoir capacity of a water-bearing material.

Permeability depends more on the size of the pore openings than on the percentage of pore space. A gravel with 20 per cent of pore space is much more permeable to ground water than a clay with 35 per cent porosity.

Laboratory measurements of permeability, as well as the distribution and yield of springs and wells, indicate that there is a fairly definite minimum size of pores—about 0.05 millimeters—through which water will move freely. Sediments with smaller pores, such as clay, silt, and shale, are impermeable even though their porosity may be high. Larger pores are present in sand and gravel and also in many incompletely cemented sandstones, conglomerates, jointed lava flows, and porous limestones.

Aquifers

A body of rock or loose surface material that is permeable as well as porous, and so can yield water to wells, is called an *aquifer*. Some aquifers may be empty of water, at least temporarily. Even in an aquifer, water movement is extremely slow, usually a fraction of a foot or a few feet per day, except in the immediate vicinity of a well or spring.

Most aquifers are sheets of sand or gravel, or beds of sandstone, limestone, or other permeable rock. A few aquifers are narrow sinuous bodies of gravel that fill former stream courses, but these are by no means as common as the popular expression "underground stream" indicates.

Water Witching

Before the Nineteenth Century men thought that ground water flowed in definite underground streams just as surface water does. They reasoned that when one dug a well, if he were lucky, his well intersected one of these streams and produced a fine flow of water. A dry well, or one that produced only a little water, supposedly had failed to intercept any of the underground channels. Since one cannot see beneath the surface of the ground, the drilling of wells was always an uncertain operation, and in their doubt about where to dig for water farmers often consulted "water witches" or "dowsers" (Fig. 14-3). Such people were supposedly endowed with supernatural powers that enabled them to discover the location of underground streams. A dowser generally walks about with a forked stick (Fig. 14-3), tightly held, which is supposed to dip violently when he crosses the position of an underground stream. His success, if any, has no known scientific basis.

Perched Water

An aquifer may rest on an impermeable substratum that overlies unsaturated material above the normal water table. The water in such an aquifer

FIGURE 14-3. *Water witch or dowser of the Sixteenth Century. (Styled after old woodcuts.)*

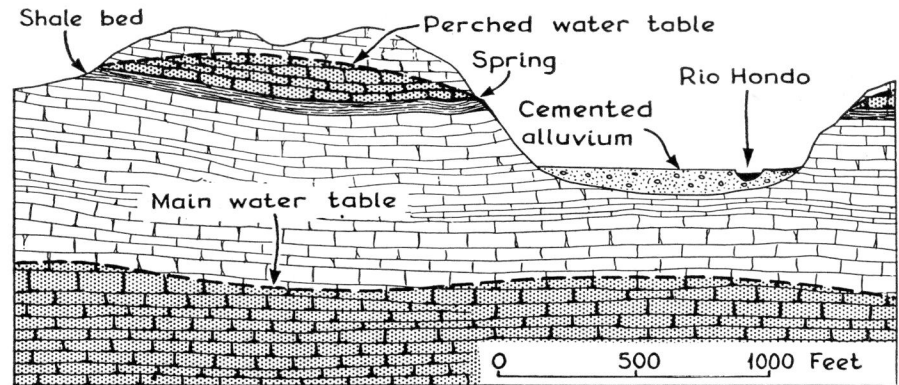

FIGURE 14-4.   *Cross section of aquifers in porous limestone, perched on an impermeable shale bed, southeast New Mexico. Note that the Rio Hondo, also, is perched above the main water table. (After A. D. Fiedler and S. S. Nye, U. S. Geological Survey.)*

is *perched*. It is prevented from percolating downward to the normal water table by the impermeable material beneath. Bodies of perched water are especially common in arid or semi-arid regions (Fig. 14-4).

Confined Water; Artesian Wells

A permeable sediment, such as an uncemented sandstone, may be overlain by an impermeable one, such as shale. If the rocks have been tilted and eroded, the permeable sediment may be exposed at the surface in hills or mountains that rise above the level of the overlying impermeable shale nearby. Under these conditions the ground water in the aquifer may be both *partially confined* (in the area where it is overlain by shale) and *easily replenished* (in the area where it appears at the surface). Such a situation is shown in Figure 14-5. Water enters the aquifer at A where it is exposed at the surface—the intake area. The water table (t-t') within the aquifer of the intake area is higher than the ground surface at B, where the aquifer is confined beneath overlying shale. If a well (W) is put down to this aquifer in area B, the confined water in the aquifer will rise through the well under the pressure of the head of water from the intake area and flow out on the surface of the ground, making this an *artesian well*.

The name "artesian" was originally restricted to flowing wells. It is now applied, however, to any well in which the water rises above the elevation of the aquifer penetrated.

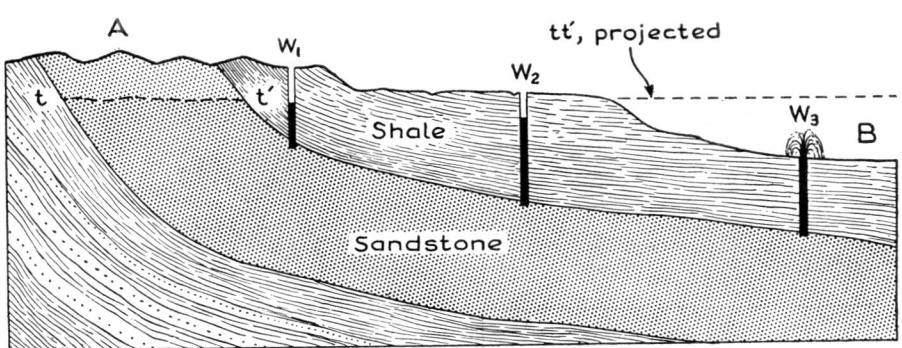

FIGURE 14-5. *Cross section showing a series of wells—W_1, W_2, W_3—penetrating a confined aquifer. The water table in the intake area is t–t'.*

Ground-Water Movement

The water that fills the pores below the water table is not static. It moves slowly under the influence of gravity. The direction of movement is controlled by the form and slopes of the water table. From the higher points on the table the water moves laterally, vertically, or obliquely downward. Similarly, it flows from all directions toward troughs or low points. The overall tendency is to lower the high points on the water table and either to raise the low points or else discharge water from them.

Points of surface discharge, located where the water table intersects the ground, are called *springs* (Fig. 14-6). Most perennial streams also mark areas of discharge in the water table. They lie in troughs on the water table, and the ground-water flow is toward and into them. Such streams are called *effluent* with respect to ground water.

Streams that flow from mountains or other well-watered areas into deserts or semi-arid regions, however, may lose water to the water table. Such *influent streams* lie on ridges in the water table, as shown in Figure 14-1.

If permeability and other factors are uniform, movement between two points on the water table is determined by the *hydraulic gradient*, which is the ratio between the difference of elevation, or head (H), and the horizontal distance between the two points (L). Ground-water gradients are usually low, such as a fall of 1 foot per 1,000 feet (0.001) or 10 feet per 1,000 (0.01).

Effect of Variations of Intake on Water Levels

For any particular permeability, the hydraulic gradient adjusts itself to the supply of water. If the discharge, as into streams, is temporarily greater

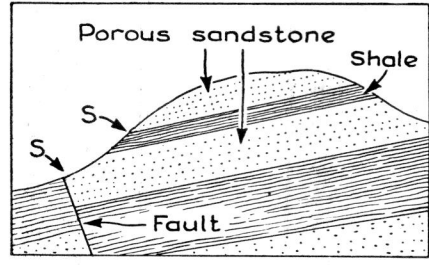 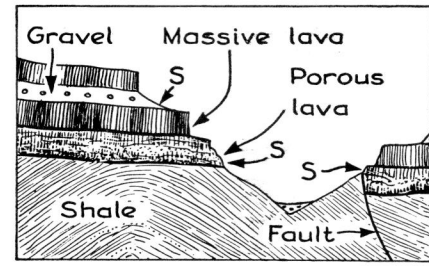

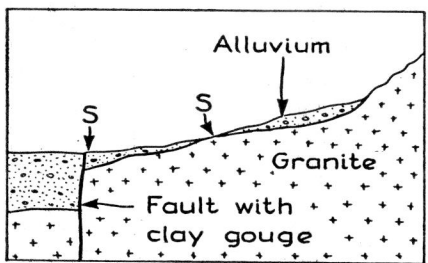 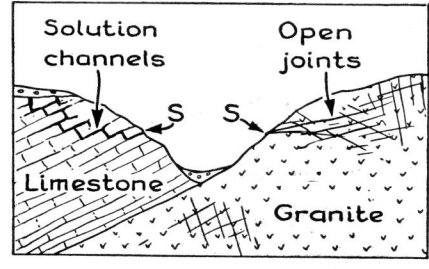

FIGURE 14-6.   *Cross sections showing likely locations for springs* (S).

than the supply into the intake area, the water table flattens. Thus, in dry spells the water table sinks farther below the surface under the ground-water divides, reducing the hydraulic gradient and the discharge.

In semi-arid western Texas, eastern New Mexico, and adjacent parts of old Mexico the water table is relatively flat and, in many places, lies 500 to 1,000 feet below most of the land surface. Much of the region either lacks permanent surface streams or else the streams are on perched ground-water bodies.

Darcy's Law

The modern concepts of ground-water movement were discovered early in the Nineteenth Century. Almost all movements of ground water are so slow that they occur by laminar flow, whereas practically all movements in surface streams are turbulent. Flow lines in laminar flow (see Chap. 12 and also Fig. 12-1) are smooth, continuous, and traceable. Investigation of the flow of water in pipes shows that for turbulent flow the velocity and discharge are approximately proportional to the square root of the hydraulic gradient, whereas *in laminar flow the velocity and discharge vary directly as the hydraulic gradient.* This fundamental law, as it applies to ground-water movement, was discovered and formulated in the 1850's by the French hydrologist Henry Darcy, during his study of the water supply of the city of Dijon. Darcy's law may be stated as:

$$V = PI$$

where: V = velocity of ground water

P = coefficient of permeability—a constant determined by the character of the material through which the water moves—a measure of the ease with which water moves through a subsurface material

I = hydraulic gradient—the slope of the water table

Geologists are usually more interested in the quantity of water moving than in its velocity. Hence, we usually replace the velocity term in Darcy's law by an equivalent term which states the quantity of water moving per unit of time (Q) through a given cross-sectional area (A).

$$V = \frac{Q}{A}; \text{ and we write Darcy's law as } \frac{Q}{A} = PI \text{ or}$$

$$Q = PIA$$

where: Q = quantity of water moving per unit of time (measured in gallons per day, cubic meters per day, etc.)

A = cross-sectional area through which water moves

If A is expressed in square feet, if I is made 1 (100 per cent gradient), and if Q is measured in gallons per day, then P comes out in *Meinzer units*. Those aquifers that have been investigated in this way have permeabilities that range all the way from 10 to 90,000 Meinzer units.

The approximate yield of a well may be predicted if the permeability of the aquifer is known, together with the hydraulic gradient under pumping conditions, and the cross-sectional area (the area of perforated casing through which water is entering the well). Similarly, the amount of ground water passing through a cross section of an aquifer can be calculated.

Rates of Ground-Water Movement

The movement of unconfined ground water through uniformly permeable material is shown diagrammatically in Figure 14-7, a section showing both ground-water divides and effluent stream levels. Some flow lines go deep, but little water follows these. The maximum velocities (and hence the greatest volumes of water transferred) are near the streams where the hydraulic gradients are steepest.

The actual velocities of ground-water movement, though almost every-

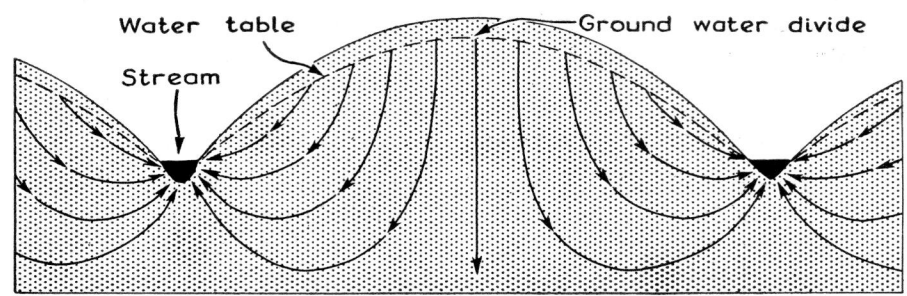

FIGURE 14-7.   *Approximate flow pattern of ground water in uniformly permeable material. (After M. King Hubbert,* Jour. Geology, *1940.)*

where low as compared to stream velocities, are highly variable. They vary in different parts of a single body of water moving through rock of uniform permeability, as shown in Figure 14-7.

Mean or average rates of movement can be calculated most easily from Darcy's Law, if the coefficient of permeability is known. The late leader of American hydrology, O. E. Meinzer, considered the flow of 50 feet per year in the Carrizo sandstone of Texas to be typical of many aquifers. In places much more rapid movements have been demonstrated by use of dyes or salts, introduced at one observation well and recovered or measured indirectly at another. Movements of 10 or 20 feet per day are sometimes attained in highly permeable materials, and one extreme velocity of 420 feet per day has been reported.

Balance Between Discharge, Permeability, and Hydraulic Gradient

According to Darcy's equation, in materials of low permeability the gradient of the water table increases steeply as intake water is added locally to the mass of ground water. In materials of high permeability the water table is relatively flat, and hydraulic gradients may be only a few feet or a few tens of feet per mile.

Discharge of Ground Water into the Ocean

The discharge of fresh ground water directly into the ocean may produce a body of fresh water extending far below sea level beneath an oceanic island or along a continental margin. It appears as if the lighter, higher column of fresh water were in static balance with the heavier sea water like a foreign mass floating within it. A column of sea water 1,000 feet high

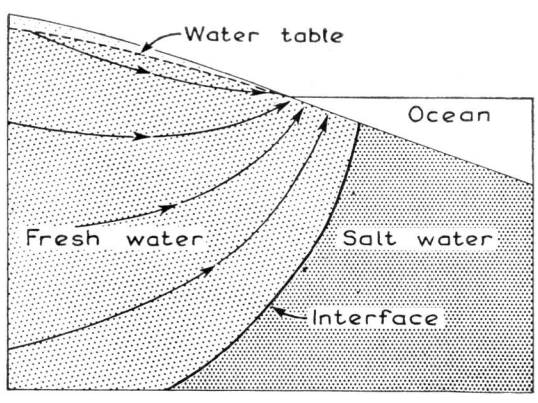

FIGURE 14-8. *Cross section showing fresh-water flow lines in relation to the contact with underground salt water. (After M. King Hubbert,' Jour. Geology, 1940.)*

can balance one of fresh water about 1,025 feet high. Thus a column of fresh water rising to an elevation of 25 feet above sea level, near the ocean shore, might indicate that fresh ground water could be recovered to a depth of 1,000 feet below sea level. Such a water body is, however, not static; it is constantly discharging into the sea. The friction of flow through the pores retards the spreading out of the fresh water, but if it were not constantly replenished from rainfall, the vertical separation between salt and fresh water would soon disappear. The probable relationships and flow lines in such a water body may be deduced from Darcy's Law (Fig. 14-8).

The body of fresh water under Oahu, one of the Hawaiian Islands, though hundreds of feet thick, is thinner than anticipated on the static flotation hypothesis. Furthermore, the fresh water is underlain by a thick transition zone of brackish water. Perhaps the mixing is due in part to oscillations caused by the intermittent pumping from many large wells. The original accumulation of fresh water extending far below sea level is presumably the result of flow following the lines induced by the hydraulic gradient of the fresh-water surface.

Drawdown by Pumping

A pumped well producing from an unconfined aquifer disturbs the water table by setting up a point of artificial discharge. We have already seen how the water table becomes adjusted to points and lines of natural discharge, such as springs and effluent streams. Similarly, removal of water through a well draws the water down adjacent to the well to produce a *cone of depression* in the water table (Fig. 14-9) that greatly increases the hydraulic gradient close to the well.

In the example shown in Figure 14-9 the pumping well removed water from uniformly permeable alluvium in the valley of the Platte River. The undisturbed water table sloped eastward 6 or 7 feet per mile. The lowering of the water levels was determined in more than 80 observation wells drilled

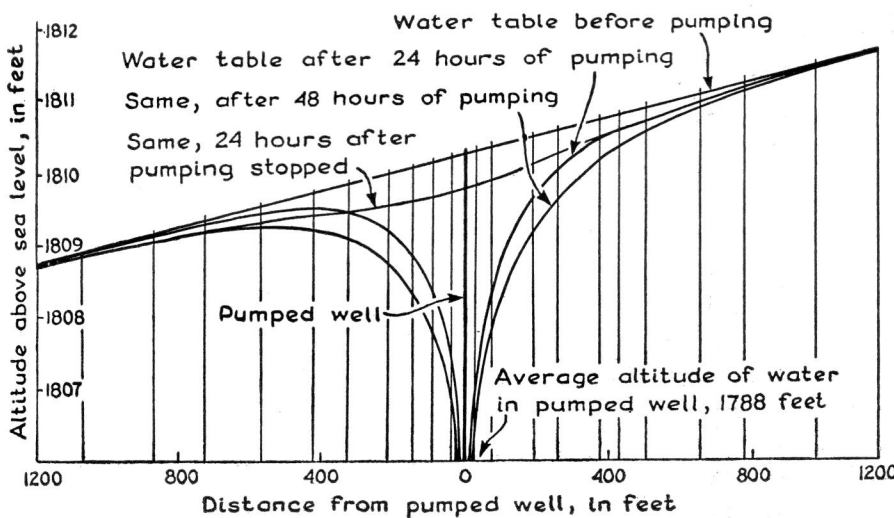

FIGURE 14-9. *Cross section showing the water table before, during, and after pumping from a well. Some of the wells used to observe the fluctuations are indicated by vertical lines. (After L. K. Wenzel, in* Hydrology, *courtesy of Dover Publications.)*

in 8 radial lines extending out to about 1,200 feet from the pumping well. Figure 14-9 is a section across the "cone" of depression in the direction of slope of the undisturbed water table. Note that a lowering, or *drawdown,* of 32 feet at the pumping well after 48 hours of pumping was accompanied by a lowering of 1 foot in the water table at a distance of 250 feet, and that measurable lowering extended at least 1,200 feet from the pumping well. Flow lines toward the well must have extended equally far laterally, and also far below the water table (cf. Fig. 14-7).

Effects analogous to the cone of depression may also be observed in wells that tap a confined aquifer. If a group of artesian wells is capped so that no water escapes, the water in each well will exert a definite pressure against the top of the casing. If the cap of one well is opened so that the water begins to flow from it, a decline in pressure will soon be noted at all the other wells.

Composition of Ground Water

Rain water, if it passes slowly through limestone, or even through decaying moderately calcium-rich rocks, such as granite, may pick up enough calcium ion—($CaCO_3 + H^+ \rightarrow Ca^{++} + HCO_3^-$)—to become *hard water.* The amount of calcium ion in hard waters in humid regions is only a small fraction of one per cent. Such waters may contain as much sodium ion as calcium ion, but the sodium ion is not easily precipitated and ordinarily goes unnoticed.

In arid regions, however, water within a few feet of the earth's surface may be partially or completely evaporated after each rain, thus progressively concentrating the water-soluble salts in the soil water. The commonest salts thus concentrated are the carbonate, chloride, and sulfate of sodium. The ground water in many arid regions is, therefore, more or less *alkaline*. Alkaline water is very toxic to plants, and the "alkali soils" are nearly useless agriculturally. If such areas are drained, and the alkaline water flushed downward by heavy irrigation, they may be reclaimed for agricultural use.

Still other ground waters are *salty*. Such waters contain enough sodium chloride to make their taste unpleasant and to make them injurious to plants. Some salty ground waters are in whole or part sea water, infiltrated directly from the ocean. Others have dissolved salt from salt beds in the rocks. Some water in deeply buried marine sedimentary rocks is supposedly sea water entrapped at the time of their deposition. Such entrapped sea water is called *connate water*. Connate waters rarely have exactly the composition of the present ocean. Variations are probably mostly due to dilution since burial, to concentration by evaporation previous to burial, and to chemical reactions with the enclosing rocks.

Ground Water in Carbonate Rocks

Solution Passages

Rain water is so effective a solvent of carbonate rocks that in limestone regions it enlarges cracks and pores and dissolves tunnels, irregular passages, and even large caverns along joints or other openings in the rock. Solution of the rock may become so extensive that much of the surface drainage goes underground through vertical tubular passages called *sinks* (Fig. 14-10) and discharges through *caves* (Fig. 14-11).

The development of sinks and caves is, of course, a very slow process. Water percolating down a crack in limestone enlarges the opening by solution. Weathering, rainwash, and collapse of the walls by gravity widens the opening at the surface, thus enabling it to trap more and more surface water which, in turn, causes more rapid solution of the limestone walls. The vertical sink thus formed may extend completely through the limestone bed, discharging at its base into other caves and caverns formed by solution of the limestone. In time, the entire bed of limestone may thus become honeycombed with interconnected sinks and caverns. Within the caves a part of

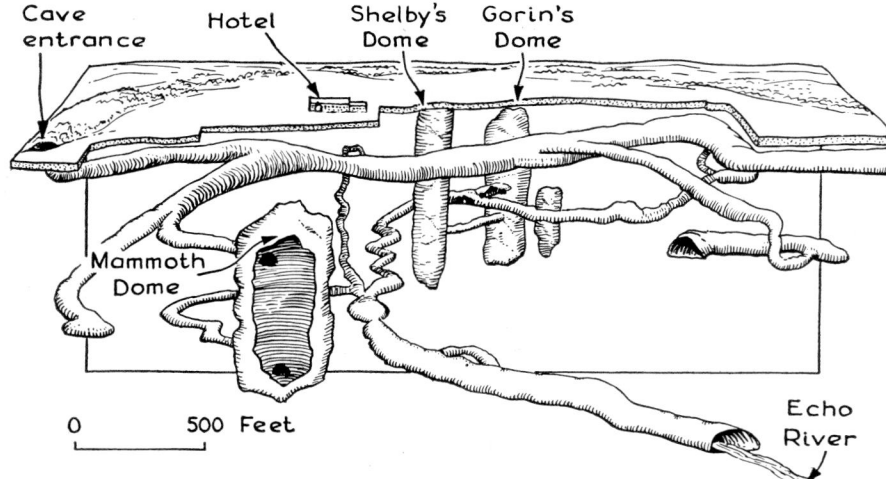

FIGURE 14-10. *A large sink, with alluvial floor, in limestone, Karst region of Yugoslavia. (Photo from Th. Benzinger, Stuttgart.)*

the dissolved calcium carbonate may be redeposited, often in bizarre icicle-like forms, producing masses of dripstone (Fig. 14-12).

The easy movement of water through a cavernous limestone aquifer and the flatness of its water table have been often demonstrated by pumping tests in mines. At the Los Lamentos mine, in Mexico, about 100 miles south-

FIGURE 14-11. *Diagram showing part of the Mammoth Cave system, Kentucky. (After A. K. Lobeck,* Geomorphology, *McGraw-Hill Book Co., 1939.)*

east of El Paso, Texas, two years of constant pumping did not lower the water table appreciably, and the mining of a rich ore body was necessarily stopped at the water table.

Karst Topography

In some limestone or dolomite regions there are few or no surface streams, and the water table is flat and far below the surface. The rainfall passes underground at once through sinks and joints widened by solution, and flows in underground streams with low gradients through large and small caverns, cascading at intervals to still lower levels, until, finally, it reappears at the surface as giant springs on the walls or floors of deep valleys. Such an area of sinks and underground drainage is called a *karst* region, from the name of such an area in Yugoslavia. Other well-known karst regions are the Causses of southern France west of the Rhone River and parts of the Cumberland Plateau in Kentucky and Tennessee.

The topography of a karst region differs from that of an area with normal surface drainage. Instead of a system of slopes definitely integrated to surface streams, a karst region is pock-marked by a maze of large and small depressions (Fig. 14-13). The smaller depressions are the upper ends of

FIGURE 14-12. *Dripstone in an Indiana limestone cavern. (Photo by Arch Addington.)*

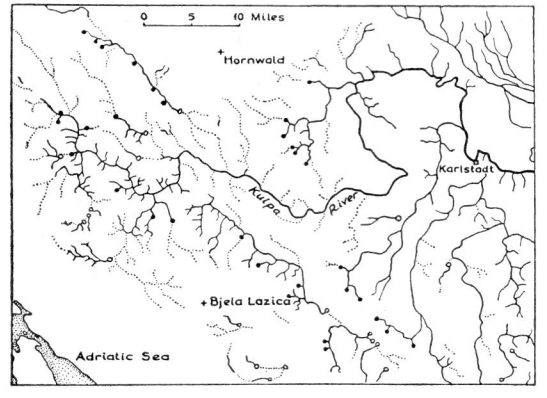

FIGURE 14-13. *Map of part of the karst region in Yugoslavia. The drainage is partly underground, as shown by stream sinks in small, partially dry valleys, and great springs in larger, deeper valleys. (After N. Krebs, 1928.)*

↢ Stream ⋯⋯ Dry valley • Karst spring ○ Stream sink

sinks and other solution passages, enlarged at the surface by weathering, rainwash, and downslope movements. Larger openings appear where the roofs of caverns have collapsed. Through such an area the main rivers flow at the bottoms of deep valleys, gaining most of their increase in volume from numerous large springs rather than from tributary streams. The largest springs arise from solution channels or even extensive caverns, which, in turn, are fed by surface drainage that goes underground. A stream may enter a sink in one valley and emerge through springs in a neighboring valley. As a result, many valleys are dry for short or long stretches.

Karst drainage has influenced men's actions for ages. Life is hard on the plateaus, easy in the well-watered valleys. Near the Causses of southern France great springs that emerge at the edge of the Rhone Valley have determined the sites of towns. Many of these, such as the ancient Roman city of Nîmes, have flourished since remote antiquity.

Ground Water in Volcanic Rocks

The cracks and cavities characteristic of many masses of volcanic rocks, as in the Hawaiian Islands or the Columbia Plateau of the Pacific Northwest, make these rocks almost as permeable as cavernous limestone or dolomite. The most productive group of springs in the United States is in volcanic rocks. The springs are distributed along a 50-mile stretch of the Snake River, Idaho, and discharge 14 million cubic meters of water per day, or 5,000 cubic feet per second, contributing to the Snake a volume of water two-thirds as great as the average discharge of the Mississippi River at St. Paul, Minnesota.

Solution and Cementation by Ground Water

In general, solution is more important above the water table; deposition and cementation below. The extreme effects of solution in carbonate rocks have already been described. Solution is not confined to carbonate rocks; in the long lapse of geologic time even minerals that the chemist regards as highly insoluble are appreciably dissolved. Grains of garnet in sandstones may be pitted and etched by ground water, and such highly insoluble minerals as augite and hornblende may be completely dissolved. Fossil shells composed of calcite are commonly leached out of shales and sandstones, leaving open cavities that may be filled to form "casts" which preserve faithfully the fossil form. Casts of soluble minerals such as halite or pyrite also are common.

In Chapter 5 the cementation of sand to form sandstone was mentioned. Calcite cements many sandstones but the reason it deposits in the sandstone pores is obscure. Evaporation of water and resultant deposition of calcite— the mechanism of dripstone formations—can hardly operate below the water table. Possibly one factor is the release of pressure on rising bicarbonate-rich ground water allowing the escape of carbon dioxide and the resulting precipitation of calcite. Around shell fragments, some sandstones are locally cemented, forming ball-like masses called *concretions*. This suggests that the shell fragments were nuclei upon which calcite from the saturated ground water precipitated.

Ground water may also deposit other substances. Silica, as opal or quartz, and iron oxide, as limonite or hematite, may fill or coat cavities in rocks. At depths of two or three miles temperatures are close to the boiling point of water at the earth's surface, and many other minerals may form in the pores.

Ground-Water Supplies in the United States

Water supply is a primary concern of farmers and inhabitants of arid regions. Even in the humid but populous areas of western Europe and eastern North America industrial demands put a heavy strain on water supply. The total quantity of recoverable fresh water underground at any one time is a matter of interest, especially in regions dependent on ground water for their supplies. All estimates indicate that the total amount of ground water is vastly less than the amount of water in the oceans, but much more than that in the atmosphere, or than that which falls as rain and snow in a single year. In almost any area some water can be obtained from wells, but in non-porous rocks, even where shattered, yields are relatively small, and of course yields will be negligible from nonpermeable rocks even if they are porous.

Abundant yields of ground water come chiefly from unconsolidated surface formations, mostly of Pleistocene and Recent age, and secondarily from older consolidated, but still permeable, sediments or from lavas. The principal unconsolidated aquifers are: (1) *alluvial gravels and sands that fill deep interior basins*, (2) *glacial outwash sands and gravels*, (3) *the coarser parts of deltaic and other coastal plain deposits, and* (4) *sands and silts beneath river floodplains.*

Interior basins filled with unconsolidated sediments are common in the western third of the United States. Ground water from many of them has been extensively developed because of the general aridity of the region. They normally supply about one half the ground water used in the United States. Several large basins are in California, including the large central valley (basins of the Sacramento and San Joaquin rivers), which is 450 miles long and 35 miles wide; and the much smaller valley of southern California, which includes the coastal plain near Los Angeles. In 1948 the ground water extracted from these basins in California was enough to cover 9 to 10 million acres to a depth of 1 foot—over 35 per cent of all the ground water used in the United States.

Glacial outwash sands and gravels are interbedded with, or lie slightly south of, the relatively less permeable Pleistocene till sheets. Glacial gravels grade into river floodplain and deltaic gravels and sands, notably along the Mississippi River. Large floodplain aquifers are also found on the Great Plains and permeable coastal plain deposits extend from New Jersey to Texas.

Among older, more consolidated rocks the chief aquifers are: (1) *permeable sandstones,* (2) *well-jointed volcanic rocks,* (3) *cavernous limestone or dolomite,* and, more rarely, (4) *fissured crystalline rocks* such as quartzite or shattered granite. Sandstone aquifers supply large amounts of ground water in the Dakotas and elsewhere in the northern Mississippi Valley but are important also in Texas, Michigan, and other regions. Basalt flows are important aquifers in the Pacific Northwest and in Hawaii. Cavernous limestones yield abundant ground water in Florida and in parts of the Cumberland Plateau.

The aquifers of three highly productive areas will be described in the following pages. These examples include the glacial outwash materials and Cretaceous sands of Long Island, the sands and gravels in the valley of southern California, and the Dakota sandstone of the northwestern Mississippi Valley.

Ground Water of Long Island

Long Island has no large streams, partly because of its size and partly because infiltration is so easy. The 40 to 50 inches of annual precipitation in west central Long Island, it has been estimated, is taken to surface runoff, ground water, and transpiration 20, 40, and 40 per cent, respectively.

Lacking adequate local surface supplies, the four million Long Islanders

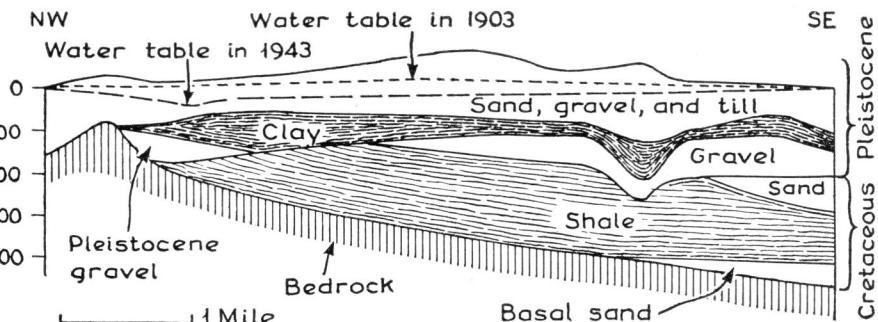

FIGURE 14-14. *Section through Brooklyn, showing the southeast slope of the bedrock (about 100 feet per mile), the Cretaceous and Pleistocene aquifers (white), and the water table in 1903 and 1943. (After C. E. Jacob, 1945; and W. de Laguna, 1948.)*

depend on ground water and surface water imported from the mainland to the Brooklyn and Queens sections of New York City. In addition, about 300,-000,000 gallons of ground water per day have been pumped in some years—a quantity about equal to that imported from the mainland. About three-fourths comes from glacial outwash sands and gravels, and one-fourth from unconsolidated Cretaceous sands (Fig. 14-14).

The most productive Cretaceous sand is the basal bed, 100–250 feet thick. It is a clean quartz sand, practically uncemented and confined by an overlying shale. The Cretaceous sands and shales are unconformably overlain (Fig. 14-14) by Pleistocene sediments as much as 300–400 feet thick. Interbedded Pleistocene till, sand, and gravel form two ridges, one near the northwest shore and the other along the middle of the island. The outwash plains of sand and gravel lie between and south of the ridges.

In a small Brooklyn area (Fig. 14-14) excessive pumping lowered the water table until it was below sea level. Along the section of Figure 14-14 the water table was 0–15 feet above sea level (0–50 feet below the ground surface) in 1903, but had been drawn down to a maximum depth of 34 feet below sea level by 1943. This reversed the hydraulic gradient, and salt water from the sea invaded the margins of the aquifer. The State of New York thereafter has required water pumped from new cooling and air-conditioning wells to be returned to the aquifer from which it was withdrawn. By 1946 more than 200 recharge wells were returning water underground in the urban portion of Long Island. In the rural areas several large surface recharge basins, where storm and industrial waste water could be stored until it would seep into the ground, had also been established. Total recharge in the summer of 1944 amounted to 60,000,000 gallons a day. The warmer recharge water raises the ground-water temperature a very few degrees Fahrenheit, making the water less valuable for cooling. The recharge wells and storm-water basins have served their main purpose, however; the water table is no longer falling, and salt-water inflow has decreased.

Ground-Water Basin in Southern California

In semi-arid southwestern California, several large interior basins have been filled with alluvium, across which the intermittent surface streams flow to the sea, charging or recharging the alluvium with ground water on the way.

A section through one of the basins is shown in Figure 14-15. The basin is filled with moderately well-sorted alluvium, 500 to 1,200 feet thick. Most of the alluvium came from the northeast and has been deposited in alluvial fans that have united to form a compound apron extending almost entirely across the whole basin. Coarse gravels are present at the heads of the fans, where the slope of the surface and initial dips of the strata are as high as 9°. Farther out on the fans the gravels grade into, and are interbedded with, sands and silts. Near the fan heads relatively impermeable soils have formed, several of which have been buried by upbuilding of the fans. The permeable gravels and coarse sands make a rather complex set of aquifers. On the lower portions of the apron some wells were originally artesian, the aquifers being confined by finer sediments or soil zones, and the hydraulic head being the result of the initial dip.

Many ground-water basins in California are broken by faults so recent that they cut through the alluvium. In this area the faults are effective barriers to ground-water movement because impervious clay gouge has been smeared along the fault planes. At one of the faults, shown in Figure 14-15, the water table drops 400 feet—a vital difference to those who must obtain their water from wells.

In this populous but arid region an attempt is made to keep the winter

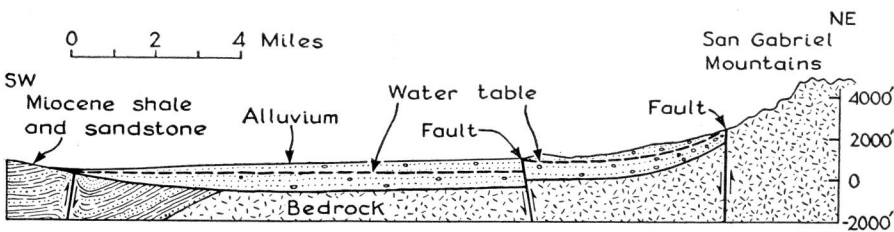

FIGURE 14-15. *Section across the Santa Ana ground-water basin, California, showing the water-bearing Pleistocene alluvium and the effect of a fault on the position of the water table. (After California Div. of Water Resources Bull. 45, 1934.)*

flood waters from reaching the sea by diverting them onto complex spreading grounds of coarse gravel which provide opportunity for the water to infiltrate. These spreading grounds serve the same purpose as the recharge wells and basins on Long Island.

It should be noted that lowering water levels in a basin by pumping is not wholly a misfortune. Surface or subsurface outflow from the basin is decreased or even stopped, and the subsurface reservoir is partially emptied, and so prepared to take in a much larger recharge.

Dakota Sandstone Artesian Aquifer

The great Dakota sandstone artesian basin is the largest and most important source of artesian water in the United States. It extends over much of North Dakota, South Dakota, Nebraska, and parts of adjacent states. At least 15,-000 wells have been drilled into this Cretaceous sandstone, which is generally somewhat less than 100 feet thick, and is overlain by hundreds or even thousands of feet of other sediments, mostly impermeable shales. As shown in Figure 14-16, the basin form resulted from folding. The principal intake areas are at the west, in the upturned zone along the edges of the Black Hills and Rocky Mountains, 2,000 or more feet higher than the sandstone on the east side of the basin. In general, there is a hydraulic gradient from west to east. Pressures have decreased progressively since the first well was drilled in 1882. The sandstone is, however, not a simple aquifer. It includes a widespread shale bed near its eastern margin, dividing the sandstone into upper and lower portions, and the compositions and pressures of the artesian waters are in detail rather complexly variable. Movement of ground water from the intake area appears to be interrupted locally by relatively impervious portions of the formation. Despite these complicating details, the Dakota may be considered a single unit, and an exceptionally important producer of ground water.

The Dakota sandstone is not the only productive aquifer in its area. One

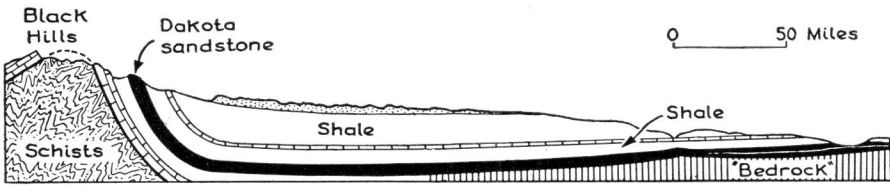

FIGURE 14-16.   *Section through the Dakota artesian aquifer, from the intake area in the Black Hills of western South Dakota to northern Iowa. The vertical scale is tremendously exaggerated. (After N. H. Darton, U. S. Geological Survey.)*

early forecast of production from a deeper sand is still remembered. The early railroads crossing this arid region were large users of water. Near the Black Hills the Dakota sandstone was an inadequate source. In 1905 N. H. Darton of the U. S. Geological Survey, who had just completed an investigation of ground water in the Great Plains region, recommended that the Burlington Railroad go deeper for water in the area near the Black Hills. The structure sections and geologic maps he had made indicated that by drilling at the town of Edgemont a probably productive Paleozoic aquifer would be encountered at a depth of about 3,000 feet. After almost three years of old-fashioned impact drilling a well flowing more than 400,000 gallons a day was developed at a depth within 31 feet of that predicted. Forty years later a new well, completed in the same aquifer after 50 days of rotary drilling, had an initial flow of almost 1½ million gallons per day, an unusually large yield from a Paleozoic sand, and very precious in this dry country.

Economic and Legal Aspects of Ground-Water Use

Where there is not enough surface and ground water to go around, disputes have arisen between individuals, communities, and states, and these disputes have been carried to the courts. Communizing the rule that the owner of the land surface also owns everything beneath the surface, all the owners of land above a ground-water basin are considered to have joint ownership of the basin. Water may not be exported without good reason beyond the basin's surface area. In addition, the *principle of best use* has been established, as in a dispute between cattlemen, who wish to preserve feeble springs, and truck gardeners, who wish to put the abundant ground water of the same area to more productive use. In some states, such as New York and Maryland, the permission of state authorities is required for the drilling of large wells, and return of used waters to the aquifer may be required. Ground water is an important public commodity, and its use now requires regulation by well-informed public officials.

Facts, Concepts, Terms

GROUND WATER IS RAIN WATER THAT FILLS OPENINGS IN SOIL AND ROCK

THE WATER TABLE SEPARATES THE
    Zone of aeration, from the
    Zone of saturation

WATER TABLE INTERSECTS SURFACE STREAMS, LAKES, AND MARSHES

POROSITY AND PERMEABILITY

PERCHED WATER; CONFINED WATER

GROUND WATER MOVES UNDER THE INFLUENCE OF GRAVITY
  Hydraulic gradient
  Darcy's Law: $Q = PIA$

NATURAL DISCHARGE OF GROUND WATER
  Springs
  Effluent and influent streams
  Discharge into the ocean

ARTIFICIAL DISCHARGE OF GROUND WATER
  Cone of depression around pumping wells
  Pressure drop in artesian wells analogous to cone of depression

SOLUTION AND PRECIPITATION BY GROUND WATER
  Solution channels, caves, and caverns in carbonate rocks
    Karst topography
    Formation of dripstone
  Slow solution of relatively impermeable minerals
  Cementation of sandstones

PRODUCTIVE AQUIFERS OF THE UNITED STATES
  In unconsolidated deposits
  In consolidated but permeable rocks

Questions

1. Draw sketches showing several geologic conditions that could result in the formation of a spring.
2. Draw one well-labeled cross section showing all of the features listed below:
   a. An area of rounded hills with two through-flowing streams, one of which has a floodplain, the other is downcutting far above grade.
   b. Two deep, but dry ravines
   c. The position of the normal water table
   d. A perched water table
   e. A swamp
   f. Two wells of equal depth, one of which has water, the other is dry.
3. A group of large fresh-water springs emerge on the sea floor about ½ mile off the mountainous coast of Ecuador. Show by a well-labeled diagram how this is possible.
4. Hornblende and augite grains are abundant in well-cemented concretions from a sandstone, but the remaining poorly cemented sandstone contains only a few etched grains. How do you account for this difference?

5. Compare the drawdown at the test well near the Platte River (p. 338) with that at the Los Lamentos mine (p. 340) and account for the difference.

6. How could you tell an area of karst topography from the hummocky surface of a large landslide or debris flow?

7. Large springs are common in areas underlain by basalt, but almost nonexistent in areas of granite. Why?

8. How can ground-water basins be artificially recharged from waste water at the surface?

9. If water is neither moving through nor being discharged from an aquifer, can there be a hydraulic gradient? Explain, using a diagram.

10. If a mass of gravel filling a valley has a cross section of 10,000 square feet, a hydraulic gradient of 5 per cent, and a permeability of 1,000 Meinzer units, how many gallons of water per day move through the cross section of the valley? How many cubic feet per second?

Suggested Readings

1. Hubbert, M. King, "The Theory of Ground Water Motion," Journal of Geology, Vol. XLVIII, 1940.

2. Meinzer, O. E., "Ground Water in the United States—A Summary," U. S. Geological Survey, Water Supply Paper 836-D, pp. 157-229, 1939.

3. Meinzer, O. E. (Editor), Hydrology: Physics of the Earth, No. 9, National Research Council, Dover Publications, Inc., New York, 1942.

4. Tolman, C. F., Ground Water, McGraw-Hill Book Co., Inc., New York, 1937.

5. Veatch, A. C., et al, "Underground Water Resources of Long Island, New York," U. S. Geological Survey, Professional Paper 44, 1906.

15. *Deserts*

Deserts are regions that are barren because they are dry. Precipitation is insufficient to support a continuous cover of vegetation. The rugged angular hills, cliffed canyons, and pebble- or sand-covered plains of the desert contrast sharply with the smoothly rounded hills and curving transitional slopes familiar in more humid country (compare Fig. 15-1 with Fig. 12-23). To a traveler from a well-watered region the desert landscape seems at first to have been molded by forces different from those of his homeland. The contrasts, however, express not different agencies, as in glaciated areas, but a difference in the results achieved by streams and downslope movements operating under different climatic controls.

About one-sixth of the land area of the earth is desert. The greatest deserts lie in the subtropical belts of high atmospheric pressure ("Horse latitudes," see Chap. 6, Fig. 6-2). Here the winds are dry, clouds are few, precipitation is low, and evaporation high. Some great deserts, like those of central Asia or parts of the American West, lie in "rain shadows" behind high mountain ranges.

Climatic Controls

In the Sahara and other great subtropic deserts the average rainfall is probably less than four inches per year. A year or more may pass without any rainfall at all. But the absolute amount of rainfall does not determine desert conditions. For example, the 5 or 6 inches of annual rainfall at Point Barrow on the Arctic coast of Alaska is nearly as low as that at Yuma, Arizona. Yet the ground at Point Barrow is sodden with water and matted with vegetation, whereas at Yuma the soil is parched and the sparse plants are highly specialized in drought resistance. The contrast is largely due to the low evaporation in the Arctic, though other factors, such as the permeability of the ground and depth to water table, also modify plant growth.

We have already noted (Chap. 6) how vegetation influences soil develop-

FIGURE 15-1. *Broad alluvial plains between desert mountains, Salton Desert, California. Belt of small sand dunes in foreground. The straight black line is a railroad. (Photo by Robert C. Frampton, Claremont, California.)*

ment. Many of the contrasts between desert and humid erosion stem directly or indirectly from the differences in soil cover and vegetation.

Some writers divide arid regions into *steppes,* where scattered bushes and short-lived grasses furnish a scanty pasturage, and *true deserts,* where vegetation is sparse or absent. On this basis the deserts of North America would be mostly steppes. There are, of course, continuous transitions from extreme deserts, through steppes, to humid regions.

Interior Drainage

Only the greatest rivers, like the Nile, Indus, Colorado, and Niger, persist through deserts to the sea. Most streams in the desert soon sink into the ground, dwindle to a series of stagnant pools, or end in small salt lakes or alkali mud flats. The drainage of the desert is therefore not integrated into larger and larger tributaries feeding one or more trunk rivers, as in humid regions. Instead, it generally consists of many small stream systems, each of which ends in a closed basin or disappears on a desert plain. Such unintegrated *interior drainage* is characteristic of deserts.

The water table generally lies far deeper in deserts than in humid regions. Furthermore, rainfall, just as in most humid areas, is generally greater at high elevations than at low. Thus after a desert rain, rills and even rushing torrents rise in the desert mountains but they quickly shrink, sink into the ground, and disappear on the desert plains below. Yet, although most stream courses are dry except for a few hours or days a year, all but the most arid deserts show unmistakable evidence that stream action dominates in molding the landscape. Graded stream courses and accordant tributaries are the rule within the small desert stream systems, just as they are in the more extensive stream systems of humid countries. Barren desert mountains scarred by stream gullies and canyons are interspersed with plains built by stream deposits.

Stream deposits are conspicuous in deserts (Fig. 15-1). Many desert storms are local, and the streams they generate run only for a few hours. Hence the sediment is not transported to the sea as in a humid area, but is dumped as alluvial cones at the mouths of mountain canyons. The cones grow until they join those heading in adjacent canyons to build great alluvial aprons along the mountain bases (Fig. 12-26 and Fig. 15-3). These compound alluvial aprons, or bahadas, flatten gradually valleyward and merge imperceptibly with the valley floor, to which the streams, decreased by evaporation and infiltration, can carry only the finer material.

Closed depressions on the desert surface are generally not filled to over-

FIGURE 15-2. *North Alvord playa, southeastern Oregon. The irregular dark patches on the white playa surface are wet ground. The straight mountain fronts to the upper left are fault scarps. (Photo by Richard E. Fuller.)*

flowing as are the lake basins in a humid country. Water gathers in them after periods of heavy rains and forms temporary lakes called *playas*. Most playas dry up during the dry season, but a few may persist for several years after an unusually wet season. Typical playas are the Black Rock Desert of northwest Nevada, the floor of Death Valley, and others in the Great Basin (Fig. 15-2). Although Great Salt Lake is perennial and thus not strictly a playa, it fluctuates widely with wet and dry seasons, so that the flat western part of its bed—the Bonneville Salt Flat—has all the features of a typical playa. When a playa dries up, the material dissolved in the water is, of course, deposited as crystallized salts, among which halite and various carbonates and sulfates of sodium are most abundant. The "alkali flats" of many arid regions are characterized by such deposits.

Thus geologic processes in deserts differ from those in humid regions primarily because of (1) the lack of vegetation, which profoundly influences erosion and especially the details of land sculpture; and (2) the intermittent stream flow and general lack of through drainage. We will now turn to a brief analysis of the influence of the desert climate on the normal geologic processes.

Geologic Processes

Weathering

Rocks disintegrate and decompose in arid regions just as in humid, though more slowly because of the lack of moisture and the paucity of organic acids in the soil. We have already seen that the Egyptian obelisk moved to New York has weathered more in 50 years than it had in 3,000 years in Egypt (Chap. 6).

Because of differences in weathering and transportation in deserts fine-grained residual soils are rare. Indeed, much of the rich soil of Egypt, flooded annually by the Nile, and of Iraq, where the alluvium of the Tigris and Euphrates have supported civilization for centuries, is transported. Most of the weathering that produced the soil minerals took place in the more humid headwater regions.

We have already mentioned (Chap. 6) that limestone, which dissolves readily and generally forms lowlands in humid climates, commonly makes bold ridges in deserts. Partly this may be because joint blocks spalled from its outcrops only slowly dissolve and thus protect the slopes, and partly because the openings in the rock are soon sealed by the redeposition of dissolved calcite when percolating ground water evaporates.

Rainsplash and Rillwash

Besides the few permanent or intermittent streams, the principal mechanisms of rock transport in the desert are rainsplash, rillwash, and sheetfloods; these give the primary stamp to the desert landscape.

Desert plants are so scattered that their roots hold only a small part of the surface material. Little humus or plant debris protects the surface; rain drops strike it directly with unbroken impact. They splash the finer fragments into the air, including even coarse sand grains. The dislodged particles tend to move downhill. Anyone who has noticed the mud and sand splashed onto a board resting on a garden plot after a light rain can readily imagine the effects of rain erosion where the whole landscape is barren and the slope steep. Fine flakes from disintegrating shale, for example, move downhill even on very gentle slopes. Steeper slopes are needed for coarse particles from disintegrating gneiss or granite.

Although heavy downpours of several inches of rain per hour may fall in even the most arid regions, the normal rains are light. The rainsplash affects the whole slope. Even though small rills may form, they grow but slowly. Though they carry silt, sand, and even gravel, they generally leave this material stranded after a short journey, partly as fillings of previously eroded channels. Newly exposed portions of the bedrock, or boulders derived from it, are made accessible to slow weathering and shed new chips and grains into the adjoining lower rill courses, tending to obliterate them in the long intervals between major storms.

Many desert slopes are gullied. A gully in the desert, however, has little advantage over the ungullied slope nearby, as both are eroded entirely by runoff, and not partly by effluent ground water, as may be the case in a humid region. Because of its steeper slopes the gully is also likely to be filled by joint blocks, chips, and grains weathered from nearby outcrops. If no rain heavy enough to produce a running stream falls for some years, the gully may lose its identity by being blocked with boulders and filled with finer debris.

Mudflows and Sheetfloods

From time to time, perhaps at intervals of years or even centuries, intense rains may come—the so-called "cloudbursts" during which several inches of rain may fall in an hour. Most are very local; a few miles away no rain, or at most a light sprinkle, may fall, while over an area of a few square miles a tremendous downpour is taking place. Such a rain may rapidly gully a long-stable slope, stripping off the loose debris, carrying away all the stranded sediment, and racing down the canyons in a wall of debris-laden water so charged with mud and sand that it forms a fluid far denser than water alone —a fluid capable of carrying huge blocks and boulders buoyed up in the viscous mass. Such a cloudburst flood, pouring down Cajon Pass, California, overwhelmed a freight train, carried the engine more than a mile down the canyon, and buried it so deeply beneath sediment that it could only be found by using a sensitive magnet! Such *mudflows* may travel completely out of the area of heavy rains before they reach the alluvial fans at the foot of the mountain. The water sinks into the fan and the mudflow comes to a halt, sometimes with a steep frontal scarp several feet high.

Less intense rains may not supply enough water to yield the violent turbulence required to move pebbles and boulders. Loose silt, sand, and gravel are so abundant on desert fans that any flowing water is soon loaded to capacity with sediment and is, therefore, unable to cut deeply into the surface. Diverted by cobbles and boulders and by jams of floating plant fragments and scattered clumps of vegetation, the water spreads widely in a film a few inches deep, forming a *sheetflood* covering the whole surface. When it sinks in or evaporates, it leaves a layer of mud and silt to dry in the desert sun.

Downslope Movements

Downslope movements operate somewhat differently in deserts. Although weathering is relatively slow, many joint blocks have weathered so long that they are "rotted" and disintegrate completely when they fall from cliffs; hence great talus piles are rare beneath steep cliffs. Both steep and gentle slopes may be mantled with fallen boulders and chips, but these are only "one boulder thick," with the bedrock often visible beneath them. The boulders lie in place until they are swept away in cloudbursts or reduced to grains small enough to be carried away by rills.

Even though every desert occasionally undergoes violent downpours— perhaps only once in a decade, or even in a century—the sparse vegetation and thin soil cover permit great runoff. Therefore great masses of water-soaked soil and rock like those responsible for the Gros Ventre slide (Chap. 11) rarely develop in deserts. In fact, the presence of large ancient landslide blocks in Arizona has been considered strong evidence of a former more humid climate.

Relation of Slopes to Structure

The lack of soil and vegetation affects downslope movement and other kinds of desert erosion in still another way. Because the soil is not bound effectively by roots, it does not slowly creep as a mass and thereby soften contrasts between surface slopes, as in moister climates. The mountain slopes are those appropriate to the size of the joint blocks yielded by the bedrock —steep on bedrock furnishing large boulders, more gentle on bedrock shedding smaller ones. Sandstones and shales that break down to individual grains may have gentle but almost barren slopes unless masked by boulders that have rolled down from above. Such differences in particle size are accurately reflected by differences in surface slopes developed on various rocks (see Figs. 7-6 and 10-6). No long-sweeping transitional slopes link the hillsides with the valley floors and round off the summits of the hills as they do in humid lands. Abrupt changes in slope at the boundaries of different rock masses are the rule in desert hills and mountains, in marked distinction with the blurred and transitional slope changes characteristic of humid regions (compare Fig. 10-6 with Fig. 12-23), where soil creep is active and streams grow larger in their lower courses.

Thus each desert hillslope tends to erode more evenly than that in a humid region, and to show much closer adjustment of slope to the structure of the underlying rock. The desert landscape is generally composed of rather even slopes, each nicely adjusted to the average grain size of the particles furnished by disintegration of the underlying rock. Steep slopes remain steep, and even on the wide desert plains the isolated hills left after erosion of great volumes of rock characteristically rise abruptly, and with steep unsoftened slopes adjusted to the coarse-sized particles weathered from them (Fig. 15-3, bottom).

Evolution of Desert Land Forms

The deformation of the earth's crust may disrupt pre-existing drainage courses in deserts just as in other regions; so, also, may rapidly growing alluvial fans, mudflows, volcanic eruptions, or even roof-collapse of limestone caverns. Basins lower than their borders are thereby formed, ranging from enormous areas like the Caspian depression, the Dead Sea trough, or the basin of Great Salt Lake, down to wind-carved hollows a few feet across. Because the runoff in many deserts is not enough to fill such basins to overflowing, they remain areas of internal drainage.

FIGURE 15-3. *Three stages in the erosion of desert mountains.* Top, *Panamint Range, California, showing alluvial fans at the foot of only moderately eroded fault block;* center, *Ibex Mountains, California, showing broad pediment embaying deeply eroded range;* bottom, *Cima Dome, California, showing residual "inselbergs" on a broad graded surface of stream erosion. See Figure 15-4. (Photos by Eliot Blackwelder.)*

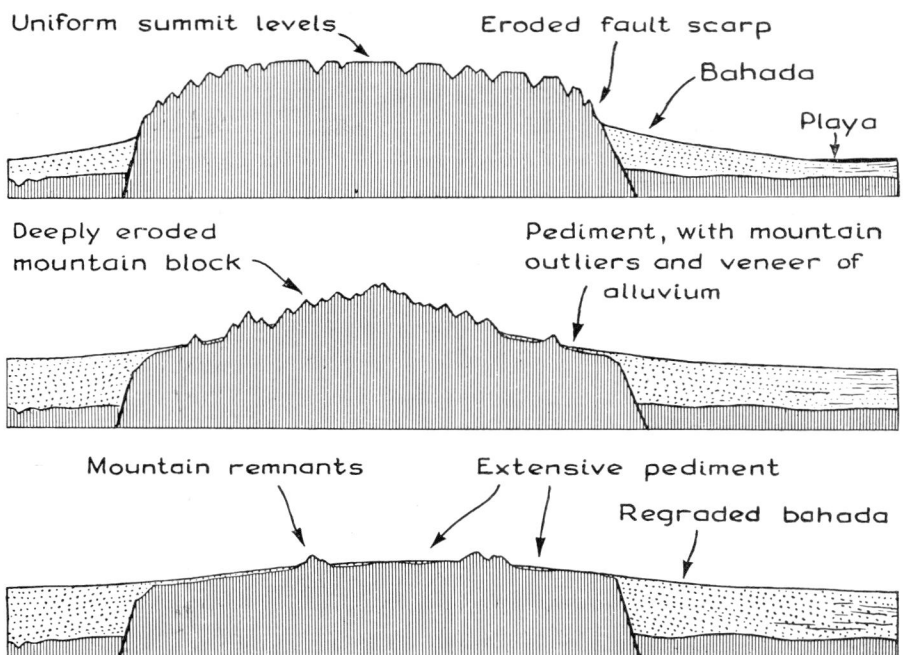

Uniform summit levels — Eroded fault scarp — Bahada — Playa

Deeply eroded mountain block — Pediment, with mountain outliers and veneer of alluvium

Mountain remnants — Extensive pediment — Regraded bahada

FIGURE 15-4. *Diagrammatic cross sections of the three stages of erosion shown in Figure 15-3.*

Every such closed basin, whether occupied by a permanent saline lake or a playa, is a local base level for the tributary area. Sediment brought to it accumulates and slowly builds up the basin floor, thereby raising the base level. Hence the streams entering the basins acquire flatter gradients and can carry less sediment. To counter this they tend to steepen their gradients by depositing material upstream, on the lower parts of their fans. But because their headward portions are also being lowered by erosion, they tend to cut down into the heads of their fans. Thus the topography of many desert areas where crustal deformation has been geologically fairly recent (as shown by deformation of late Tertiary or Pleistocene deposits, for example) falls into three kinds of land forms: (1) the relatively *steep mountain slopes* of bare rock and talus, (2) the *bahada slopes* made up of coalescing alluvial fans along the mountain bases, and (3) the *playa floors* covered with finer silts, clays, and various salts left behind by evaporation of the water upon it (Fig. 15-3, *top*, and Fig. 15-4, *top*). This stage is common in the Great Basin part of Utah, Nevada, and southeastern Oregon, where all these forms abound.

But, although the basins build up by sedimentation, thus raising the local base level, erosion continues in the mountains. The slopes developed on the mountains are those appropriate to the particles supplied by the bedrock. As we have noted, these slopes are little modified by soil creep and remain steep, each segment of slope adjusted to the size of the joint blocks or other particles loosened by weathering. On the other hand, the foot of the slope is slowly weathered back as the bedrock breaks down to sizes that can be moved by streams on the bahada below. Thus the base of the mountain retreats from its original position, but the sharp break from bahada to mountain slope remains.

As the base level rises the ephemeral streams build their courses above the neighboring land and break out into adjoining lower basins. Sheetfloods build up the fans in their lower parts while rainwash and rills slowly lower their upper parts. Lateral swings of the streams emerging from the mountain canyons sap the spurs of the interstream ridges and plane off the higher parts of the fans as they cut down. The lower ends of the interstream ridges are also reduced by rillwash as the mountain front retreats and are regraded to slopes practically identical with those of the frontal fans. A broad, gently sloping surface, called a *pediment,* is thus carved on the bedrock at the foot of the mountain slope. This surface is more or less completely covered with a thin veneer of gravel in slow transit toward the valley. At such a stage the desert landscape is composed of four main elements: (1) *the mountains,* whose slopes are about as steep as in the earlier stage described above, (2) *the pediment,* or planed-off surface of bedrock, which is separated from the mountains by an abrupt break in slope, (3) *the bahada,* composed of the old fan deposits regraded to a lower slope that blends imperceptibly with the pediment upslope and with (4) *the playa* downslope (Fig. 15-3, *middle,* and Fig. 15-4, *middle*). This stage of development of the desert landscape is widely represented in eastern Arizona and southern New Mexico, areas in which geologically recent crustal disturbances have been mild.

As erosion continues, streams draining to lower closed basins or to the sea continue to extend headward, as they do in humid regions. As a result, the basins whose floors are at higher elevations are successively captured by these extending streams and are eroded and ultimately regraded to lower levels. The main streams thus grow longer and longer and, in time, their long profiles become graded throughout, even though no single storm may produce enough rain to cause the main stream to flow throughout its full length at any one time. It is clear that this process of drainage integration and regrading of higher basins to lower levels must result in the scouring out of old basin fill. With respect to the lower base level, such materials are like the bedrocks of the original topography and are eroded in slopes appropriate to their component rocks. Generally, they are poorly consolidated and readily eroded. As a result, pediments develop rapidly across the beveled edges of the old basin deposits and extend widely at the expense of older bahada slopes. In this stage of desert erosion playas are absent (if the drainage is external), or much fewer (if sea tributaries have not succeeded in extending through the region); bahadas are relatively much less extensive; mountains have shrunk; and pediments cover much of the area (see Fig. 15-3, *bottom*, and Fig. 15-4, *bottom*). This is the stage represented in the United States by large parts of southwestern Arizona, where the Gila and Colorado furnish a slowly lowering base level for the local streams, and there are no present-day playas (cf. Fig. 15-1). Nevertheless, as shown by wells drilled for water, large parts of this area are underlain by thick alluvial fans and playa clays. The surfaces of these areas blend indistinguishably into the larger areas of pediment in which only a thin veneer of gravel masks hard bedrocks, and small mountain masses rise abruptly from the smooth surfaces at their bases.

In a still later stage, apparently represented by parts of the Kalahari Desert in South Africa, the mountains, though they have retained their steep slopes, have shrunk to small isolated hills, rising abruptly above a rock floor like islands from the sea. Such mountain remnants are called *inselbergs,* from the German word for "island mountains." Presumably, if no structural or climatic change intervened, these would eventually be reduced by continued erosion to produce a wide rock plain subject primarily to wind erosion. Small parts of the Kalahari and other deserts approach this condition, but no large area has been recognized as representing such a final stage in desert landscape evolution.

Work of the Wind

The desert land forms just described indicate the dominance of water-molded surfaces in the landscape. Many people, however, have the impression, perhaps fostered by movies and romance, that deserts are chiefly great wastes of sand dunes. These, they suppose, furnish the chief contrast with the humid landscape. As we have just noticed, for most deserts this is not true. Nevertheless, because vegetation is sparse or absent in deserts, wind erosion is much more prominent than in humid regions, and in parts of the desert its effects may be very noticeable.

Even a gentle wind wafts dust along a city street. On hot summer days dozens of "dust devils" (whirlwinds) swirl fine debris high above the plains, and occasionally tornadoes uproot trees, loosen the soil, and carry the debris of houses for hundreds of yards. The amount of material transported by the wind is difficult to measure, but even in humid countries it is considerable, as shown by the dust accumulated in a closed room within a few days.

Sorting by the Wind

Anyone who has allowed a handful of dry soil to dribble slowly from his hand during a wind has noticed that some of the material falls nearly vertically, but much of the pile strings out to leeward. If there are some fine particles of dust, these will be carried away completely. By repeated trials, the coarse grains of the soil can be rather cleanly winnowed from the fine, even in a gentle breeze.

This example illustrates a general condition. If any object is dropped through a fluid less dense than itself, it falls at a speed that increases at first but eventually becomes constant—the so-called "terminal velocity" of fall for the object. Two forces are acting on the body: (1) the pull of gravity acting downward, and (2) the resistance of the fluid to the passage of the object. At any particular point on the earth's surface the pull of gravity is practically constant and depends only on the difference between the mass of the body and that of the volume of fluid it displaces (Archimedes' Principle). But the resistance of the fluid to the movement depends on the viscosity of the fluid, on the diameter of the body about which the fluid must be displaced, on the shape ("streamlining") of the body, and on its speed through the fluid. The resistance also increases with increasing speed. Thus the force of gravity, which in a vacuum would produce constant acceleration, is ultimately balanced by the increasing resistance of the fluid. There is then no net downward acceleration of the body—its speed is constant at the terminal velocity.

Experiments have shown that the terminal velocities of different-sized spheres falling in air vary tremendously. When the particles are smaller than about 0.01 mm., their terminal velocities vary almost exactly with the square of their diameters, conforming to the law deduced by Sir G. G. Stokes in 1851. The speed of such particles is slow enough for the air to accommodate their passage by laminar flow. The velocities of larger spheres are not so simply related to their diameters because their greater volumes displace more air, and the inertia of the air displaced in their passage becomes important. The air around, and especially behind, them becomes turbulent during their fall. Figure 15-5 shows in a general way the relations between size and terminal velocities of particles in air, though irregular grains do not have quite as high terminal velocities as spherical grains of the same mean diameter. The fine particles in our handful of soil thus are blown farther by

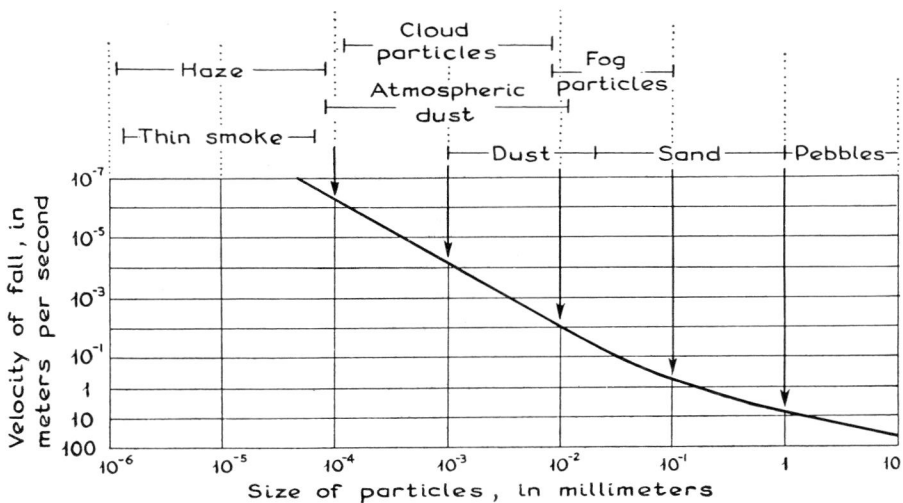

FIGURE 15-5. *Graph showing the variation of terminal velocity of fall with grain size of falling particles. (After R. A. Bagnold,* The Physics of Blown Sand and Desert Dunes, *William Morrow and Co., 1942.)*

the wind before sinking to the ground than are the coarse particles that fall more rapidly.

These relations help us to understand the transporting power of the wind. Winds are always turbulent; gusts and eddies that swirl in every conceivable direction are superposed on their general motion. Close to the ground the vertical components of these gusts are naturally less than those in other directions. The ratio of speed of up-or-down gusts to the average forward velocity of the wind ranges widely but averages about 1/5. Hence, if there are particles in the air whose terminal velocities are lower than 1/5 of the wind speed, some of them will be carried upward by gusts and remain suspended in the air for a while until either they are caught in a compensating downdraft or sink with their ordinary terminal velocity to earth. In the time they are in suspension they will, of course, travel along with the wind. Particles with terminal velocities greater than 1/5 of the wind speed are not wafted aloft; they remain on the ground. Measurements of wind speed in dune areas indicate that sand motion begins when the average velocity reaches about 5 meters per second (11 miles per hour). Assuming the maximum updrafts to have 1/5 of this speed or 1 meter per second, we would expect from the curve of Figure 15-5 to find grains smaller than about 0.2 mm. diameter to be winnowed out of the surface material—those much finer to be carried away as dust—while the grains remaining behind should be coarser.

This deduction from the curve is amply confirmed by observation of actual dunes. When dune sands are shaken through sieves graded in size, it is found that grains of the size range 0.3 mm. to 0.15 mm. in diameter greatly predominate. Even the finest dune sands contain few grains smaller than 0.08 mm.

Wind-blown sand rarely rises more than 6 feet, even in a severe storm, and most moves along within a few inches of the ground, as shown by abrasion of the bases of telegraph poles. The great clouds that blacken the sun in areas like the floodplain of the Nile and the "Dust Bowl" of Oklahoma, Kansas, and Texas, are clouds of dust, not sand.

Hence, in sandy deserts away from floodplain or playa silts the air is generally clear above a carpet of moving sand only a few feet thick during even high winds.

Motion of Particles with the Wind

Sand grains may be momentarily suspended by gusts. On falling they strike the surface at a low angle, and, if the surface is rocky, bounce again into the air and journey along by a series of hops. If the surface is loose, such a falling grain may eject others as it "splashes" into the surface, so that, though a particular grain may make only one jump, it is replaced by another or by several jumping grains. Even grains somewhat too large to be ejected from the surface may be pushed slowly along by the successive impacts of many smaller sand grains. Thus a layer of sand whose thickness depends on wind speed and grain size creeps downwind. If the wind acts on a pebbly sand, it may winnow the sand grains away and leave the pebbles behind. They accumulate over the desert surface as a residual layer, one pebble thick, to form a so-called "desert pavement" (Fig. 15-6). The more exposed pebbles generally have smooth surfaces cut on them by the sandblast. Some pebbles are eventually so worn that they may be rolled over by the wind, exposing new sand beneath. This is then blown away, allowing neighboring pebbles to be undermined and overturned. Sandblasted surfaces can then develop on the new upper sides. In this way pebbles may ultimately become faceted by two or more flat surfaces meeting at sharp angles like those of a Brazil nut, as in several pebbles shown in Figure 15-6.

The surface of a sand dune is rough enough to make the air film directly in contact with it turbulent. The most exposed grains are whipped aloft in momentary whirls. However, the matter is somewhat different with grains whose average diameters are less than about 0.03 mm. (below sand size). When grains so fine have settled from the air and accumulate, even the most exposed grain will project so slightly above the general surface that it

FIGURE 15-6.   *Desert pavement in Death Valley, California. Note how several of the pebbles have been faceted by sandblast. (Photo by Eliot Blackwelder.)*

fails to deflect the air into turbulent eddies except at very high wind speeds. A surface composed exclusively of grains as fine as this requires a very high wind to set it in motion.

In a wind-tunnel experiment, the British military engineer Bagnold showed that a surface of loose, dry portland cement was stable, and the air above it dustless, even though the wind was strong enough to move pebbles 1/6 inch in diameter. This stability of even-surfaced fine material accounts for the general lack of dust storms on large playas whose surface material is both fine grained and well sorted. The small size of the grains and their strong cohesion when capillary water is present stabilize such material. Only when a playa surface has recently dried up and is scored with curled-up flakes of dried mud does it furnish much dust to even a strong wind. After these flakes have been blown away (locally to accumulate as dunes of clay chips to leeward) the playa surface is nearly dust-free except in places disturbed by animals or wheels.

Velocity measurements during sand drifting show that wind speed near the ground is much less over a loose sandy surface than over a rocky floor, even though the wind at the height of 6 feet is identical in the two localities.

A study of the grain movements suggests why this is so. Momentarily suspended grains merely bounce along the rocky floor and are so highly elastic that relatively little energy is needed to keep them going. But grains moving over an incoherent surface lose much energy in splashing into and disturbing the surface. More energy is needed to keep such grains moving, and, as the wind is the only available source, grains traveling over loose sand exert more drag on the wind than those traveling over rock. Hence, grains traveling over a rocky floor slow or stop when passing over a sand accumulation. This accounts for the peculiar ability of sand dunes to collect grains from intervening barren areas into themselves, rather than permitting the sand to spread evenly over the whole surface.

Wind Erosion

Unlike streams and glaciers, winds are not constrained within banks but blow freely over the whole surface of the earth. Dust clouds raised by the wind may be blown far outside the area before they settle. This process is called *deflation* (Latin: "to blow away"). The only base level for wind erosion is the local water table, and even this may be slowly lowered as wider areas are eroded down to the capillary fringe.

Most of the large undrained depressions of the deserts of North America and Asia—those of Death Valley and the Dead Sea, for example—have been formed by crustal movements rather than erosion. However, in Wyoming, Texas, New Mexico, and Colorado, some hollows carved by wind are several hundred feet deep and several square miles in area. In the Kalahari Desert of South Africa there are many shallow "pans" on flat erosion surfaces carved across granite. These are undrained and hence could not have been formed entirely by running water.

Perhaps the most striking depressions in whose formation wind erosion must have played a major part form a chain of oases extending about 400 miles west-southwestward from the Nile Delta into the Libyan Desert (Fig. 15-7). Although the depressions may have been formed originally through some other process (ground-water solution, for example), there is strong evidence that they have been enlarged and deepened by the wind.

The northern margins of these depressions are steep escarpments, much dissected by stream-carved ravines. The bottoms of the depressions range from well below sea level to several hundred feet above. From their floors the southeastern slopes rise gradually to the general level of the desert plain, a few hundred feet above the sea. On these slopes, and extending in long chains for hundreds of miles to the southeast, are innumerable sand dunes, representing part of the material excavated in forming the basins. The general desert floor is formed of flat-lying sandstones, and there is no evidence that the basins are fault troughs. The dunes to leeward are strong evidence that the wind has been a major factor in excavating these depressions, but the water-scoured slopes, draining down into the depressions, show that the wind need have done little actual wearing away of rock. Sand carried into the basins by rainsplash and rills was already of about the right size for the wind to transport. The wind has probably enlarged the basins not by abra-

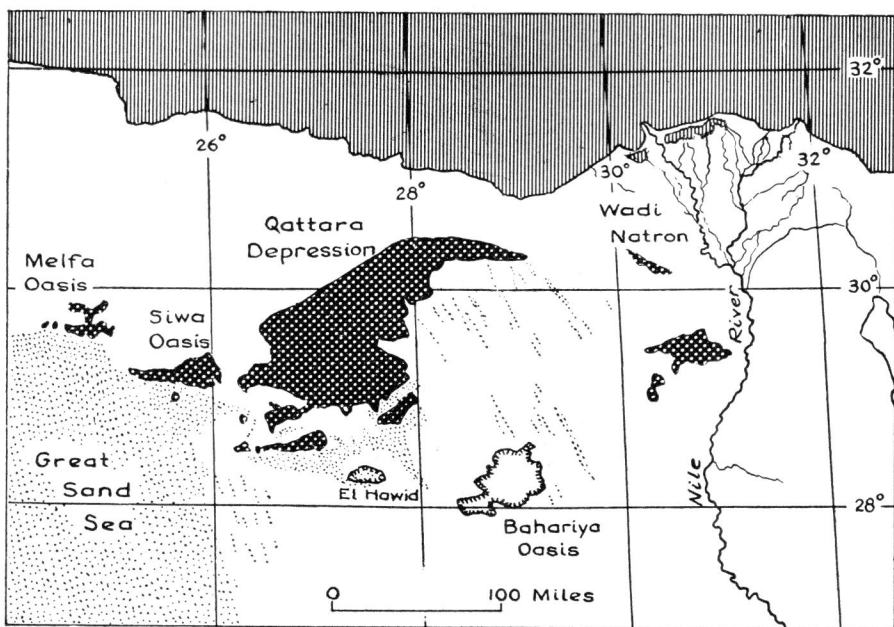

FIGURE 15-7. *The large depressions and sand dune areas (dotted) of Egypt and Libya. The dark-patterned depressions are below sea level. (After AAF Aeronautical Charts.)*

sion but by merely removing material already prepared for transport by other agencies.

When such basins as these are cut to the local water table, moist ground and vegetation prevent further lowering. As the side slopes weather back and larger areas are exposed, the water table itself may be further lowered by evaporation. Many of the Egyptian oases have springs of fresh water around a central depression filled by a salt marsh or playa whose floor is sealed from the main body of ground water by clay.

Although such depressions are impressive, they are rare. Most deserts afford little direct evidence of deep wind erosion. Shallow grooves, a few feet deep and a few hundred feet long, are locally carved from unconsolidated silty sediments. In topographic gaps, where wind armed with sand is locally concentrated, the bedrock may be smoothed and polished or etched by shallow grooves, testifying to the ability of the wind actually to abrade rocks. However, the main role of the wind is in removing unconsolidated material from the sandy and silty surfaces of fans and other deposits left by streams, the very presence of which testifies to the dominance of running water in the making of the desert landscape.

Surface Forms of Moving Sand

Small-Scale Features

A flat surface of sand is unstable in the wind. As soon as the wind rises to the speed at which grains begin to leap along, the surface is bombarded by grains that differ in size, paths through the air, and speed. What happens when a high-speed grain splashes into the sand and ejects other grains from the surface?

Figure 15-8 represents a nearly flat surface of sand, much magnified. A small hollow has developed at B. Though the size and paths of the leaping grains range greatly, there will be a mean size and angle of fall characteristic of many. Hence at first most grains will strike the surface at roughly the same angle and momentum. In the figure they are represented by the equally spaced parallel lines. The forward motion of the sand grains, both by leaping through the air and by impact from other grains, should be roughly proportional to the number of grains striking any particular area. Thus on the upwind AB side of the hollow, the impacts per unit area are few relative to those on an equal area of the downwind side BC. More grains will be driven up the slope BC than down the slope AB, and the hollow will get bigger. Also, there are more impacts per unit area on the slope BC than on an equal level area immediately to leeward. Grains carried up the slope will, therefore, accumulate at the lip of the hollow, C. A second slope, CD, is thus formed. On it, as on AB, the grain motion is at a minimum; a second hollow must form farther to the right. In this way the originally flat surface will become rippled. The ripples grow until the wind speed over the crests is enough greater than that over the hollows to prevent their further growth; their forms and dimensions are then stable for that particular wind speed, though they will continue to migrate downwind.

Once the rippling begins the grains no longer fall at random. Though individual grains travel very different distances and paths, there is a characteristic path determined by grain size and wind speed about which the individual paths fluctuate. Thus when rippling begins, more grains will be

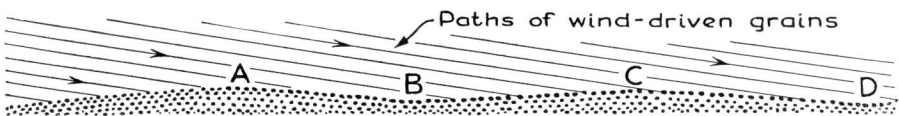

FIGURE 15-8. *The initiation of rippling on a sand surface.* (*After R. A. Bagnold,* The Physics of Blown Sand and Desert Dunes, *William Morrow and Co., 1942.*)

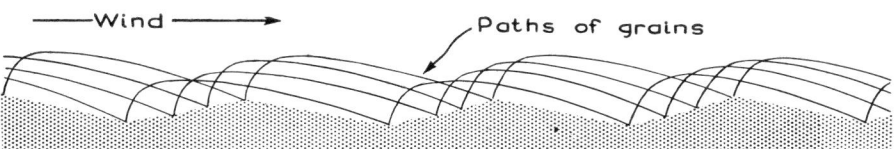

FIGURE 15-9. *Uniform transfer of grains on wind-formed ripples. (After R. A. Bagnold,* The Physics of Blown Sand and Desert Dunes, *William Morrow and Co., 1942.)*

ejected from slopes facing the wind than from sheltered slopes; on the average they travel a characteristic path and their impacts will be correspondingly concentrated (Fig. 15-9). The ripples thus tend to become identical in size and spacing. Furthermore, as their crests rise into layers of stronger wind, only the larger heavier grains can remain on them, for the lighter are more readily moved by impact and by gusts. This detailed sorting of coarser grains on crests than in troughs is the opposite to that in water-formed ripples. When preserved in consolidated rocks, it helps to distinguish rocks of one depositional environment from those of the other.

When wind speed rises enough (in wind-tunnel experiments to about three times that needed to start grains moving), the ripples are destroyed, apparently because the difference in speed over crests and hollows becomes negligible. Likewise, when the wind dies away, the hollows tend to fill because the wind is too feeble in these protected places to maintain them. Hence winds slowly slackening over a long time may also leave a nearly flat, though mildly rippled, surface. But if wind speed slackens systematically in a particular area (perhaps because of topography) so that more sand is brought to it than is carried away, a smooth, unrippled sand sheet may form.

Large Accumulations of Sand

When sand travels across country and enters an area where either the nature or configuration of the ground or the vegetation interferes with the wind, it accumulates. Two kinds of accumulations are obviously related to topography: *climbing dunes,* which form where the wind must rise over a sharp topographic break (an example is the sand sea banked against the northeast wall of Panamint Valley, California) and *falling dunes,* which form where sand is swept over a cliff and falls into the wind shadow below. Almost as readily understood are the dunes and sand sheets formed where the wind, after sweeping sand through a topographic gap, enters a wide plain and diverges, with consequent slackening of speed. Sand also accumulates on wide, flat plains and forms great persistent dunes that slowly travel across country for scores or even hundreds of miles. Though the factors involved are complex, we may understand some of them if we briefly consider the mechanism of dune advance. Among the factors influencing dune motion, vegetation is of paramount importance. Other important influences are the effect of sand accumulation itself on the pattern of wind currents over it, and the relation between sand supply and the prevailing winds.

Dunes in Barren Deserts.   Although no desert is free of vegetation, in many the plants are so small and widely scattered as to exert trivial influence on wind speeds. When a sand accumulation grows high enough to affect the air flow over it, the wind speed will obviously become greater on the windward than on the leeward side (Fig. 15-10). Sand will thus be selectively removed from windward and carried to the leeward side. When the differential speed becomes great enough so that the average path of grains in momentary suspension does not carry them all the way to the foot of the leeward slope, the sand tends to accumulate high on the leeward slope in the wind shadow.

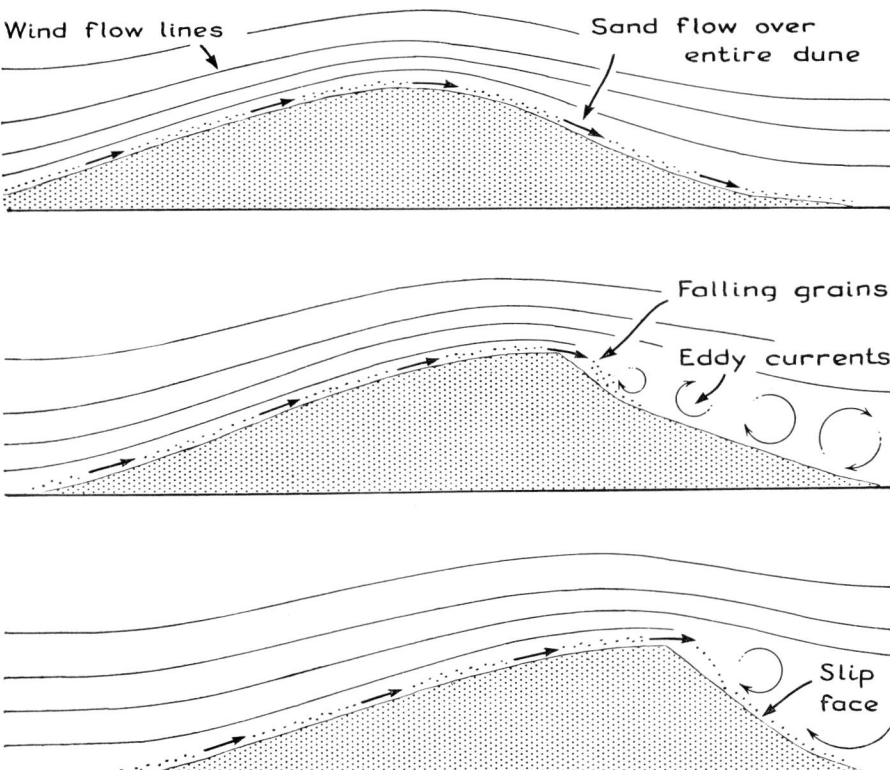

FIGURE 15-10.   *Evolution of a sand dune with a slip face. (After R. A. Bagnold, The Physics of Blown Sand and Desert Dunes, William Morrow and Co., 1942.)*

Eventually, the grains pile so high that they become unstable. The loose pile then slides down to form a *slip face*, thus building a still more efficient wind shadow than before.

Dune accumulations a few feet high rise into air streams speedier than those near their bases. They therefore become increasingly unstable. Any sags along their crests funnel the wind through the gaps, counteracting the tendency of grains to roll sidewise into the depression. Hence a large transverse dune of irregular height may split in two.

Barchans   In deserts of extremely sparse vegetation the most common dunes are of the crescentic variety called *barchans* (Fig. 15-11). The bow of the crescent faces the wind, the middle of the lee face is a slip face. When the wind is blowing there is relative quiet in the lee of the dune. All the sand swept over the top of the slip face, therefore, accumulates. At the edges of the dune, on the other hand, sand streams away to leeward in great quantities. The slip face commonly covers about two-thirds of the dune length at right angles to the wind. Since no sand escapes here, all sand arriving from upwind must either be carried away from the wings or the dune will grow larger. If the barchan remains of the same size, the sand streams from the tips must carry three times as much material as reaches an equivalent length of the dune from upwind. These streams thus make especially favorable places for new dunes to form farther to leeward. Field observations indicate that barchans form where the wind is almost uniform in direction throughout the year.

In general, the larger a barchan the slower it migrates. A small dune will

FIGURE 15-11.   *Barchan dunes near Laguna, New Mexico. The dunes are several hundred feet long. (Photo by Robert C. Frampton, Claremont, California.)*

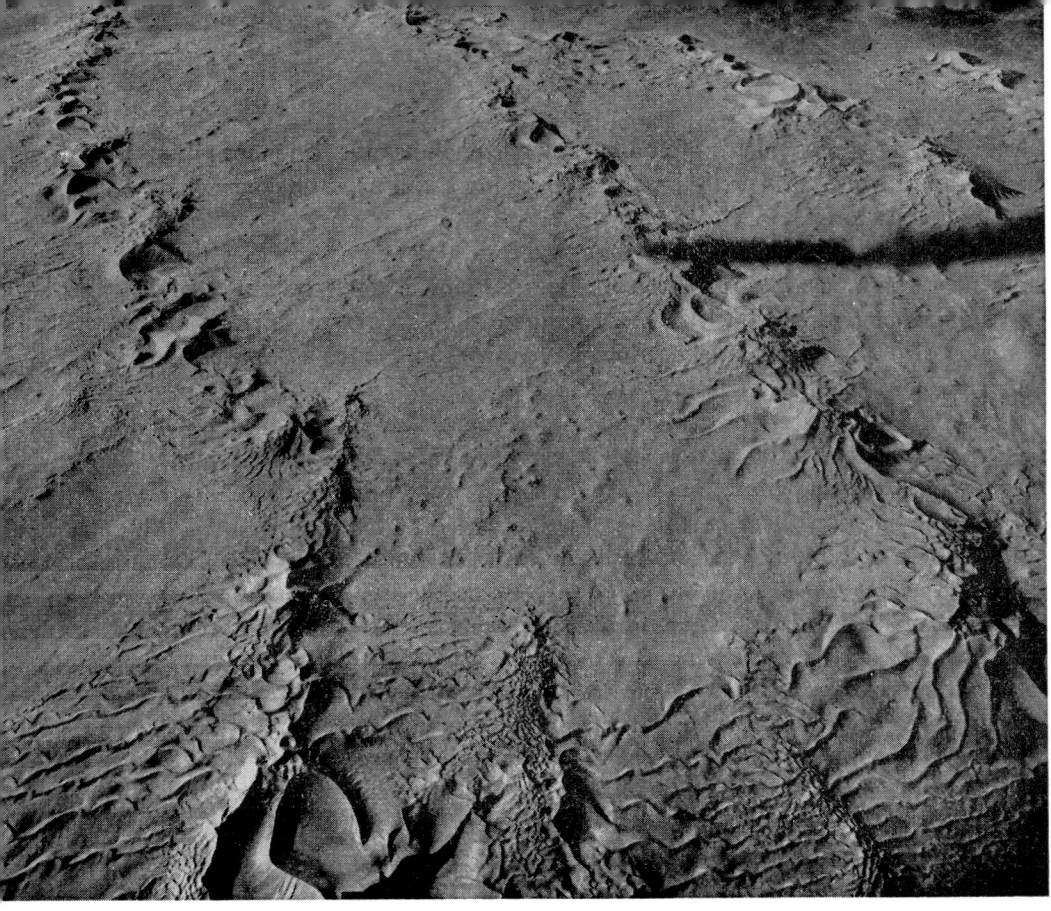

FIGURE 15-12. *Seif ridges of the Sahara, Africa. (Photo by U. S. Air Force.)*

thus overtake a larger one downwind, enclosing a hollow between its wings and the leeward dune. This, in turn, alters the wind pattern and hence the dune shape. Barchan fields can thereby become very complex, especially where their advance is also slowed by vegetation or topography.

Seifs Where wind direction varies considerably, the barchan shape is unstable. If the sand source is in one direction so that winds from that direction furnish a large proportion of the total sand for the dunes, whereas the winds from another direction are more powerful, the sand movement becomes highly irregular, and long dune chains are formed (Fig. 15-12). For this variety of dunes the name *seif* (Arabic: "sword") has been suggested. These dune ridges seem to be able to grow to great size. Some in Iran rise more than 700 feet above their bases and are ¾ of a mile wide. Individual seifs as much as 60 miles long are known, and chains of seifs extend for more than 200 miles in western Egypt.

Dunes in Conflict with Vegetation. Even an open vegetative cover greatly influences dune forms. Plants establish themselves more readily in

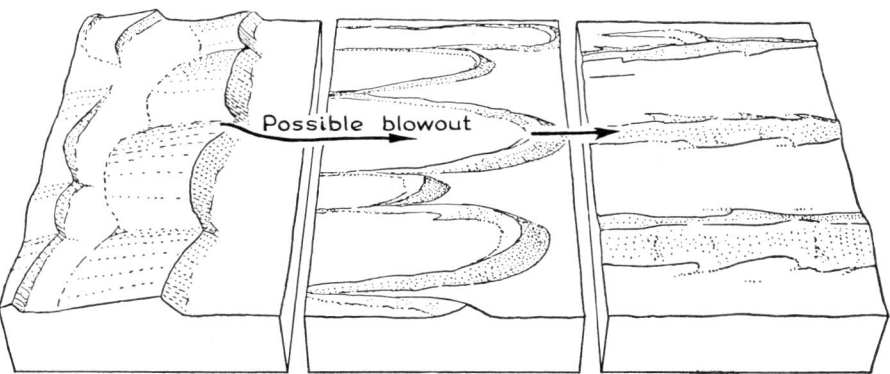

Possible blowout

FIGURE 15-13. *From left to right, transverse, parabolic, and longitudinal dunes; the arrows indicate a possible transition between the three forms. (After J. T. Hack, 1941.)*

sags in the dunes than on the more active crests. Here they are less abraded by moving sand, and their roots are more likely to reach ground water. Hence *transverse dunes* are not so unstable at low heights as they are in barren deserts, for the plants in the sags gather sand and the dunes do not so readily split up. These dunes often attain lengths of half a mile or more at right angles to the wind, and reach heights of 10 or 15 feet before breaking up into barchans. Where vegetation is able to establish itself on the sand, however, long transverse sand waves do not form; instead, two other varieties of dunes predominate: parabolic or "blowout" dunes and longitudinal dunes. The relations between the three forms are shown in Figure 15-13.

Parabolic Dunes   Dunes of parabolic or even elongate "hairpin" shape, with the points facing the wind may form, both by blowouts of older stabilized sand and by accumulation of sand downwind from patchy sources. A parabolic dune may begin where sand is swept from a dry stream bed up gullies in a bordering cutbank onto a brushy terrace, overwhelming the plants in the line of maximum supply. The scoop shape funnels both wind and sand grains so the center advances faster than the sides (Fig. 15-13). This extends the dune front, which becomes a slip face, though the sides are rounded by the stilling effect of the vegetation on the wind eddies. The wings may lag enough so that vegetation develops over their outer slopes and ultimately anchors them completely. In extreme cases the lengthening is so great that the dune becomes a longitudinal dune elongated directly downwind. Parabolic dunes also form where excessive cultivation or trampling by animals destroys the plant cover to furnish patchy sand supplies, as in the "Dust Bowl."

Longitudinal Dunes   Where the sand supply is spotty in distribution and amount, and the wind is of constant direction, a common form of dune is a long ridge parallel to the wind. These *longitudinal dunes* also form typically where climbing dunes reach a cliff top and sand emerges from gullies in the crest (Fig. 15-14), or where other topographic gaps—even wind-made ones like the lee ends of parabolic dunes—channel the sand.

Such ridges are the dominant dune forms of the Navajo country of northeast Arizona, where many of them attain lengths of several miles and heights as much as 30 feet. They require extremely constant wind direction and are evidently favored by a rather sparse sand supply and a climate so dry that only a little sand motion is needed to overwhelm the vegetation. The pattern of these dunes is remarkably constant over hundreds of square miles.

Summary of Factors Influencing Dune Shapes.   We have suggested some of the many factors that influence the formation and shapes of dunes. Doubtless, others are involved, and much remains to be learned. However, it seems clear that, if we have wind from a constant direction, three factors are highly important in shaping dunes: wind speed, sand supply, and vegetation. Abundant sand and strong winds will produce transverse dunes either in barren or brushy deserts; with less sand and weaker vegetation, longitudinal dunes. Moderate winds with little or much sand will produce parabolic dunes, if the vegetation is sufficiently aggressive to partially anchor the slowest moving parts.

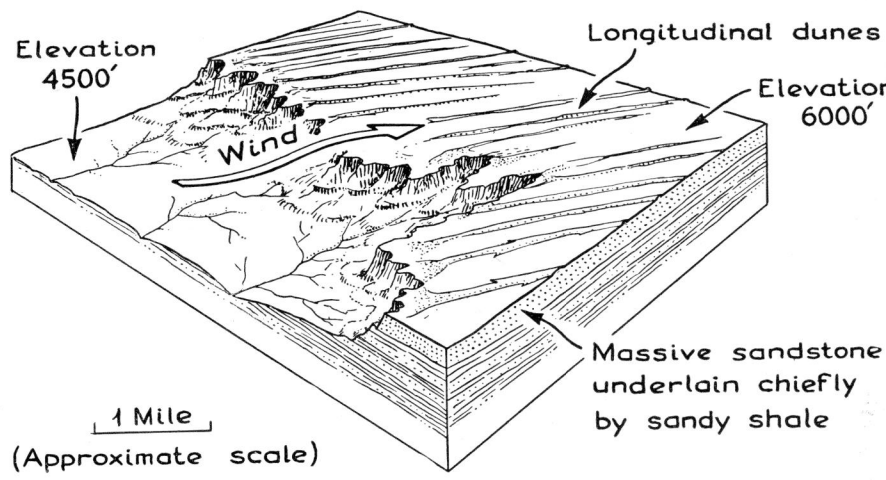

Figure 15-14.   *Longitudinal dunes on the Moenkopi Plateau, Arizona, formed where sand released by weathering of sandstone cliffs is drifted across the plateau by the wind. (After photo of the U. S. Soil Conservation Service.)*

Characteristics of Ancient Dune Sands

Study of dunes and of highway and railroad cuts through them shows that dunes differ from other sand accumulations in several ways. One of the most conspicuous is the type of cross-bedding they exhibit. The internal stratification of dunes is extremely complex, as suggested by their diverse forms, their slip faces, and their complicated progress as they move downwind. A sandstone that contains these complex cross-bedded patterns is illustrated in Figure 15-15. This particular sandstone, a part of the Navajo sandstone of Jurassic age, contains other features that suggest its dune origin. Among these are the excellent sorting, the rare occurrence of wind-carved pebbles, and the "frosted" or sandblasted surfaces of the grains. Such frosted surfaces on sand grains are common in wind-blown but not in water-laid sands, both because small grains move faster in wind than in streams and because their effective mass is far higher, since the density of the fluid in which they are suspended is much lower. Hence, the strong impacts, though not enough to shatter the grains completely, do tend to pit or "frost" the surfaces of wind-blown grains in contrast to those of water-borne sands of the same size. Such features have enabled ancient dune sands to be recognized in many geologic formations from pre-Cambrian age onward.

FIGURE 15-15. *Cross-bedding in the Navajo sandstone, Kane County, Utah. (Photo by H. E. Gregory, U. S. Geological Survey.)*

Loess

Great areas of southern Germany, Russia, Turkestan, and China in the Old World (see Fig. 13-12), and of the Mississippi Valley in the New, are blanketed with fine-grained, loosely coherent material called *loess*. Many of the loess-covered areas are very fertile farm lands, as in the Palouse country of eastern Washington and in the Mississippi Valley. Some loess deposits are hundreds of feet thick, as in western China, but a more usual thickness is a few feet.

Microscopic studies show that loess is composed of angular particles, mainly less than 0.02 mm. in diameter, of quartz, feldspar, hornblende, mica, pieces of fine-grained rocks, and some clay. The bulk of the grains are fresh or only slightly weathered.

All these features suggest that the loess is a deposit of dust that settled from the air in grassy country, where it would be protected from being blown away again. Most loess lies in the lee of areas that were glaciated during the Pleistocene period (see Chap. 13), but some deposits lie in the lee of deserts. The great loess deposits of China, for example, lie downwind from the Gobi and other deserts, and are probably being added to at the present time, just as dust from the "Dust Bowl" in the early 1930's must have added to the soil of the more humid lands of the eastern United States.

Facts, Concepts, Terms

INTERRELATIONS BETWEEN CLIMATE, SOILS, AND LAND SLOPES

INTERIOR DRAINAGE; BAHADAS; PLAYAS

PEDIMENTS

SHEETFLOODS AND MUDFLOWS

RELATION OF GRAIN SIZE OF WEATHERED PARTICLES TO DESERT SLOPES

EVOLUTION OF DESERT LANDSCAPES

SETTLING VELOCITY AND WIND SORTING

ORIGIN OF RIPPLES ON SAND SURFACES

BARCHANS, SEIFS; DESERT PAVEMENT; WIND-CARVED PEBBLES

DUNES IN CONFLICT WITH VEGETATION: PARABOLIC AND LONGITUDINAL DUNES

Questions

1. Why are the southwestern slopes of the Hawaiian Islands arid?
2. Why do dune sands range so slightly in grain size? Why are they generally free from clay?
3. Why do low winds disperse sand and higher winds cause it to accumulate?
4. Draw a cross section through the area shown in Figure 15-1 and label the following features on the section: (a) area that is being reduced in height by rainwash and gullying, (b) pediment, (c) area of stream deposition and braided streams, (d) area of active sand dunes, (e) area where wind-carved pebbles and desert pavement might be found, (f) area from which water might be recovered from wells.
5. Why are pediments not formed in humid regions?
6. Why are wind-borne sands more likely to be frosted than stream sands?
7. What differences can you find between pediments and stream-cut terraces?
8. The so-called "cloudbursts" rarely exceed 3 or 4 inches in total rainfall over a period of an hour or two. Heavy rains that last for much longer periods of time are common in humid regions. Explain the cause of the generally more drastic results associated with the desert cloudbursts.
9. Rainfall is higher along the recently uplifted shores of the Baltic Sea than it is along the Italian coast. Why, then, are sand dunes so much more abundant on the North German and Polish coasts than near Naples?
10. Why do talus piles fail to accumulate at the foot of the high sandstone cliffs of the Navajo country of northeastern Arizona? Compare with the abundant talus at the foot of the limestone cliffs along the Grand Canyon.

Suggested Readings

1. Bagnold, R. A., *The Physics of Blown Sand and Desert Dunes*, William Morrow and Co., New York, 1942.
2. Bryan, Kirk, "Erosion and Sedimentation in the Papago Country, Arizona," U. S. Geol. Survey, Bull. 730, 1922.
3. Gautier, E. F. (translated by D. F. Mayhew), *Sahara, the Great Desert*, Columbia University Press, New York, 1935.
4. Hume, W. F., *Geology of Egypt*, Vol. I, Government Press, Cairo, 1925.

16. *The Oceans*

PART 1

Oceanography

Fossil corals weathered from outcrops of limestone in the desert and marine shells in the rocks at the summit of high mountain ranges prove that the relative positions of land and sea have changed in the geologic past. If we are to read the history of the earth from the record of the marine rocks, we must know something of the sea, the processes that go on within it, its organisms, and its varying environments of erosion and sedimentation.

The study of the sea—the science of *oceanography*—is young. Its modern development began with the great expedition of the British research ship "Challenger" in 1873–1876. Rapid advances in our understanding have been made, but the vastness of the oceans and the complexity of their problems still leave many puzzles unsolved—puzzles whose solution would greatly aid our interpretations of the geologic record.

Composition of Sea Water

Except where great rivers pour into the sea, the chemical composition of sea water is almost uniform. The concentration of the dissolved ions varies slightly from place to place because of differences in evaporation and precipitation, but their relative proportions are everywhere nearly constant. Mixing and diffusion, although sluggish and by no means complete, have been very effective during the long history of the oceans.

Average sea water contains about 3.5 per cent of dissolved matter—35 parts per 1,000. More than 40 elements have been detected in sea water. A few of these overwhelmingly predominate (Table 16-1), but several minor constituents, such as phosphorus and nitrogen, deserve special mention because they are important to life.

It is assumed that the dissolved solids were brought to the sea by rivers.

The runoff from the lands has been carefully estimated, and many analyses of river waters have been made (see Chap. 7). From these data, the American chemist F. W. Clarke has computed that about 3 billion tons of dissolved solids are brought yearly to the sea. These solids, however, are very different from those composing the oceanic salts, as is shown in Table 16-1.

TABLE 16-1. *Composition of Dissolved Solids in River Water and in the Sea*

ION	RIVER WATER (WEIGHTED AVERAGE)	SEA WATER
CO_3^{--}	35.15%	0.41 (HCO_3^-)%
SO_4^{--}	12.14	7.68
Cl^-	5.68	55.04
NO_3^-	0.90	—
Ca^{++}	20.39	1.15
Mg^{++}	3.41	3.69
Na^+	5.79	30.62
K^+	2.12	1.10
$(Fe, Al)_2 O_3$	2.75	—
SiO_2	11.67	—
Sr^{++}, H_3BO_3, Br^-	—	0.31
	100.00	100.00

Either the river waters of today differ greatly from those of the past, or else oceanic processes are selectively removing certain constituents from the sea. River waters differ from oceanic waters chiefly in their great excess of carbonate, calcium and silica and lower content of sodium and chlorine. The calcium contributed by the rivers would double the concentration in the sea in about 100,000 years, if it were not precipitated. Obviously, silica must also be quickly precipitated. These substances are extracted by marine organisms to form extensive deposits of rock (Appendix IV).

Circulation of the Sea

Water, like all fluids, has no strength; it flows in response to the slightest differences in unbalanced pressure. If we heat a pan of water, the warmed water expands—that is, it becomes less dense—and rises, whereas the cooler, heavier water sinks and flows beneath it. Similarly, if we place pure water in one compartment of a vessel with a removable partition, and salt water at the same temperature in the other compartment, on removing the partition the heavier salt water will flow beneath the fresh. For a time the water will be layered according to density, but diffusion ultimately makes it uniform, and the layering disappears. Differences in temperature and salinity cause differences in density in the ocean and set up a circulation of the ocean water.

Though less effective than that of the air, the circulation of the sea notably affects climates. Changes in the configuration of land and sea, such as are implied by the distribution of marine rocks, must have affected ancient climates also, with all that this implies in respect to erosion, deposition, and life environments.

Density Variations in the Sea

Oceanographers have found that the major circulation of the sea is a response to two main factors: (1) the drag of the prevailing winds and (2) the differences in density of different water masses. Both factors largely depend on the sun's radiation. Density differences are generally controlled by temperature and salinity (concentration of dissolved salts), but the local content of micro-organisms or of suspended sediment also modifies these effects. The oceans are so vast that diffusion only slowly equalizes differences in salinity brought about by evaporation and precipitation. Similarly, the heat conductivity is so low that temperature differences persist. Hence the oceans are made up of great water masses within each of which the temperature and salinity vary but little, though the boundaries against neighboring masses with differing characteristics may be relatively sharp. The differing densities of these water masses and the systematic drag of wind friction on the surface of the sea sets up a slow motion and causes the oceanic circulation.

Variations of Density with Latitude. The transfer of heat from low latitudes to high has been described in Chapter 6. As there mentioned, most is carried by the winds, but in both the North Atlantic and North Pacific oceanic currents carry about 10 per cent of the total heat transported.

Water strongly absorbs solar radiation; even the clearest ocean water absorbs more than 80 per cent within 33 feet of the surface (more than 60 per cent in the first 3 feet). Water turbid with silt, air-bubbles, or micro-organisms absorbs even more—over 99 per cent within the first 33 feet. The sun thus warms only a thin surface layer of the oceans.

Warming the surface water layer makes it less dense, but warming also increases evaporation and thus tends to increase its salinity and, hence, its

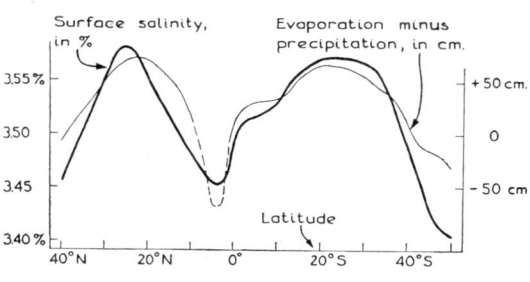

FIGURE 16-1. *Graph showing the close correlation between changes of surface salinity and evaporation minus precipitation when these are plotted against latitude. Average for all oceans. (After Wüst, from H. U. Sverdrup, M. W. Johnson, and R. H. Fleming, The Oceans, Copyright 1942 by Prentice-Hall, Inc.)*

density. In the equatorial belt evaporation is high, but wind speeds are low, cloud cover is dense, and rain abundant. Rains dilute the surface water, counteracting evaporation in its effect on density. Farther from the equator in the trade wind belts, despite lower temperatures, evaporation is higher because here are strong winds, few clouds, and little rain. Here, a surface layer about 2,000 feet thick is so warm that it floats on colder water despite its high salinity.

The interplay of evaporation and precipitation on salinities is clear from Figure 16-1. Because of differences in temperature and salinity the warm water of the equatorial and trade wind belts tends to spread north and south over the colder water near the poles. Conversely, the cold polar water sinks and flows equatorward. These movements have been traced by measuring salinities and temperatures at many places in the ocean. In the Pacific Ocean water that sinks in the Antarctic can be traced by its distinctive temperature and salinity even to the equator; cold Arctic water can also be traced far to the south in the depths of the Atlantic. The slowly drifting water masses are so large that mixing fails to disperse them completely.

Warm surface waters, however, do not by any means flow directly from the equator toward the poles. It is clear that density differences alone do not explain their circuitous paths. Deflection of the masses by wind friction is one obvious factor; another is the rotation of the earth.

Effect of the Earth's Rotation

We have seen in Chapter 6 that the tendency of a moving mass to persist in its line of motion (its inertia) causes the trade winds to be deflected to the right (westward) in the northern hemisphere and to the left (also westward) in the southern. The same relations prevail with ocean currents, but because the sea, unlike the air, is confined to depressions on the earth's surface, and also because its density and hence its inertia are higher compared to the driving force, its movements show a somewhat more stable pattern. There are seasonal displacements of currents, but these are much less capricious than those of air masses. In general, the density differences in the sea arise in the same places, and the motion that tends to equalize them becomes a steady flow.

Pressure Differences in the Sea

The fundamental cause of movement of the large water masses in the sea is the differences in density set up by variations in temperature and salinity. Such differences cause unbalanced pressures that operate in the vertical direction, at right angles to the earth's surface. On the other hand, the force set up by the rotation of the earth operates in the horizontal direction and it changes in amount with differences in latitude. Wind friction also sets up horizontal forces. From detailed observations of temperature and salinity, made at measured depths and at points where longitude and latitude are known, it is possible to calculate the resultant pressure existing at any definite level beneath the surface from all of these forces. *Isobaric maps* (isobars are lines of equal pressure) can thus be prepared for the sea, just as they are from barometric readings of the air.

Since water has no strength, it is also possible to calculate the currents that will be set up and maintained by the pressure differences. The calculations involve hydrodynamical principles beyond the scope of this book, but one of the triumphs of modern oceanography has been the close agreement between the current motion as determined by measurements and the motion as calculated from the theory of flow. The method has, therefore, become standard practice in oceanography, because it is much easier to measure temperature and salinity than the velocity and area of the moving water masses.

One general rule emerges from the analysis of pressure differences: In the Northern Hemisphere, if the observer looks in the direction that the current is traveling, he sees lighter water on his right and denser water on his left—the isobaric surface slopes upward to the right. In the Southern Hemisphere, the lighter water lies to the left, and the denser water to the right. As the density of the water near the ocean surface is more dependent on temperature than on salinity, "lighter water" may, in general, be replaced by "warm water," in the rule stated above.

Let us turn to a description of one of the largest and best known of the oceanic currents, the Gulf Stream.

The Gulf Stream System

"There is a river in the sea—the Gulf Stream," said Maury, the great American oceanographer of a century ago. This statement was regarded by some as a typical American exaggeration, but modern studies essentially confirm his picture of a well-defined stream of warm water, with relatively sharp boundaries, coursing with the speed of a river across thousands of miles of ocean (Fig. 16-2).

The Gulf Stream system consists of three segments: the Florida Current, the Gulf Stream proper, and the North Atlantic Current. The Florida Current pours through the Florida Strait with a speed of nearly 3 miles per hour, carrying as much water out of the Gulf as 14 Mississippi Rivers would

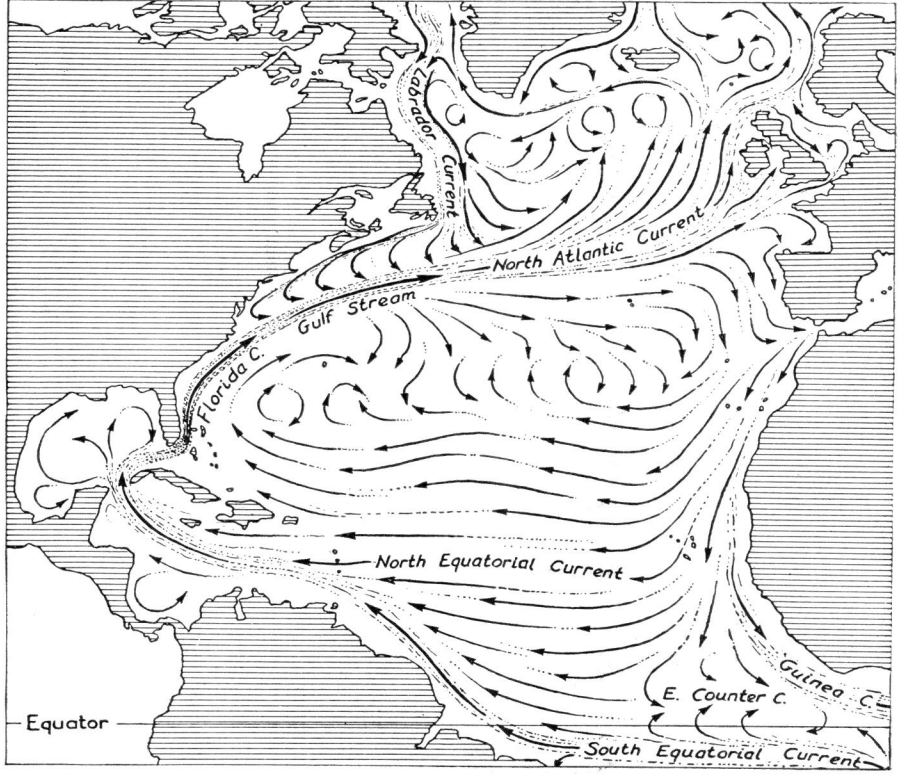

FIGURE 16-2. *The currents of the north and equatorial Atlantic Ocean in February and March. (After H. U. Sverdrup, M. W. Johnson, and R. H. Fleming, The Oceans, Copyright 1942 by Prentice-Hall, Inc.)*

bring into it. Theoretically, sea level should stand about 18 inches higher on the Cuban coast than at Key West, Florida. Of course, this cannot be measured, but measured current speeds in different parts of the channel agree with the speeds computed from the measured temperature and salinity. Temperature and salinity determine the water density from which velocity is computed by physical theory.

Precise leveling across Florida shows that the Gulf of Mexico stands about 7 inches higher than the Atlantic—this is the immediate cause of the Florida Current. The higher level in the Gulf is, in turn, caused by the westward deflection by the trade winds of the warm surface waters of the Atlantic. Most of the water feeding the Florida Current flows northward into the Gulf through the Yucatan Passage between Yucatan and Cuba.

As the Florida Current emerges from the Gulf of Mexico it is joined by a current which flows north along the east side of Cuba. As the combined stream sweeps north along the continental shelf it is continually augmented by water from great eddies in the Atlantic. Deflected to the right by the earth's rotation, the stream leaves the continental shelf and courses northward as a well-defined broad current. Its speed is as great as 2½ miles per hour, and its depth, hitherto about 2,600 feet, becomes as much as 6,500 feet. As it sweeps far offshore past Virginia, it is about 60 miles wide and in volume about three times that of the Florida Current—in other words, that of 40 Mississippis. Inshore from the Gulf Stream, great counterclockwise eddies are thrown off, giving rise to a southward drift along the Atlantic shores that shifts sand southward along the beaches, and may help to shape the cuspate forms of the sandy capes from Hatteras southward.

The Gulf Stream spreads out to the northeast and becomes less definite. It gives off great eddies both to the north and south and encloses still smaller ones. Here, it is called the North Atlantic Current. Its northern branches spread far into the Norwegian Sea and have been traced beyond North Cape into the Arctic Ocean. They notably affect the climate of the British Isles and Scandinavia, for they transport warm water far to the north.

Effects of the Shape of the Coast on Currents

Coastal configuration strikingly affects ocean currents, especially along the eastern shores of the Atlantic and Pacific. The Benguela Current along the west coast of Africa, the Peruvian Current, and the California Current are all in part maintained by the prevailing winds which blow equatorward, either parallel to the shore or offshore. The wind drift of surface waters is, therefore, offshore along all three coasts, and, as no surface flow is available to replace the water lost, each of these currents is replenished by upwelling of colder water from the depths. Such deep water is rich in plant nutrients and supports an abundant marine life.

With climatic fluctuations the currents are occasionally disturbed, as has happened several times with the Peru Current. The upwelling may then cease for a time. As a result, the marine life may be destroyed on a whole-sale scale, and a rain of dead organisms falls to the sea floor. Perhaps some fossil-rich strata record similar disturbances of oceanic currents in the geo-logic past. Such disturbances of currents may, in turn, give rise to local cli-matic changes. When the Peru Current shifted, torrential rains wrought tragic destruction to the normally arid west coast of Peru and Ecuador. Changes in configuration of sea and land during the geologic past must also have affected the ocean currents. For example, if the Isthmus of Panama were submerged (as we know from marine sediments that it has been), or even if the eastern point of Brazil were several degrees north of its present position, the power of the Gulf Stream would be greatly changed. Water warm enough for the growth of coral reefs now bathes the Bermuda shores; it may be doubted that a feebler Gulf Stream would allow these animals to grow so far north. It must be kept in mind, then, that the environment of modern marine organisms depends not only on climates determined by lati-tude, but also on the courses of ocean currents.

Shore Processes

The great ocean currents intimately affect climates and marine life, but the oceanic agencies that most directly affect the shore are the waves, tides, and longshore currents associated with them. Waves and wave currents wash every shore on earth. Tides, though locally spectacular, are generally far less effective in shaping the shores, though incomplete information suggests that they are important in controlling the distribution of marine sediments at considerable depths.

Waves

Geologic Effectiveness. Anyone who has systematically watched the never-ending play of waves on the shore—whether the gentle ripples that carry grains of sand up and down the beach or the great rollers that pound against the cliffs—cannot but be impressed with the power of the sea in molding the shape of the coast and in moving detritus. Historic records prove that for decades, or even for centuries, some shorelines have been driven back several feet per year by wave attack. From repeated soundings and studies of bottom sediments we also find that waves profoundly affect the movement of sediments at shallow depths.

The Theory of Waves.   The restless waves of the sea seem almost to defy analysis, and, indeed, no theories yet proposed explain all their vagaries. Nevertheless, the need—vital alike to the naval architect designing ships, the civil engineer building docks and jetties, the military scientist planning amphibious landings, and the geologist studying sedimentation and erosion —has led to decades of intensive study and many useful conclusions.

The mechanism by which wind forms water waves is complex and not completely understood. Theoretically, waves should not arise until the wind speed is 11 to 15 miles per hour, yet measured winds of only about 2½ miles per hour produce waves on ponds. Clearly, wind motion must be turbulent, that is, some air particles move at much higher speeds than the current as a whole. These variations in wind speed produce corresponding variations in surface pressures and friction; it must be these that first ruffle the surface.

Now, it has long been known from watching floats that the water particles in surface waves move up and down in orbits, but retain their average positions. Indeed, if the water masses actually moved forward at the same speed as the wave, no ocean would be navigable.

Wave Characteristics.   A wave is described in terms of its length, L (the distance from crest to crest or from trough to trough); its height, H (the vertical distance from trough to crest); and its period, T (the time interval between the appearance of two consecutive crests at a particular position). The velocity, V, of a moving wave is the distance the wave form travels in a given time; thus, $V = \frac{L}{T}$. See Figure 16-3.

The motion of the water particles, and hence the wave form, depends on the wave length, the ratio of height to length, and the depth, d, of the water. Waves in water that is deep compared to wave length (d is greater than $\frac{L}{4}$) are called *deep-water waves*. Their speed is independent of the depth.

The water particles in a deep-water wave move approximately in circles, the radii of which decrease very quickly with depth (Fig. 16-4). For waves 330 feet long and 16 feet high, traveling about 28 miles per hour, the parti-

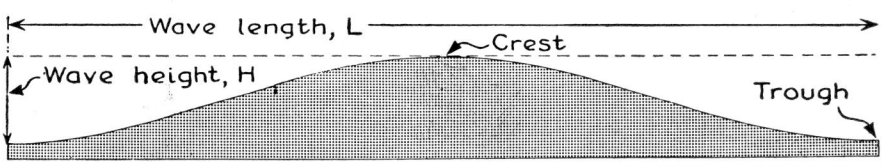

FIGURE 16-3.   *Schematic diagram of a progressive wave.*

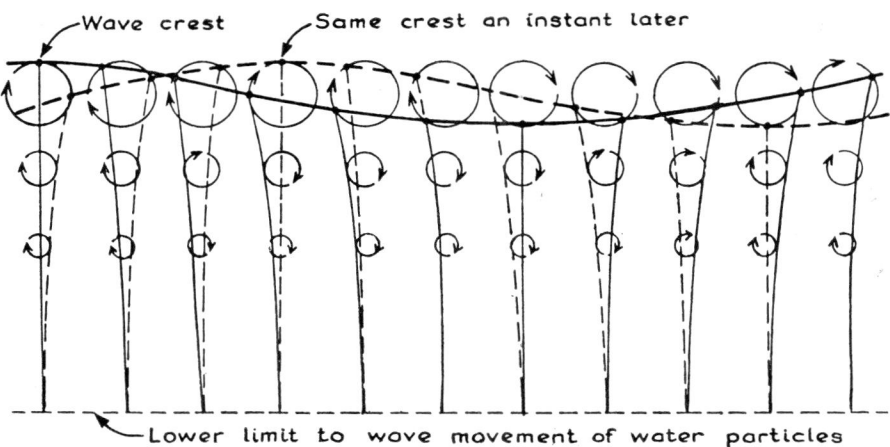

FIGURE 16-4. *Circular movement of water particles in a deep-water wave of small height. Full lines indicate the position of the water particles at one instant, dashed lines the same particles 1/4 period later. (After U. S. Hydrographic Office, Publ. No. 11.)*

cle velocity at the surface is about 4.4 miles per hour; at a depth of 65 feet about 1.2 miles per hour; while at 330 feet the particle motion is negligible. This quick falling off of orbital velocity with depth gave Vening-Meinesz' submarine enough stability for gravity observations in moderate seas when it was submerged less than 100 feet. But in great swells 1,250 feet long and 33 feet high, traveling at 56 miles per hour, though the particle velocity at the surface is again only about 4.4 miles per hour, the motion persists to greater depth and at 330 feet is still nearly 1 mile per hour.

The water particles do not move at exactly uniform speed in their orbits; they are slightly faster when at the top of the path. Here they are moving in the direction of wave progress, so there is a slight net forward movement of water with the wave. For low waves of long period this net movement is slight but it is noticeable in high steep waves.

Under steady winds the waves grow in size and speed, up to a limit. The wind gives them energy in two ways: by its push against the wave crests and by the frictional drag of the air along the water surface. The first process depends on the difference between wind speed and wave speed; when wind speed is greater the waves are pushed, but when wave speed is greater the wave encounters air resistance like that against a moving car. On the other hand, frictional drag depends not on differences in the speed of air and the wave *form*, but on the differences of speed between air and water *particles*. As we noted above, the particle speed is moderate even in the greatest waves. Hence wind can speed up waves that are moving faster than itself, although this speeding up is opposed by the wind resistance of the wave crests. From measurements it appears that the two tendencies balance when the wave speed becomes 1.45 times that of the wind. Beyond this there is no increase in wave speed, no matter how long the wind may blow.

Thus wave size and speed depend on several factors: wind speed, wind duration, the *fetch* (the distance across which the wind blows with the same direction and speed) and, finally, the state of the sea at the time the wind began to blow. Fairly good formulas involving these factors have been worked out to predict the height, speed, and period of waves. These were used with remarkable success in prophesying surf conditions for the Normandy Invasion and other amphibious operations during World War II.

Waves generated in an area of wind may progress into a region of calm or of different wind direction. Here they are called "the swell." They lose height and energy because of air resistance to the wave fronts and internal friction. Somewhat surprisingly, they also become longer and speed up. Roughly, the swell loses one-third its height in traveling a distance in sea miles equal to its wave length measured in feet. Nevertheless, some of the greatest waves that pound the coast of southwest England have traveled thousands of miles from their source area in the South Atlantic.

When either "forced" waves, which are still growing with the wind, or "free" waves, like the swell, which are no longer growing, enter shallow water, several things happen. When the depth becomes less than $\frac{L}{4}$, the bottom interferes with the orbital motion of the water particles, causing the wave to slow and steepen. Where the water shallows to $\frac{L}{20}$, the wave speed no longer depends on L but becomes proportional to $\sqrt{d}$. Such waves are called *shallow-water waves*, as distinguished from the deep-water waves whose speed is independent of the depth.

Wave Refraction.   A deep-water wave thus begins to "feel bottom" where the depth is about a quarter or, at most, half the wave length. Commonly, this is less than 500 feet, but for the largest storms it may be nearly three times this depth. Though the wave period remains the same, the wave length shortens—in other words, the wave slows down. This is not because of friction but because the water particles at the bottom can no longer move in circles; their paths become ellipses, and directly at the bottom are reduced to straight lines, to and fro, in the direction of wave advance. Where the wave approaches shore obliquely, the inshore end feels bottom sooner than the offshore parts. It therefore slows sooner and the wave becomes bent, for the offshore part continues to advance as a deep-water wave (with greater speed) after the inshore end has begun to change to a shallow-water wave. Along a straight, evenly sloping coast, then, no matter what the origi-

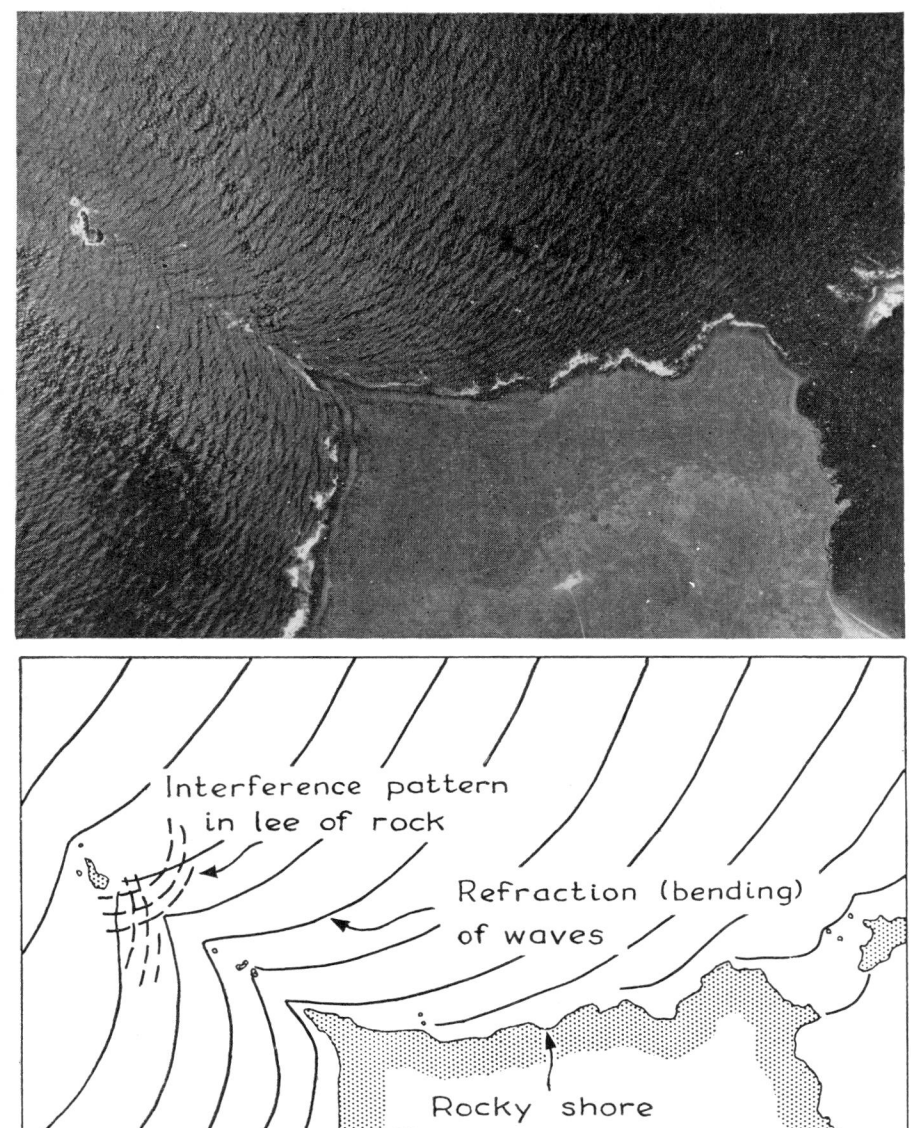

FIGURE 16-5. *Vertical airplane photo* (top) *and diagram* (bottom) *showing bending of the waves* (*wave refraction*), *around a point of land.* (*After U. S. Hydrographic Office, Publ. No. 234.*)

nal angle of approach may have been, the waves tend to be bent (refracted), so that they approach shore nearly straight on—that is, with their crests nearly parallel to the beach (Fig. 16-5).

Wave refraction modifies the distribution of the energy in the waves. The energy which was uniformly distributed along the wave length in the deep-water wave becomes concentrated on the headlands and diminished in the coves. Hence the waves on the headlands are more powerful and higher, those on the shores of the coves less powerful and lower, than they would be on a straight coast. We shall see that this selective attack on the headlands makes the waves tend to shorten and straighten the shoreline by eroding the headlands and filling the coves with the transported debris.

Breakers.   In deep-water waves only half the energy is carried forward with the speed of the wave. The other half is used in the oscillation of the water particles above and below their quiet-water positions. On advancing into shallow water little of the energy is used up in friction; the wave is, nevertheless, slowed up by interference with orbital motion near the bottom and, since its energy is nearly the same, its height must theoretically increase. On the other hand, when the wave has finally been transformed to a shallow-water wave, all of its energy is being carried forward with the wave. Theoretically, this should lower the wave height, since a smaller mass of water can carry the same energy as the larger mass of a deep-water wave, which is only 50 per cent efficient in transporting energy. These theoretical conclusions are borne out by observations of real waves.

Waves whose ratio of height to length in deep water is more than 1/100 steepen on entering shallow water until the crest angle is about 120°. When they are this steep they become unstable and begin to break but may continue for hundreds of feet without curling over of the crest. This is the ideal wave for the surf-board rider—it is often only the exceptionally high wave that will do this. On the other hand, waves in deep water that are lower relative to length carry most of their energy in the wave crest. As these slow and steepen in shallow water they no longer find water available to fill their wave forms. At this point the wave curls over, breaks with a single great crash, and the water tumbles forward onto the beach. This is the breaker. Depending on the wave period, breakers occur where the still-water depth is from one to two times the wave height.

When a wave breaks, the energy it carried is dissipated in collision with the bottom and in turbulence of the water. Bottom materials in the breaker zone are thrown violently into suspension, as any surf swimmer can testify. Because of the concentrated turbulence, a trough is often scoured in the bottom along the breaker line. Material thrown into suspension sweeps forward with the rush of the water up the beach. As the water shallows and becomes less turbulent, the detritus is sorted according to its size, shape, and density.

Wave Erosion

Waves erode the shores in three ways: by their impact and *hydraulic pressure,* by *corrasion*—the sawing and grinding action of the sand, gravel, and cobbles hurled against the cliffs or rolled and dragged across the foreshore —and by *solution,* which is a minor process even along limestone shores.

Hydraulic Action. Waves contain astounding energy. Even moderate waves, 10 feet high and 100 feet long, exert a pressure (theoretical) of 1,675 pounds per square foot when they strike an obstruction (Fig. 16-6). Great storm waves, 42 feet high and 500 feet long, should produce a pressure of more than 3 tons per square foot. On the east coast of Scotland pressures as great as this have actually been measured by dynamometers.

At Wick, in the north of Scotland, waves of a great storm in 1872 tore a concrete monolith (45 feet long, 26 feet wide, 11 feet thick, and bound by 3½-inch iron rods to the breakwater foundation) from its place, together with a great mass of the foundation, and dropped the whole unbroken inside the harbor. The mass weighed 1,350 tons. The block was replaced by a much larger mass of concrete weighing 2,600 tons; five years later this, too, was carried away by the waves. At Tillamook Rock lighthouse on the Oregon Coast, waves have repeatedly hurled stones large enough to break the plate glass of the light 132 feet above the sea. One fragment, weighing

FIGURE 16-6. *Wave breaking on rocky coast near Carmel, California. (Photo by M. R. Campbell, U. S. Geological Survey.)*

FIGURE 16-7. *Sea caves and sea stack at low tide, near Santa Cruz, California. (Photo by Eliot Blackwelder.)*

135 pounds, was thrown through a roof more than 100 feet above sea level, breaking a hole 20 feet square. At Dunnet Head, north Scotland, the lighthouse windows 300 feet above the water are occasionally broken by storm-driven stones.

The tremendous wave pressures act not only directly, but also by compressing air driven into crevices in the rocks. Such pressures suffice to pry huge blocks from cliffs exposed to full-wave attack. At Ymuiden, Holland, a 7-ton block in the breakwater was seen to move forward—toward the sea —presumably because of compression of the air in crevices behind it.

Thus storm waves seek out weak strata or joint cracks in a cliff and dislodge the blocks between them. Sea caves so formed may be scores or even hundreds of feet long; some penetrate entirely through small promontories and produce spectacular arches. The roofs of others collapse, leaving isolated *stacks* in front of the cliff line (Fig. 16-7).

Corrasion. As we have seen, great storm waves can tear huge blocks from the cliffs and hurl them against the shore. Even moderate waves can move boulders and cobbles, and feeble ones sand. In a Cornish mine that extends beneath the sea, it is reported that the grinding of boulders is easily heard through a 9-foot roof of rock, as the waves beat against the cliff overhead. The shore zone is a veritable grinding mill. On the beach at Cape Ann, Massachusetts, angular fragments from the granite quarries have become rounded within a single year. The sea is a great saw; armed with cobbles, pebbles, and sand, it works away at the beach, undercutting cliffs and wearing a notch in every rock exposed at this level. In relatively unconsolidated material, such as the glacial gravels making up the Holderness coast of Yorkshire, the sea has eaten back the cliffs at a measured rate of

7 to 15 feet per year for more than a century. Shakespeare's Cliff, several hundred feet high, a part of the "Chalk Cliffs of Dover," becomes undercut so as to produce frequent large landslides. One of these, in 1810, was so great that the vibrations were felt as a strong earthquake at Dover, several miles away. These are spectacular examples, but even in hard rocks, like the rhyolite shown in Figure 16-8, wave-notching is impressive.

Corrasion is not confined to the water's edge. By far the greatest energy is spent between the breaker line and the shore; nevertheless, particle motion at greater depth is considerable, especially if the length of the waves is great. Waves with a period of ten seconds that are 500 feet long and 22 feet high should, theoretically, produce water particle speeds of 10 inches per second at 300 feet depth; this is enough to move fine sand. When account is taken of tidal and other currents, it appears that the bottom water is commonly agitated enough to keep clay particles in motion to depths of 600 feet and on some coasts to 900 feet or even more. These conclusions from wave theory are partly confirmed from the character of the sediments found by dredging. The depth to which wave motion is perceptible is called *wave base*.

Now, it is true that the particle speeds mentioned reverse their direction with every wave. But most of the shallow sea floor slopes seaward, so that

FIGURE 16-8. *Cliff in rhyolite, undercut by the waves. Kindall Head, Moore Island, Maine. (Photo by E. S. Bastin, U. S. Geological Survey.)*

a particle moving to and fro on the bottom is moving downhill seaward and uphill landward; outside the breaker zone, where these oscillations are about equal, the sediment tends ultimately to progress downhill (seaward) in response to the pull of gravity. As the motion becomes less and less with greater depth, the coarser particles eventually reach positions where agitation is too feeble to keep them moving. Here they come to rest. But finer particles can still be moved and will continue to travel seaward until they, in turn, are sorted out in accordance with their size. During all this agitation the particles of sediment wear against each other and against the bottom. They gradually become smaller and smaller until they can be moved by successively feebler oscillations. Dragged back and forth with each passing wave, they also wear away the bottom, so that not only the breaker zone, but all the coastal zone to the depth of wave base, is corraded.

Inshore from the breaker zone, where the waves have become forward-moving masses of water, the situation is somewhat different. The breakers dashing onshore may carry coarse fragments to heights where the feebler backwash cannot carry them back, even though it is favored by a steep slope. Although as much water must flow away from shore as toward it, the shoreward movement is generally concentrated in a shorter time, so that the velocities in the breaking waves are greater than in the flow of the backwash. A coarse pebble or cobble *storm beach* may thus be built above the reach of the normal waves. Such a beach is steep, for coarse material can be moved seaward only on steep slopes.

Eventually, a profile is developed for any particular size of waves at which swash and backwash are about equally effective. Even after this profile is developed, the beach material is gradually ground down in size so that slow seaward transport continues. As the waves fluctuate in size, the profile and the grain size of the material continually readjust to conform to them.

The Profile of Equilibrium

Thus, partly from wave theory, partly from observing the movements of detritus in waves, partly from dredgings and soundings, and partly from study of large, dried-up lakes (see p. 314, Lake Bonneville), the concept of the *marine profile of equilibrium* has been developed (Fig. 16-9). This profile is a smooth, sweeping curve, concave upward. It is steep in the breaker zone and flattens seaward. Its slope is adjusted to the average particle size of the detritus moved by the waves at each particular place along it. The slope is that which just suffices to keep the material in slow transit toward deep water. As the waves forever change in size, the profile constantly re-

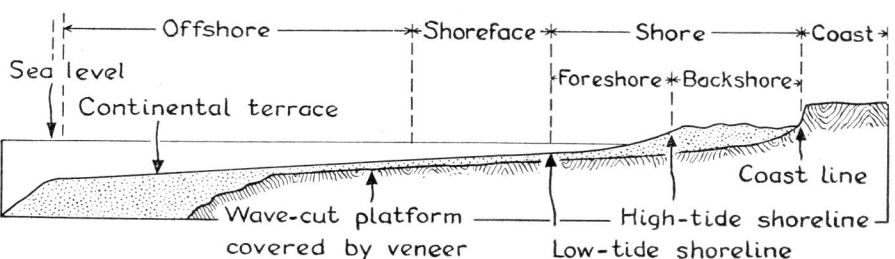

FIGURE 16-9. *Elements of the shore zone and profile of equilibrium.* (*After D. W. Johnson,* Shoreline Processes, *John Wiley and Sons, 1919.*)

adjusts, flattening and shallowing with feeble waves, steepening and deepening with more powerful ones. But for any coast there should be, theoretically, a general average slope about which the profile fluctuates.

The concept of the profile of marine equilibrium is not purely hypothetical. Soundings commonly show that the sea floor near shore is concave upward, despite the complications to be expected because of the world-wide changes in sea level that occurred during the Pleistocene. As shown by Figure 16-10, the slope and depth of the offshore profile is more or less closely adjusted to the fetch and power of the waves. The reality of such a profile—at least of its shoreward portions—is still more convincingly shown by exam-

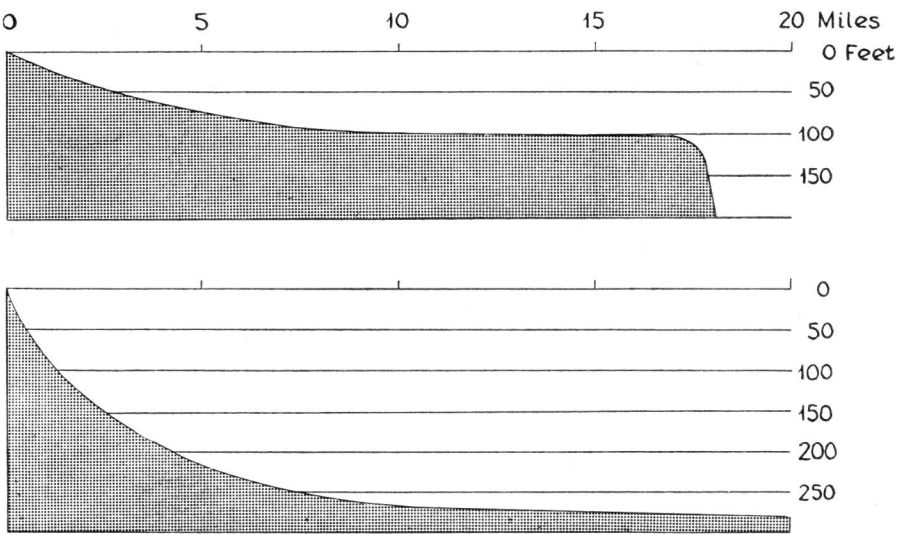

FIGURE 16-10. *Average profile of the sea floor off Madagascar. Note the striking difference between the profile of the protected west coast* (top) *and the exposed southeast coast* (bottom). (*Redrawn from D. W. Johnson,* Shoreline Processes, *John Wiley and Sons, 1919.*)

FIGURE 16-11. *Uplifted beaches on Middleton Island, Alaska. The lines of low bluffs are old sea cliffs. Note how the uplifted wave-cut platform truncates the inclined beds. (Photo by S. R. Capps, U. S. Geological Survey.)*

ination of uplifted sea floors in many places (Fig. 16-11). Such graded sea floors, uplifted to form *marine terraces,* are found in many places. They show some or all of the features seen and deduced from modern coasts: cliffs and cliffnotches, stacks and caves, storm beaches, gravel and sand beaches, and a sedimentary veneer resting on a smoothly truncated surface of bedrock whose profile is gently concave upward but retains a general seaward slope. It is convenient, then, in geology, to refer to the profile of equilibrium in discussing shore processes. But it should be clearly understood that the idealized profile is rarely achieved, partly because of the continual changes in wave power and the complicating effects of tidal and other currents, and probably mainly because of the fluctuations in sea level and slow movements of the earth's crust (Chap. 9) that have interrupted its development. Nevertheless, profiles approaching the ideal are common, and beach materials can be seen to move in sensitive response to natural and artificial changes in the profiles.

Tides

The ancients knew that the ebb and flow of the tides varied with the phases of the moon. So complex is the real earth as compared with the idealized earth assumed by the astronomers and physicists that we have, as yet, no general theory that permits tidal forecasts for any point on an ocean. Tides

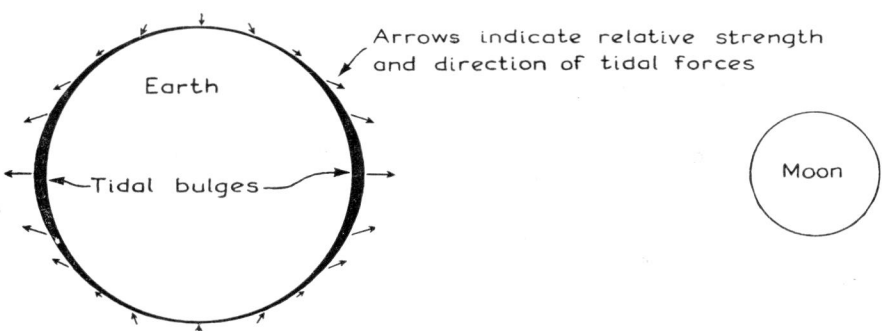

FIGURE 16-12. *The lunar tides. The distance between the moon and the earth should be much greater than shown in the diagram. The tidal bulges are greatly exaggerated. (In part, after, H. U. Sverdrup, M. W. Johnson, and R. H. Fleming, The Oceans, Copyright 1942 by Prentice-Hall, Inc.)*

are, of course, predicted with great accuracy for all the principal ports; these are not computed from general theory, however, but from analysis of tidal records over a long period of years at the particular port concerned.

Since Newton's day the tide-producing forces have been visualized much as shown in Figure 16-12. If D is the distance from the center of the earth to that of the moon, M is the mass of the moon, and r the radius of the earth, the moon's attraction for a unit mass, m, at the earth's surface on the side nearest the moon is greater than its attraction on a unit mass at the earth's center: $\frac{Mm}{(D-r)^2}$ is greater than $\frac{Mm}{D^2}$. Theoretically, if this surficial mass is water, it will be bulged up. Even the rocky crust also shows tides, though these are much smaller because of the rigidity of the rocks. A unit mass on the earth's surface directly opposite the moon is less attracted than an average like mass for the earth as a whole: $\frac{Mm}{D^2}$ is greater than $\frac{Mm}{(D+r)^2}$. Theoretically, a water mass here is "left behind" in a corresponding bulge. At all other points on the earth's surface there is also a difference in the moon's attraction for surface particles and for the "average" particles. Thus there is a resultant force acting on each particle along a line inclined to that joining the centers of the earth and moon, as illustrated in Figure 16-12. In general, there are both vertical and horizontal components of the tide-producing force. Because the earth rotates with respect to the moon once in 24.84 hours (not 24, because the moon advances eastward in its orbit), two tidal bulges pass over a particular point on the earth's surface during this time.

The sun's attraction acts like that of the moon. Despite its far greater mass, it is so much farther away that its maximum tide-producing force is only 0.46 that of the moon. Twice during the lunar month, at new and full moon, both sun and moon lie on a straight line that also passes through the earth and their influence is greatest. These are times of extreme tides (*spring tides*); at all other times their tide-producing forces tend in some degree to neutralize each other. At the first and third quarters their influences are directly opposed and tides are smallest (*neap tides*).

Though the simple diagram of Figure 16-12 gives the essentials of the tide-producing forces, it certainly fails to explain fully the vagaries of the local tides. For example, many ports have but one tide in a lunar day; in others the lag is many hours from the time the moon is at zenith; in still others the two daily tides are of greatly different heights. They also vary with the seasons. These and many other facts make it clear that the tides are not a simple direct response to the vertical component of the moon's gravitational pull, which is really far too small for effective lifting of the water masses anyway. Tides are really great shallow-water waves, induced by periodically shifting tidal forces acting on the water particles of the sea in much the same way that wind turbulence produces water waves. And, like other shallow-water waves, they "feel bottom" and are influenced by its configuration. Depending on this configuration, and on still other factors only little understood, they produce currents that vary greatly in strength and direction at different points and at the same point in different seasons.

To the geologist the principal interest of the tides is in these currents. Some are prodigious, particularly in estuaries with converging shores. In the Bay of Fundy, between New Brunswick and Nova Scotia, the extreme tidal range is as much as 70 feet vertically, and the possibility of harnessing the power of the huge water masses involved in this vertical movement has been seriously studied. Thus far, the great cost of the dams needed and the fluctuation in amount of power available from hour to hour and day to day have made the project seem uneconomic. The tidal currents attain speeds of 9 miles per hour during both rise and fall of the tide. In places they have scoured basins on the bottom more than 150 feet deep. In St. Malo Bay, Brittany, the tidal range is 40 feet, and currents reach speeds of 8 miles per hour. Between the Orkney and Shetland Islands the tidal currents move at 12 miles per hour. In some rivers the tide enters as a high wave of water called the *bore*. The bore of the Hangchow River in China is as much as 16 feet high and speeds upstream at 16 miles per hour.

Since experiments have shown that water moving at ½ mile per hour will transport medium sand grains, and at 3 miles per hour gravel an inch in diameter, it is clear that tidal currents are important agents of erosion and transport of sediments. Dredgings off the Mull of Galloway in Scotland show that coarse gravel is moved by tidal currents at depths of more than 800 feet.

Because the wave length of the tide is so great, it always acts as a shallow-water wave—that is, the greatest ocean depth is only a small fraction of the wave length. (See Fig. 16-12 which is, of course, oversimplified). Tidal waves must everywhere reach bottom and produce some water motion even at great depths. Because of the density stratification of the sea, secondary internal waves are formed at the boundaries of water masses of different density, and bottom motion is probably negligible in many places. The currents should be strongest near the edges of the shallow continental shelves because, here, more water must flow in and out (in proportion to depth) than elsewhere. This may account, at least in part, for the sediments near the outer edge of the shelf being somewhat coarser than those farther inland. The stronger currents should winnow out the finer sediment and drop it off the shelf onto the continental slope.

Over submarine ridges, regardless of depth, tidal currents must be perceptible where the ridge rises steeply from the bottom. Many ridges yield bare rock to the dredge, and nearly all show coarser sediments than adjoining basins or depressions.

In summary, we have reason to think that the winnowing action of tidal currents significantly affects the size, sorting, and distribution of sediment on the sea floor. But only locally are they important in shaping the shore itself and, in general, their effects are subordinate to those of wave currents.

Two wholly different kinds of waves, neither of them tidal, have, unfortunately, become popularly known as "tidal waves." One, the sea wave caused by earthquake, *seismic sea wave,* is considered in Chapter 18; the other is a high-water wave caused by prolonged and unusually violent on-shore winds.

Wave Currents

Despite refraction, most waves approach shore at a slight angle, so that, on breaking, the swash has a component of motion parallel to the shore. This produces a current called the "longshore drift," which, depending on the exposure of the shore, may consistently flow in one direction or reverse with differences in the wave approach. Such a drift may become swift, especially if there is a longshore wind, or if tidal currents reinforce it. The breaking wave carries sand and gravel obliquely up the beach—the sand particles farther than the gravel because they move with a feebler current—and even the backwash may retain a longshore component. Each particle thus moves

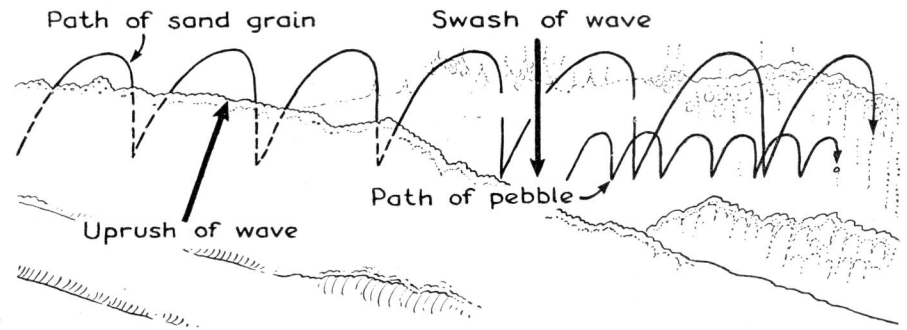

FIGURE 16-13. *Longshore drift* (to the right) *resulting from oblique wash of the waves. Each arcuate segment of the particle's path represents the movement due to one wave.*

in a series of sawtoothlike oscillations down the beach (Fig. 16-13). Marked pebbles made from bricks have been traced along the upper beach as much as half a mile in a single day; similar movements doubtless occur in the whole bottom zone agitated by the waves. Vast amounts of sand are transported parallel to the shore by these longshore currents.

Artificial Interference with Shore Processes

Artificial interference with waves and currents furnish many examples of the nice adjustment of beach profiles and shore outlines to the average power of waves and currents. Even a slight obstacle to movement of detritus often causes drastic changes in the shape of the beach. Dredging of sand and gravel offshore has frequently been followed by erosion of the shore; the material left behind the excavation is swept seaward to restore the slope that had been modified by removal of the dredged material. Still more striking results have come from the building of groynes on many bathing beaches. Groynes are low walls extending from the high-tide line out into the sea with the purpose of preventing the removal of sand by longshore drift. Where they have been successful in holding the fine sand, the down-drift beaches, being deprived of their normal supply of traveling sand, have often been severely scoured and changed from sand to gravel or cobble beaches.

Breakwaters, walls of rock or concrete built to protect anchorages from storms, have often greatly modified beaches, as at Santa Monica, California. A 2,000-foot breakwater was built parallel to the shore and about 2,000 feet away from it, in order to furnish anchorage for small boats. The general longshore drift here is to the southeast. The breakwater, of course, reduced the power of the waves striking the shore behind it. The detritus moving along the beach was no longer kept in suspension in the shore zone, and the beach in the lee of the breakwater began to fill in. In less than 20 years it had advanced more than 500 feet. Despite being robbed of their normal quota of drifting sand from the northwest, the beaches to the southeast of the breakwater have not been eroded inland very far, although it is noteworthy that in time of great storms there is considerably more damage to the southeast of the breakwater than to other nearby parts of the shore (Fig. 16-14).

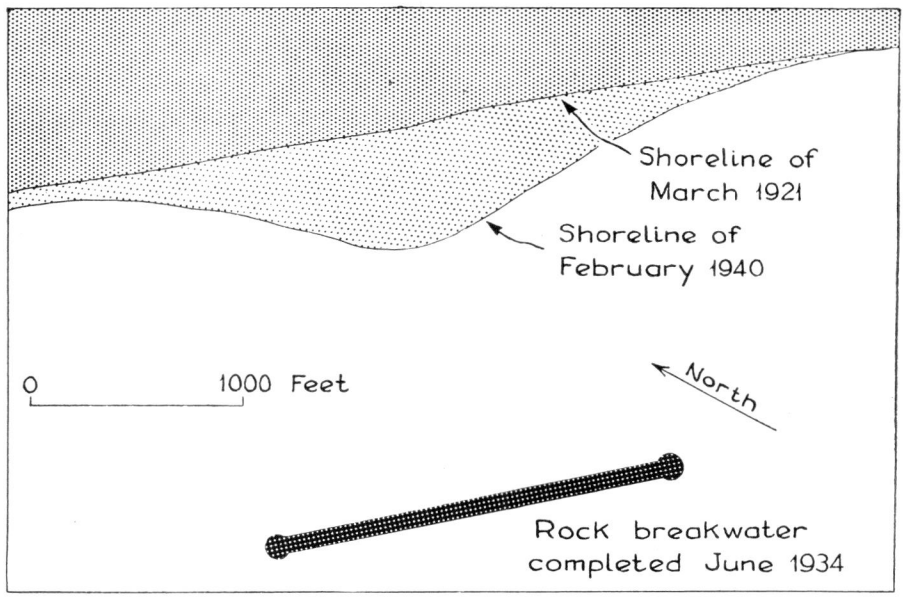

FIGURE 16-14. *Map of Santa Monica beach before and after building of breakwater. Prograding has continued since 1940, but no accurate surveys are available since that year. (The direction of longshore drift is from left to right.)*

Erosion and Deposition by the Oceans

The Shores of Mountainous Coasts

THE SHORES of mountainous or hilly coasts are more varied than those of plains. Waves attack exposed promontories, producing cliffs; the coarse and fine debris eroded from them is distributed by longshore drift to make characteristic shore features—bay-head and bay-mouth bars, land-tied islands, spits, and bars.

The wave refraction on the shores of a bay produce currents that tend to sweep material to the head of the bay, where it accumulates as a bay-head bar (Fig. 16-15).

Material carried by longshore drift tends to continue straight across indentations in the coast. Debris from headlands may thus tend to close a bay by forming a sand spit that continues in a smooth curve from a cliffed point. Toronto Harbor is protected by a long sand spit growing westward. As such spits are built out into deep water, wave refraction and, in the sea, local tidal currents, cause their ends to become curved (Fig. 16-15). Where the outflow from a bay is small, such spits may build completely across the bay, forming a bay-mouth bar (Fig. 16-16). The lagoon behind may be brackish

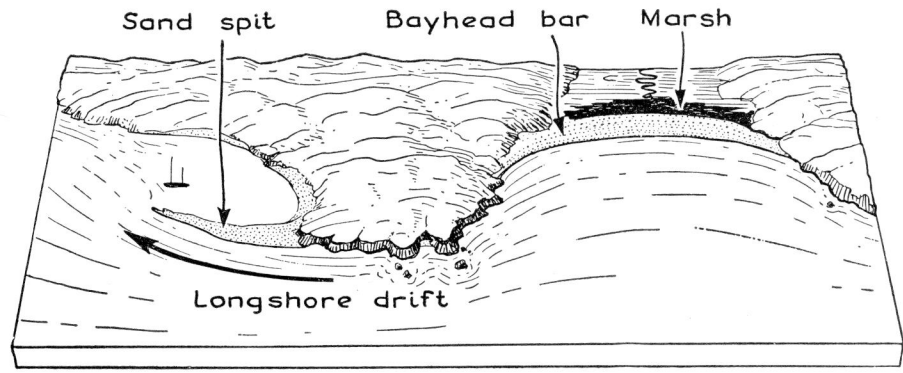

FIGURE 16-15. *Two common depositional features on an embayed coast.*

FIGURE 16-16. *Bay-mouth bar and lagoon, St. Mary's Lake, Glacier National Park. (Photo by E. Stebinger.)*

or fresh and may be slowly filled by silt from the land or by sand blown inland from the beach.

Islands refract the waves and protect the beach behind them from the full force of the sea (see Fig. 16-5). Hence the longshore drift cannot effectively carry material behind an island. Therefore the mainland beach tends to be built out, diverting the longshore currents toward the island. Eventually, a spit may extend to the island, tying it to the shore. Many islands thus become joined to the coast in both directions, enclosing triangular lagoons (Fig. 16-17).

Even on straight coasts the drift may be deflected by bottom irregularities and build forward into the sea. The beach formed varies in height with the detritus supplied and the height of the storm waves. As the beach advances seaward it leaves behind a series of older beach ridges—some 20 feet or more high—separated by swales. Perhaps the best known of these advancing beaches is that of the Dungeness, in southeastern England, which has grown more than a mile into the sea since the time of Elizabeth—about 6 yards per year.

The Shores of Plains

Along low-lying coasts such as those of the Gulf of Mexico and the southeastern Atlantic states, the wave attack is spread widely instead of being con-

FIGURE 16-17. *Land-tied island, Hanock County, Maine. (Photo by E. S. Bastin.)*

centrated upon headlands. The sea deepens gradually offshore, hence the waves feel bottom far out, and the material they stir up is built into an *offshore bar* just inland from the zone of greatest breakers (Fig. 16-18). As wave agitation must continually grind down the bar material and slowly move it seaward, offshore bars must advance slowly landward in order to survive with a stable sea level. Most offshore bars along plains do touch the mainland locally. Here the shore is undercut to produce low cliffs. Part of the material eroded is carried by longshore currents and added to the bar.

The lagoon behind an offshore bar is, of course, salty, but rivers may freshen the water considerably. A large river may raise the lagoon level to such a height, especially at low tide, that the bar is broken, forming an inlet. Tidal currents help to keep the inlet open. Inlets shift from time to time because of shore drift, and new ones are sometimes formed. In the protected lagoon, salt grass grows. Gradually the lagoon changes to a marsh as plant debris accumulates along with the silt brought by streams, tidal currents, and winds.

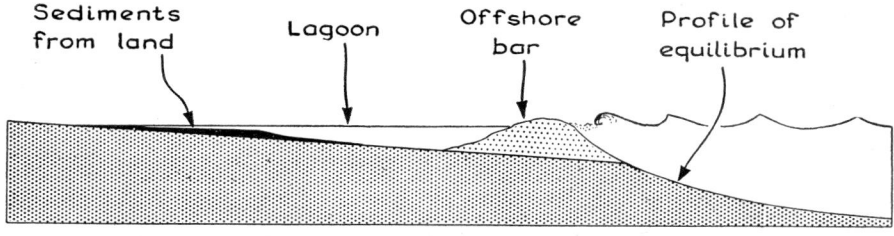

FIGURE 16-18. *Cross section showing the features of the shore zone of a low-lying coast. The vertical scale is exaggerated.*

The famous Florida and New Jersey beaches are such offshore bars. Another striking example is the long sand bar that fringes the southern coast of the Baltic from Danzig to Memel, enclosing an almost continuous brackish lagoon. Figure 16-19 illustrates a plains shoreline along which some offshore bars still enclose lagoons, though others crowd meandering tidal rivers behind them.

Magnificent examples of offshore bars fringe the Carolina coast. Their cuspate shape has been attributed to the junction of eddies thrown off by the Gulf Stream (Fig. 16-2) but it may be partly controlled by wave refraction from bottom irregularities offshore.

Beaches that dry out between tides furnish sand which may be blown inland by winds. Because of the absence of coarse shingle and the prevalence of sand beaches along plains coasts, the country inland is likely to be overrun by sand dunes wherever onshore winds are common.

FIGURE 16-19. *The west coast of Cape York Peninsula, Australia, showing features of a plains coast. The curved white stripe in the upper half of the view is an offshore bar; the ridged beach in the foreground is probably an older offshore bar. (Photo by U. S. Air Force.)*

Sediments of the Sea

The Sources of Marine Sediments

Except for a trivial amount of meteoritic dust and the much greater increment from submarine volcanoes, all marine sediments come ultimately from the land. Even the material of marine shells has been leached from rocks of the lands, though it is extracted from solution in the sea. Sluggish or swift, the rivers sweep the land waste seaward. Detritus from rivers and rills, from downslope movement, from glacial plucking and abrasion, along with wind-blown volcanic ash and desert dust, are carried to the sea, there to mingle with the detritus won by the ocean itself in its attack upon its shores.

The rivers bring to the oceans about twenty billion tons of rock waste every year. More than four-fifths of this is clastic material carried in suspension or rolled along the stream bottoms; the rest is dissolved. The fate of the clastic material is clear. Samples dredged from the sea floor show partly consolidated sediments identical in grain size and mineral content with river sediments. Stratification is more common, though not universal, and sorting more complete, than in sediments deposited by rivers. Furthermore, the widespread marine sedimentary rocks that are uplifted to form part of the lands show that the solid particles carried to the sea by streams accumulate on the bottom. The fate of the dissolved material is not so obvious. We know that some of it is deposited, for there are great marine beds of limestone, dolomite, salt, and anhydrite, but we do not know whether deposition exactly keeps pace with replenishment and, hence, whether the sea is becoming more or less salty. As was previously noted, river waters have a much larger proportion of silica and calcium than sea water, so that if these substances were not withdrawn from solution in the sea, its composition would change greatly within a short time, geologically. Life processes regulate this; limy and siliceous shells of various organisms represent much of the material brought to the sea in solution.

Marine organisms are geologically important not only because they contribute their shells to sediments but also as indicators of the environment of deposition. An understanding of the main features of marine life is essential for interpreting the sedimentary rocks and the historical record they contain.

The Life Zones of the Sea

Life is everywhere in the sea. Even the greatest depths—far below levels to which light penetrates, and where the water temperature is only a little above freezing—yield living organisms to the trawl. Although marine environments are less variable than those on land, they, and the organisms living within them, do differ from place to place. Marine biologists classify the different parts of the sea as shown in Figure 16-20.

The two main divisions are the *benthic*, or sea-bottom environment, and the *pelagic*, or open-water environment. The benthic environment is divided, at a depth of 200 meters—about the depth of the edge of the continental shelf—into a *littoral* system above and the *deep-sea* system below. The depth of 200 meters is critical because it is near the limit to which light can penetrate. On the deep-sea floor there is neither light nor seasons, and hence no photosynthesis by green plants. The bottom life is confined to scavengers living on organisms sinking from above and to a few bacteria, nearly all of which depend on organic compounds formed ultimately by photosynthesis but brought to this depth by currents or by sinking.

The life of the sea, both plant and animal, is classed in three large groups: the *benthos* (Greek: "deep" or "deep-sea"), or bottom dwellers; the *nekton* (Greek: "swimming"), or swimming forms; and the *plankton* (Greek: "wanderer"), or floating and drifting organisms. Of these, the benthos includes all the attached, creeping or burrowing organisms of the bottom: seaweeds and grasses, sponges, barnacles, clams, oysters, corals, bryozoa, worms, lobsters, crabs, and many other animals. Among the inconspicuous but geologically important organisms of the benthos are the minute single-celled animals called *Foraminifera,* most of which secrete shells of calcium carbonate. The nekton includes the squids, fishes, seals, whales, and many other animals which are economically important but of little geological significance. The plankton includes all the surprisingly varied organisms, chiefly microscopic, that float with the ocean currents. Among them are several

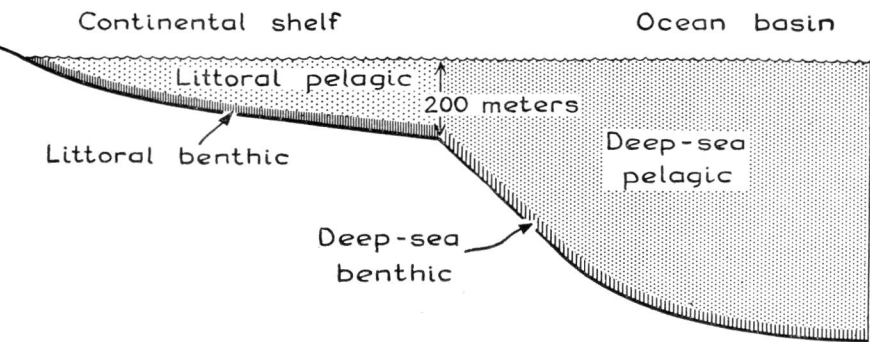

FIGURE 16-20. *Diagrammatic cross section showing the four major divisions of the life zones of the sea. (After H. U. Sverdrup, M. W. Johnson, and R. H. Fleming,* The Oceans, *Copyright 1942 by Prentice-Hall, Inc.)*

kinds of diatoms with shells of opal, the *Coccolithophores,* a group of algae
that secrete rounded shells made up of calcite in thin plates, many individual
Foraminifera (though but few species), and the silica-secreting *Radiolaria.*
As many as 100,000 diatoms have been found in a quart of water from the
open North Pacific, though the usual abundance is far less.

Except for a few very specialized bacteria, all life depends on photosynthesis—the process by which plants utilize radiant energy from the sun to
make compounds of carbon, hydrogen, and oxygen. These compounds constitute the bulk of the plants, and are used directly or indirectly by animals
that eat the plants or other animals. Marine plants, like land plants, form
organic compounds directly from carbon dioxide and water, but only in the
presence of light. Plants require small amounts of many substances, among
them phosphates, nitrogen, iron, and manganese. These factors of food supply that limit plant growth are partly reflected in the very different densities
of population in different parts of the sea, and in the comparable variations
in the animal life that feeds on them.

When an organism dies, unless it is eaten by another animal, the organic
compounds comprising it ultimately sink below the lighted zone, perhaps to
feed some scavenger in the depths. In sinking, much of the organic matter is
decomposed into carbon dioxide and other constituents by bacteria. This
decomposition both uses up the oxygen dissolved in the water and adds to
the carbon dioxide in it—two factors that strongly influence the kinds of
sediments of the deep sea. Decomposition at any depth returns the phosphorus, nitrogen, and other nutrient elements to solution as simple compounds once more available for plant growth, if the water mass containing
them returns to the lighted zone. As we have seen on p. 384, this return of
deep water to the surface is by slow mixing or by upwelling. These factors
account for the flourishing plant life in parts of the sea and explain why the
Newfoundland Banks are such prolific fishing grounds, as are similar areas
of upwelling along the west coasts of California, Chile, and Africa.

Realms of Sedimentation in the Sea

The "Challenger" expedition proved that sediments are being deposited
virtually everywhere in the sea except where currents scour the bottom or
where slopes are too steep to retain them. Even where depth, exposure, distance to shore, submarine slope, and other obvious factors seem identical,
the local sediments vary greatly. Despite much careful work, we are far from
understanding all the controlling factors. Sir John Murray, the leader of the
"Challenger" expedition, classified the sediments into two great groups: *terrigenous* (derived from land), and *pelagic* (of the sea). As we shall see, the

red clays, some of the most representative of his pelagic sediments, are actually composed largely of land-derived material. Also, the bottom mud at a depth of 33,000 feet in the Mindanao Trough is indistinguishable from many continental shelf deposits; it is a "terrigenous mud." Despite these anomalies, deposits of different realms vary broadly and some of them sharply. Modern oceanographers call sediments "pelagic" if they are thoroughly oxidized—a condition that indicates long contact with deep water and hence slow deposition. Deposits of any depth that show lack of this oxidation, and hence accumulation rapid enough to prevent it, are called "terrigenous" even though found at greater depth.

We accordingly consider the *terrigenous sediments of the strand zone,* the *terrigenous sediments of the shelf zone,* and the *pelagic sediments* as different classes. Because of their geologic importance and the peculiar problems they offer, we shall consider separately the deposition of carbonate rocks (limestones and dolomites).

Sediments of the Strand Zone.　The strand zone is the area immediately along the shore. Its sediments differ considerably from place to place. The most important kinds are beach deposits, deltaic deposits, and tidal-flat deposits.

Beaches　Beaches are deposits of cobbles, gravel, and sand on and near the shore. During great storms much or all of this material is moved about or even stripped away by the waves, but usually only the superficial layers are agitated. Beaches vary greatly in size; some between cliffs and headlands are only a few feet long; others on plains coasts stretch uninterruptedly for miles. They range from a few feet to several thousands of feet in width. Most beaches of coarse material are narrow because steep slopes are necessary for coarse material to move seaward. Even in sand beaches the steeper ones tend to be made of coarser sand than the flatter ones, though the correlation of slope and grain size is not perfect (Fig. 16-21). Most beaches are concave upward because the coarsest material is nearest the shore.

The waves not only wear down and round the grains, but also sort them according to size. Variations in the waves are reflected in the size of grains they can move. Thus the beach develops a laminated structure. The perfection of sorting and lamination depends on the available material. It is much poorer along coasts of high relief (which generally supply heterogeneous material) than along plains coasts (where the source material has already been sorted by sluggish streams or the waves of a former sea).

Beach sands may be largely of quartz if they come from deeply weathered granitic regions or from reworking of sedimentary rocks, for such reworking means, of course, that any feldspar originally accompanying the quartz will have been exposed to weathering at least twice. But in regions of rapid

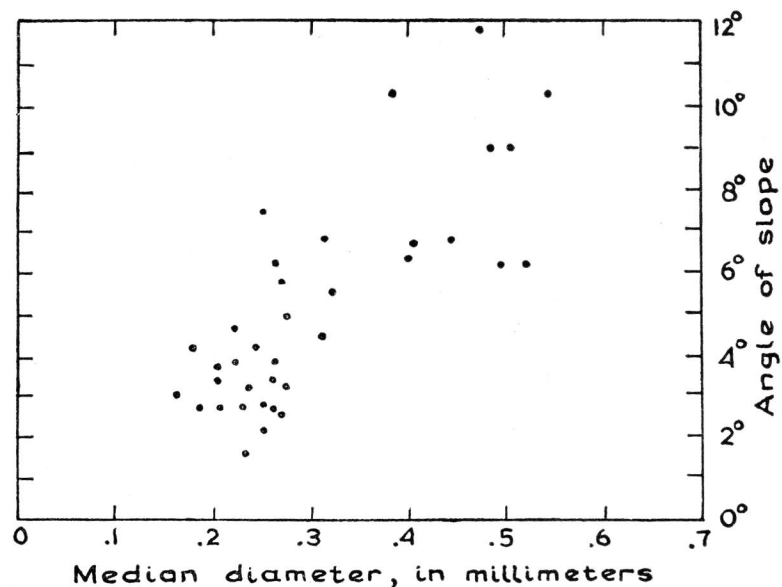

erosion or of ineffective chemical weathering, feldspar abounds on sand beaches. In Hawaii, whole beaches are made up of grains of olivine from nearby basalts. Some beaches, particularly in the tropics, are composed largely of shell fragments. Along the Brazilian coast these are cemented almost as fast as they are being formed because the spray dissolves a little calcite at the surface and redeposits it below. Many sand beaches contain concentrations of heavy minerals ("black sand"), such as magnetite, chromite, and garnet, sorted out by the wash because of their high density as compared with other minerals. At Nome, Alaska, gold thus concentrated has long been worked.

Few beach deposits have been recognized among the rocks, though recently uplifted beaches are abundant (Fig. 16-11). Beaches are quickly destroyed because uplift exposes them to erosion, whereas submergence to a depth less than wave base subjects them to reworking by waves and tidal currents. Nevertheless, Cambrian beaches, with associated cliffs and tumbled stacks, have been recognized in the Montana Rockies, Carboniferous and Jurassic beaches in Great Britain, and other examples elsewhere.

Deltas   Streams generally form deltas where they enter the sea, but if the stream load is small or the coast exposed, the detritus may be widely scattered by waves and currents, and no delta is formed. Deltas furnish another example of the difficulty of geologic classifications, for all ocean deltas contain some beds that are truly marine, others that are brackish lagoon deposits, and still others that are fluviatile—all complexly interfingering.

Where a stream enters the sea the surface slope that gives it a current vanishes. If wave action is strong, river water and sea water quickly mix, but along quiet coasts the fresh water may float for awhile above the salt water. The powerful flood of the Amazon extends far into the Atlantic, and fresh water may be dipped from the surface scores of miles out to sea.

The sediment suspended in the river water is thus distributed widely over the sea floor. The water of some rivers is so heavily laden with sediment that it sinks under the ocean water and flows down the slope of the sea bottom. Ultimately, of course, such sediment-laden water will deposit its load, either in a submarine delta or widely over the sea floor. There is growing evidence that similar sediment-laden "density currents" may form independently of rivers within the sea itself (see p. 432).

Rivers vary widely in the grain size of their sediment and all carry coarser material in flood than in low water. Some, like the Colorado, carry much coarse sand; others, like the Mississippi and the Rhine, discharge little but clay, silt, and fine sand. At the river mouth, the coarsest material settles out of suspension first, followed progressively by finer and finer grains, as can be seen in the Mississippi delta (Fig. 12-15), where coarser silt and fine sand extend on either side of the distributary channels out into the Gulf. Traced landward, these submarine ridges emerge and become natural levees. Like all natural levees, they are made of coarser sediment than is found behind them.

The activities of delta distributaries vary with river discharge and with shifts to other newly formed distributaries. Thus some segments may advance into the sea while other segments are being destroyed by the waves. The shore processes are like those of a plains coast. Offshore bars sheltering tidal lagoons are common; the waves may push the bars inland to the delta front and then rework the stream-laid beds, concentrating the coarser material on the beach and strewing the finer over the sea floor.

The factors just outlined condition deltaic sediments; the coarser material comes to rest chiefly on the stream bed or on beaches; the next finer forms natural levees and their submarine extensions; the finest sediment is deposited in back of the levees during floods or is swept out to sea.

The clay particles in river water have very large surface areas relative to their bulk. They are dispersed as individual particles, largely because the unsatisfied valence bonds of the surface ions attract negatively charged ions such as O^{--} and OH^- and thus prevent one particle from coming in contact with another (Chap. 4). On entering the sea, with its abundant ions of Na^+ and Mg^{++}, the surface shells of attracted ions are stripped away, and the clay particles come into contact and adhere, forming larger aggregates which quickly sink.

Clay should, therefore, settle quickly when it enters the sea. Actually, few careful studies of this process have yet been made, though the above idea conforms with laboratory experiments. In one study off the California coast, Revelle and Shepard (1938) found that, despite thorough mixing of stream and ocean water, much clay was carried by eddies and currents as far as 150 miles offshore before settling to the bottom, though most was deposited much nearer the coast.

Some deltas, like that of the Mississippi, are known from surveys to be subsiding, locally as much as 8 feet per century. It has been suggested that this is due to the load of sediments added to the crust. As with glacial loading, this seems a necessary consequence of isostasy, although the sinking of the Mississippi delta seems too great to be explained in this way alone. The unconsolidated sediments are much less dense than deeper-lying rocks and hence should not cause their floor to sink as much as their own thickness. But, whatever the cause may be, parts of the subaerial delta are carried below the sea, forming tidal lagoons or even open channels where fine marine sediments accumulate. Perhaps this subsidence gives the Mississippi delta its peculiar bird-foot form by submerging all the outer surface except the natural levees. As distributaries shift, deltaic marine beds may be buried by river deposits. If subsidence continues, the delta may remain nearly in place and not advance into the sea. Thus interlayered marine, fluviatile, and brackish-water deposits may accumulate to great thickness, thinning into beds completely fluviatile landward and marine seaward. Many deltas have been recognized among the ancient sediments by the intimate association of these varied deposits. In New York and Pennsylvania a delta formed during the Devonian shows strata more than 15,000 feet thick; borings for petroleum indicate a comparable thickness of the Mississippi delta along the Louisiana coast.

Tidal-Flat Deposits Tidal-flat deposits accumulate in estuaries of sluggish streams or behind offshore bars. Many fill quickly; at Wilhelmshaven harbor and Cuxhaven, both on the North Sea coast of Germany, 8 to 10 feet of mud accumulates in a single year. Much of this mud and silt is brought by rivers, the rest must be swept in by the tide. Some of the mud from the German and Dutch coasts is so full of organic matter that it is used as agricultural fertilizer. Shallow parts of tidal lagoons commonly support a luxuriant growth of brackish-water plants. This vegetation speeds up the silting of the lagoon, for it acts as a filter to screen out sediment.

Like beaches, lagoonal deposits are only exceptionally preserved among the older rocks. Nevertheless, some of the Devonian shales of Germany, as well as other examples, have so many of the characters of modern lagoonal deposits that they are thought to represent this environment.

Sediments of the Shelf Zone. As was pointed out earlier in the discussion of shore processes, the "profile of equilibrium" is an ideal concept. Soundings and samplings of the continental shelf sediments seem to show actually as much deviation from a smooth profile as agreement with it. Nor is there a notable correlation of grain size of sediments with either distance from shore or depth of water. It does appear that the immediate shore zone generally has both a flattening profile and finer sediments seaward, but within a short distance from the shore both these regularities usually disappear. Such deviations may be partly explained by the rise in sea level of perhaps 300 feet since melting of the Pleistocene glaciers—a time so short

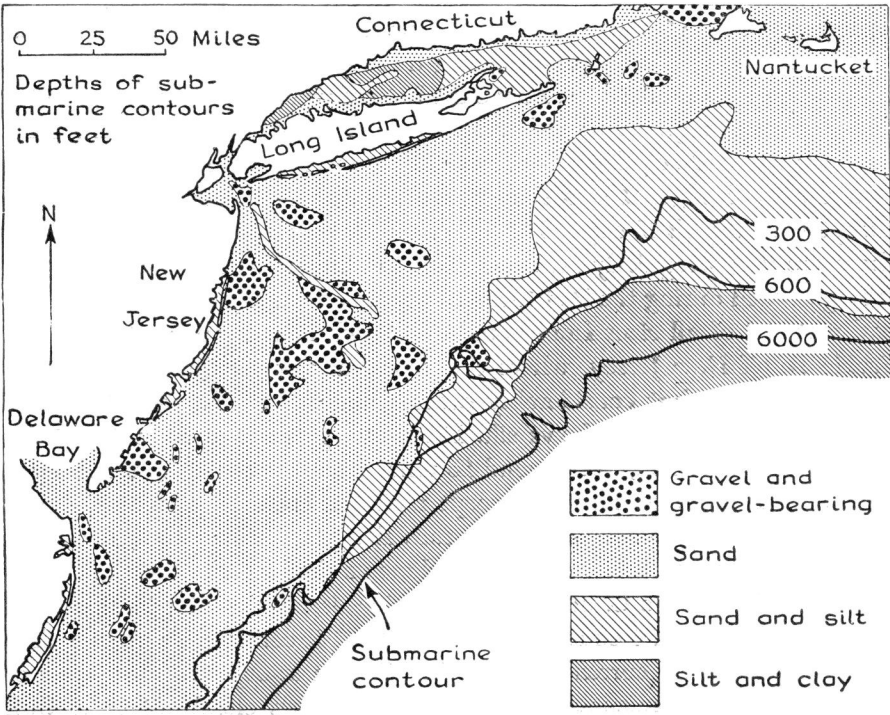

FIGURE 16-22. *Size-distribution of sediments on the continental shelf, off the mid-Atlantic states. (After F. P. Shepard and G. V. Cohee, 1936.)*

that the sea has not yet had time to adjust its submarine profile and the grain size of its sediments to its new level. Or perhaps crustal movements have similarly interfered with the development of a smooth slope and effective sorting. It seems, though, more likely that the currents set in motion by the winds, waves, and tides act in a more complex way than the idealized theory envisages. Whatever the reason, the fact remains that many recent studies have shown that the simple graded slope, steeper inshore and flatter offshore, and covered with sediments that become regularly finer seaward, is far from characteristic of many of our present seas.

Sediments of the Atlantic Continental Shelf   Figure 16-22 shows the distribution of bottom sediments on the continental shelf off the mid-Atlantic states, as reported on the official hydrographic charts. The material reported is largely that adhering to the tallow on a sounding lead. The charts show that, far from having a simple gradation from coarse to fine seaward, the sediments are very irregularly distributed. There is actually more sand nearshore and more gravel in deep water. The sediments of the outer edge of the shelf are coarser than those inshore. Silt and clay, however, are rare at depths less than 600 feet and extend from here down the continental slope.

Figure 16-23 shows the profiles along three lines surveyed across the shelf off the eastern United States. Above the bottom profile is plotted the median diameter of the grains of sediment, as determined by sieving and other methods. Profile A, off Cape Cod, shows a bottom slope rather regular except near the shelf edge; it also shows finer sediments—largely silty—offshore except near the shelf edge. The coarser materials within 20 miles of the shore are very irregularly arranged. This profile has been developed on an area glaciated during the Pleistocene, and the advancing sea has, therefore, had material of all sizes from boulders to rock flour available for sorting. Profile B, off New Jersey, where the floor was not glaciated, shows finer material inshore but considerably coarser material at the same depth as that at which Profile A shows silt. Here, too, grain sizes of sediment are irregularly distributed, but beyond 25 miles from the coast the tendency is toward finer material seaward except near the shelf edge, where, again, coarser material prevails. Silt may be rare along this profile because the sea is here working on old coastal plain sediments that contain little but sand-sized material. Profile C, off northern Florida, is very regular in slope, but even here the material outside the 25-mile limit is consistently coarser than that inshore. Strong currents of the Gulf Stream sweep the outer edge of this profile. The coarseness of sediment here shows clearly that the Gulf Stream is swift enough to carry fine sediment right across the shelf, down the slope, and even across the Blake Plateau, whose outer edge is marked

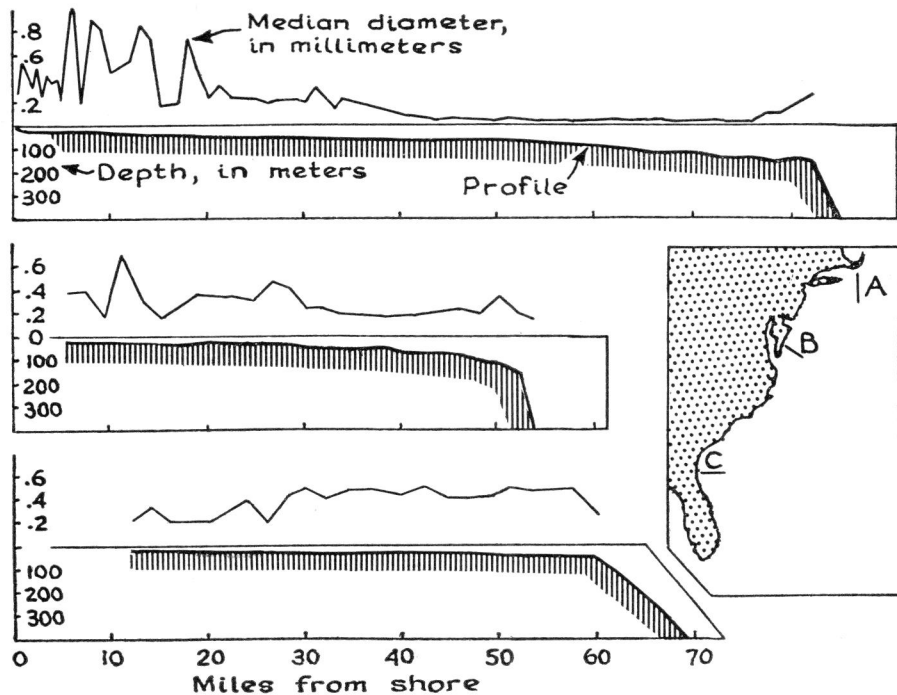

FIGURE 16-23. *The distribution of sediment sizes along three profiles of the Atlantic shelf of the United States. (After H. C. Stetson; redrawn from H. U. Sverdrup, M. W. Johnson, and R. H. Fleming,* The Oceans, *Copyright 1942 by Prentice-Hall, Inc.)*

by the 2,000-meter (6,500-foot) depth line. It is concluded that nearby a fairly coarse sand is being deposited at depths greater than 6,500 feet.

Sediments of the Dutch East Indies The submarine topography of the Dutch East Indies is highly irregular, with deep basins, shallow marine plains, and submerged mountain ridges. Some of the basins—the Sulu, Celebes, and Banda seas—are more than three miles deep, separated from the ocean basins by much shallower ridges and saddles. The situation is further complicated by the many volcanoes, of which more than 50 have been active at some time since 1600. Although only a few hundred samples of the sediments of this great region of more than a million square miles have been studied, some extremely interesting results have been gained, chiefly by the Netherlands "Snellius" Expedition of 1929–30.

When Tamboro, on an island east of Java, erupted violently in 1815 (see Chap. 17), the ash was thrown high in the air and carried by the winds for tremendous distances. This ash is readily recognizable from its distinctive minerals. In places 240 miles distant from the volcano as much as 10 inches of this ash was deposited. Even more striking than the prodigious volume of the ash is its presence as the topmost layer in submarine cores collected from the basins of the region. Although it has been suggested that some of the very finest grains of this ash are actually still being deposited—reaching bottom for the first time 115 years after the eruption—it seems more likely that reworking of the sediment by benthonic animals may account in part for its abundance in the surface layer.

Rock bottom was found on many of the ridges between the East Indian basins at depths of several thousand feet. Fine muds were found at far shallower depths in less-exposed places; thus we have evidence of strong currents sweeping the bottom clear of sediments at depths far below the zone of agitation by surface waves. Land-derived mud was cored fully 300 miles from the nearest land and at depths of 16,000 feet; volcanic muds from still greater distances and depths. From the Philippine Trench, east of Mindanao, 50 miles from land and at a depth of 33,000 feet, the core recovered silty and clayey sediments with a little sand that included fragments of metamorphic rock like that known from eastern Mindanao. This core, 20 inches long, showed no signs of sliding sediment.

These indications of long suspension and wide dispersal of land-derived sediments confirm the evidence of strong currents at great depth. It appears that the currents cross the ridges and saddles, winnow out the fine material settling there, and waft it along until it falls either into one of the basins or off the continental slope. On many of the steep slopes bare rock has been dredged at intervals. Nearby the sediments may be coarse and unsorted at depths of thousands of feet. Gentler slopes at the same depths have much finer sediments and no bare rock; it thus appears that some of these coarse sediments may have slid down the steep submarine slopes from higher levels, exposing the rock and coming to rest at depths where the normal sediment is far finer. Submarine slumping like that inferred here is well shown in consolidated rocks by contorted bedding and sliding surfaces, as in the Silurian rocks of parts of Wales, in Tertiary rocks of Peru and Ecuador, and in many other places. We refer to this again in discussing submarine canyons later in this chapter.

On the divide between two of the basins, the "Snellius" made current readings at depths ranging from the surface to 6,500 feet. Currents ranged in direction from NE to E to SE and even SW at different depths; in speed they ranged from 1/3 mile per hour at the surface to 1 mile per hour at 600 feet, whence they fell off to 1/14 mile per hour at 6,500 feet. From experiments it has been found that particles of quartz of various diameters settle in water at speeds indicated in the accompanying table. Miss Neeb, a Dutch geologist who studied the sediments, computed the distance they would be transported by currents of only 1 centimeter per second (1/44 mile per hour) during the time required for them to settle 1,000 meters (3,300 feet). In this calculation she assumed no turbulence, which would, of course, further prolong the suspension of some particles, as would also the higher viscosity and density of sea water, at the lower temperatures which prevail at depth (Table 16-2).

TABLE 16-2. *Theoretical Dispersal of Sedimentary Particles by Currents (After Neeb)*

DIAMETER OF PARTICLE (MM.)	RATE OF SETTLING IN WATER AT $27°$ C. (CM./HR.)	DISTANCE TRANSPORTED WHILE SINKING 1,000 METERS AND BEING CARRIED BY CURRENT OF 1 CM./SEC. (KM.)
0.100 (fine sand)	2000	1.8
0.030 (coarse silt)	180	20
0.005 (fine silt)	5	720
0.001 (clay)	0.2	18,000

These computations, though they neglect flocculation and therefore cannot be literally accepted, make it appear probable that the finest silts and clays can be carried almost indefinitely by the currents actually found in the East Indies; in fact, the clay particles in the sediments of a basin may have sources quite different from those of the coarser silts and sand grains with which they are deposited.

Still another result of the studies of the East Indian sediments is noteworthy. The drowned river plain of the Sunda Shelf, discussed in Chapter 9 (Fig. 9-10), hardly exceeds 200 feet in depth. The bottom muds near Java are made up of volcanic debris from the nearby island; farther north, the sediments are coarser quartz sands, undoubtedly derived from the deeply weathered granites of Borneo and the islands north of Sumatra. Because they are separated from these quartz-bearing islands by a belt of much finer and presumably younger quartzose sands, they have been considered as residual from the drowned land surface, relatively undisturbed by marine processes. Although the erratic size-distribution of many of our Atlantic shelf deposits makes this deduction somewhat doubtful, it may be that the

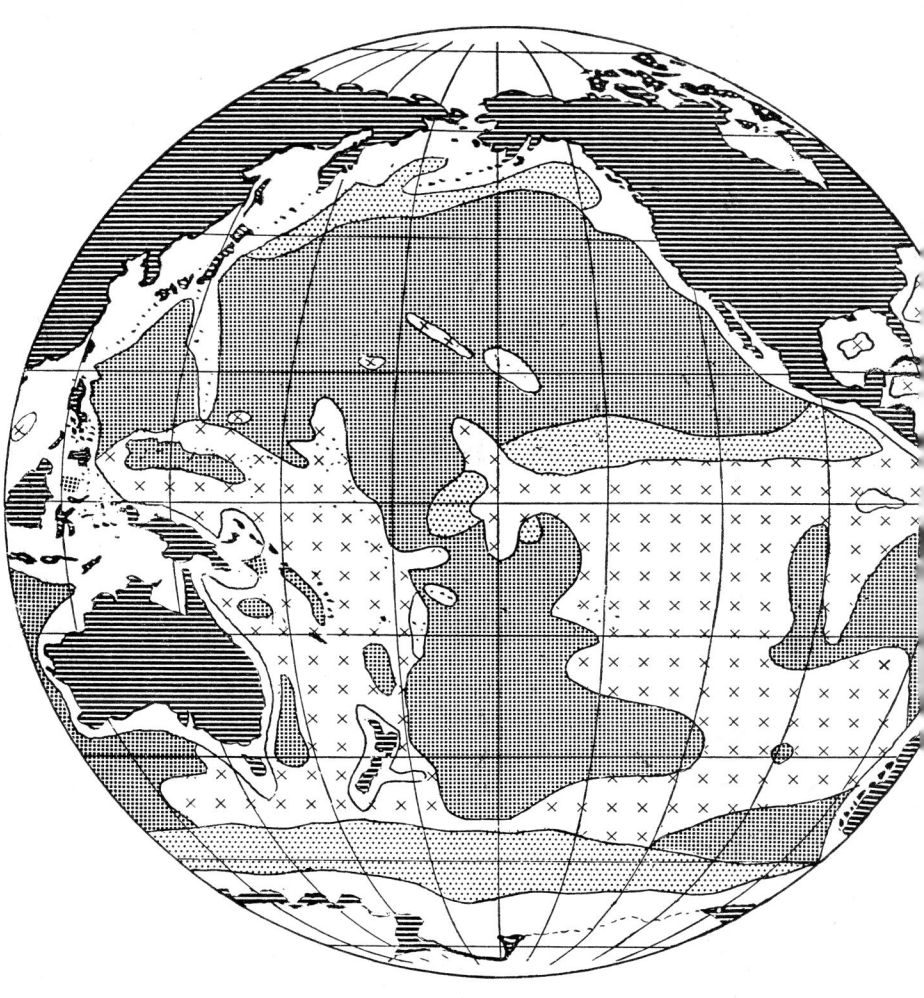

Terrigenous sediments   Red mu

FIGURE 16-24.   *The distribution of the terrigenous and pelagic sediments. (After H. U. Sverdrup, M. W. Johnson, and R. H. Fleming, The Oceans, Copyright 1942 by Prentice-Hall, Inc.)*

sea, during its rise since the Pleistocene glaciers began to melt, has indeed only slightly reworked these sands.

Comparison of Ancient and Modern Terrigenous Offshore Sediments Only a few thousand samples of modern marine sediments have been carefully studied, probably less than one sample for each 100,000 square miles of area. Clearly, any general conclusions about modern terrigenous sediments

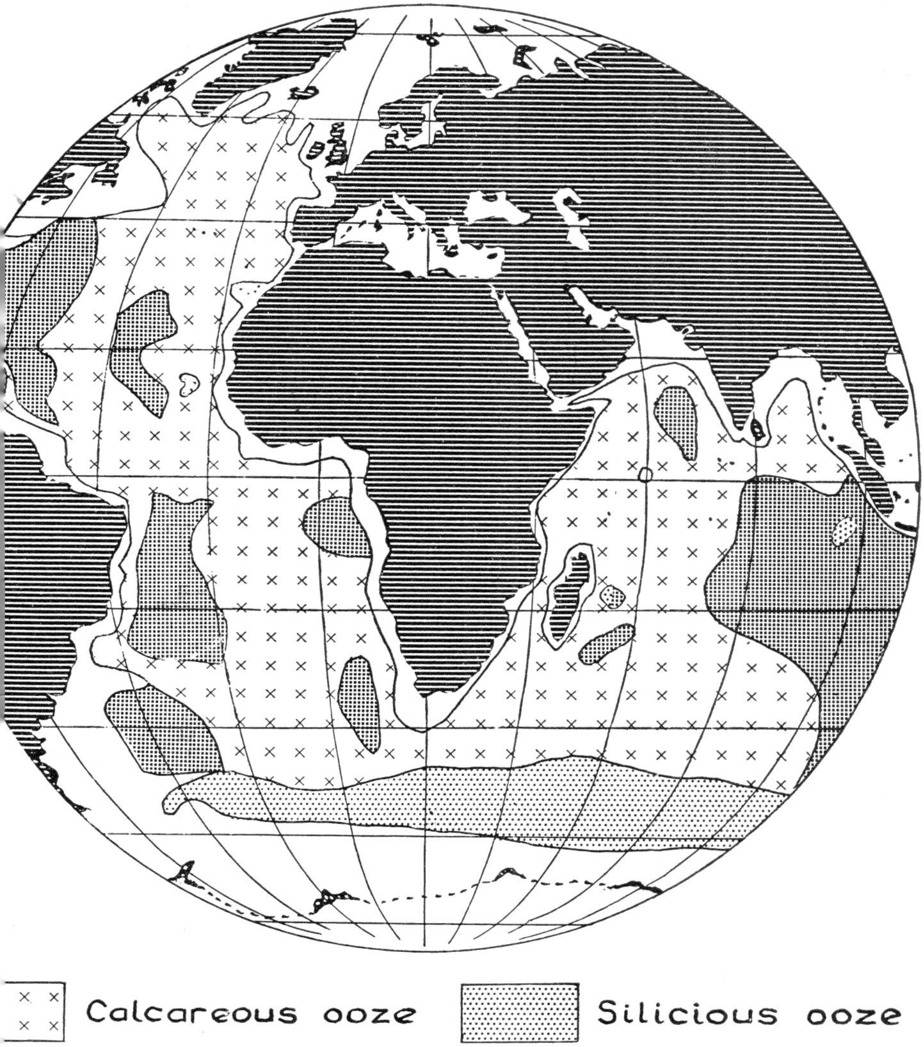

Calcareous ooze   **Silicious ooze**

must be very tentative. Our data suggest that relatively little silt and clay are now being deposited on the continental shelves—most of the fine sediment appears to be suspended until it drops into a basin or over the shelf edge. Locally, where strong currents sweep the sea floor, as off Florida or on the interbasin ridges of the East Indies, even sand is kept moving until it, too, reaches a deep basin. Beyond the immediate beach zone, the grain size of the sediment seems not to diminish regularly with either depth or distance from shore, but to be more closely related to the submarine topography.

The most widespread ancient sediments are marine shales, clays, and mudstones. They generally contain fossils of shallow-water organisms. Many of these rocks can be traced laterally into sandstones in one direction and into limestones in the other; both contain similar shallow-water fossils. We know of no present marine environment where muds and clays are accumulating in shallow water over areas of thousands of square miles, yet geologic mapping shows inescapably that they did so accumulate at many times in the geologic past.

A few formations, notably the silty *Repetto formation* (Pliocene) of California, contain deep-water fossils. It seems entirely probable that the Repetto was deposited in a deep basin like some of those off the present California coast and in the East Indies. The basin was ultimately filled completely by clay, silt, and winnowed sand dumped from nearby high topography onto the sea floor. However, such deposits seem rare among rocks of the continents; pending further studies of marine sediments now forming, our inferences as to the conditions under which most shales were formed must be made from the intrinsic characters of these rocks and their fossils, and not from close analogy with similar deposits now forming.

Pelagic Sediments.   Pelagic sediments cover nearly three quarters of the ocean bottom, as the "Challenger" discovered. Deep-sea deposits are of three main types: the *calcareous oozes,* the *red clays,* and the *siliceous oozes.* Of these, the calcareous oozes are most widespread, covering nearly 48 per cent of the sea floor, followed by the red clay (38%) and siliceous oozes (14%) (Fig. 16-24). As we have noticed (p. 416, Mindanao Trough) these sediments are not found near land, even in extremely deep water, because their characteristic materials are masked by a deluge of terrigenous debris. According to recent compilations, the calcareous oozes (812 samples) have been dredged from depths of 2,300 to nearly 20,000 feet, with the average less than 12,000 feet, the siliceous oozes (only 37 samples) from depths between 3,400 and 26,500 feet (average about 14,000), while the red clay (126 samples) has nowhere been found at depths less than 13,000 feet and extends to as much as 27,000 feet.

Calcareous Oozes   Calcareous oozes cover most of the floor of the Atlantic and Indian oceans and of the equatorial and southern Pacific as far as latitude 50° S. They consist largely of remains of planktonic organisms having calcium carbonate shells. Shells of *Foraminifera* and plates from calcareous algae are most abundant (Fig. 16-25). Minute particles of clay and subordinate siliceous shells are present. As noted above, the calcareous oozes are, on the average, the shallowest of the pelagic sediments. They are considered to be forming where, though some terrigenous clays are concurrently settling, the calcareous plankton can rain down on the ocean floor

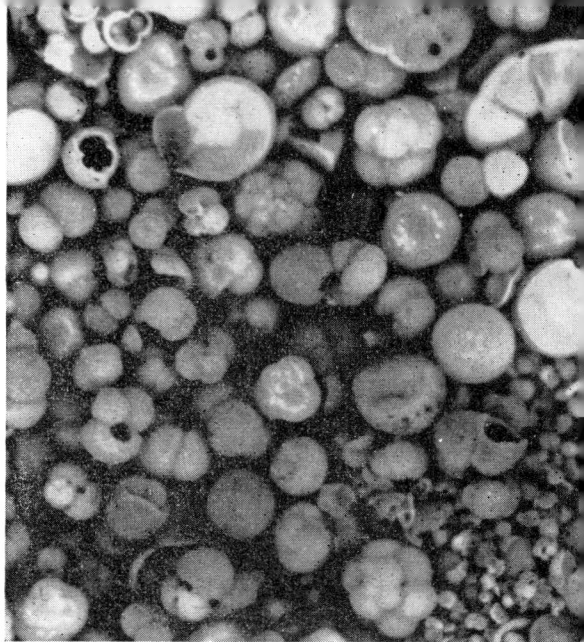

FIGURE 16-25. *Foraminiferal ooze dredged from a depth of 9,900 feet in the Gulf of Mexico. Enlarged about 15 times. (Courtesy of The Humble Oil and Refining Company.)*

in sufficient quantity to mask the clay. The relatively high content of carbon dioxide below the zone of photosynthesis (where it is largely consumed by plants) makes deep-sea water a slow solvent of the carbonate shells. The ooze, nevertheless, accumulates because the falling material buries the underlying sediments and protects them, for the solution is slow and the interstitial water diffuses so sluggishly that it soon becomes saturated with carbonate.

Red Clay   Red clay covers most of the deep-ocean floor. Most of it is too fine grained to be identified microscopically, but X-ray study has shown that the chief minerals are quartz, mica, and several kinds of clay minerals. All are generally stained red by iron rust. None of the truly oceanic islands has any quartz rocks, and the clay minerals are those found in soils formed on land, hence much of the material must have been derived from the continents. Volcanic dust may also be an important source. The red clays contain a few siliceous organisms—diatoms or radiolaria—and also a few of those so prominent in calcareous oozes. In fact, all three types of pelagic deposits intergrade, and their distinctions are arbitrary.

It has been noticed that red clay accumulates either where few calcareous planktonic organisms can thrive (in high latitudes) or where such organisms must sink through vast depths to reach bottom. The deep water is cold and carries considerable carbon dioxide in solution. Thus most of the calcareous materials dissolve during sinking, and not enough reach the bottom to mask the clay that is everywhere accumulating. There is little organic matter, so that the oxygen in the water can combine with iron to give the characteristic red hue. Manganese oxide is abundant in some of the red clay, and its deposition is not fully understood, but it is thought to be somehow related to the oxidizing condition at the bottom.

Siliceous Oozes   Siliceous oozes carpet much of the polar seas and an equatorial strip across most of the eastern Pacific. The equatorial oozes are largely composed of shells of radiolaria, minute single-celled animals, whereas those of high latitudes are chiefly shells of the single-celled plants, the diatoms. Radiolaria flourish in the nutritious water of the equatorial drift. In high latitudes the mixing of deep water with cold surface water and the excessively deep turbulence caused by the prevailing westerlies furnish abundant food for plankton. Furthermore, the waters are cold and rich in carbon dioxide, so that most limy shells are dissolved before reaching bottom. Diatoms, therefore, accumulate here so abundantly as to mask other constituents of the ooze.

Pelagic Sediments in the Geologic Record   Claystone deposits so similar to pelagic red clays (manganese nodules, very fine grain, rich in iron, with fragmental foraminifera and other features) that some geologists consider them deep-sea deposits, have been described from the Alps, from Barbados in the West Indies, and from Timor in the East Indies. These rocks are so intimately associated with coarse clastics and rocks with other shallow-water characteristics, however, that a deep-sea origin seems doubtful, though not impossible. The Cretaceous chalk of England and France was at one time also considered a consolidated deep-sea ooze. Its fossils are largely shallow-water forms, however, and it is now thought that the chalk was probably deposited in clear water at depths not exceeding a few hundred feet.

The Cretaceous rocks of the Caspian region include a great thickness of coccolith-bearing chalks, associated with narrow zones of terrigenous ("blue") mud shales, and, apparently, shallow benthic fossils are lacking. Whether or not these particular beds do, indeed, represent a fossil ooze, it seems certain that deep-sea sediments are extremely rare on land.

Deposition of Limestone and Dolomite

Limestones and dolomites make up about 20 per cent of the sedimentary rocks. Today, as we have just noted, calcareous oozes cover nearly half the ocean bottom. These pelagic oozes, however, are only subordinately, if at all, represented on land, and certainly most limestones have been formed in shallow water, as is shown by their fossils. Among modern sediments we find many analogs to these shallow marine limestones but not to the dolomites.

Chemical Conditions Affecting Calcite Deposition.   Calcite is among the most soluble of the common minerals. Though only slowly soluble in pure water, it readily dissolves in the presence of carbon dioxide (Chap. 6).

The solubility of calcite in natural sea water is difficult to determine because of the interference of so many other dissolved ions. Despite generations of chemical work, recent studies still differ by a factor of 6: according to one, sea water in contact with air, and both at a temperature of 68° F., is about 6-fold oversaturated; according to another, it is about saturated. The latter result seems far more credible, for oversaturation could hardly continue in a sea constantly stirred and with many calcite shells available as nuclei of precipitation.

Despite the complexity of the problem, all chemists agree that the warming of sea water, which consequently decreases its carbon dioxide content and, usually, increases salinity, is favorable to precipitation of calcite. On the other hand, decomposition of organic matter, which yields carbon dioxide, increases the solubility of calcite. As we have seen, the composition of the pelagic sediments suggests that the great bulk of the ocean is slightly undersaturated, but that the warm surface waters of low latitudes are nearly or quite saturated with calcite.

Life processes affect marine deposition and solution of calcite by modifying the carbon dioxide content of the water. Many organisms secrete shells of calcite or aragonite (a different crystalline form of calcium carbonate). Photosynthesis by benthic and pelagic plants extracts carbon dioxide from the water; in the light zone of the warm tropical seas the water is practically saturated. Over the shallow Bahama Banks (Appendix IV), many processes unite to reduce the carbon dioxide content: The water is warmed as it flows over the bank to replenish that lost by evaporation; flourishing plants withdraw much more carbon dioxide; bacteria in the muds include species that turn nitrogen into ammonia. Ammonia, being a weak base, neutralizes some of the free hydrogen ions in the water and thus further reduces the solubility of calcium carbonate. As a result of all these processes, the water is saturated, and fine needles of calcium carbonate are constantly being formed and settling to the bottom as lime muds. It seems likely that some of the extremely fine-grained limestones of the geologic past were formed similarly.

Many lime-secreting plants and animals flourish in the littoral zone of tropical seas. Shells and their broken fragments are so abundant in many places as to make up whole beaches. They also accumulate in shallow offshore waters and are slowly cemented to limestone.

Coral Reefs Conspicuous among calcareous accumulations are the coral reefs. They are confined to waters at least as warm as 68° F., for reef-building corals do not thrive in colder water. Not only must the water be warm, but it must also have normal salinity and be nearly free from silt; coral reefs are therefore absent near the mouths of muddy rivers. The coral animal, anchored in his limy case, depends upon food brought to him by the waves and currents and therefore thrives better on the windward and offshore sides of the reefs. Many species of corals live on the reefs, but reef-builders will not grow below a depth of about 150 feet nor much above low-tide level. Because of their narrow depth range, coral reefs are delicate markers of crustal movement, as noted in Chapter 9.

Though corals form the conspicuous framework of the reefs, algal remains bind them together, and many other calcareous organisms such as mollusks, worms, foraminifera, and a host of others contribute to the reef and may constitute far more of its bulk than coral.

Coral reefs are of three kinds: *fringing reefs, barrier reefs,* and *atolls.* Fringing reefs are confined to the very border of the land. They may be thousands of feet wide, but most are only a few score feet. In the aggregate, however, they are undoubtedly much bulkier than the more spectacular and individually larger barrier reefs and atolls.

The Great Barrier Reef of Australia roughly parallels the Queensland Coast at distances ranging from 25 to nearly 200 miles and extends for 1,200 miles south from Torres Strait. The barrier reef is separated from shore by lagoons which may be shallow but are in places hundreds of feet deep.

Many of the volcanic islands of the Pacific and Indian oceans are surrounded by barriers (Fig. 16-26). In atolls there is no central island, only a ringlike reef enclosing a central lagoon. The enclosed lagoon is dotted with

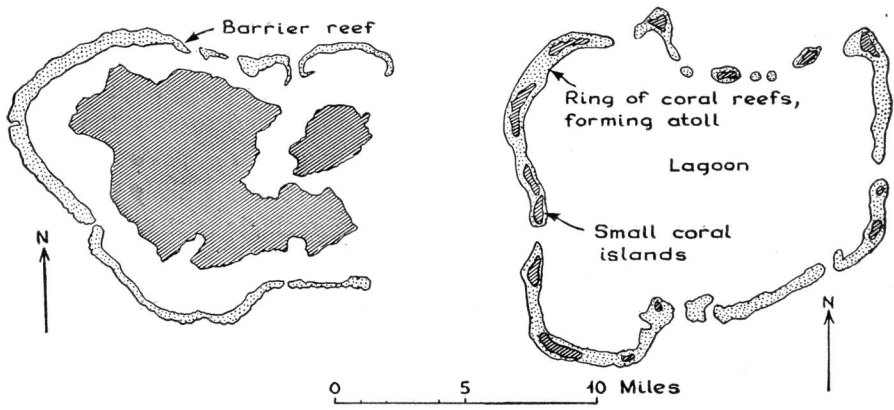

FIGURE 16-26.   Left, *the barrier reef of Vanikoro Island, Caroline Archipelago;* right, *Peros Banhos Atoll, Chagos Archipelago. (After Charles Darwin, 1842.)*

isolated coral heads and floored with algal mud and some sand derived from broken reef fragments hurled over by the waves.

The Origin of Atolls. The origin of barrier reefs and atolls has fascinated students for more than a century, since Darwin's "Voyage of the 'Beagle'." Darwin considered fringing reefs, barrier reefs, and atolls to form a sequence (Fig. 16-27). He thought that if an island fringed by a reef subsided slowly, the reef would grow upward essentially in place, keeping pace with the subsidence to become a barrier reef, and when subsidence had drowned the central island, an atoll. In support of this, it has been pointed out that the shore forms of many islands with fringing reefs are indented like those that might be developed by submergence of a stream-dissected mountain. Furthermore, the outer slopes of many atolls and reefs descend steeply to great depths, and at least two borings on reefs show their structure to be chiefly coralline to depths of many hundred feet.

An alternate hypothesis was suggested by the American geologist Daly in 1910. Daly noticed that charts of the lagoons of many barrier reefs and atolls show a remarkably uniform depth, ranging from about 150 to 250 feet. He also noted that some of the higher peaks of the Hawaiian Islands had been glaciated, presumably during the Pleistocene, and inferred that the water along the shore below must have been too cool for coral growth at that time. If so, the present fringing reefs of Hawaii must all date from postglacial

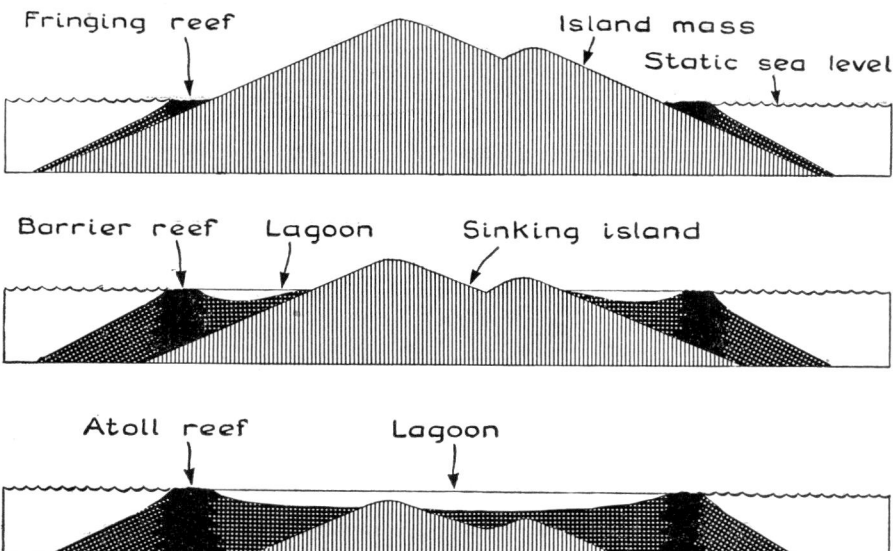

FIGURE 16-27. *Cross sections showing three stages in the origin of an atoll, according to the sinking-island hypothesis. (After Charles Darwin, 1889.)*

time. He then computed the volume of sea water that must have been locked up in the ice sheets during their greatest expansion (Chap. 13). From this he inferred that sea level was then lowered about 300 feet beneath its present stand. If so, the water in which the corals now flourish must have been not only colder, but also siltier, because the waves would be working on unconsolidated sediments of the sea floor. Therefore, Daly concluded, the reefs would not have been able to withstand the attack of the sea—they should have been planed off at a level appropriate to the lowered sea of the time. When the ice sheets melted, and sea level gradually rose, the surviving corals would find smoothly planed banks to colonize. According to this theory, then, the atolls and barrier reefs have been built on smoothly leveled surfaces of abrasion as the water gradually rose. No subsidence of the land is needed to explain the lagoons. The present lagoons vary in depth because of postglacial sedimentation, the smaller lagoons are shallower than the larger because their areas are proportionately less, relative to the reefs over which sediment is washed during storms. Banks and sea mounts that have not been colonized by reef-builders since the rise of sea level are simply the planed-off islands of preglacial time (Fig. 16-28).

Daly's theory fails to explain the indented shores of many islands within

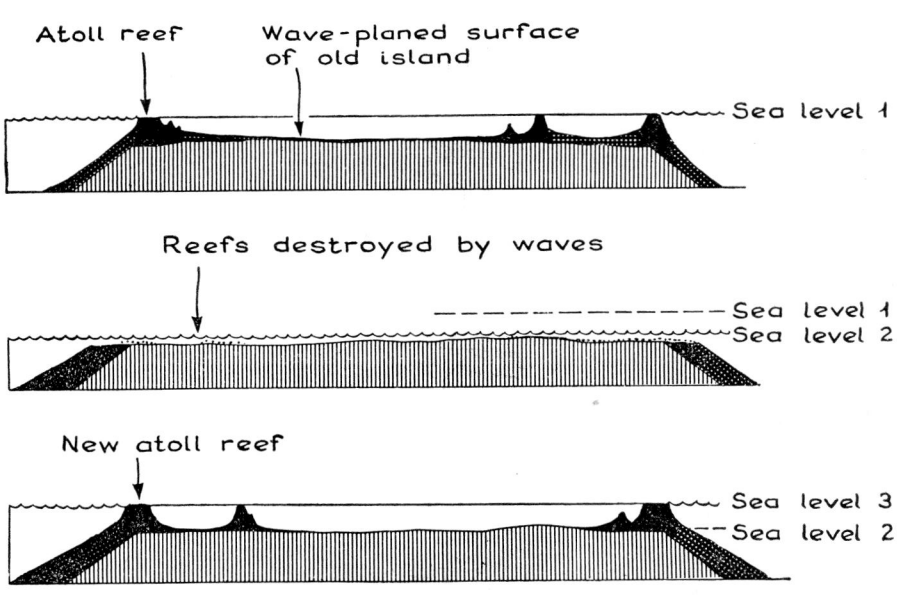

FIGURE 16-28. *Cross sections showing three stages in the origin of an atoll, according to the glacier-control hypothesis. (After R. A. Daly,* The Floor of the Ocean, *University of North Carolina Press, 1942.)*

barrier reefs, for the relief of the valleys and ridges is far greater than can be explained by a mere 300-foot drop in sea level. On the other hand, it does satisfactorily explain the uniform depths of lagoons. Probably both Darwin and Daly are partly correct in explaining the origin of fringing reefs and atolls, for certainly not all parts of the ocean floor have had the same history, even since the Pleistocene. For instance, we know of both drowned reefs with perfectly preserved atoll shapes near the Philippines and of greatly elevated reefs in Fiji and Samoa.

Dolomites. Dolomites offer one of the greatest of geological puzzles. Very few organisms secrete shells with appreciable magnesium carbonate content, and in the few that do, the ratio of magnesium carbonate to calcium carbonate is far short of that in dolomite. We know of no deposits of inorganic dolomite on the present sea floor. Yet dolomites, though practically absent among the younger rocks, are both abundant and widespread among Mesozoic and older rocks.

Locally, around certain igneous intrusions, limestone has been altered over considerable areas to dolomite. This dolomite shows, however, definite association with faults and fissures and is clearly a product of contact metamorphism. Such an origin is not possible for the great sheets of early Paleozoic dolomite that can be traced for scores or even hundreds of miles, some of them interbedded in the closest association with unaltered limestones.

A possible clue to the origin of these widespread dolomites comes from a boring on the atoll of Funafuti, in the Ellice Islands, not far from Tarawa of World War II fame. In this boring, which reached a depth of about 1,100 feet, all the material greatly resembled that at the surface—that is, coral heads, algal deposits, and shells of many other animals. However, although the structures of these organic remains were well preserved, the composition of the material at depth differed notably from that at the surface. The content of magnesium carbonate irregularly increased with depth, and considerable dolomite was present near the bottom. It was inferred that part of the calcium in the calcite shells had been replaced by magnesium ions from the sea water, which, of course, thoroughly soaks the atoll structure. Magnesium is five times as abundant as calcium in sea water. This hypothesis may be correct, but, if so, other factors are locally involved that prevent its occurrence everywhere, for a core hole drilled on Bikini atoll to a depth of 2,200 feet found no dolomite or any significant increase in magnesium content, even though the lower beds penetrated had fossils of Miocene age. These beds must have been in essentially continuous contact with sea water for more than 15 million years. The details of the origin of dolomite offer some of the many geologic problems for the future.

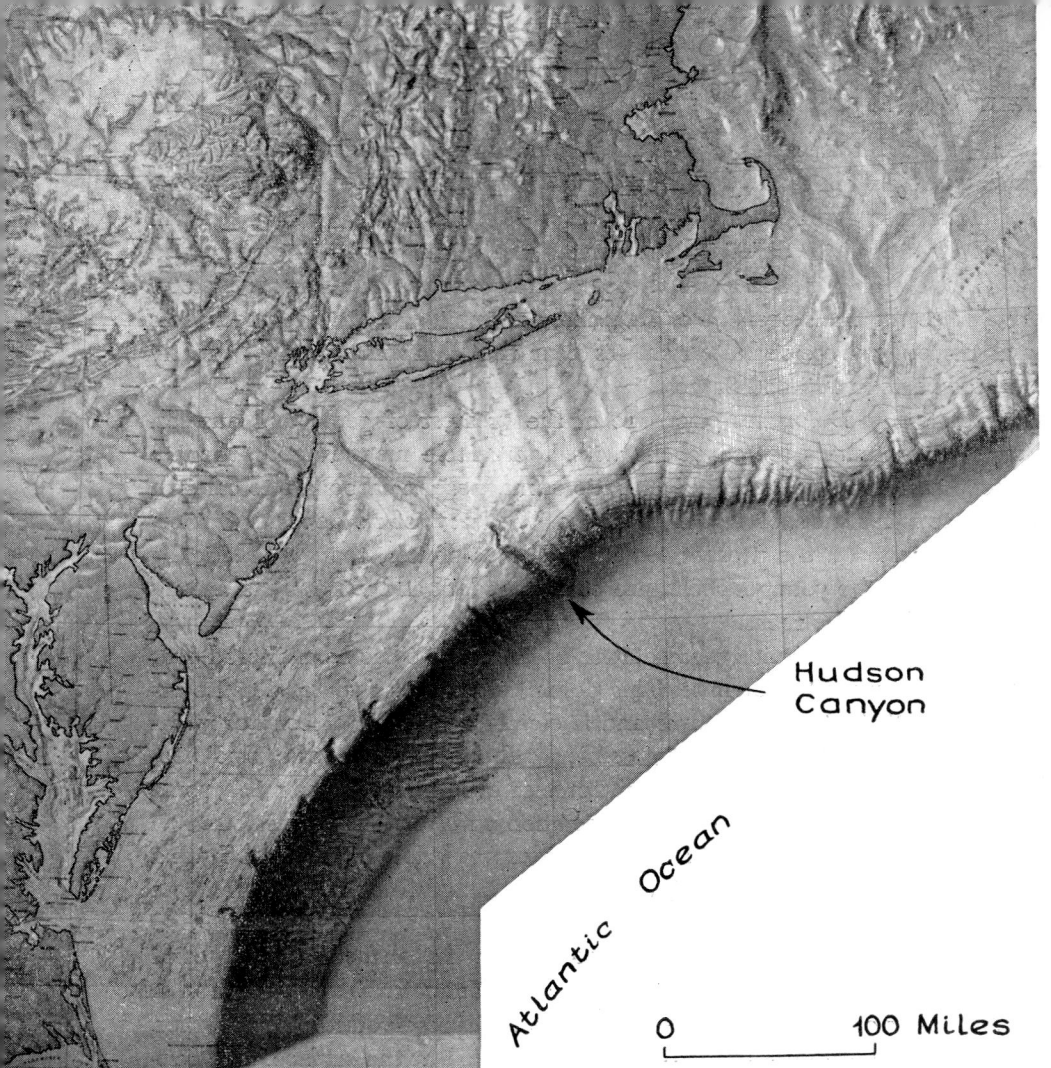

Hudson
Canyon

Atlantic Ocean

0   100 Miles

FIGURE 16-29. *Relief model showing the land and submarine topography of the northeastern United States. Submarine canyons are abundant on the steep continental slope (partly in shadow). The irregularly ridged submarine topography in the upper right corner is submerged glacial topography. (Courtesy of the Aero Service Corporation, Philadelphia.)*

Submarine Canyons

Terrigenous sediments are mostly deposited nearshore, as shown by the great thicknesses—measured in thousands to tens of thousands of feet—of Cenozoic and Cretaceous shale and sandstone penetrated by many coastal wells, and by the fineness and thinness of the clastic layers in deep-sea deposits. The continental shelves appear to be primarily constructional terraces, built up through the later part of geologic time; and, secondarily, erosional features cut on rocks of varied kinds and ages. We see in this combination of deposition and marine erosion a principal reason for the upper dominant level at the earth's surface noted in Figure 2-9.

Submarine erosion may not, however, be limited by wave base. Since 1890 increasing knowledge of the sea floor has revealed more and more signs of deeper erosion, notably, valleys cut sharply into the shelf. Figure 16-29 shows many such canyons indenting the margin of the shelf off the northeastern United States.

Note that the only river mouth clearly connected with a submarine canyon is the Hudson's. This exceptional trench extends from the estuary of the Hudson River almost across the shelf and, then, after fading out on the shelf, reappears as a deep trench in the continental slope, extending clear to the floor of the Atlantic. Its deep-sea part is not well shown in Figure 16-29.

Other stretches of continental shelf, notably those bordering California, Alaska, and southwestern Europe, are cut by similar gorges. Most submarine canyons are short and steep features of the continental slope, heading far from shore except where the shelf is very narrow or interrupted by deep basins, as off southern California. A few canyons are 70 to 150 miles long and 3,000 to 5,000 feet deeper than the adjacent sea floor, thus comparable to the Grand Canyon in size. Among the great submarine canyons are the Monterey Canyon off California that heads near the mouth of Salinas River, the Congo Canyon extending into the Atlantic from the mouth of the Congo River, and the Nazaré Canyon off Portugal that heads far from any existing river mouth.

Monterey Canyon's long profile is shown in Figure 16-30, together with that of the Salinas River. The irregularities in the submarine profile may be caused by errors in the soundings, but the general form is trustworthy. The average grade of the canyon bottom is almost 4° for the first 40 miles from the shore, that of the lower Salinas River less than 0.1°. The submarine canyon is probably cut almost entirely in Miocene, Pliocene, and Pleistocene sediments, though near mid-course granite has been found on one wall. The tributary Carmel Canyon may be excavated chiefly in sheared granite. The head of Monterey Canyon is just outside the beach zone, apparently cut into latest Pleistocene or even recent marine and river deposits.

In general, submarine canyons are marked by steep gradients, roughly concave long profiles, few tributaries, and they head within a few hundred feet of sea level. The relatively small number of canyons whose heads are near the mouths of land rivers include most of the very large examples. The canyon mouths have not been adequately surveyed. Some are at the levels of local basins at various depths below 1,200 feet, or at the base of the steep continental slope 7,000 to 11,000 feet subsea, or, in the case of Hudson Canyon on the deep-sea floor, at even greater depth. The walls are composed chiefly of the younger sedimentary rocks, in part, fairly well consolidated.

Hypothesis of Subaerial Origin   The many hypotheses advanced to account for submarine canyons belong to two groups, the subaerial and the submarine. The principal subaerial hypothesis makes use of the sea level lowering during the Pleistocene glacial epochs, discussed in Chapter 13. The question here is the amount of water removed from the oceans and locked up in glacial ice. Although the areal extent of the Pleistocene glaciers is fairly well known, their thickness is uncertain, so estimates of the ice's volume vary.

Perhaps the lowering of sea level can be determined more easily from direct topographic evidence. As sea level is both the locus of wave erosion and the base level for stream erosion, there would be two sets of topographic effects. After a lowering of sea level, wave erosion would have had new and weak materials on which to work. Wave-cut terraces would therefore be ex pected on the continental slope. All trunk rivers would have been rejuvenated. The Mississippi, Colorado, Yangtze, Ganges, Indus, and La Plata would have been among the great rivers especially effective in cutting valleys into the newly exposed weak sediments of their lower courses. Even the Salinas River (Fig. 16-30) could have cleaned out its lower valley with ease for the first 3,000 feet of depth, and with little difficulty down to 8,000 feet subsea. The valleys so excavated might have been filled with continental or marine sediments as base level rose again, but the deep fill would then tell

FIGURE 16-30.   *Profile of Salinas River, Monterey submarine canyon, and its tributary, Carmel Canyon. The vertical scale is exaggerated 20 times.*

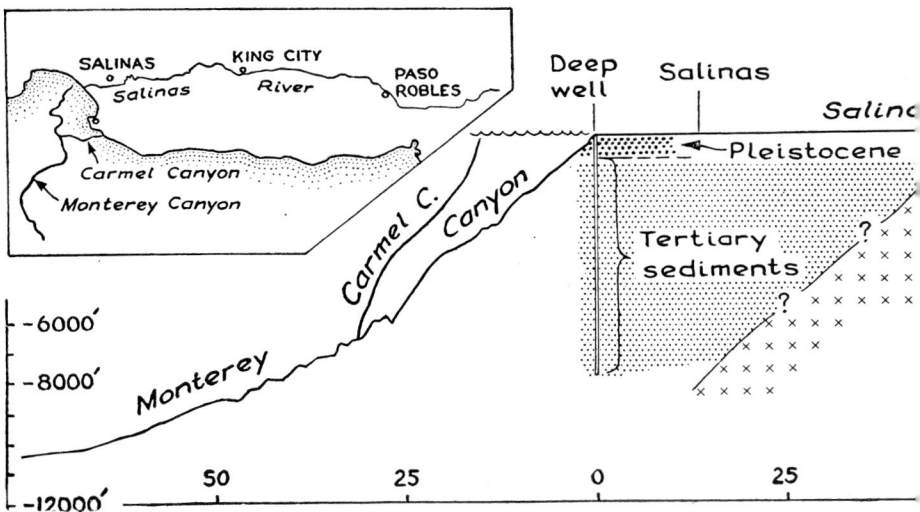

the tale. The long profile of the part of the canyon still unfilled should pass beneath the base of the fill in a smooth curve. This smooth profile curve might extend inland 1,000 miles for the Mississippi and perhaps 50 miles for the Salinas River.

What are the actual topographic facts? Terraces submerged several thousand feet may have been detected off Angola, west central Africa, but they have not been found elsewhere. Two isolated sea mounts 600 and 800 miles off the California coast have flat tops 1,700 and 2,400 feet below sea level. There is no known example of a smooth long profile with a 6,000-foot, 7,000-foot, or other similar deep base level, extending inland along an estuary or along the base of Pleistocene alluvial or marine fill, though the Congo may be a possible imperfect example. On the other hand, all the topographic features that would be caused by submergence have been found, related to a base level only a few hundred feet below present sea level. We have described in the previous section of this chapter the floors of atolls 200 or 300 feet subsea, and submerged drainage systems, deltas, and even beaches at about the same depths. Valleys floored with Pleistocene or Recent alluvium 200 to 600 feet deep are known along many coasts (cf. Fig. 16-30) and may be generally distributed. After allowances for local deformations and for the depths of great rivers, the evidence seems to indicate that the lowest Pleistocene sea levels were about 300 feet below the present one, in good accord with reasonable estimates of the volume of the ice.

Another effect of a theoretical drop of sea level by 6,000 or 7,000 feet deserves mention. The volume of the ocean would be reduced about one-half, with a corresponding doubling of the salinity. The more sensitive forms of marine life, including many shelled animals, would be destroyed. However,

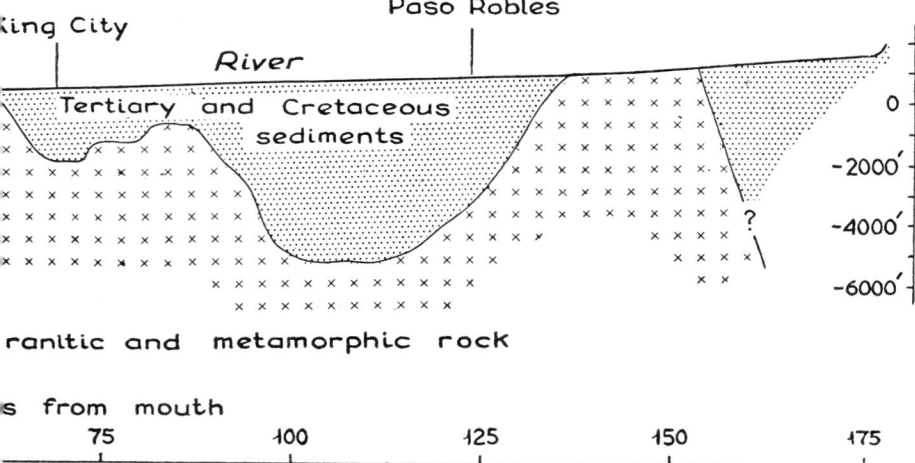

only a few kinds of shellfish became extinct during the Pleistocene, and the number of these extinctions appears to be merely normal, as it corresponds nicely with estimates of the period's length based on other grounds. Apparently, not one of the corollaries of this subaerial hypothesis is actually true.

Hypothesis of Submarine Formation by Density Currents   Turbid water is denser than clear water. A turbid water mass moves along a sloping bottom as a density current. These currents have been studied experimentally, especially by Kuenen of Holland. The motive power of such a current is gravitational, of course, and is measured by the difference between its density and that of the main water body, and by the slope. Since the difference in density is usually much less than that between water and air, a greater slope is required to produce a given velocity. Even boulders, however, if suspended in a muddy fluid, move rapidly down 3° or 4° slopes.

The best historical evidence for erosion by currents comes from Swiss lakes. The subaqueous flow of 1887 in the Lake of Zug has been described in Chapter 11 (see especially Fig. 11-16). The mean gradient of 4° is similar to that of many submarine canyons, and is much higher than most stream gradients on land. The largest trench in a Swiss lake is that cutting the delta of the upper Rhone in Lake Geneva. American reservoirs, such as Lake Mead on the Colorado River, also have their density currents, which travel tens of miles along the lake floor.

In a fresh-water lake a density current sinks to the bottom if its density exceeds 1. In the sea the density must be greater than 1.025, the value for clear sea water. Bottom density currents in the sea are obviously hard to discover. The evidence for their effectiveness in cutting canyons must be mostly circumstantial. We would expect steep gradients; canyon heads just below sea level; mouths at various levels, perhaps especially at the foot of the continental slope; and a distribution of canyons determined by the sources and movements of abundant supplies of sediments. A large muddy river would be one good source of sediment, the meeting place between two longshore currents another, the outer edge of the continental shelf a third—especially good during the Pleistocene stages of lowered sea level and abundant debris from melting glaciers. Once started, erosion of the canyons themselves would furnish additional sediment.

If density currents cut deep submarine canyons, they must also be responsible for some marine deposits, and ancient deposits of this sort might, after elevation and erosion, be exposed to our view. Just such contorted and tumbled, medium-to-coarse fragmental deposits have been found in Italian and Californian Cenozoic strata and are less certainly represented in many other formations of all eras, including the pre-Cambrian.

Nearly all the known facts about submarine canyons seem to fit the density current hypothesis. Nevertheless, two or three of the best-informed American students of the subject still favor a modified subaerial hypothesis. One modification involves deformation after canyon formation. The canyons are thought to have been bowed down as the continents spread laterally under their own weight. The resulting long profiles would be convex. No such bowing is suggested by the data shown in Figure 16-30.

We cannot feel really satisfied with any hypothesis yet; more information is needed. Can hard rocks be eroded by density currents? What is the form of the deep-sea floor? What is the significance of the flat-topped sea mounts? Are there terraces at or near the base of the continental slope? How thick is the sedimentary layer on the deep-sea floor? Do debris deltas extend out from the mouths of submarine canyons? Some of these questions have been answered partially and locally, but many more sonic and seismic surveys are needed, as well as more long cores from the uppermost deep-sea sediments. Perhaps trough experiments, or instruments planted in the canyons themselves, will measure density-current erosion. Ultimately, the solution of this one problem will give us firm ground for further advances in our knowledge of both oceans and rocks.

Facts, Concepts, Terms

HEAT TRANSFER

WIND FRICTION

RELATION BETWEEN EVAPORATION, PRECIPITATION, SALINITY, TEMPERATURE, AND
    DENSITY

WATER MASSES, OCEAN CURRENTS

DEFLECTIONS OF CURRENTS BY THE EARTH'S ROTATION

COASTAL EFFECTS ON CURRENTS

EFFECTS OF CURRENTS ON CLIMATES AND MARINE ENVIRONMENTS

WAVES—DEEP-WATER WAVES, SHALLOW-WATER WAVES, WAVE LENGTH, PERIOD,
    HEIGHT

FACTORS IN WAVE GROWTH AND SPEED

ENERGY DISTRIBUTION IN WAVES

ORBITS OF WATER PARTICLES

WAVE REFRACTION

BREAKERS

WAVE EROSION; HYDRAULIC ACTION, CORROSION

PROFILE OF EQUILIBRIUM

TIDES, WAVE CURRENTS, LONGSHORE DRIFT

COASTAL TOPOGRAPHY AND DEPOSITS

LIFE ZONES; THE ROLE OF PHOTOSYNTHESIS; THE ROLE OF OCEANIC CHEMISTRY

TERRIGENOUS AND PELAGIC SEDIMENTS

THE CHARACTER OF BEACH, LAGOONAL, SHELF, AND DEEP-SEA DEPOSITS

EVIDENCE FOR STRONG CURRENTS AT DEPTH

LIMESTONE DEPOSITION ON BANKS

CORAL REEFS—FRINGING REEFS, BARRIER REEFS, AND ATOLLS

GLACIAL LOWERING

DARWIN'S THEORY OF ATOLLS; DALY'S THEORY

DOLOMITE

DENSITY CURRENTS

SUBMARINE CANYONS

Questions

1. Why is the systematic distribution of sediments in lakes not paralleled by that in the oceans?
2. Why is the salinity of the Arctic Ocean near the surface lower than that of average sea water?
3. If the effect of wave refraction is to concentrate attack on headlands, why is a cuspate form of sand spit such as those off Cape Hatteras not destroyed?
4. If you knew only the most general facts about ocean currents, could you predict the course of the Gulf Stream? How?
5. If a wave 400 feet long has a period of 8 seconds, what will be its velocity in feet per second? In miles per hour?
6. What are the swiftest marine currents, and where are they found?
7. Roughly estimate the ratio between the velocity of a swiftly moving wave far from land, and the velocity of the swiftest marine currents.
8. What would be a reasonable difference in elevation between the seaward and landward edges of the rock floor of an erosional marine terrace 2,000 feet wide? How was this difference calculated?
9. Assuming that density currents build deltalike deposits on the floor of the ocean at the mouths of submarine canyons, how would you expect the form of such a delta to differ from a normal delta? How would its composition and internal structure differ?
10. The Congo submarine canyon heads in the estuary of the Congo River, which

is studded with alluvial islands. The depth of sedimentary fill in the estuary is not known. Draw two long profiles of the base of this fill, (1) assuming that the estuary is the slightly emergent head of a canyon formed beneath the sea, and (2) assuming that the whole submarine canyon was formed subaerially, but has been warped below sea level. Explain.

Suggested Readings

1. Sverdrup, H. U., Johnson, M. W., and Fleming, R. H., *The Oceans, Their Physics, Chemistry, and General Biology*, Prentice-Hall, New York, 1942.
2. Ommanney, F. D., *The Ocean*, Oxford University Press, New York, 1949.
3. Shepard, F. P., *Submarine Geology*, Harper & Bros., New York, 1948.
4. Kuenen, P. H., *Marine Geology*, John Wiley & Sons, Inc., New York, 1950.

17. *Igneous Activity*

Igneous Rocks

IGNEOUS ROCKS, as explained in Chapter 5, congeal from molten magmas generated within the earth. Some, the *volcanic rocks,* have been erupted to the surface; others, the *plutonic rocks,* have solidified underground and are revealed only where unroofed by erosion or penetrated by wells and mines.

From the earliest times men have been interested in the igneous rocks. The spectacular eruptions of Vesuvius, Etna, and other Mediterranean volcanoes strongly impressed the ancients. Our industrial civilization values the ores of gold, silver, copper, lead, uranium, and other elements concentrated near the borders of intrusive igneous masses.

The igneous rocks are wonderfully diverse. Microscopic study reveals thousands of different varieties, and even at a glance many different textures, minerals, and other physical characters are apparent. Yet despite this diversity, two varieties of igneous rocks tremendously preponderate. These are the dark-colored, quartz-free volcanic rock called *basalt* and the light-colored, quartz-bearing intrusive rock called *granite.* Why should this be? Only a partial answer can be given in this chapter, but more evidence will be developed later.

Volcanoes

Volcanoes are mountains or hills formed by eruption of materials from the earth's interior through a central vent, or, more commonly, a group of vents. Most of the material ejected is either liquid *lava* or fragments of partially or completely solidified rock (*pyroclastic debris,* see Chap. 5). Every active volcano also gives off *gas,* chiefly water vapor, often in astounding amounts. During one phase of the 1906 eruption of Vesuvius vast quantities of steam blew from the crater throat as from a gigantic safety valve. The steam jet eroded the vent walls and greatly enlarged the crater.

Shield Volcanoes

One characteristic type of volcano is built of coalescing and overlapping lava streams. The hot lava wells out either from the central conduit or, more commonly, from fissures on the flanks of the cone. It pours down the cone (Fig. 17-2) in thin interlacing streams. The lava is of low viscosity and, before congealing, generally flows out to a gently sloping surface. There is little explosive activity, and pyroclastic material is scarce or absent, so that these volcanoes are low and flat, shaped like an inverted dinner plate in contrast to the steep-sided pyroclastic volcanic cones (Fig. 17-7). Such flat cones of superposed flows are called *shield volcanoes*. Most shield volcanoes are composed of basalt, though a few are andesite. Many contain a few small bodies of rhyolite or obsidian either as intrusive masses or as flows and pyroclastics.

The summit of a typical shield volcano usually contains a conical *crater*, though some bear a much larger depression called a *caldera*, which may be several miles in diameter. In an active cone the crater may contain a lake of molten lava.

Mauna Loa and Kilauea. Despite their gentle slopes, the world's highest and largest volcanoes are shield volcanoes. The Hawaiian Islands are essentially shield volcanoes, some of which rise more than 13,000 feet above the sea. As the nearby Pacific Ocean is more than 15,000 feet deep, some of the volcanoes must be more than 28,000 feet high (Fig. 17-1).

Mauna Loa (Fig. 17-2), on the island of Hawaii, rises nearly 14,000 feet above sea. At the summit is a caldera 2 miles wide and about 1,000 feet deep. In historical times eruptions have come mostly from fissures on the flanks of the cone, or from vents on the caldera floor.

On the southeastern slope of Mauna Loa, about 20 miles from the summit and nearly 9,000 feet lower, lies a second volcano, Kilauea, with a caldera about 1½ miles in diameter. It contains a lava pool that has been crusted over during most of historic time, but in a pit in this crust lies Halemaumau, a pool of molten lava. From time to time the lava rises and floods the floor of the main Kilauea sink. In 1924 it withdrew deep into the crater, and

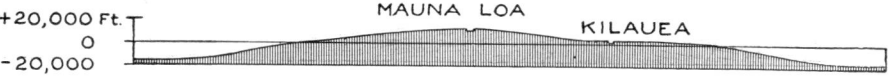

FIGURE 17-1. *Profile of Hawaii. (After H. T. Stearns and G. A. MacDonald, Hawaii Div. of Hydrography, 1946.)*

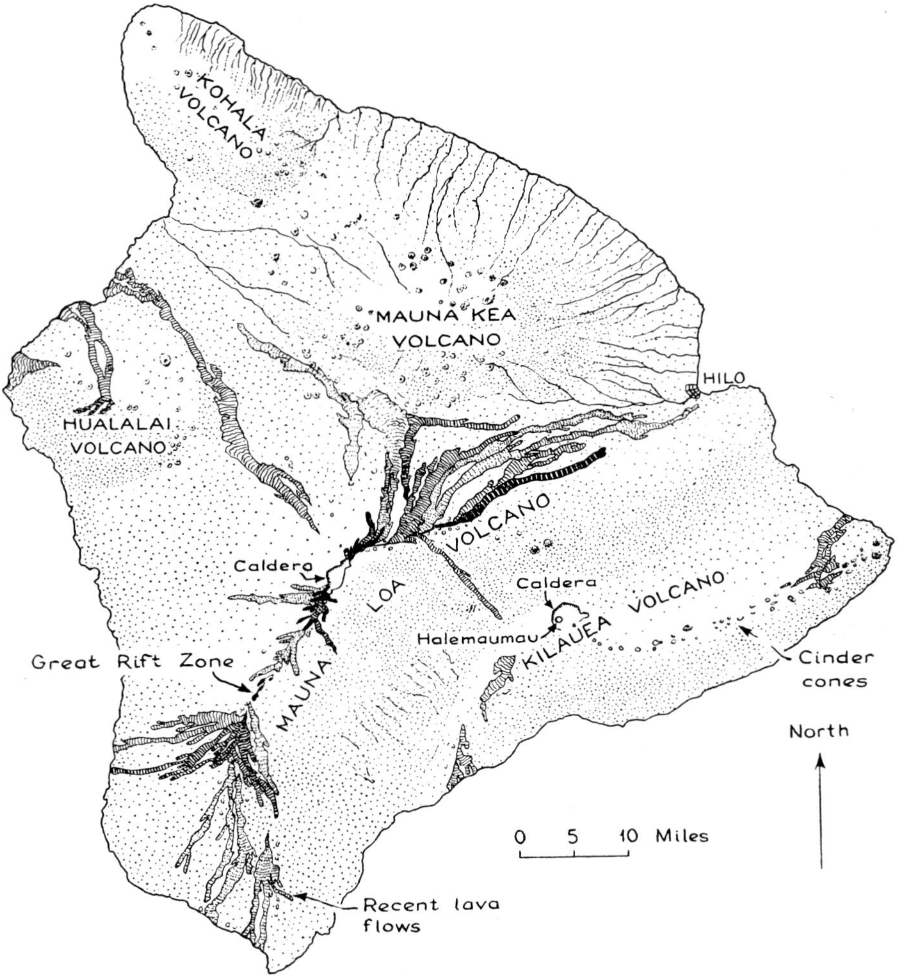

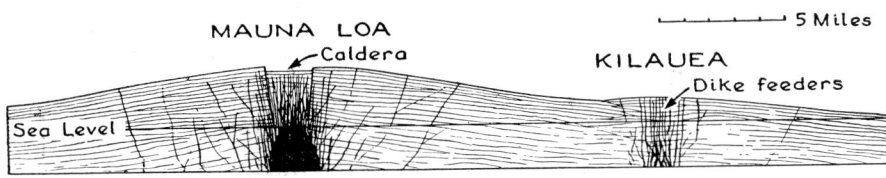

FIGURE 17-2a.   *Generalized map of Hawaii. Note the alignment of cinder cones and lava flows on the fissure zones of Mauna Loa and Kilauea. (After H. T. Stearns and G. A. MacDonald, Hawaii Div. of Hydrography, 1946.)*

FIGURE 17-2b.   *Cross section through Hawaii, showing the concentrations of dike feeders that make up the rift zones. (Same source as Fig. 17-2a.)*

ground water poured into the hole, causing a violent steam explosion. Later the lava lake reappeared.

Surface crusts form on the Halemaumau lava pool but are soon broken up and engulfed by slow convection currents. Gases, largely water vapor, continuously emerge and in places, especially near the center of the pool, they escape so violently that the molten rock boils up in "lava fountains." Some of the gaseous sulfur and hydrogen, present in small amounts, burns when it reaches the surface.

As the lava of Mauna Loa stands nearly 9,000 feet higher than that in Halemaumau, it seems that the two lava columns cannot now be directly connected. Otherwise the lava in Mauna Loa would drain down to the level of Kilauea.

Fissure Eruptions on the Flanks of Shield Volcanoes. Within historic times several lava flows have been added to Mauna Loa (Fig. 17-2). These have emerged, not from the Halemaumau and Mauna Loa "fire pits," as must sometimes happen, but from fissures that opened lower on the shield. Many came from a northeast-trending zone of roughly parallel cracks—the "Great Rift Zone"—crossing the apex of the shield (Fig. 17-2). Some flows are very small; others, such as that of 1855, contain over 4,000 million cubic feet of rock. Hundreds of such lava streams must slowly have built up the massive pile of the shield volcano.

On Mauna Loa the sources of only a few fissure flows can be seen, but younger flows must conceal hundreds of dikes that fed these and older flows.

Innumerable dike feeders are exposed in the deeply eroded shield volcanoes of nearby Oahu.

Radial Dike Swarms. In many eroded volcanoes such dikes radiate from

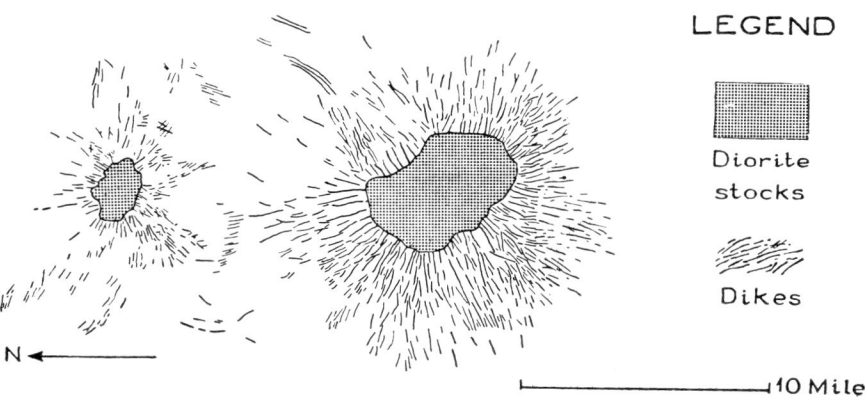

LEGEND

Diorite stocks

Dikes

N ←

10 Miles

FIGURE 17-3. *Geologic map of stocks and dikes, Crazy Mountains, Montana. (After W. H. Weed, U. S. Geological Survey.)*

the central vent. How would the roots of such a volcano appear if erosion had removed all the lavas of the shield, exposing the floor on which it was built? In many regions geologists have mapped isolated, rudely circular masses of igneous rock from which swarms of dikes radiate. Others show swarms of parallel dikes, recalling the Mauna Loa "rift zone." Some retain a few patches of lava, perhaps all that remains of a once-voluminous volcanic shield, but in others deep erosion has removed all traces of any former volcanic pile. Figure 17-3 is a geologic map of two such masses in central Montana.

Parasitic Cinder Cones. Small cinder cones (Fig. 17-4), which are steep-sided accumulations of pyroclastic debris (chiefly bombs and lapilli, Appendix IV) heaped around a central vent form on the surface of many shield volcanoes. These "parasitic" cinder cones are generally about a mile across and 50 to 500 feet high. Mauna Loa has several, and Mount Newberry, a shield volcano in central Oregon, is embellished with over 150. Commonly, several cones are grouped along a definite line, indicating that they formed along a fissure, perhaps while the dike occupying the fissure was congealing.

Calderas. Much larger parasitic cones, composed either of lava or pyroclastic products, or mixtures of both, may also appear near the summit of a shield volcano, particularly along the rim or else on the floor of summit calderas. *Calderas* are large volcanic depressions, steep-sided and more or less circular in form. They are commonly from three to nine miles in diameter, whereas volcanic vents and plugs average less than 1,000 feet across. Although some calderas have been interpreted as pits made by gigantic explosions, it is generally agreed that most have been formed by the subsidence of the top of the volcano into the underlying magma chamber following voluminous eruptions. At many volcanoes the area of subsidence is

FIGURE 17-4. *Cinder cone with lava flow in the middle distance. Lava Butte, Oregon. (Photo by H. A. Coombs.)*

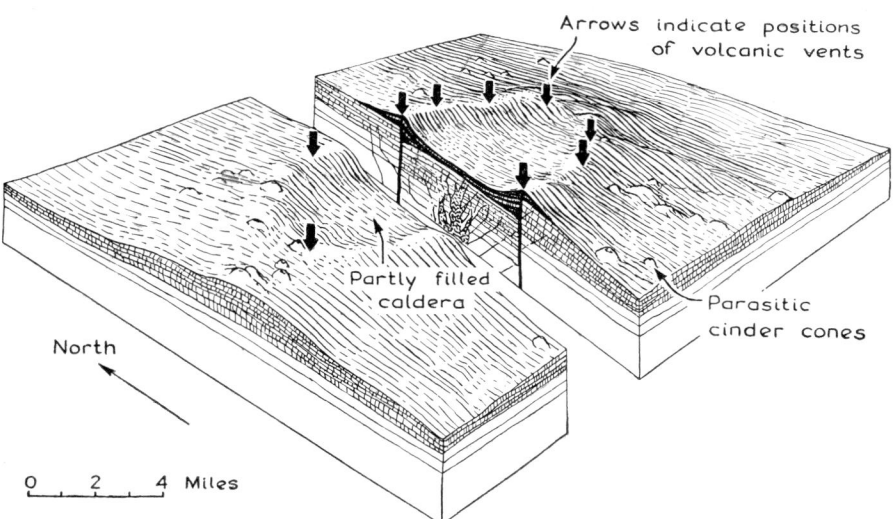

Arrows indicate positions
of volcanic vents

Partly filled
caldera

Parasitic
cinder cones

North

0   2   4 Miles

FIGURE 17-5.   *Relief diagram of Medicine Lake volcano showing the arrangement of vents along the border fracture of an old caldera. The cross section through the caldera shows its collapsed center and the infilling by later lavas. (After C. A. Anderson, 1941.)*

bounded by a roughly circular fault. The rocks on some caldera floors can be matched with those hundreds of feet higher on the walls, thereby proving central subsidence. A few calderas have collapsed so recently that the concentric faults are still visible along their margins, but, more commonly, the actual faults are hidden by talus cones or by lavas and pyroclastics extruded along the rim faults.

Ring Dikes.   As the top of a volcano slowly subsides, magma from the underground chamber may rise along the ringlike bounding fracture and solidify as an arcuate or circular dike called a *ring dike* (Fig. 17-6). Lava from a ring dike may well out and flood the caldera floor or, more often, may feed several parasitic volcanoes along the caldera rim. At Askja volcano, in Iceland, a caldera about 4½ miles in diameter was formed by downsinking along a ring fracture, then several rim volcanoes developed along the margin of the caldera. Basalt flows and scoria erupted from them have partially buried the caldera floor and walls.

Another example is the Medicine Lake volcano of northeastern California (Fig. 17-5). Here, a great andesitic shield volcano is capped by nine steep-sided lava cones arranged in an ellipse. Presumably, this ellipse marks an old caldera rim completely buried by lava and ash from the rim volcanoes.

Eroded volcanic centers tell us still more about ring dikes and calderas

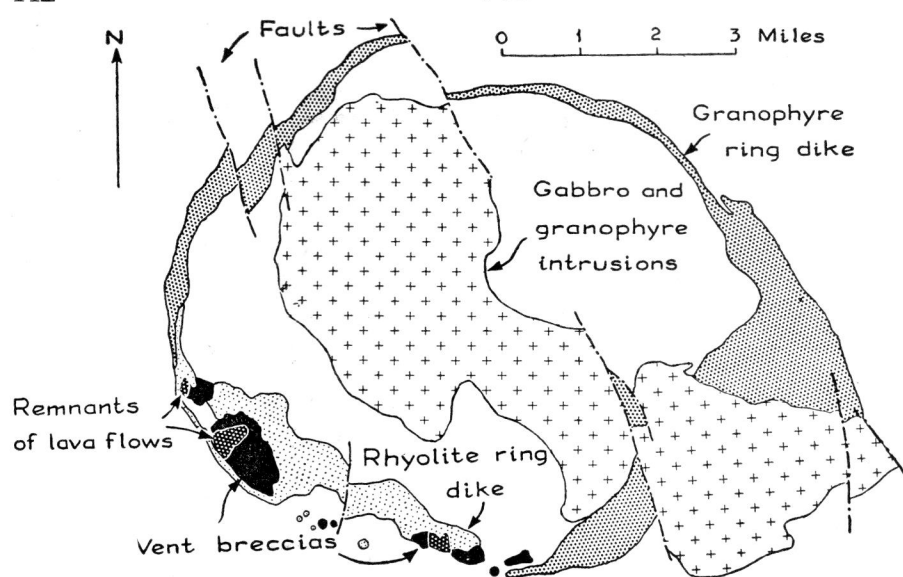

FIGURE 17-6. *Geologic map of an eroded volcanic complex, Slieve Gullion, Ireland.* (*After J. E. Richey, 1932.*)

due to subsidence. The deeply eroded volcanic center of Slieve Gullion in northern Ireland (Fig. 17-6) was probably once surmounted by a volcano very similar in form to that at Medicine Lake. At Slieve Gullion few patches of surface lava have survived and these only because they subsided into the depths of the volcano. Today the most obvious feature is a ring dike of granophyre (a fine-grained variety of granite), about 7 miles in diameter, which encircles a down-dropped block riddled with intrusions of gabbro. The gabbro intrusions are probably a part of the congealed magma which fed the volcano. A second ring dike of aphanitic rhyolite clings closely to the first for about one-fifth of its circumference. But perhaps the most interesting feature disclosed by the geologic mapping is that, at intervals along one side of the complex, the main ring dike grades into or is cut across by pipelike masses of explosion breccia. Obviously, these mark the plugs or vents of explosive volcanoes built along the caldera rim, much like those at Medicine Lake.

Composite Volcanoes

Composite volcanoes differ from shield volcanoes chiefly in their strong explosive activity. The beautiful steep-sided cones of snow-capped Fujiyama in Japan, Mount Rainier in Washington, Vesuvius in Italy, and Cotopaxi

(Fig. 17-7) in Ecuador have been built by a combination of relatively quiet outpouring of lava flows and violent explosions that heaped pyroclastic fragments about the vent. In most of them layers of pyroclastic debris alternate fairly regularly with lava flows. Others are great chaotic piles of pyroclastic rock with relatively little interbedded lava.

A graceful steep-sided cone such as Cotopaxi or Fujiyama appears at first to have little in common with a low flat-topped lava shield like Mauna Loa. Yet there are close similarities. Most of the lava may come from fissures low on the flank of the cone. Subsidence of the top to form large calderas is common. Such subsidence may be along ring fractures as in shield volcanoes, or by the piecemeal crumbling and sinking of a chaotic jumble of fault-bounded blocks into the magma chamber immediately following violent eruptions of pumice lapilli (Fig. 17-8). As in shield volcanoes, large or small parasitic cones may appear. Furthermore, owing to a change from quiet lava outflow to strong explosions, some composite volcanoes have grown directly on the top of old shield volcanoes.

Varieties of Explosive Activity. Explosive activity may take many forms. The Italian volcanoes, Stromboli and Vulcano, which have been observed for many hundreds of years, and three or four other volcanoes that have erupted violently in recent centuries, illustrate the range of activity.

Stromboli At Stromboli, called the "Lighthouse of the Mediterranean,"

FIGURE 17-7. *Cotopaxi volcano, Ecuador. (Photo by courtesy of E. Lewis.)*

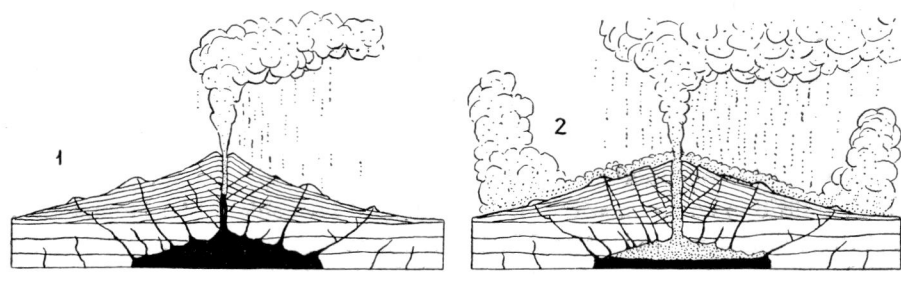

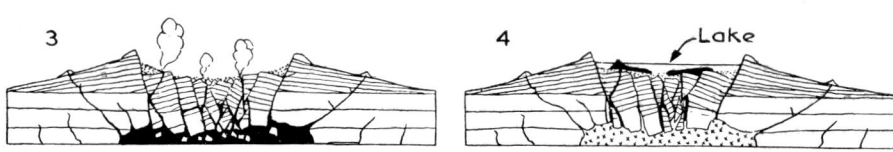

FIGURE 17-8. *The evolution of the caldera at Crater Lake, Oregon. (1) early eruption cloud; (2) great pumice eruptions, with mountain beginning to founder; (3) caldera resulting from collapse into magma chamber; (4) crystallization of magma after minor eruptions. (After Howel Williams, 1942.)*

mild explosions spaced at intervals of a few minutes is the characteristic activity. With each explosion, showers of glowing lava belch from the crater and rain down as red-hot pasty clots. Many are shaped into spindle or tearlike forms (*bombs*) during their flight through the air. Rarely, during longer quiet intervals the lava in the throat crusts over. Such quiet periods are ended by a stronger explosion that blasts the solid crust into fragments and expels it, together with many gas-charged lava clots, onto the sides of the cone. Few of the quiet periods last long enough, however, to allow much magma to congeal.

Strombolian explosions generally build huge cinder cones composed almost entirely of lapilli, bombs, and blocks, but some contain lava flows either from the crater or from fissures (Fig. 17-4).

Cinder cones may be small, like those decorating Mauna Loa or they may grow into towering cones such as Stromboli. Parícutin, the celebrated new volcano in Mexico that first erupted in a cultivated field in 1943, is a typical cinder cone.

Vulcano   The mountain called Vulcano is another Italian composite cone. Unlike the almost continuously active Stromboli, Vulcano has much stronger and more irregularly spaced explosions. Vulcano expels mostly solid material. During the long quiet periods the magma in the throat freezes to considerable depths, then is shattered and blown out by explosions to form great seething eruption clouds of mainly solid fragments. These are mixed with some clots of pasty glowing lava from farther down in the throat. These black clouds are often followed by tremendous masses of steam that boil from the crater and hang as white, swirling clouds above the cone. Some of the activity at Parícutin has been similar to that at Vulcano.

Katmai The strongest known volcanic explosions have occurred in long-dormant composite volcanoes. Some of these volcanoes seemed dead, not having erupted within historic times. In many the cones had been partially destroyed by erosion.

In 1912 Mount Katmai in Alaska, not previously known to be active, exploded with great violence, changing its steep-sided but deeply eroded cone into a shattered irregular-topped mountain capped by a pumice-filled basin. Bits of pumice and broken lava rained down from the eruption cloud, burying forests on Kodiak Island 60 miles away to a depth of 10 feet or more.

Tamboro and Krakatoa A tremendous explosion at the volcano Tamboro, on an island in the East Indies, occurred in 1815. It is estimated that over 30, and possibly as much as 50, cubic miles of material was shattered into fragments and blown into the air, scattering a thick coating of pyroclastic debris over a radius of 200 miles and producing heavy dust falls 500 miles and more from the volcano.

Krakatoa, in the strait between Java and Sumatra, exploded violently in 1883, blowing about a cubic mile of shattered rock into the air. In prehistoric time a large composite volcano had been built up and had later subsided, after great eruptions of pumice, leaving only an arc of islands to mark the rim of the caldera. (Compare with Crater Lake volcano, Fig. 17-8.) Subsequent eruptions built several parasitic cones within the ring; of these the highest was Krakatoa, which grew to more than 2,600 feet. Then, in 1883, after emission of some steam and dust, a violent steam explosion blew nearly all the islands to bits and blasted out a great hole reaching 1,000 feet below sea level. The eruption cloud from this gigantic explosion rose 17 miles upward and rained millions of tons of angular rock fragments, dust, and pumice lapilli over the sea. Indeed, the finer ash remained suspended in the atmosphere for years and was wafted all over the earth.

Fortunately, few people lived within the killing range of the eruption cloud, but sea waves as high as 100 feet were generated, which overwhelmed the low-lying coasts of Java and Sumatra, killing thousands of persons and causing tremendous damage.

Pelée In 1903 Mount Pelée, on the island of Martinique, was strongly active. Fortunately, many stages of this activity were observed by the great French geologist A. Lacroix. Lacroix described how, during the strongest explosions, immense clouds of dust, pumice lapilli, and fragments of red-hot rock shot to great heights, boiling, eddying, and swelling into a gigantic black mushroom of seething particles high above the volcano. Then, slowly, gravity overcame the upward force, the cloud of fragments fell back and, on hitting the steep slopes of the volcano, shot out laterally, rolling down the slope like an avalanche. The mobility shown by these racing hot avalanches is astonishing. Some traveled down the mountain side at more than 60 miles per hour, and, of course, completely devastated everything in their path. Their great mobility is explained by the fact that each bit of pumice in the cloud continuously emits water vapor and other volatile substances; the pressure of these expanding gases buoys up the particles and makes the mixture so mobile that it flows at appalling speeds.

The town of St. Pierre, containing about 28,000 persons, was caught in the path of one of these glowing avalanches. Everyone perished except for one prisoner who was protected by his underground dungeon.

Spines and Domical Protrusions. Steam explosions like those described, though generally smaller and less destructive, may accompany the rise of almost solid lava into the crater. At Mount Pelée such a *domical protrusion* of glassy lava rose into the crater, and, being too viscous to spread laterally, it punched up as a gigantic *spine* (Fig. 17-9), much like a cork emerging from the mouth of a bottle. At intervals steam exploded violently at the edges of the spine, shooting out eruption clouds of gas-charged pumice fragments, ash, and even large blocks. These clouds did not rise high but almost immediately began to pour down the sides of the cone.

Spines and domical protrusions seem to be characteristic of waning activity at composite volcanoes. They may rise in the volcanic throat, as at Mount Pelée, or punch up through the lower flanks like the newly formed Showa Shin-zan protrusion at Usu-Yama volcano, Japan. This dome, in 1944,

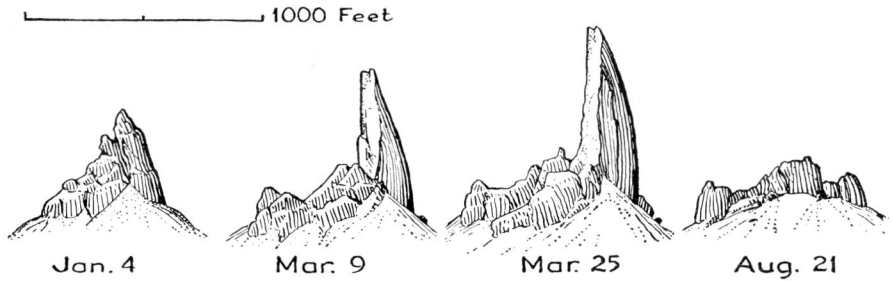

1000 Feet

Jan. 4   Mar. 9   Mar. 25   Aug. 21

FIGURE 17-9. *Four stages in the evolution of the Mount Pelée spine.* (*After A. Lacroix, 1904.*)

pushed up through wooded ground and a cultivated field, lifting and heaving the broken rock and soil aside, and finally emerging as a steaming blocky mass of practically solid lava.

Many domical protrusions are not as viscous as the two examples given. On approaching the surface they swell into rounded masses shaped like an inverted bulb or may even flow down the slopes for a short distance as rounded steep-sided flows. Although these viscous masses of glassy lava are generally associated with composite volcanoes, they locally appear in shield volcanoes, and some even rise independently to form isolated volcanoes such as the Mono "Craters" in California.

Most domical protrusions consist of obsidian. As they swell into bulbs at the surface, the chilled crust breaks into great angular blocks, mantling the edge of the dome in talus.

Pyroclastic Flows. Glassy lava may be so highly charged with gases, however, that, instead of rising as a dome or spine, it froths violently into bits of pumice on approaching the surface. This violent frothing generally produces seething eruption clouds like those of Mount Pelée. At some volcanoes the gas-charged glassy lava boils more quietly from the crater in a pumiceous froth that breaks into small gas-emitting particles. These discharge over the lip of the crater and down the sides of the volcano in hot, highly mobile *pyroclastic flows* ("sand flows"). Some sand flows are still so hot when they come to rest (perhaps because of chemical reactions between the contained gases) that the particles weld together again, forming a rock (welded tuff, Appendix IV) that superficially resembles obsidian or rhyolite. Most welded tuffs have boiled out of fissures (dikes) but some come from central vents. Fenner, an American geologist who first comprehensively described this type of eruption, showed that the "sand flows" that rolled down the Valley of Ten Thousand Smokes soon after the Katmai eruption came from fissures on the valley floor. Since his work, many rocks formerly mapped as flows of rhyolite or obsidian have been recognized as welded tuffs, deposited as hot avalanches of pumice fragments.

Volcanic Mudflows. Volcanoes occasionally explode through a crater lake, as did the Javanese volcano Klut in 1919. Or the gas-charged avalanches of pumice may melt snow on the volcano and pour down the slopes as a mudflow, as in the 1877 eruption of Cotapaxi in Ecuador. Eruption clouds may also roll into a river, and the mixture of water and volcanic debris may travel far down the valley. Such deposits are called *volcanic mudflows*. Heavy rains may also wash the loose pyroclastic debris down steep volcanic cones in mudflows. Mudflows are very common in volcanic assemblages of all ages.

FIGURE 17-10. *Plateau basalt flows, near Vantage, Washington. (Photo by courtesy of the Washington Dept. of Conservation and Development.)*

Plateau Basalts

In some of the world's most extensive volcanic areas true volcanoes are rare or absent. Instead, great sheets of lava have welled out through fissures in no way connected with volcanic cones. Such *plateau basalts* cover vast areas in the northwestern United States, western India, the Parana basin of South America, and the Iceland-Northern Ireland-Hebrides region. Old plateau basalts, now folded and metamorphosed, are found in many other areas.

Basalts of Miocene age cover approximately 150,000 square miles on the Columbia River plateau of eastern Washington (Fig. 17-10) and northern Oregon. Slightly younger basalts underlie vast areas in southern Idaho and southeastern Oregon. In western Washington and Oregon Eocene basalts are perhaps even more voluminous. Hundreds of dikes of basalt and dolerite, marking the fissures through which the lavas rose, can be seen in central Washington and in the Blue Mountains of northeastern Oregon. Such floods of basaltic magma are the most extensive of all volcanic deposits.

Pillow Lavas

Basaltic flows in Samoa have been observed to subdivide into irregularly rounded and ellipsoidal masses from several inches to a few feet in diameter on entering the sea. Such closely packed pillow lavas (Fig. 17-11) with peculiar cavernous spaces between them are common the world over. They are commonly interstratified with sedimentary rocks of great thickness. Many of the Eocene basalts of western Washington and Oregon are pillow lavas that were erupted into shallow seas. Each pillow commonly shows a thick rind of glass formed by chilling of the surface by water. Many inter-

FIGURE 17-11. *Pillow lavas, near Ellensburg, Washington.*

stices between the pillows are partly filled with altered fragments of glass, or with mud squirted up by hot water and steam from the sea floor. The Miocene basalts of the Columbia River plateau also contain abundant pillow lavas, but the fossils in the associated sediments show that these flows entered shallow fresh-water lakes.

Intrusive Igneous Masses

Volcanic phenomena require us to infer the presence of masses of magma underground that supplied the lavas and pyroclastic products. We now describe these "root zones" of the volcanoes and also other igneous masses, large and small, that have congealed within the earth's crust but may never have fed volcanoes.

Plugs and Stocks. The pipelike masses of igneous rock that solidify in the conduit of a volcano are called volcanic *plugs* or *necks*. They are commonly rounded or elliptical in outline, but may be quite irregular. They average only about 600 feet in diameter but range up to well over a mile.

Most plugs are composed of porphyritic rock that may be medium-grained in the center, but is commonly chilled to glass or fine-grained rock at the borders. Some are entirely glassy and may have once fed domical protrusions or boiling pyroclastic flows. Many, particularly in their upper parts, are made of explosion breccia. As might be expected from observed volcanoes, many plugs are *composite,* consisting of several kinds of igneous rock, differing in texture and perhaps in composition, emplaced at slightly different times.

The walls of most plugs are nearly vertical, but some are known to flare outward with depth and pass into larger igneous masses, called *stocks,* that are a few miles in diameter. Not all stocks, however, show evidence of any connection with volcanoes; many may never have reached the surface, and have been exposed only by deep erosion of overlying rocks.

Most plugs and stocks are more resistant to erosion than their surroundings; they rise as steep-sided buttes above the surrounding country. In some areas that retain few volcanic rocks today, plugs are thickly clustered, indicating a former volcanic landscape very different from the present one. Over 150 plugs, marking the sites of vanished volcanoes, rise above the plateaus of northwestern New Mexico and nearby Arizona. Over 100 plugs appear in an area of less than 150 square miles near Neuffen, Germany. Most of the steep-sided buttes in the Midland Valley of Scotland, along the Rhine in Germany, and in the John Day basin of Oregon mark the position of former volcanoes.

Dike Swarms.   Dikes (see p. 79 for definition) are among the most common igneous forms. They are associated with volcanoes and plateau basalts, and also with the borders of intrusive masses. Dikes range in size from fillings of microscopic cracks to bodies like the Great Serpentine Dike of Rhodesia, which is over 300 miles long and as much as 7 miles wide.

Many regions contain *dike swarms*. Some dike swarms are obviously related to volcanoes; among them are the linear dike swarm at shallow depth that must be inferred from the rift zone of Mauna Loa (Fig. 17-2), and the radial swarms of Oahu and of the Crazy Mountains in Montana (Fig. 17-3). Others, of which the dike swarm shown in Figure 17-20 is a good example, appear to be offshoots of large plutonic masses deep beneath the surface.

Some large dike swarms appear to have been the feeders for plateau basalts; others for sill complexes, lopoliths (see p. 454), or other igneous masses. One of the most interesting is the great series of dolerite dikes of western Scotland and northern Ireland (Fig. 17-12). The dikes cluster thickly near the major volcanic centers such as Skye, Mull, and Arran, but many, such as the 110-mile-long Cleveland dike of northern England, extend so far they cannot be considered "parasitic" fissures from the volcanic plugs. The map in Figure 17-12 has been greatly generalized from many detailed geologic maps. The reduced scale permits only a few of the largest dikes to be shown; for each line on Figure 17-12 there are scores or even hundreds of dikes visible in the field.

Similar swarms of dolerite dikes are present in areas of widely varied structure and geologic ages. Yet, large areas of the earth's surface show none. It appears that in places deep fissures have cut through the crust to a world-wide source of basalt beneath, and that the basalt has erupted through the fissures in large amounts to produce plateau basalts, sill swarms, or lopoliths.

Sill Swarms.   In areas of flat-lying sedimentary rocks magma may in-

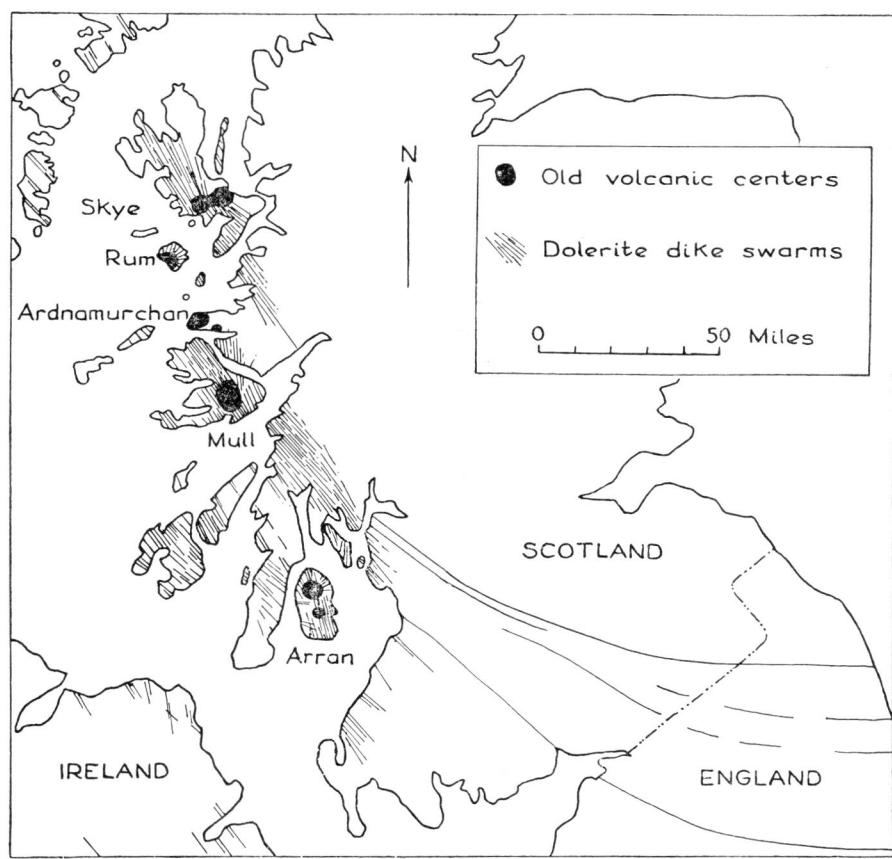

FIGURE 17-12. *The dolerite dike swarms and eroded volcanic centers of Scotland and adjacent parts of England and Ireland. (After J. E. Richey, H. H. Thomas, and others, Geological Survey of Great Britain.)*

sinuate itself between the strata to form flat pancakelike bodies called *sills* (see Chap. 5, especially Figs. 5-2 and 5-3). Sills, like dikes, vary greatly in size, in the kind of rocks composing them, and in their associations with other igneous masses. Sills of dolerite or its coarse-grained equivalent, gabbro, are most common, and, like dikes, appear in great swarms.

A tremendous sill swarm in South Africa—the Karoo dolerites—consists of hundreds upon hundreds of dolerite sills, ranging from a few feet to over 1,000 feet thick. They inject the relatively flat sedimentary rocks of the Karoo series in Basutoland and nearby Cape Province. The aggregate volume of these sills and related dikes appears to be nearly as great as that of the Columbia River plateau basalts.

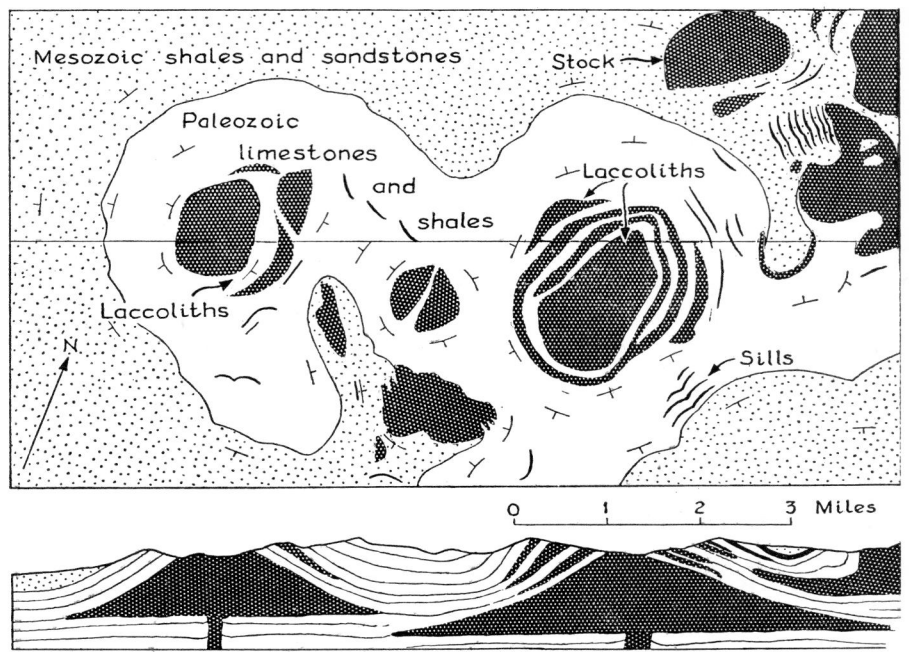

FIGURE 17-13. *Map and cross section of a group of laccoliths and stocks in the Judith Mountains, Montana. (After W. H. Weed and L. V. Pirsson, U. S. Geological Survey.)*

Laccoliths.   Magma may spread sill-like between sedimentary layers, and then lift the roof rocks into a dome, forming an underground magma body with a flat floor and an arched roof, called a *laccolith* (Fig. 17-13). Few laccoliths exceed a few miles in diameter. Like sills, they tend to form groups (Fig. 17-13) and are generally intimately associated with stocks, dikes, and sills. Laccoliths grade into sills, and also into stocks and irregular igneous masses.

Differentiated Sills and Laccoliths.   The great bluffs that form the Palisades across the Hudson River from New York City mark the eroded edge of a dolerite sill locally more than 900 feet thick. This large sill is not uniform dolerite from top to bottom, but consists of layers of differing composition. Along the base of the sill, and in places along the top, there is a zone of fine-grained rock containing scattered olivine phenocrysts. Most of the rest of the sill consists of a mass of relatively uniform dolerite containing no olivine. Pyroxene tends to be more abundant in the lower part of the uniform dolerite than in its upper portion. Finally, just above the basal fine-grained zone there is a layer 10 to 20 feet thick containing up to 25 per cent of olivine crystals (Fig. 17-14).

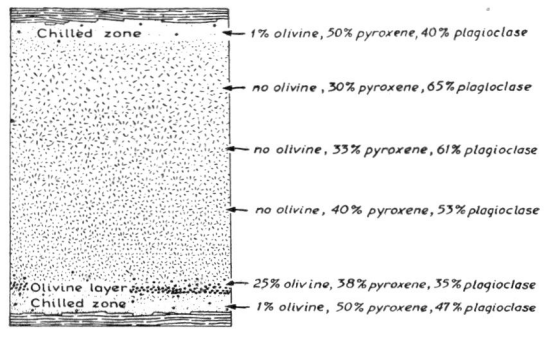

FIGURE 17-14. *Diagrammatic cross section of the Palisades sill, New Jersey, showing the vertical variation in texture and mineral composition. (After Frederick Walker, 1940.)*

The proportion of olivine to other minerals in the fine-grained border rock is such that, if the olivine of the basal olivine-rich layer were uniformly distributed throughout the much thicker mass of olivine-free dolerite, the entire sill would have the same composition as the fine-grained zone. It is inferred that this is a chilled zone representing the bulk composition of the magma of the sill at the time of injection. Evidently, during the slow cooling of the magma the early-formed olivine crystals had time to settle through the magma and accumulate on the top of the basal chilled layer.

The Palisades sill illustrates the phenomenon of *differentiation,* which means the process whereby a once-uniform magma may yield rocks of differing compositions upon solidifying. The special kind of differentiation illustrated by the Palisades dolerite is *gravitative crystallization differentiation.* Many thick sills and laccoliths show ferromagnesian-rich rocks of high specific gravity at their bases; apparently most of their minerals have sunk during crystallization. Doubtless, gravitative differentiation also occurs in some volcanic plugs and stocks, perhaps accounting for the rise of domical protrusions of obsidian or rhyolite of low density into volcanic throats that formerly erupted much more dense basalt and andesite.

Deep Intrusive Masses

Many sills, dikes, and larger intrusive masses were doubtless intruded at much greater depths than the shallow forms just described. In general, even small deep intrusions are coarse-grained and show little chilling at contacts, indicating slow cooling against hot wall rocks. The wall rocks are not sedimentary or volcanic but are generally metamorphic or deep-seated igneous rocks.

Characteristic of these deeper zones of intrusion are very large igneous masses called plutons or batholiths; these, and the rarer lopoliths, will now be described. It must be remembered, however, that, although the lower parts of these great masses were undoubtedly formed at abyssal depths, the tops of some extended nearly to the surface and locally show features typical of shallow intrusions.

Lopoliths: Stratiform Complexes.   Among the deep intrusive masses are the highly differentiated igneous rock complexes, called *lopoliths* (Greek: "basin"). These are shaped like a basin—both floor and roof of the chamber are bowed downward like a saucer (Fig. 17-15).

The Duluth gabbro mass which underlies Lake Superior is a lopolith. It is about 150 miles across, is estimated to be about 10 miles in maximum thickness and to contain about 50,000 cubic miles of rock. Thus lopoliths differ from sills and laccoliths in their enormously greater size, and in their downsagging centers.

Most lopoliths, some large and deep sills, and even a few small intrusions are marvellously differentiated into many thin sheets and bands of contrasting mineral composition. In the Bushveld complex, a large lopolith in South Africa, this banding is so striking that outcrops of the igneous rock resemble thin-bedded sediments when viewed from a distance. The origin of the layering in such *stratiform complexes* has been debated. Since, in general, the heavier rocks are toward the base and the lighter toward the top, crystal settling by gravity probably plays a part, as it almost certainly did in the Palisades sill. Nevertheless, the hundreds of thin layers of alternating light and heavy minerals are difficult to explain by crystal settling alone. Perhaps convection currents aided in sorting the sinking crystals into layers.

In Greenland, a small stratiform complex of gabbro, the Skaergaard intrusion, has the form of a tilted cone with the apex downward. Its strikingly layered structure is well exposed on the walls of the deep Greenland

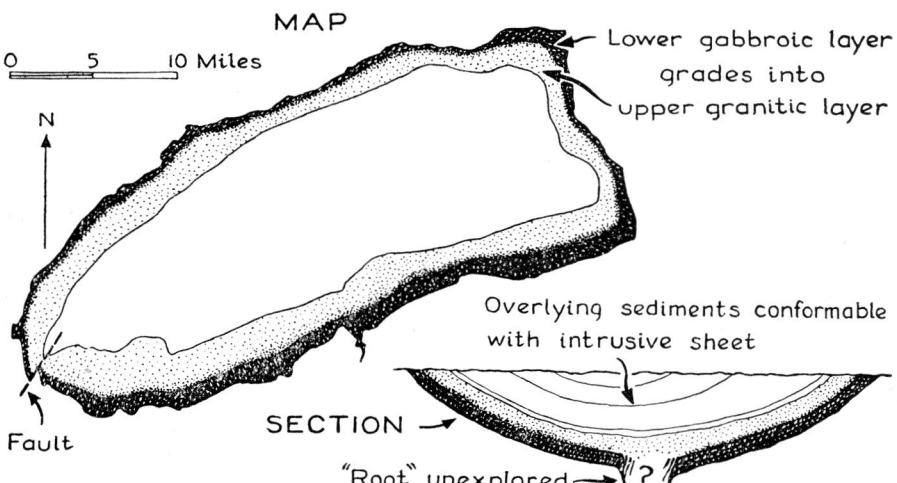

FIGURE 17-15.   *The Sudbury lopolith, Ontario, Canada. (After A. P. Coleman and E. E. Moore, 1929.)*

fiords. From the way the stratiform layers abut against the contacts, and from other evidence, the English geologists Wager and Deer concluded that convection currents sweeping slowly across the floor had sorted the crystallizing minerals into thin layers.

Batholiths, or Plutons. The largest igneous masses are the great bodies of granite and granodiorite that extend along the cores of most major mountain ranges and underlie vast areas of the ancient shields (see Chap. 19). The body of granodiorite and related quartz-bearing plutonic rocks in the Coast Range of British Columbia (Fig. 17-16) is at least 1,000 miles long and 20 to 150 miles wide. Uniform granitic rocks are exposed on the canyon walls through a vertical range of as much as 7,000 feet without indication of any bottom. Similar great granitic masses are exposed in the Sierra Nevada, the Patagonian Andes (Fig. 17-16), and along the cores of other great mountain chains. Granitic masses of various sizes—some rounded in plan, some elongated, and some irregular—form intricate patterns with the metamorphic rocks that underlie most of eastern Canada, Scandinavia, and Brazil.

Masses of granitic rock with a surface area of more than 40 square miles are called batholiths; if less, they are called stocks. In most definitions bath-

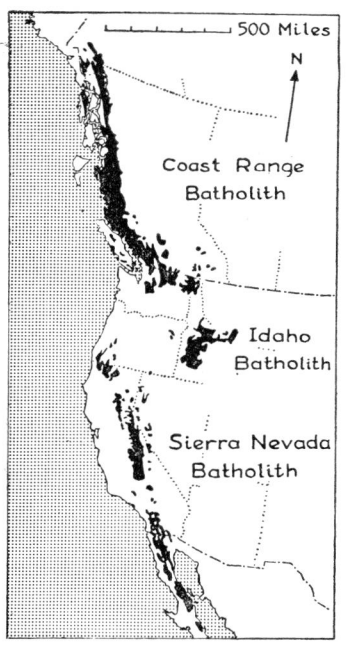

FIGURE 17-16.   *Large granitic intrusives (black) of western North America* (left) *and the Patagonian batholith. (After Geologic Maps of North and South America, Geological Society of America.)*

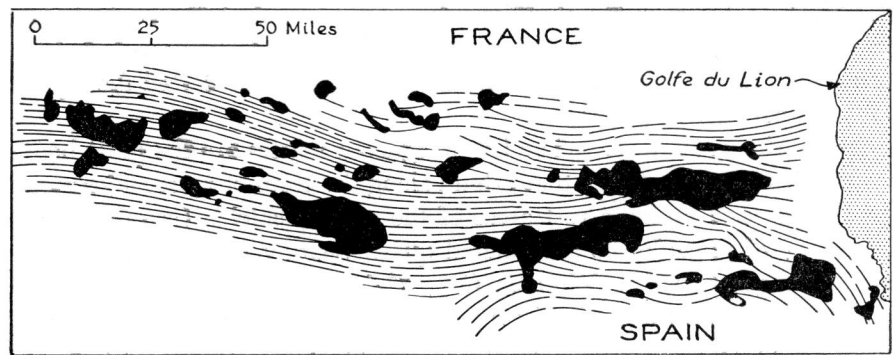

FIGURE 17-17. *Map of cross-cutting batholiths* (black) *in the Pyrenees. The lines are trends of folds.* (*After R. A. Daly,* Igneous Rocks and the Depths of the Earth, *McGraw-Hill Book Co., 1933.*)

oliths are said to flare outward with depth and to be "bottomless," in contrast to intrusions with a definite floor such as lopoliths and laccoliths. Although anything resembling a floor can rarely be seen in the field, many students of batholiths have advanced indirect evidence that they taper downward or else terminate in other ways at depths of only a few miles. Isostatic and thermal evidence for this conclusion is given in Chapter 19. From surface exposures alone it is impossible to infer the form of an igneous mass at depths of 10 or 20 miles. Because of the uncertainty regarding the form of the lower parts of granitic masses many geologists prefer to use the term "pluton" instead of batholith. A *pluton* is defined as any very large igneous mass, irrespective of form.

Contact Relations of Plutons   Plutons vary in their roofs and walls. Some of them break across the bedding, folds, and other structures of the surrounding wall rocks (Fig. 17-17). Obviously these acted as liquids, engulfing numerous fragments of wall rock in the granite and penetrating their walls in large and small dikes (Fig. 5-5).

But in other areas the metamorphic wall rocks bend concordantly around the pluton, and the granite itself may show a marked flow banding, or even a gneissic structure parallel with the wall, along which it may be streaked out, granulated, and even partially recrystallized. The structural pattern indicates that both pluton and wall rock have undergone slow laminar flow as pasty masses of only slightly different viscosity (Fig. 17-18). Such structural relations seem best explained by the idea that the pluton injected hot wall rocks and moved into place as a pasty mass, either as an almost completely crystallized magma, or as a heterogeneous mass of "*migma*" that had only partially melted into *magma* before it was forced upward.

In still other plutons no definite contact separates granite and wall rock;

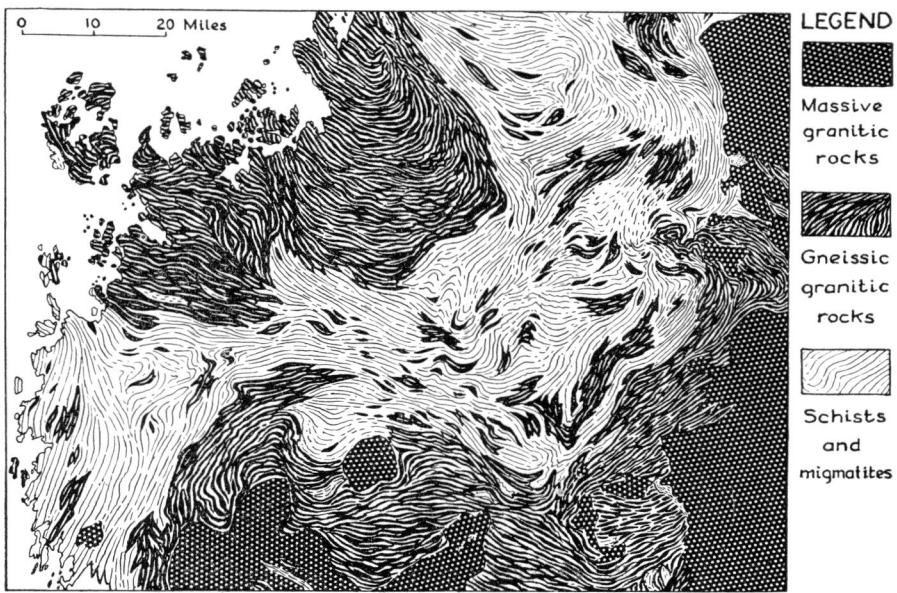

LEGEND

Massive granitic rocks

Gneissic granitic rocks

Schists and migmatites

FIGURE 17-18. *Concordant batholiths of southern Finland. The lined patterns indicate the strike of foliation in the gneissic granites, schists, and migmatites. (After Martti Saksela, 1935.)*

the granite appears to fade gradually into the wall as though it had been formed by metamorphism of the wall rock into granite by introduction of feldspar and other minerals characteristic of granite.

Convincing field evidence from several areas indicates that metamorphic rocks—particularly mica schists, hornblende schists, and quartzites, derived respectively from normal shales, basalts, and sandstones—have been transformed into granite. When traced toward the granite mass, they begin to lose their typical foliation or other structural features; feldspar and other granitic minerals appear, first in small isolated crystals, then in increasingly abundant clots until, finally, only granitic minerals are present with an occasional tell-tale streakiness to give a hint of the original metamorphic foliation. Less easily replaced beds such as marbles and schistose conglomerates may survive in the granite as faint ghostlike relics.

Still another common kind of contact is the "injection gneiss" or "mixed rock" type. This consists of intimately interpenetrating masses of granite and metamorphic rock in small layers, bands, dikes, or irregular masses (Fig. 17-19).

Such complex small-scale mixtures of granite and wall rocks are called *migmatites* ("mixed rocks," see p. 612). Migmatites abound in many areas; in Finland they make up 26 per cent of the surface rocks (Fig. 17-18).

FIGURE 17-19. *Migmatite, showing gneissic structure and cross-cutting veins. Near Storm Lake, Montana. The pencil indicates scale. (Photo by F. C. Calkins, U. S. Geological Survey.)*

Migmatites are very puzzling. Do they represent an intimate injection and soaking of schists and other metamorphic rocks by granitic magma in small sills, tongues, and dikes? Or do they represent the selective melting of the wall rocks, the granitic parts being accumulations of the most easily fusible constituents which have melted and run together into small masses, leaving the dark-colored, less fusible metamorphic parts of the migmatite unassimilated because cooling ensued before they could be melted into the magmatic solution? Or are the granitic parts due to differential replacement in the solid state of parts of the metamorphic wall by the minerals of granite? Or do all these processes occur together when complex migmatite zones are formed?

It is now well established that granite may be of either igneous or metamorphic origin. But how can we certainly discriminate between the two modes of origin either in the small granitic layers of migmatites or in the great relatively uniform batholiths? These questions and the related one of how granitic masses make room for themselves are among the most difficult problems in geology. They will be mentioned again in connection with the structural history of mountain belts (Chap. 19).

Satellite Dikes   Although some plutons show no dikes along their borders, others are cut by hundreds of dikes that appear to be differentiates of the granitic mass (Fig. 17-20). Four chief kinds of dike rocks are commonly

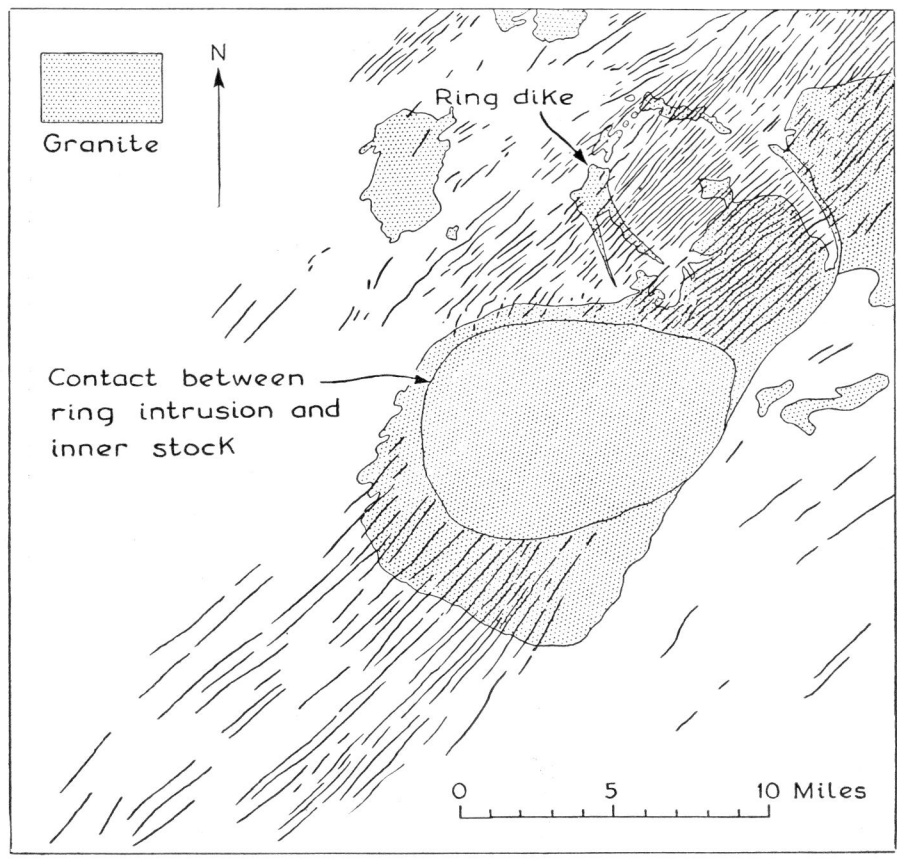

FIGURE 17-20.   *Dike swarms at the margin of granite stocks, Ben Nevis, Scotland. (After E. B. Bailey, Geological Survey of Great Britain.)*

associated with plutons. All may occur together, or only one kind may appear. Mineralogically, *granite porphyries* (or *granodiorite porphyries*) closely resemble the rock of the pluton, but the grain size is smaller and the rock is strongly porphyritic. *Lamprophyres* are dark, generally porphyritic dikes, made up chiefly of the ferromagnesian minerals found in the granite and considerable feldspar. *Pegmatites* consist mostly of quartz and feldspar in very coarse grains that may be a foot or more across; they also often contain rare minerals as minor constituents. *Aplites* are like pegmatites in composition, but are of slightly finer grain than granite.

   Schlieren   Inclusions of wall rock, some sharply angular and little modified by the surrounding granite, and others softened, half-assimilated, and pulled out into long dark-colored streaks—called *schlieren*—are characteristic of the edges of plutons. Many schlieren—perhaps nearly all—are stretched-out, half-assimilated inclusions of hornblende schist or similar refractory ferromagnesian-rich rocks, but some may be segregations or clots of ferromagnesian minerals crystallized out of the granite magma itself. Schlieren may be difficult to distinguish from lamprophyre dikes—indeed the two have been observed to grade into one another. In some plutonic border zones inclusions make up as much as 40 per cent of the total volume of rock.

Space and Time Relations   The distribution of batholiths in space and time is significant. Granite has not been found on any of the isolated Pacific Islands, and it is probably not present in the crust beneath them, since granitic rock fragments are never found in their volcanic ejecta. Also, the travel time of earthquake waves across the Pacific basin (Chap. 18) is such that no significant amount of granite or granodiorite can be present. Quartz-bearing plutonic rocks thus appear to be confined to the continents, whereas the more widespread basalt is intrusive and extrusive in both continental and oceanic areas. Apparently, large masses of granite are formed mainly in the "root zone" of mountain belts (Chap. 10), and they appear at the ground surface in the cores of deeply eroded mountains, or in areas marking the site of former mountains that have been completely eroded away (Chap. 19).

Physical Chemistry of Magmatic Crystallization

For well over 100 years geologists have attempted to duplicate the processes that occur in crystallizing magmas by melting rocks in the laboratory and studying the products that separate from them. Because of the high temperatures necessary to melt rocks and because of the large number of solid-solution minerals with complicated chemical formulas that compose them, such experiments pose many difficulties.

The rapid development of physical chemistry in the last half century has given considerable impetus to such investigations. Most geological problems involve very complex chemical systems containing many components. The Norwegian geologist J. H. L. Vogt made a great step forward by applying physical chemical reasoning to the crystallization of slags from blast furnaces. He also experimented with artificial rock melts. In 1904 the Geophysical Laboratory of the Carnegie Institution was founded in Washington, D. C., and its investigators began a serious long-range attack on the physical chemistry of artificial rock melts. The details of their work are beyond the scope of this book, as they require an advanced knowledge of both geology and physical chemistry. Their experiments, however, have sharpened, clarified, and, along some lines, even revolutionized geological thinking about the origin of igneous rocks. In particular, N. L. Bowen of this laboratory has offered a comprehensive theory of the evolution of igneous rocks which is being extended, tested, and modified by field work and additional laboratory experiments.

Bowen's Theory

A greatly simplified statement of Bowen's theory is that *basalt magma is the parent of all the igneous rocks, and the many varieties have arisen through crystallization differentiation.* The great diversity of volcanic and plutonic rocks is explained by the separation of early-formed crystals from the parent basalt magma during its crystallization and the segregation of the minerals in different parts of the mass. Crystal differentiation by gravity, the process that was suggested by the olivine-rich ledge of the Palisades sill, is considered very important. Squeezing ("filter-pressing") of a mesh of early-formed crystals while the interstices are still filled with molten material is a second mechanism for separating liquid from early-formed crystals. By careful laboratory experiments Bowen and his colleagues have worked out the order of crystallization (*i.e.,* the order in which the different minerals appear) during the freezing of basalt magma. This order is summarized in the following diagram, which is called the "*Bowen Reaction Series*":

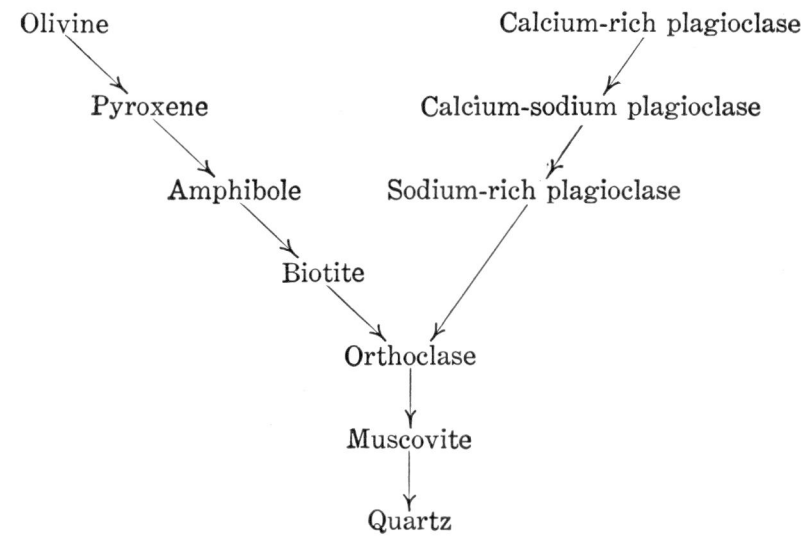

Olivine and calcium-rich plagioclase are the first of the minerals to crystallize in abundance from basaltic magma. Unless these early-formed crystals are removed from the melt by some such process as crystal settling, the olivine will react with the surrounding liquid to produce pyroxene, and the calcium plagioclase will react to produce a plagioclase richer in soda. If no crystals are removed from the melt, thus allowing these reactions to go on unimpeded, the whole mass solidifies to a mixture of plagioclase and pyroxene with a small amount of olivine—the typical minerals of basalt. If, however, the olivine and/or the calcium plagioclase is separated from the melt, perhaps by settling to the floor of the magma chamber, such reactions between crystals and liquid cannot take place. Olivine contains a higher proportion of magnesium and a lower proportion of silica than the melt as a whole; calcium-rich plagioclase a higher proportion of calcium and a lower proportion of both sodium and silica than this melt. The liquid left behind will, consequently, be relatively impoverished in magnesium and calcium and relatively enriched in silica, alkalis, and iron. Hence the liquid left behind is no longer of basaltic composition but has reached a composition approximating that of andesite. If no further separation of crystals occurs, it will crystallize to a rock of andesitic composition, but if the next crop of crystals (which are now pyroxene and calcium-sodium plagioclase) are removed as fast as they are formed so that they cannot react with the liquid, the left-over magma may continue to change in composition, ultimately reaching the composition of a siliceous alkali-rich rhyolite. Or, depending on special conditions of crystal removal, final liquids rich in iron and the alkalis and only moderately rich in silica might be formed. Only 10 per cent or less of the original basalt magma could thus be transformed to rhyolite; the rest has gone into the early crystalline phases now accumulated as peridotites, olivine-rich gabbros, diorites, and similar rocks on the floor of the magma chamber.

By this simple mechanism Bowen has offered an explanation of the great diversity of the igneous rocks. The theory is powerfully supported by certain field occurrences: for example, the common presence of olivine-rich rocks along the floors of thick sills and of quartz-bearing rocks in their tops, and the appearance of a small amount of rhyolite late in the history of a great basaltic or andesitic volcanic cone. These small siliceous flows and pyroclastics represent the last liquids remaining in the deep chamber after crystallization and gravitative separation had removed the earlier-formed crystals. Also, the stratiform complexes may well have been formed as accumulations of the earlier crystallizing minerals into sheets and layers so closely packed that the remaining magma was squeezed out and could not react with them.

Objections to Bowen's Theory

Although Bowen's theory has won wide acceptance as an explanation of differentiated sills, stratiform complexes, and some composite volcanoes, it has not been so generally accepted as an explanation of the origin of the granite and granodiorite of plutons. It is difficult to believe that these huge igneous bodies represent only 5 to 10 per cent of the original magma from which they crystallized. Their varied kinds of contacts also suggest other processes than crystal differentiation. In particular, it is well established that the replacement, partial melting, and assimilation of wall rock fragments by magmas locally play important roles in developing large granite masses. The confinement of granites and granodiorites to continental areas and the absence of granite differentiates in the Pacific Islands is also against the theory. Thus cogent evidence indicates that large masses of granite magma arise in other ways than as differentiation products of basalt. And finally, the huge amounts of both granite and basalt and the relative rarity of intermediate rocks such as diorite, which, according to Bowen's theory, should be much more abundant than granite, pose a question to those who would derive all granites by differentiation of basalt magma. Perhaps much granite has come from the remelting of crustal rocks in the roots of mountain ranges, as mentioned in Chapter 10 and discussed more fully in Chapter 19.

Role of Volatile Constituents in Igneous Activity

Several times we have mentioned the water vapor and other volatile constituents of magmas. Many characteristics of magmas and of the rocks solidified from them arise from their content of volatiles. Among the better-known effects are the following:

Mechanical Effects

Although density differences between magma and wall rock are important, a part of the energy that brings lavas to the surface, and most of the energy that causes volcanic explosions is supplied by the expansive power of pent-up gases. The expansive force of a gas-charged liquid is familiar to everyone. Mechanically there is no difference except in scale between the results of uncorking a highly carbonated bottle of pop or champagne and the flashing of underlying magma into pumice upon the blowing out of a partly solidified volcanic plug. Both liquids are driven upward and expelled by the pent-up gases within them. Even the greatest volcanic explosions such as those at Tamboro and Katmai appear to be nothing more than the bursting of the upper walls of a magma chamber by the slowly increasing steam pressure within it.

Crystallization taking place in a magma increases the gas pressure. Molecules of gas, free to diffuse through a liquid, become more and more confined as crystallization proceeds because they cannot enter the newly formed crystals. Hence, as more and more crystals are formed, the gas pressure rises. It may ultimately be able to burst the roof, allowing part of the gases to stream out and escape. Sudden removal of pressure causes the dissolved gases in the magma to separate at once into bubbles: the magma froths into pumice or scoria. Increase of temperature, perhaps caused by chemical reactions between gases, may cause further frothing, since it would expand the gases still more and also reduce the viscosity of the magma surrounding the gas bubbles. Such an increase of temperature may well play a part in the relatively uniform boiling over of pumiceous magmas into pyroclastic flows and glowing avalanches.

Fumaroles; Geysers; Hot Springs

Gases and liquids will pour off from crystallizing magmas as long as there are escape channels. Within the craters, or on the slopes of some active volcanoes such as Katmai and Vesuvius, are many openings called *fumaroles* from which issue jets of steam and other gases. From other openings, or from the same ones in the wet season, hot springs well forth. Similar features are common in volcanic fields such as Yellowstone Park, where there has been no historic volcanic activity but where evidences of recent activity are clear. *Geysers,* which periodically spout water and steam, are intermediate between hot springs and fumaroles.

The temperature at fumaroles ranges from the boiling point of water to as much as 700° C. (over 1200° F.). Chemical analysis shows that, ordinarily, more than 98 per cent of the vapor is steam, but sulfur compounds, hydrochloric acid (HCl) fumes, carbon dioxide, ammonium chloride, hydrogen, and other gases are nearly always present. Many fumaroles and hot springs bring to the surface small amounts of arsenic, mercury, antimony, copper, lead, manganese, and especially iron. Chemical studies of the gases suggest that most of these metals were carried in the gas as compounds of chlorine or fluorine, and it is known that many metallic chlorides and fluorides vaporize at relatively low temperatures. Near the surface, where air mingles with the fumarole gases, the metallic compounds react with oxygen of the air to produce oxides of metals. At the temperature of the fumaroles iron oxide cannot be vaporized; it deposits as crystals of hematite and magnetite on the walls of the vent. Rapid filling of fumarole channels by hematite and magnetite has been repeatedly observed at Katmai, Vesuvius, and other volcanoes. Chemical considerations suggest that deeper within the channel, out of contact with the air, many of the metals are deposited as sulfides— a deduction strongly supported by the arrangement of sulfides in ore deposits that occupy fissures which may once have been the channels for ascending magmatic fluids.

The close association of fumaroles with volcanoes, and the occurrence in the gases of unusual substances such as are not found in ground water in nonvolcanic regions, implies that the gases are derived from cooling magmas. That the unusual compounds found in fumarolic gases are indeed contained in magmas is shown by the occurrence of chlorides, water, and carbon dioxide as minute inclusions within the minerals of igneous rocks.

Many fumaroles and hot springs bring to the surface not only fluids expelled from crystallizing magmas, but also large amounts of ground water with which they mixed during their ascent. Fumaroles with temperatures far above the boiling point in the dry season may become mere hot springs during the rainy season, and then revert to fumaroles again in dry weather— clearly, they were cooled and diluted by surface water during the rainy season. Only rarely can the relative amounts of ground water and magmatic water be accurately determined; careful estimates at the hot springs of Mount Lassen, California, gave 10 per cent magmatic water, at the springs of Yellowstone National Park much less. Some hot springs yield no magmatic water; the ground water may penetrate along fissures far enough to be heated by the earth's internal temperature and thence rise as a hot spring to the surface. As already noted, such springs differ in composition from those inferred to be of magmatic origin.

In their passage upward toward the surface—perhaps for many thousands of feet—the magmatic fluids may react with the channel walls, forming many new minerals and drastically altering the wall rocks in places. As the ascending thermal waters approach the surface, reactions leading to the formation of clays by alteration of the feldspars and the volcanic glass of the wall rock are likely to be prominent. Ordinarily, the clay is left as a lining of soft, altered rock bordering the channel, but locally it may be carried upward, clogging and choking the channel, and eventually emerging as bubbling springs of mud. The mud is generally brightly colored from iron oxides, manganese oxides, and other impurities. Such brilliant yellow, purple, red, orange, black, and gray bubbling mud springs are called *paint pots.*

Alteration of the Parent Igneous Rock by Volatiles

Fluids escaping from crystallizing magmas may, under certain conditions, drastically change the igneous rock from which they were generated, as well as the adjacent wall rocks. Minerals that separate from a magma may be stable as long as they are surrounded by the magma, but may be readily transformed to new substances if they are steeped in the last volatile residues of a crystallizing magma. Microscopic examination of nearly any igneous rock shows a few changes of this kind. The feldspars of even the freshest granite generally contain a few flecks of muscovite or clay minerals, and the biotite may be partially altered to chlorite. The olivine of a gabbro is generally partially or completely altered to serpentine minerals.

In some rocks these reactions are so extensive as to change completely the rock composition and texture. Pegmatite, which apparently crystallizes from magmas exceptionally rich in volatile substances, shows such alterations to a marked degree. In many pegmatite masses careful field and petrographic studies have shown that most of the original magmatic minerals were replaced by new minerals as water-rich fluids passed through the still hot pegmatite after it had crystallized.

Rocks containing olivine are particularly subject to such changes because olivine is readily altered by hot volatiles. Even in basaltic lavas, from which the volatiles readily escape into the air, the rims and cracks in grains of olivine are commonly coated with decomposition products. In peridotites and other rocks rich in olivine which crystallize underground, where the hot volatiles are kept in contact with the mineral grains much longer, olivine is likely to be completely altered, and other ferromagnesian minerals partially altered. Such altered peridotites appear very different from the fresh rocks; the altered rock has been named *serpentine*. Serpentine can, of course, be formed either by magmatic fluids during the late stages of crystallization of the parental magma or, long afterward, by the action of hot water from an entirely foreign source. There is, therefore, good justification for classing serpentine as a metamorphic rock, as is done in many texts. It is probable, however, that most serpentine was formed during the last stages of magmatic crystallization, and that it differs from rocks such as gabbro only because the much higher original content of olivine emphasizes the alteration. The associations of serpentine are with the igneous rocks, not the metamorphic; it is commonly found as sills and dikes in unmetamorphosed sedimentary rocks. Therefore, like the pegmatites, serpentines have been grouped with the igneous rocks in this book (Appendix IV, Table IV-2), although some serpentines (like some granites) are undoubtedly of metamorphic origin.

Alterations and replacements by magmatic fluids, though more common in intrusives, also take place in volcanic rocks. Lapilli, bombs, and other volcanic products are commonly "burned" to a bright red along the edge of craters and at the orifices of fumaroles or other places where hot gases mixed with air are escaping. The coloration comes from oxidation of the iron-bearing minerals to bright-red specks of hematite. Similar oxidation, occurring along flow bands, shears, and brecciated areas where the gases can gain entrance, gives many rhyolites and obsidians their yellow, brown, and red colors. The welding and oxidation of pumiceous fragments in pyroclastic flows have already been commented upon. Underground, where free oxygen is not abundant in the vapors, chlorite, serpentine minerals, clay minerals, and many other "hydrothermal minerals" appear.

Facts, Concepts, Terms

SHIELD VOLCANOES
> Fissure eruptions, radial dikes, cinder cones
> Calderas, ring dikes
> Lava pools, lava fountains

COMPOSITE VOLCANOES
> Small rhythmic explosions at short intervals
>> Incandescent bombs and lapilli
> Strong explosions after periods of inactivity
>> Eruption clouds of shattered rock, pumice lapilli, and ash
>> Pyroclastic flows and welded tuffs
> Spines and domical protrusions
> Volcanic mudflows

PLATEAU BASALTS
> Extent, volume, distribution
> Submarine eruptions, pillow lavas

SHALLOW INTRUSIVE BODIES
> Plugs, dike swarms, sill swarms, laccoliths
> Gravitative differentiation

DEEP INTRUSIVE BODIES
> Lopoliths
>> Causes of layering in lopoliths and stratiform complexes
> Batholiths, or plutons
>> Size, shape, composition

Relation to wall rocks—concordant, discordant, gradational
Migmatites; schlieren; satellitic dikes
Confinement to deeply eroded mountain belts

BOWEN'S THEORY
The reaction series
Objections to the theory as applied to batholiths

ROLE OF VOLATILE CONSTITUENTS
Expansive force of gas-charged magmas
Fumaroles, geysers, hot springs
Alteration by magmatic fluids
Origin of serpentine

Questions

1. Why do composite volcanoes have steeper slopes than shield volcanoes?
2. A shield volcano has a caldera 3 miles in diameter and 1,000 feet deep. Three parasitic cones have developed along the caldera rim. A domical protrusion of glassy rhyolite and several basaltic cinder cones decorate the surface of the shield outside the caldera rim.

    Draw two sketch geologic maps, one to show the inferred geology along a horizontal surface within the volcano, but about 500 feet below the floor of the caldera; the other to show the inferred structure along a still deeper horizontal surface which lies about 1,000 feet below the base of the volcanic shield itself. (Assume that the volcano was built on a uniform basement of eroded granite.)
3. Why are cinder cones more likely to develop along a dike after it has fed lava flows and is in process of congealing than when the dike first breaks through to the surface?
4. How can you distinguish a deposit of welded tuff from a flow or domical protrusion of flow-banded rhyolite?
5. Obsidian from a Japanese volcano has a specific gravity of 2.60, pumice lapilli from the same volcano floats two-thirds submerged in water. Approximately what volume of pumice would be required to account for the formation of a caldera on the top of this volcano 4 miles in diameter and with an average depth of 1,000 feet, assuming that the caldera was the result of piecemeal subsidence after strong explosions of pumice lapilli?
6. Draw cross section diagrams of the following igneous bodies: (a) a spine, (b) a ring dike, (c) a differentiated sill, (d) a stratiform complex.
7. Draw a sketch or geologic map showing a portion of the edge of a batholith. Show and label the following typical features: (a) a discordant contact, (b) inclusions, (c) granite porphyries, (d) lamprophyres, (e) a series of dolerite sills that are older than the batholith, (f) a rhyolite dike that is younger than the batholith.

8. Give the evidence for classifying serpentine as (a) an igneous rock, (b) a metamorphic rock.
9. Explain the process of gravitative crystallization differentiation.

Suggested Readings

1. Bowen, N. L., *The Evolution of the Igneous Rocks*, Princeton University Press, Princeton, N. J., 1928.
2. Daly, R. A., *Igneous Rocks and the Depths of the Earth*, McGraw-Hill, New York, 1933. (Full discussion of igneous rock problems.)
3. Fenner, C. N., *The Origin and Mode of Emplacement of the Great Tuff Deposit in the Valley of Ten Thousand Smokes*, National Geographic Society, Katmai Series, No. 1, 1923. (A penetrating analysis of the origin of the deposits formed during the 1912 eruption of Mount Katmai.)
4. Lacroix, A., *La Montagne Pelée et ses Eruptions*, Masson et Cie, Paris, 1904. (A comprehensive memoir, beautifully illustrated, on the 1903 eruptions of Mont Pelée.)
5. Tyrrell, G. W., *Volcanoes*, T. Butterworth, Ltd., London, 1931. (Written in nontechnical language.)
6. Williams, H., *Geology of Crater Lake National Park*, Carnegie Institution of Washington, Washington, D. C., 1942.

18. *Earthquakes and the Earth's Interior*

Effects of Earthquakes

On November 1, 1755, All Saints' Day, the churches of Lisbon were thronged with worshippers. Suddenly, the ground thundered forth a terrifying roar and began to heave and writhe in horrible, jarring shocks that seemed endless. The roofs and arches of the great stone churches crumbled, crushing hundreds beneath them. Most of the buildings in the city crashed in rubble. Within six minutes, when the awful quaking stopped, nearly a quarter of the population had perished, and thousands more lay trapped beneath the wreckage. The sea withdrew from the harbor, exposing the bar at its mouth, then rushed in as a wall of water 50 feet high, drowning hundreds who had taken refuge on the open wharves, and destroying ships, docks, and nearly everything in its path. At the same time a shorter quake of great intensity came. Slides of rock cascaded down from the mountains behind the city, raising dust clouds so dense that many thought a volcanic eruption had begun. A third great shock struck the desolate city two hours later. Fires completed the destruction; by nightfall 60,000 people of a population of 235,000 had perished, and thousands of the survivors were crippled, maimed, or mad from terror. Less than one house in six was habitable.

Probably no other earthquake in recorded history has been so widely felt. Rivers as far away as Lübeck, Germany (1,400 miles) rose or fell several feet. Loch Lomond, 1,200 miles away in Scotland, was thrown into waves 2 feet high. At Fez and Mequinez, Morocco, 400 miles from Lisbon, many thousands perished in wreckage nearly as complete as that in Lisbon, and buildings were wrecked in scores of towns in Spain and North Africa. It has been estimated that the shock was felt over as much as 1/13 of the whole surface of the earth.

On December 16, 1811, at 2 A.M., the pioneers of the Mississippi Valley near New Madrid, Missouri, were thrown from their beds by a terrifying shock. In a few seconds many of their log cabins were destroyed and landslides came rushing down. Over great areas the earth cracked open. Sand, water, and mud gushed out, burying many of the fields. For miles the banks of the Mississippi caved in; islands sank, others rose. River boats were swamped or hurled over the banks. During the next 15 months came thousands of minor shocks and at least two earthquakes as severe as the first one. Large areas of land rose, draining former swamps. A huge tract nearly 150 miles long and 35 miles wide, mostly in the Mississippi floodplain, sank 3 to 10 feet, and river water rushed into it and formed new lakes and swamps, one of them being Reelfoot Lake, a water body several miles across. The shocks were strong enough to stop pendulum clocks and ring church bells as far away as Boston, and to crack plaster in Virginia. Had this earthquake come a century later the loss of life would have been appalling, but in those pioneer days few were endangered.

On September 10, 1899, after a long series of minor shocks, a great earthquake wracked southeastern Alaska, centering in the region of Yakutat Bay. Though this shock was felt as far north as the Yukon and from Sitka on the east to Cook Inlet on the west, little damage was done in this sparsely settled area. This earthquake, however, is distinguished as having accompanied the greatest vertical displacement of any yet studied: Barnacles and boring clams marking the old sea level were locally lifted more than 47 feet above the sea (Chap. 9). Nearby areas were depressed.

This earthquake affected the coastal glaciers (Fig. 13-6) in a most interesting way. Muir glacier, more than 100 miles to the east, was so greatly shaken that it shed gigantic icebergs from its snout and soon wasted away at abnormal speed. The glaciers at Yakutat Bay, near the area of greatest shaking, reacted differently. Though they, too, shook free many bergs, tremendous snowslides crashing down from the towering St. Elias Range at the head of the glaciers more than made up for the loss. The added load of new snow drove the glaciers vigorously forward—in places many hundreds of yards within less than 10 months—and even stagnant parts of the Malaspina glacier, so long inactive that a dense forest had grown on the debris-strewn ice, resumed motion and were tossed into a wild confusion of crevasses, hummocks, and pinnacles.

This account of major earthquakes could be greatly extended: The one that struck eastern Sicily and the Calabrian Coast across the Straits of

Messina on December 28, 1908, destroying the cities of Messina and Reggio, killing about 100,000 people, and leaving the coast at Messina submerged more than 2 feet below its former stand; the two great earthquakes of Kansu, in western China—one in December 1920, the other in May 1927—each of which is reported to have killed about 100,000 people, chiefly by the collapse of dwellings dug in loess; the great earthquake of Kutch, at the mouth of the Indus, in 1819, felt as far away as Calcutta, during which a great area of low land was flooded by the sea, while to the north a scarp—the "Allah Bund" or "Dam of Allah"—50 miles long, was raised as much as 20 feet; the great Assam earthquake, in the big bend of the Brahmaputra in 1897, during which buildings were destroyed in Calcutta, 200 miles away, and a scarp was raised 35 feet; the Hawkes Bay earthquake of 1931 and the Wellington earthquake of 1855 in New Zealand, both of which were accompanied by land tilting and uplift of 6 to 10 feet with respect to the sea; and many more.

The San Francisco Earthquake

The San Francisco earthquake is of special interest, for from it have developed many of our current ideas about earthquakes. At a little after 5:00 in the morning, April 18, 1906, a great earthquake struck San Francisco. Many buildings were wrecked, especially those built on marshy or filled ground, but many of those on solid rock were little damaged. Hundreds of people were killed or injured, pavements were broken, and gas and water mains and electric power lines were torn apart. Fires sprang up and burned for days—the firemen helpless because of the broken mains. The city was a shambles. Though the loss of life could only be estimated, it probably reached 700. Material losses exceeded 400 million dollars, about 5 per cent from the earthquake, the rest from the usual accompaniment, fire. San Francisco suffered most but Santa Rosa, Palo Alto, San Jose, and many other places were severely damaged.

A great fault zone—the San Andreas rift—cuts obliquely across the California Coast Ranges for more than 600 miles, from the ocean at Punta Arenas on the north, past San Francisco, and thence far to the southeast, where it is finally lost in the alluvium of the Colorado Desert (see Fig. 18-1). From Punta Arenas to San Juan—more than 270 miles—the ground was rent open along this fault, and the country on the west side moved northward with respect to that on the east. The greatest displacement, 21 feet as measured by the offset of a road, took place about 30 miles northwest of San Francisco. The displacement diminished both to north and south, though not regularly. Locally, especially toward the north, a slight vertical displace-

FIGURE 18-1. *The San Andreas fault zone, California.*

ment occurred, but it did not exceed 3 feet. For the most part only horizontal movement took place.

The San Andreas rift had been recognized as a fault for many years before the San Francisco earthquake. Not only are geologic formations cut off and the rocks much sheared and crushed along it, but it is also conspicuous in the landscape. For many miles it is marked by large or small valleys and parallel ridges. Some streams bend abruptly when they reach it, and follow it for scores or hundreds of feet before swinging again to their original trends. Small ponds are strung out along it—some of them on steep slopes—strikingly anomalous features in arid southern California. In 1857 a strong earthquake shock occurred along a segment of the San Andreas rift that crosses the Tehachapi Mountains, far to the south of the segment active in 1906. It, too, was probably accompanied by a northward movement of the west side of the fault relative to the east side.

Causes of Earthquakes

No one knows the ultimate cause of earthquakes. It seems certain that the immediate cause is the sudden movement of rock masses along faults, for such movements would be adequate to cause the shaking, and fault displacement has been observed during a large number of quakes. The great majority of earthquakes, however, are unattended by visible fault displacement, though such displacement may have occurred below the ground surface. Every fault is of finite length and beyond the last point of rupture it disappears. No movement was measured on the San Andreas fault in 1906 south of San Juan. Similar relations are known to prevail in mines and wells; many faults at depth can be observed to die out before they reach the surface.

Furthermore, the close connection between fault displacement and intensity of shaking is clear. The intensity of shaking died rapidly away at right angles to the San Andreas fault except where boggy ground rendered buildings particularly susceptible. Although, as we have seen in Chapter 9, faults like that at Buena Vista Hills may sever oil-well casings without producing earthquakes—presumably because of slow and continuous movement, rather than abrupt slipping—the converse is more often true. In the Los Angeles Basin, many wells have been cut off along faults during earthquakes even though there was no surface sign of the slipping. For these reasons all seismologists now agree that major earthquakes arise from movement on faults, even when movement is undiscernible at the ground surface.

Elastic Rebound Theory

The fault motion accompanying the San Francisco earthquake was chiefly horizontal. It therefore could not be due directly to gravity. The region has no active volcanoes, and, furthermore, no known volcanic eruption has shown energy remotely comparable to that released in moving thousands of cubic miles of rock for many feet along a break 270 miles long. What, then, was the source of the prodigious energy?

The American geologist H. F. Reid suggested an answer. Most crustal movements take place slowly. Though the fault displacement was sudden, the energy it released had probably been slowly accumulating in the rocks of the region. The storage had been as elastic strain in the rocks, built up as the crust to the west of the break slowly drifted northward with respect to that on the east. Slipping of the fault in response to these forces was long prevented by the cohesion of the rocks on the two sides. The strain continued slowly to accumulate, just as energy is stored elastically when we bend a bow. When the elastic strain reached a critical value, the friction on the fault plane was overcome, and the rocks along the fault "snapped past each other," just as a stick will snap and break if we bend it too far.

Reid showed that an analysis of the data from precise surveys of the San Francisco region supported this idea almost conclusively. Accurate triangulation surveys had been made by the U. S. Coast and Geodetic Survey in 1851–1865, 1874–1892, and also after the earthquake in 1906–1907. Though the first survey was less accurate than the later two, its data were consistent. If the stations called Diablo and Mocho, which are both 35 miles east of the rift, are considered not to have changed position, then a comparison of the computed positions of the points near and those west of the San Andreas rift showed systematic and very considerable movement with respect to these "fixed points" during the time between the last two surveys. Reid grouped the data for these points according to the average distance of the several stations from the fault. His results are shown in Table 18-1.

TABLE 18-1. *Average Displacement of Points Between Surveys of 1874–1892 and 1906–1907 (Relative to Diablo-Mocho Line)*

NUMBER OF TRIANGULATION STATIONS	AVERAGE DISTANCE FROM FAULT (MILES)		DISPLACEMENT (FEET)	
	EAST	WEST	NORTHERLY	SOUTHERLY
1	4.			1.9
3	2.6			2.8
10	.9			5.1
12		1.2	9.7	
7		3.6	7.8	
1		23.	5.8	

The relations may be visualized from the accompanying diagram, Figure 18-2. Let us assume a time when there was no elastic strain in the rocks of the region; the line AOC represents a line crossing the fault at right angles. As the regional displacement began the point A moved slowly north to A' and the point C south to C', but as there was no slipping on the fault at this time, the formerly straight line connecting them through O became strongly bent. Just before the earthquake, the point A had advanced still further, to A'' and the point C to C'', these distances being such that the total $A–A''$ and $C–C''$ amounted to 21 feet. Then the fault gave way and the stress was relieved with explosive violence. $A''–O$ straightened out to $A''–O'$ and $C''–O$ to $C''–O''$. It is easy to see that lines such as $A''–O'$ and $C''–O''$, which had been straight just before the movement, would become bent into curves $A''–B$ and $C''–D$, curves identical with $A''–O$ and $C''–O$ but bent in opposite directions. When the fault moved, it may be assumed that the elastic

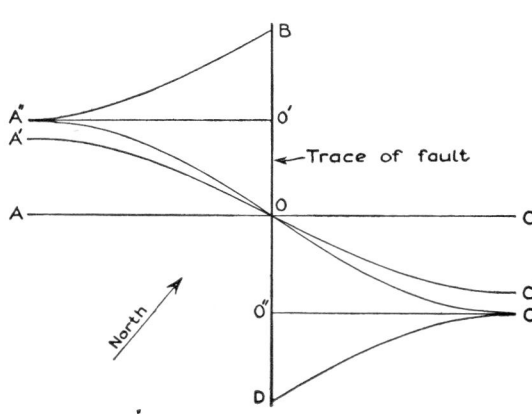

FIGURE 18-2. *Diagrammatic map showing how the 1906 offset along the San Andreas fault is explained by the elastic rebound theory. (After H. F. Reid, 1911.)*

strain (the bending of the rocks without rupture) was almost completely released; the fault movement represents only the "elastic rebound" of the rocks from their condition of strain just prior to the fault. The data in Table 18-1 conform to the pattern of lines like $A''-B$ and $C''-D$, lending strong support to the theory.

Since 1907 the Coast and Geodetic Survey has made two further systematic studies of the same triangulation points, in 1922 and 1946, respectively. These surveys confirm a relative northerly displacement of triangulation stations to the west of the assumed fixed line Diablo-Mocho. The total movement seems to be at the rate of about 2 inches per year. If we could assume, which of course we cannot, that when this displacement again reaches the value of 1906 there will be another similar earthquake, about a century must still elapse. Such a suggestion is wholly without value, of course, for we cannot be sure that the same displacement will be needed before movement is renewed, nor that another nearly parallel fault might not take up the motion instead. But the existence of the movement is strong support for the "elastic rebound" theory.

Seismic Sea Waves

The great wave that overwhelmed Lisbon harbor was doubtless caused by displacement of the sea floor just to the west, perhaps accompanied by submarine landslides. The dramatic effects of earthquakes on the sea floor have already been mentioned in Chapter 9, in connection with the Japanese earthquake of 1923. Another example was furnished by the earthquake of November 18, 1929, located by instruments as having occurred near the Grand Banks in the North Atlantic. Within a few hours 12 telegraph cables were broken in 28 different places within a triangular segment of the sea floor about 400 by 200 miles. All the breaks took place in the deep sea in a region where the bottom topography is relatively smooth and the slopes less than 200 feet per mile. The breaks spread in rough sequence fanwise from a center, and the time intervals between earthquake and breaks indicate that the disturbance of the bottom traveled between 0.4 and 30 miles per minute. Presumably, the later breaks were caused by flowing bottom sediments; and the high apparent speed indicates perhaps there was more than one such stream. Even the lowest of these speeds is amazing when we think of the very gentle slopes down which the flowing debris traveled. This occurrence makes it easy to accept the inferences (from the character of the sediments) of large-scale slumping of sediments into deep basins off California and the Dutch East Indies (Chap. 16). The slopes here are far steeper, and earthquakes more numerous than near the Grand Banks.

Few earthquakes on land are accompanied by notable topographic changes (Chap. 9) and presumably this is also true of those beneath the sea, for relatively few earthquakes are accompanied by big waves such as those that wrought havoc at Lisbon. It is true that modern, continuously recording tide-gages reveal many sea waves presumably set up by earthquakes that would otherwise go unnoticed. Waves caused by earthquakes are called *seismic sea waves* or *tsunamis* (Japanese). In many, as at Lisbon, the first movement is a withdrawal of water, followed, after an interval measured in minutes, by a great inrush of the sea. Some of these waves are truly gigantic: Perhaps the greatest on record is one 210 feet high that broke on the south tip of Kamchatka in 1737; another 93 feet high struck the city of Miyako, Japan, in 1896; another, in 1868, carried the *U.S.S. Wateree* far inland from its anchorage off the Chilean coast and left it high and dry; waves from an earthquake off Peru were 8 feet high as they were recorded on Japanese tide gages 10,300 miles from their source. These heights depend very largely on the configuration of the shores; in the open ocean the waves are rarely noticed by ships.

The speed of the waves depends only on the depth of water. It is often 400 or 450 miles per hour in the Pacific, but much less in the shallower Atlantic. In Honolulu it is now a routine matter to send coastal warnings when a seismic sea wave is to be expected, as indicated by seismograph records of earthquakes thought to be submarine.

Earthquake Waves and Their Transmission

If we break a bat while hitting a baseball, our hands are stung by the vibrations transmitted through the wood. In the same way, when two huge blocks of rock slip past each other along a fault, elastic vibrations are set up and transmitted through the earth in all directions. The nature of these vibrations may be seen from Figure 18-3, *left*.

If we imagine the point P in Figure 18-3, *left*, is a particle within a uniform mass of perfectly elastic rock, and that it is pressed over to the right toward P' by some outside force, we can see that the material to its right is compressed, while that to its left is expanded. Since the rock is assumed to be perfectly elastic, these *changes in volume* mean that potential energy is stored up in it, and the compressed material tends to expand to its original volume and the rarified material to contract. Let us now consider the material along the center before deformation, represented by the line APB. In response to the external force that displaced P to P', this line is bent to some such position as $A'P'B'$. Thereby a *distortion*, or *shear*, is set up in it; each point on the line nearer to P tending to slip parallel to PP' farther from

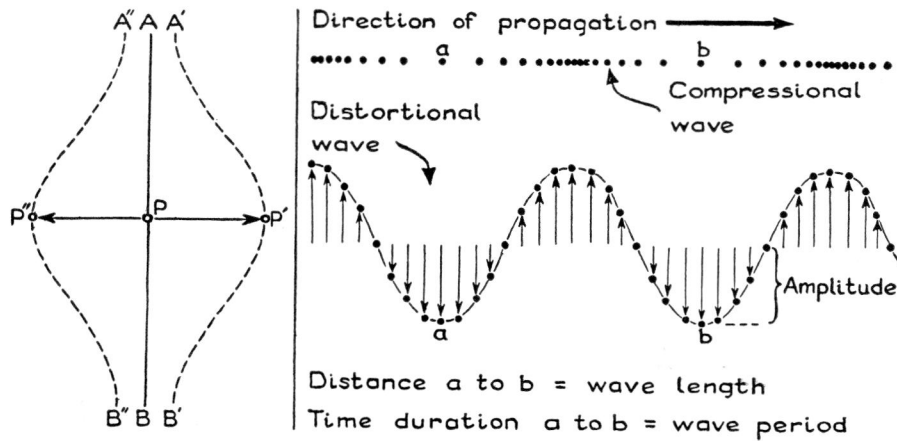

FIGURE 18-3. Left, *the distortion of line AB, during the passage of compressional waves at right angles to AB;* right, *the effect of compressional and distortional waves on the grouping of a series of equally spaced points. Note how the compressional wave affects their spacing in the direction of propagation, whereas the distortional wave offsets them as it passes from left to right.* (After J. B. Macelwane, When the Earth Quakes, *Bruce Publishing Company, 1947.*)

its original position on the line *APB* than its neighbor farther from *P*. This means that some energy is also stored up by distortion or shear.

Now let the outside force be removed. The point *P'* will not simply return to its original position *P*, but will have a momentum (like a pendulum) that will carry it on past *P* to a point such as *P''*. The part of the rock to its left will now be compressed and that to the right rarified; similarly, the shearing tendency will have been reversed in direction. At *P''*, the kinetic energy that carried it past the center will have been transformed again to potential energy; the point will again oscillate back through *P* and then to *P'*.

Of course rocks are not perfectly elastic. In nature some of the energy is consumed as frictional heat, so that each oscillation is less than the preceding one. Still the principle involved is not affected. If we consider the effects of the vibrations of *P* on its neighbors, and of these in turn on their neighbors, we see that two types of waves must emanate from the point and spread through the surrounding rock in *all* directions.

One of these waves is a *compressional* or *pressure wave* (a longitudinal wave) like a sound wave in air, the particles vibrating in the line of wave progress. It is called the primary or *P wave*. The other is a *distortional* or *shear wave* (a transverse wave) in which the particles vibrate at right angles to the direction of wave progress (Fig. 18-3, *right*). It is called the secondary or S *wave*. In the compressional wave there is alternate condensation and rarifaction in the line of progress; in the shear wave the particle motion is like that in a loosely hanging rope, one end of which is given a sharp flip. Each particle of the rope moves essentially at right angles to the length of the rope, but the wave travels from end to end of the rope. These transverse vibrations are not the visible waves in the ground surface noted in some strong earthquakes; their wave lengths are measured in tens of miles and their speeds in thousands of feet per second, far too fast for the human eye.

From the theory of elasticity the compressional waves in an ideal elastic solid travel at speeds which vary directly as the resistance of the solid to compression and shear, and inversely as its density.* The speed of shear waves depends on resistance to shear and inversely on density. If the density is the same, both waves travel the faster the more rigid and incompressible the rock. On the other hand, in two equally rigid and incompressible rocks the speed is greater in the less dense rock. As can be seen from the formulas, the compressional wave (P wave) travels faster than the shear wave (S wave).

The elastic properties of rocks, incompressibility and rigidity, can be measured in the laboratory. Assuming that the wave theories are correct, they can also be computed from the measured speeds of waves in the earth made by exploding dynamite (artificial earthquakes). In a dynamite explosion, we know the exact point at which the "earthquake" occurred, and hence can measure distance from the source accurately. By sensitive chronometers we can also time the explosion to less than $1/10,000$ second. These measurements fail to agree exactly with those made on the same rocks in the laboratory, but the deviations are generally not great and are thought to come from imperfect sampling, errors in measurement, and mainly to lack of homogeneity of the rocks in nature. When we think of all the faults, bedding surfaces, joints, gneissic structures, slaty cleavages, and other structures of rocks, this explanation of the discrepancies seems very reasonable. Certainly, rocks in the part of the earth accessible to us are far from being ideal, homogeneous solids.

The P and S waves are called "body waves" because they travel in all directions from their origin through the elastic body of the earth. In a homo-

---

* The formulas are: Velocity of P wave $= \sqrt{\dfrac{\kappa + 4/3\,\mu}{\rho}}$

Velocity of S wave $= \sqrt{\dfrac{\mu}{\rho}}$; in which $\kappa =$ bulk modulus (a measure of incompressibility or resistance to change in volume); $\mu =$ modulus of rigidity (a measure of resistance to shear or change of shape; and $\rho =$ density. It is assumed that the medium is isotropic and perfectly elastic.

geneous medium (see Chap. 3) waves travel in straight lines, but in the earth's outer parts we know that the medium is not homogeneous. When these waves pass from one rock into another with differing elastic properties, they are bent or refracted like water waves or light rays. If they meet the boundary at an angle greater than a critical value which depends on these properties, they are reflected, just as light is reflected from a water surface or sound echoes from a cliff. (The P wave is indeed a sound wave in rock.) Furthermore, when either the P or S wave strikes a sharp boundary—a *discontinuity*—it sets up new waves. This happens, for example, at every geologic contact and at the surface of the earth. These new waves generated at the discontinuity include four new body waves (a new compressional and a new shear wave corresponding to each of the original P and S waves). Besides these, there are formed at least three kinds of "surface waves," so called because they travel along, or at least close to, the discontinuity.

The surface waves have considerably lower speeds than either P or S waves. They are called *"long"* or *"L waves"* and are far more complex. They spread along the surface only, instead of through the entire mass of the earth.

From this brief discussion, it is obvious that earthquake waves are prodigiously complex. Furthermore, they are generally perceptible to ordinary observation only near the source. Clearly, delicate instruments are needed for recording them effectively. We will turn now to a brief discussion of these instruments, called seismographs, before passing to the deductions they enable us to make about the structure of the earth.

Seismographs

Recording earthquake vibrations requires the establishment of a point that can move as nearly independently of the earth as possible. We all know that we can knock the bottom book out of a pile laid on the table so quickly that the books above are hardly disturbed and so fall almost vertically, without tipping over. The inertia (tendency to resist acceleration) of the upper books prevents them from traveling along with the quickly accelerated book. If we move the bottom book slowly, the acceleration is communicated upward, and thus the whole pile goes along with the bottom book. The *seismograph* is an instrument designed to measure the displacement of the ground with respect to a mass that is, like the upper books in our example, as independent as possible of the support or ground surface upon which it rests. When the support (which is of course rigidly attached to the earth) moves quickly under the impulse from an earthquake wave, as little motion as possible is transmitted to the main mass of the seismograph. If, by a suitable de-

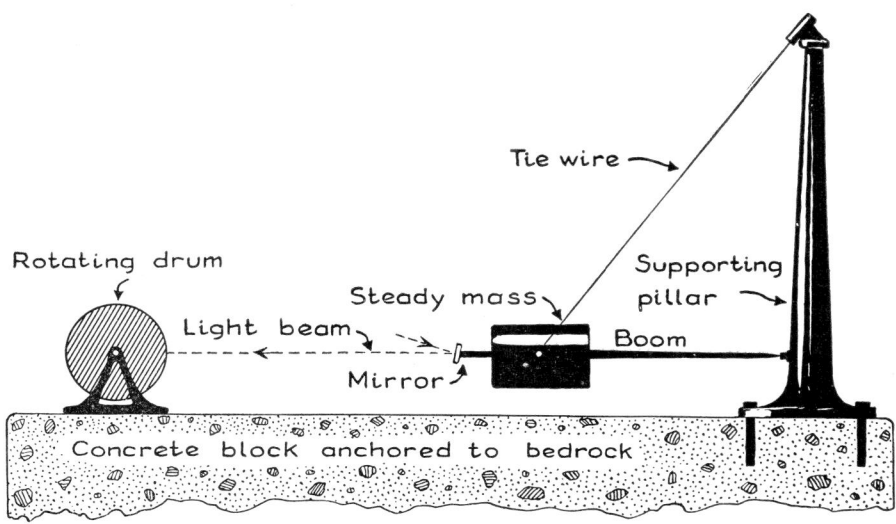

FIGURE 18-4. *Diagrammatic sketch of a horizontal pendulum seismograph.*

vice, we can measure the displacements between mass and support, we thereby measure the ground motion during the earthquake.

This problem has been attacked in many ingenious ways: Masses have been supported by springs; long pendulums have been suspended from high structures (so their time of swing would be great compared with that of earth vibrations); a third solution, and for many years the most widely adopted, is that of the so-called "horizontal pendulum." In this device a pendulum, consisting of a steady mass and a boom (Fig. 18-4), is held by a wire fastened to a supporting pillar. The steady mass thus tends to remain in its position of rest as the support vibrates during an earthquake. Either by a delicate pen attached to a lever from the pendulum weight, or by a beam of light reflected from a mirror on the pendulum to a moving strip of photographic paper, the relative motion of the mass and support is recorded. Such a seismograph measures only the components of motion at right angles to the length of the pendulum; to record the horizontal movements completely two such pendulums are needed, one supported so that it hangs north and south, the other east and west. We also need a spring suspension for measuring vertical movements, but the principle is the same. Time is marked automatically on the records, or *seismograms,* usually by an electric clock suitably linked to the device moving the record. In this way the arrival time of the several vibrations can be read from the record within a second, or, on some seismographs, considerably more accurately.

Inferences from Seismograph Records

By comparing seismograph records of the same earthquake at different stations, and of different earthquakes at the same station, seismologists are able to recognize the various kinds of waves that elastic theory predicts. In general, a seismogram shows several different kinds of waves, each of which can be related to a definite wave path through the earth. For example, in the seismogram shown in Figure 18-5 the points indicating the arrival of the *P*, *S*, and *L* waves, are labeled. The arrival of a reflected compressional wave, *PP*, and of a reflected shear wave, *SS*, are also indicated (cf. Fig. 18-8).

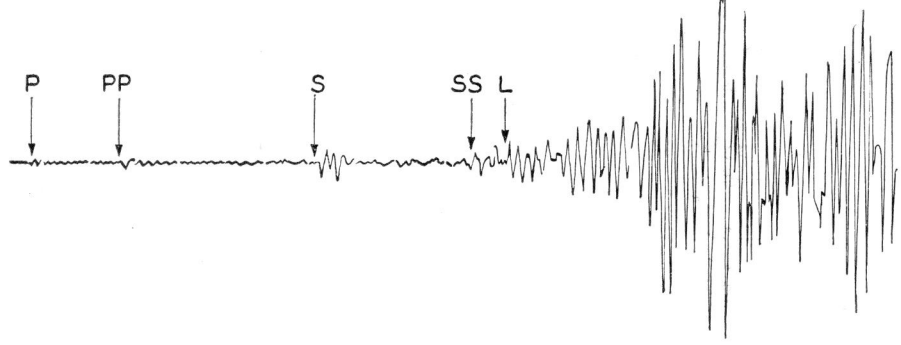

FIGURE 18-5. *Seismogram with letters indicating the arrival of the various earth-quake waves. (After L. D. Leet, Practical Seismology and Seismic Prospecting, Appleton-Century-Crofts, 1938.)*

Refraction and Reflection at Depth

By tabulating the travel times of the waves from an earthquake of known source, and by identifying the several wave groups on the seismograph records of many stations, "time-distance" tables have been made. When these are graphed, by plotting the distances from the sources (measured in degrees on the earth's surface) against travel time, we have a *time-distance curve* (Fig. 18-6). It is significant that for the *P* and *S* waves such lines are curves that are concave toward the axis of distance, that is, the farther the source from the station, the faster the apparent speed of wave travel. This means that the waves must travel faster as they penetrate deeper within the earth. For the *L* wave, which travels along the surface, the time distance curve is a straight line (Fig. 18-6); therefore its travel time is in direct ratio with distance from the source.

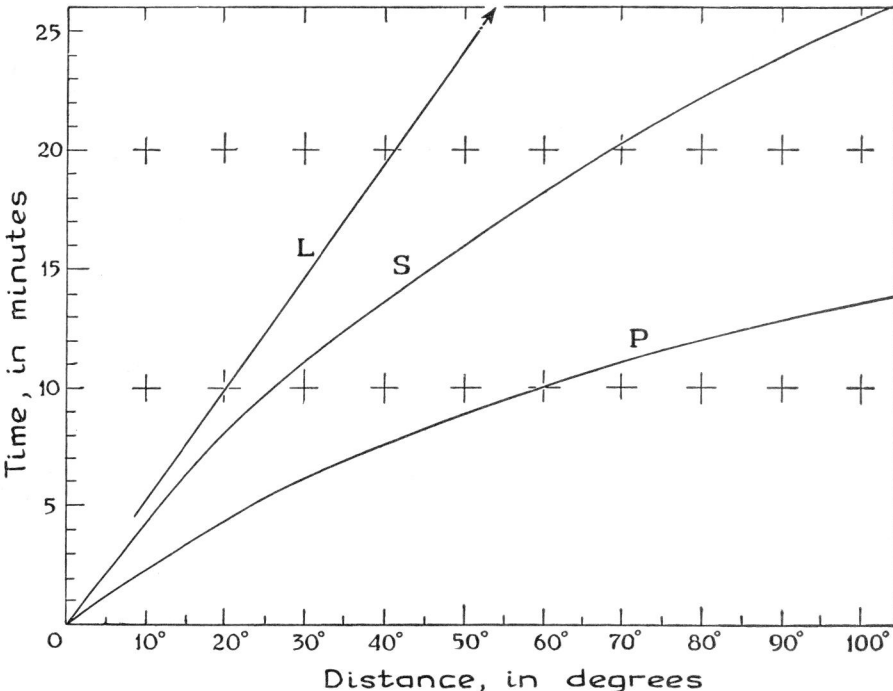

FIGURE 18-6.   *Time-distance curves for the three principal earthquake waves.*
(Data from L. D. Leet, 1938.)

Because of the increase in travel speed with distance, it is apparent that the P and S waves do not travel along the surface but through the body of the earth, and the deeper their paths, the greater their speeds (Fig. 18-7). From this it follows that their paths are curved. Therefore, at points distant from the source they emerge at the surface at higher angles than they would if they followed straight lines. When they strike the surface they are reflected back and proceed again in curved paths (Fig. 18-8). Accordingly, a whole train of reflected waves is recorded on seismograms at distant stations, as shown in Figure 18-8.

We know from astronomical data that the average density of the earth is about 5.52, whereas the average density of surface rocks is only about half this. Therefore the material making up the interior of the earth must be far denser than that at the surface. As the formulas for wave speeds show (footnote, p. 479), if only the *density* of the rocks increased with depth and their other properties remained unchanged, the waves should actually be slower, as they go deeper. Since they speed up, we can say that the elastic properties—rigidity and incompressibility—must *increase even more than*

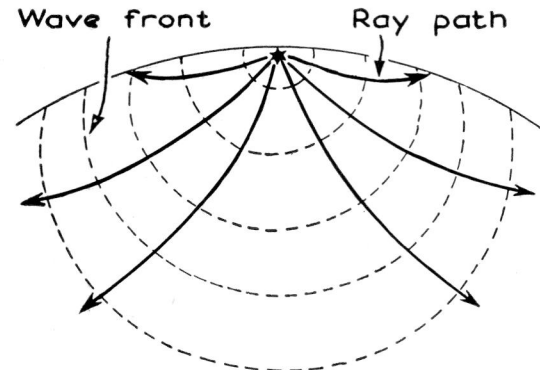

FIGURE 18-7. *The bending of rays of earthquake waves caused by the increase of wave velocity with depth. The rays must always move at right angles to any small segment of the ray front.*

the density. We shall return to this important fact later in the chapter, in summarizing our inferences as to the deep interior of the earth, but before doing so let us see what other information the seismograms reveal.

Epicentral Distance

The sharply pulselike records on seismograms suggest that most earthquakes begin in a very localized area, even though many miles of fault may ulti-

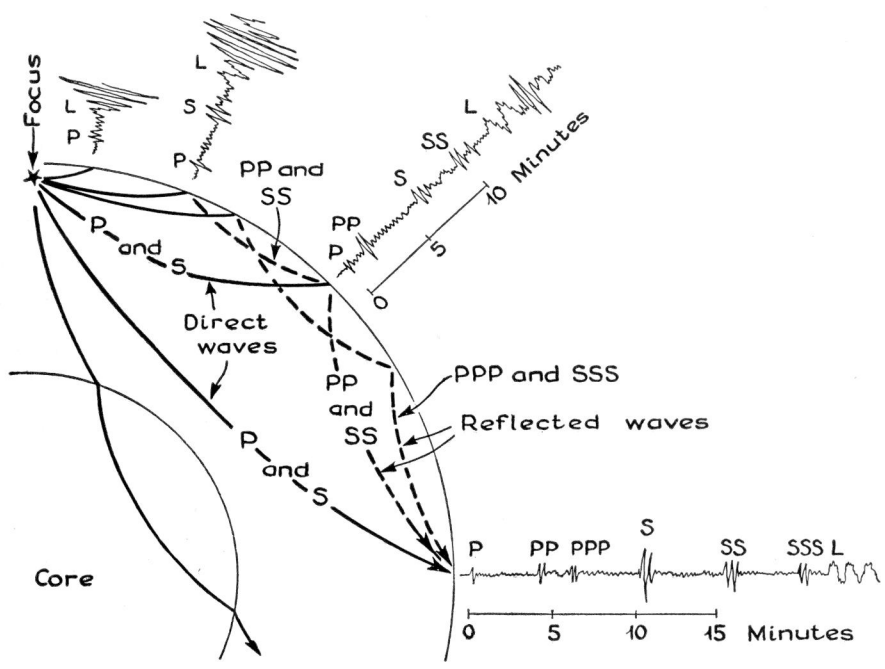

FIGURE 18-8. *Section through a part of the earth, showing the paths of a few of the many earthquake waves and the records they leave on the seismograms of four stations. Note the reflected waves PP, PPP, etc. The time scale of all the seismograms is the same. (After J. H. F. Umbgrove,* Pulse of the Earth, *Martinus Nijhoff, 1947.)*

484

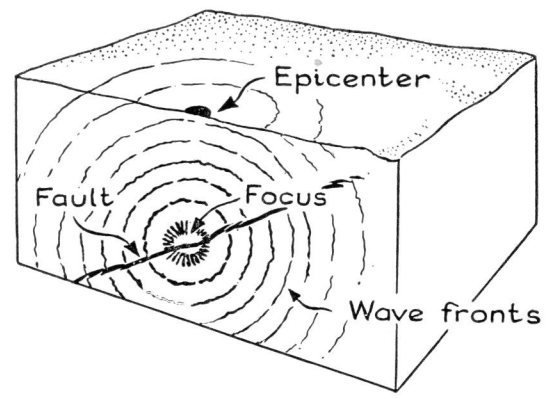

FIGURE 18-9. *Diagram showing the position of the focus and epicenter of an earthquake caused by local deep-seated movement on an inclined fault.*

mately be active, as in the San Francisco earthquake. The point from which the first movements appear to start is called the *focus*. The point on the surface vertically above the focus is called the *epicenter* (Greek: "above the center"). See Figure 18-9.

The time lag between the P and S waves shown on a seismogram enables us to read the "epicentral distance" (the distance from the station to the epicenter) from the time-distance curves. This distance is measured in degrees of arc of the earth's curvature. If we draw a circle on a globe, using the seismograph station as the center and the epicentral distance as the radius, we know that the shock must have occurred somewhere on the periphery of this circle. If we have adequate records from at least three stations, it is possible to fix the location of the epicenter at the point where the three circles drawn about the three stations intersect (Fig. 18-10). In fact, it is sometimes possible to find both the distance and direction of an earthquake from a single station by observing the direction of first movement and the relative amplitudes as shown on the vertical and two horizontal records.

Isoseismal Lines

Near the most intensely shaken area we ordinarily have no quantitative way of measuring the force of the tremors. We must rely on qualitative information such as the destruction caused or the perceptibility of the shock to persons. Many "scales of intensity" have been suggested by students of earthquakes, but all are highly influenced by the psychology of the people and the methods used in questioning them, the density of population, the nature of the ground, and the quality of the building construction in the area. Most scales now in use are modified from one proposed by the Italian seismologist Mercalli. The most widely used scale in America is the "Modified Mercalli" or Wood-Neumann scale (1931).

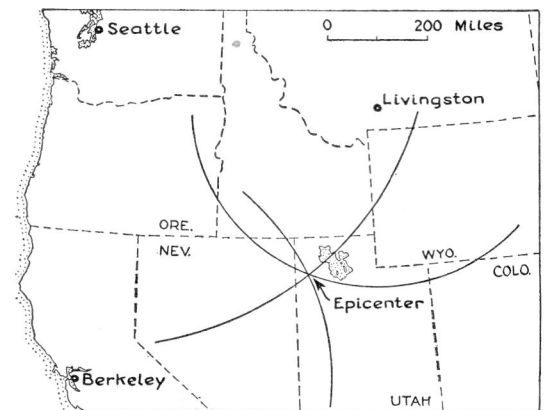

FIGURE 18-10. *Locating an epicenter in northwest Utah from seismograms recorded at Seattle, Berkeley, and Livingston stations.*

Abridged Wood-Neumann Scale of Earthquake Intensities

I. Not felt except by very few, favorably situated.

II. Felt only on upper floors, by a few people at rest. Swinging of some suspended objects.

III. Quite noticeable indoors, especially on upper floors, but many people fail to recognize it as an earthquake; standing automobiles may sway; feels like passing truck.

IV. Felt indoors by many during day, outdoors by few; if at night wakens some. Dishes, windows and doors rattle, walls creak; standing cars may rock noticeably. Sensation like heavy truck striking building.

V. Felt by nearly all; many wakened; some fragile objects broken and unstable objects overturned; a little cracked plaster; trees and poles notably disturbed; pendulum clocks may stop.

VI. Felt by all; many run outdoors; slight damage: heavy furniture moved, some fallen plaster.

VII. Everyone runs outdoors; slight damage to moderately well-built structures, negligible to substantially built, but considerable to poorly built; some chimneys broken; noticed by automobile drivers.

VIII. Damage slight in well-built structures; considerable in ordinary substantial buildings with some collapse; great in poor structures. Panels thrown out of frame structures; chimneys, monuments, factory stacks thrown down; heavy furniture overturned; some sand and mud ejected, wells disturbed; automobile drivers disturbed.

IX. Damage considerable even in well-designed buildings; frame structures thrown out of plumb; substantial buildings greatly damaged, shifted off foundations, partial collapse; conspicuous ground cracks; buried pipes broken.

X. Some well-built wooden structures destroyed; most masonry and frame structures destroyed, with their foundations; rails bent, ground cracked; landslides on steep slopes and river banks; water slopped over from tanks and rivers.

XI. Few if any masonry structures left standing; bridges destroyed; underground pipes completely out of service, rails bent greatly, broad cracks in ground and earth slumps and landslides in soft ground.

XII. Damage total: waves left in ground surface and lines of sight disturbed; objects thrown upward into the air.

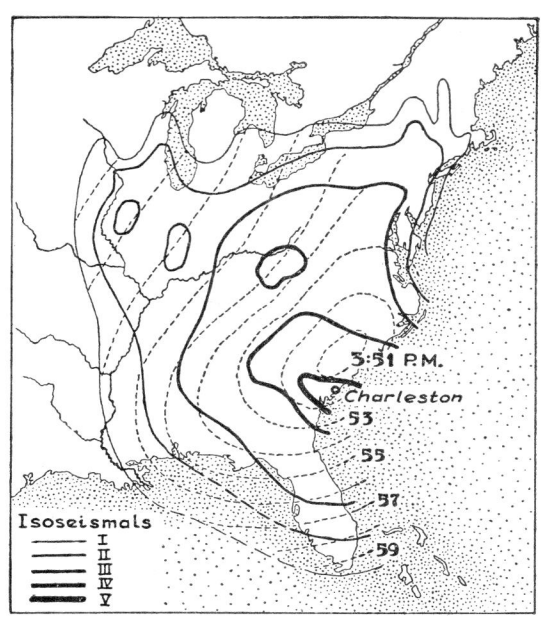

FIGURE 18-11. *Isoseismals of the Charleston earthquake, 1886. The fine dashed lines (called coseismals) connect points where the earthquake struck at the same time; the heavy numerals indicate their actual arrival time. (After Dutton; redrawn from J. B. Macelwane*, When the Earth Quakes, *Bruce Publishing Company, 1947.)*

Local factors vary too much for such estimates of intensity to be either quantitatively accurate or directly comparable from one locality to another. Nevertheless, much information valuable to engineers, insurance underwriters, civic planners, and others can be derived from a systematic study of the way intensities are distributed. In the United States, immediately after a destructive earthquake, the Coast and Geodetic Survey carries out such studies by distributing large numbers of postcards to postmen for delivery to the public along their routes. These cards contain lists of characteristic earthquake effects which the people receiving them are asked to check to correspond with their own experiences during the shock. When the returned cards are tabulated by locality, a map can be compiled showing the intensities according to the Wood-Neumann scale.

Lines drawn on such a map through points of equal intensity are called *isoseismal* lines (see Fig. 18-11). The isoseismals generally lie in rough ovals about a center which either coincides with or is close to the position of the earthquake epicenter as determined from seismograph records. The fact that the instrumentally determined epicenter commonly does not quite coincide with the "field epicenter" may be accounted for by differences in surficial geology. Differences in the nature of the underlying rock not only influence the distribution of earthquake damage and therefore the isoseismals, but also control the speed of transmission of elastic waves to the nearby seismographs and thus affect the instrumental records. A complex pattern of damage characterized the San Francisco earthquake, because hills of hard rock alternate abruptly with small basins filled with unconsolidated muds and silts.

Earthquake Magnitude

The intensity scale gives only a crude idea of the effects as they are observed locally and does not give a clear measure of the actual amount of energy released by the shock. How can we find some way of comparing the amounts of energy released *at the source* in different earthquakes? To answer this question, we need a scale that does not depend on the type of building construction or the nature of foundations.

Earthquake body waves carry energy away from the source in all directions; therefore it is clear that the amplitude of any wave must diminish with distance from the source. This enables us to build a better scale by which to estimate the energy released at the source.

The method is by no means accurate. No single seismograph can be sensitive to all of the waves of different periods in which earthquake energy is carried. Also, no two seismographs will react identically, because of differences in the foundation on which they are placed, although most seismographs are placed on bedrock to avoid extreme differences in stability. Nor is it likely that the same proportion of the total energy of every quake is always carried in waves of a particular period. Nevertheless, if these limitations are disregarded, it is possible to compare source energies carried by waves of particular periods by examining their amplitudes as recorded on seismographs from instruments of identical mechanical properties. Even if the method is not precise, the tremendous range in energy of earthquakes makes even a rough scale useful.

The American seismologist C. F. Richter has developed such a scale. It is based on the logarithm of the amplitude of the largest horizontal trace made by a seismograph of standard mechanical properties 100 km. from the epicenter. Records from such standard instruments at several stations make it possible to compute what the amplitude *should have been* if one of them had happened to be located at this standard distance.

Thus, on this scale Nevada's Sonoma Range earthquake of 1915 (Chap. 9) had a magnitude of $M = 7.75$, only a little less intense than the San Francisco earthquake of 1906 ($M = 8.25$) or the Tokyo earthquake of 1923 ($M = 8.1$), although its damage to life and property was negligible.

The highest magnitude yet determined is about 8.5. Magnitude $M = 2$ corresponds to a shallow shock barely perceptible near the epicenter; magnitude $M = 7$ is the lower limit of a major earthquake. Each unit on this scale of magnitudes corresponds to an energy release about 60 times that of the preceding. Thus a magnitude 8 shock releases as much energy as 200,000 shocks of magnitude 5.

Depth of Focus

Since the L wave cannot start until the P wave hits the surface, the lag of the L waves gives us a clue to the depth of the earthquake focus. So, too, does the time lag between P and S waves. From these data the depth of focus of most earthquakes can be computed.

The majority of earthquakes—the great bulk from which the time-distance graphs have been computed—have focal depths between 5 and 20 miles. If the depth of focus is less than about 5 miles, the earthquake is rarely felt for any great distance, though it may be strong at the epicenter. Such shallow earthquakes are abundant near volcanoes, especially just before and during eruptions, and thus were formerly grouped as a separate class whose origin supposedly was different from that of "normal" or tectonic earthquakes. However, their records have been found to be similar to those of normal shocks; and hence they are now thought to originate in the same way, by movements of rocks along faults, even though the forces that bring about the faulting may be the result of a bursting magma chamber rather than tectonic forces.

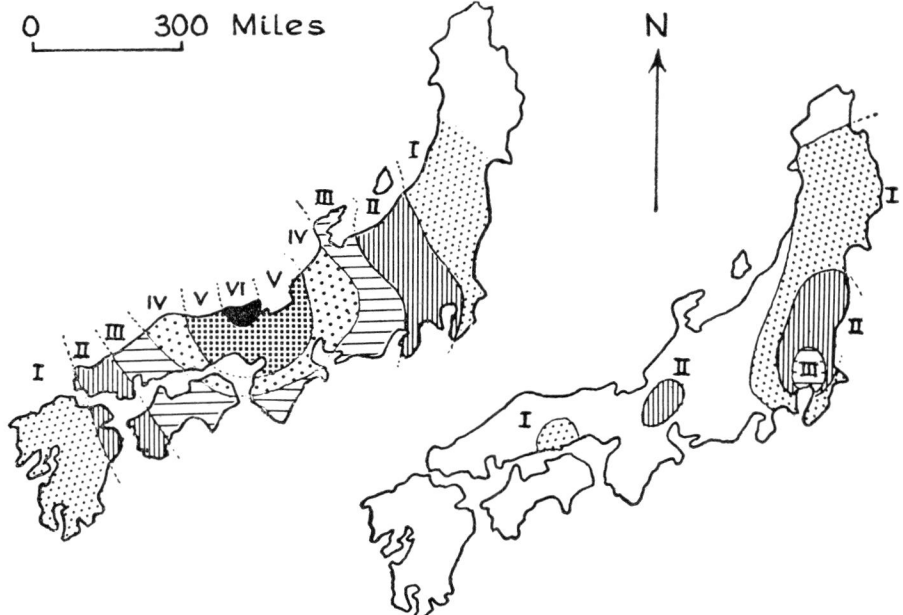

FIGURE 18-12. *Maps of Japan showing the arrangement of isoseismals of the normal North Tazima earthquake* (left), *and of the deep-focus earthquake of March 29, 1928* (right). (*After Wadati; redrawn from J. B. Macelwane,* When the Earth Quakes, *Bruce Publishing Company, 1947.*)

Deep-Focus Earthquakes. About 4 per cent of the recorded earthquakes differ from "normal" in having few or no L waves recorded, and also, when the normal time-distance curves are used, they arrive too soon at distant stations, as if they had taken a "short-cut" from their origin. Furthermore, when the isoseismals for some of these are drawn, the areas of perceptibility show very erratic patterns, in notable contrast with the fairly systematic arrangement of the isoseismals of a "normal" earthquake (see Fig. 18-12). All of these features are taken to indicate that the focus lies very much deeper than in normal earthquakes. Some of the focal depths determined for such shocks range down as far as 700 kilometers (435 miles), many are deeper than 100 km. These shocks are not due to mysterious explosions within the earth, as has been suggested. If they were, the first P motion at all stations should be a compressive movement. This is not the case, as seen by their records which, except for the anomalies already mentioned, are identical with those of earthquakes known to be connected with faulting. We have, then, a further clue to the conditions at great depth in the earth—the rocks are strong enough to depths as great as one-eighth of the earth's radius to accumulate elastic strain and to fault like those at shallow depth.

Practically all of these very deep earthquakes are associated with the island arcs of the Pacific or with the Andean chain of South America; none has been recorded from beneath the North American continent.

When the depths of focus of the earthquakes along an island arc are plotted on a vertical section at right angles to the trend of the arc, some strikingly systematic associations are found. The deeper shocks invariably originate farther within the concave side of the arc; in fact, the foci rise successively in depth along a plane that intersects the sea floor at the boundary of arc and foredeep (Fig. 18-13). In South America, too, the deeper shocks center farther from the Pacific.

Some geologists have suggested that these shocks arise on thrust faults along which the island arcs and South America are overriding the floor of the Pacific. For some shocks at least, this theory is supported by the direction of the first motion recorded on seismograms from variously located stations. Certainly, the systematic distribution of the shock centers on planes dipping away from the Pacific Basin supports this idea.

Distribution of Earthquakes

Seismologists estimate that every year the earth shakes with more than a million earthquakes which are potentially strong enough to be felt ($M = 2$). Clearly, only a very small number are strong enough to be recorded for

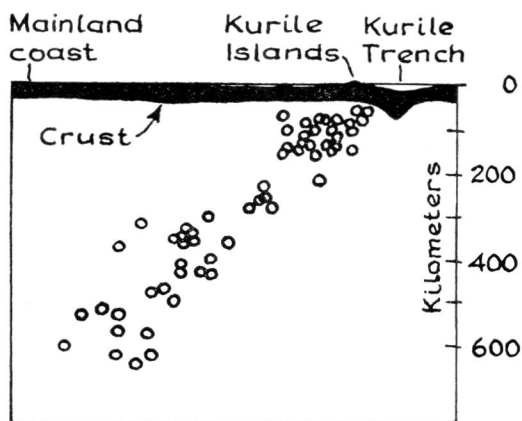

FIGURE 18-13. *Cross section of the outer earth, at right angles to the Kurile Island Arc, showing the foci of deep earthquakes. There is no vertical exaggeration. (After H. H. Hess, 1948.)*

any distance from their sources. It has been estimated that about 220 great shocks ($M > 7.75$) and about 1,200 other strong earthquakes ($M = 7.0$ to 7.7) occur per century over the entire earth. Figure 18-14 shows the distribution of epicenters of recorded earthquakes.

Small shocks ($M = 5$ or less) of shallow depth apparently occur nearly everywhere over the earth, but the larger shocks are very differently distributed. The crust of the earth seems to consist of large blocks, notably the Pacific Basin (except for the neighborhood of the Hawaiian Islands) and most of the shield areas of the continents which are essentially free from large earthquakes. Between these blocks are the zones of major seismic activity. These include:

1. The Circumpacific belt of young mountains, including its branches

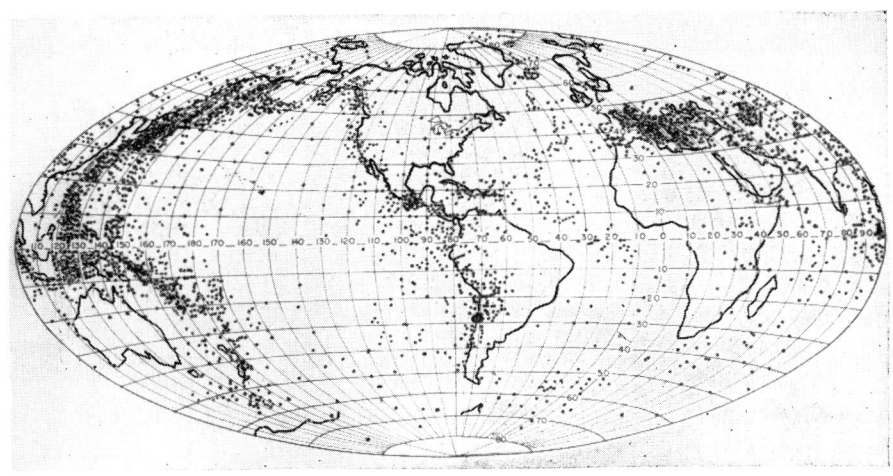

FIGURE 18-14. *Epicenters of most of the earthquakes recorded between 1899 and 1930. (Reproduced by permission from L. D. Leet,* Practical Seismology and Seismic Prospecting, *Appleton-Century-Crofts, 1938.)*

through the Antilles and through the island arcs south of Japan. Here occur about 80 per cent of all shocks with foci at less than 60-km. depth.

2. The Mediterranean and trans-Himalaya zone, in which occur nearly all the remaining large shallow shocks.

3. The Mid-Atlantic and Mid-Indian ridges, along which many shallow shocks occur.

4. The Hawaiian Islands and African rift valleys, with which moderate activity is associated.

The deep-focus shocks are still more localized: 90 per cent of those whose origins lie between 60 and 300 km. and all that have thus far been recognized from depths greater than 300 km. are in the Circumpacific belt.

The Crust of the Earth

The Yugoslav seismologist Mohorovičić (pronounced Mōhōrōvee'cheech) in 1909 found certain features in seismograms that indicate a layered structure for the continental segments of the outer part of the earth. He saw that earthquakes occurring less than about 800 km. (495 miles) from the recording station gave records of two compressional and two shear waves instead of one. By comparing travel times to more distant stations, the smaller pair could be identified as the normal P and S waves long known on records of distant shocks. The larger pair traveled more slowly but seemed to have started earlier. They were therefore received first on nearby stations (up to about 150 km.) but lagged more and more behind P and S at more distant stations, finally becoming unrecognizable at 800 to 1,000 km. from the source. Mohorovičić showed these facts could be explained on the assumption that the earth has a layered structure, with an outer layer, in which speeds are relatively slow, resting on the deeper body of the earth, where speeds are higher. A shock originating in the upper layer will send waves directly to a nearby station. These waves will be powerful and will arrive before other waves that penetrate to the deeper layer, even though the travel speed in the deeper material is high. But at more distant stations the waves that penetrate to the lower layer travel within it at speeds high enough to more than make up for the time they took to travel from the focus down to the lower layer (see Fig. 18-17). Hence they will arrive first.

Later studies with better records than those available to Mohorovičić have shown that still a third pair of compressional and shear waves can be recognized at stations between about 100 and 1,000 km. from the epicenter of shallow shocks. These suggest two layers rather than a single one (Fig. 18-15). The *upper layer* seems usually to be rather sharply separated from the *intermediate layer*, which, in turn, is even more distinct from the *lower*.

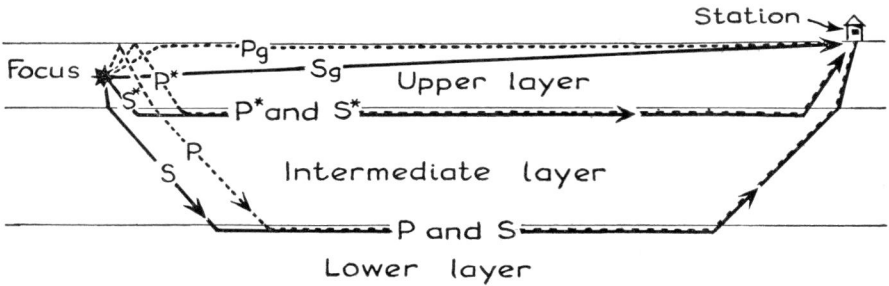

FIGURE 18-15. *Diagrammatic cross section of the continental crust, showing the paths of the three sets of P and S waves. The Pg and Sg waves traverse the upper (granitic) layer; the P° and S° waves travel chiefly through the intermediate (basaltic) layer. (After H. Jeffreys,* The Earth, *Cambridge University Press, 1929.)*

The name "layer" seems hardly appropriate for this deeper zone, for it grades gradually into the deeper body of the earth. The discontinuity between the intermediate and lower layer—a fundamental feature of earth structure—is called the *Mohorovičić discontinuity. It marks the base of the crust of the earth.*

The speed of the various waves in the continental crust varies somewhat but on the whole is surprisingly constant in each layer (Table 18-2). The

TABLE 18-2. *Wave Speed in Continental Crustal Layers (km. per sec.)*

	UPPER LAYER	INTERMEDIATE LAYER	LOWER LAYER (TOP)
Compressional (P-type) wave	5.5	6.2 to 7.2	7.8 to 8.1
Shear (S-type) wave	3.3	3.6 to 3.8	4.3 to 4.8

apparent delay in starting time of the waves in the deeper layers gives a way of measuring the thickness of the layers. Though the thickness varies from place to place (and in many places a sedimentary blanket rests on the upper layer and further complicates the interpretation) the upper layer generally ranges between 10 km. and 30 km. in thickness. The intermediate layer is between 20 and 40 km. thick. The combined thickness of the two layers above the Mohorovičić discontinuity averages roughly 30 to 40 km. (18 to 25 miles). In places, as beneath the Sierra Nevada and the Alps, they are nearly twice as thick.

The Continental Crust.   To summarize the facts disclosed by these studies of crustal structure, we can say that the crust beneath the continents has a surprisingly uniform structure. From geologic mapping and from seismic and gravity studies we of course know that in many parts of the earth there are deposits of sedimentary rocks, ranging in thickness up to perhaps 15 km. (10 miles). Compared to the whole body of the earth, however, this mantle is extremely thin, and for North America has been estimated to average only 0.8 km. (about 0.5 mile) for all Cambrian and younger rocks, and is almost surely less than 2 km. (1.5 miles) on the average for all sediments including the pre-Cambrian. Beneath this pellicle of sediments the continents seem to have an "upper layer" in which the wave speeds are remarkably uniform. This is surprising, for wherever we see the basement rocks that underlie the sediments they are highly varied in composition, including granite, migmatites, and many other intrusive rocks, as well as many kinds of schists and gneisses.

Perhaps the nearly uniform speeds in the upper layer are averages of more diverse speeds in local rocks, distributed evenly enough so as to give about the same properties over large areas. These speeds are nearly identical whether measured in Europe, Japan, or North America. Furthermore, they approach closely those computed from the elastic properties of granite, the dominant rock of the continental basement. It seems reasonable to conclude that the average rock of the "granitic layer," though varied and largely not a true granite, resembles granite in that it has mostly minerals rich in silica (Si) and alumina (Al). Following a suggestion of the Austrian geologist Suess, the material of this layer is called by the coined name, *sial.*

Beneath the sial, or granitic layer, is the "intermediate layer." As we can see from Table 18-2, the wave speeds recorded in this layer are by no means so uniform as those from the granitic layer. Local deviations of elastic properties (and hence perhaps of rock type) may be considerable. Some seismologists suggest that a less distinct horizontal layering can be made out within this layer, and that it becomes more highly elastic toward the base. Despite these complexities, it is significant that on all the continents a fairly distinct layer 20 to 40 km. thick can be recognized beneath the granitic layer and above the Mohorovičić discontinuity. The elastic properties do not seem to correspond precisely with those to be expected in any particular rock thus far examined in the laboratory, but they do not differ greatly from those of gabbro or basalt. As there is independent evidence of widespread basalt in the earth, the layer has been called the "basaltic layer" and because magnesia is a prominent component, the material of the layer is also referred to as *sima* (Si = silica; Ma = magnesia).

As we have mentioned above, the rocks immediately beneath the Mohorovičić discontinuity constitute the "lower layer" of the seismologists. There seem to be no sharp discontinuities within this layer for long distances down into the earth. The elastic properties of the lower layer vary slightly from place to place, at least near its top, but at depths of 60 km. (or at most 100 km.) they appear to be uniform over the entire earth. The rock whose elastic properties as measured in the laboratory seem most nearly comparable to those of the lower layer is peridotite, perhaps a variety especially rich in olivine. We must, however, maintain a lively skepticism about the exact composition of all rocks below the zone of observation, and refrain from taking a positive position about a peridotite layer.

The Crust Beneath the Oceans. Seismograms suggest that the crust beneath the oceans differs from that beneath the continents. The difference is especially notable for the Pacific. The evidence is of two kinds: (1) from the energy of reflected body waves, and (2) from the speeds of surface waves.

The proportion of the wave energy reflected back into the earth is greater the higher the angle at which the wave strikes the surface; it is a maximum for waves striking at right angles. The significant feature in crustal studies is that earthquake waves of equal epicentral distance, one of which has been reflected from a continental surface and the other from the bottom of the Pacific Ocean, differ greatly in intensity. The oceanic reflection is the weaker. Yet, if the crustal layering were the same, they should have about equal amplitudes for shocks of equal magnitude. These differences suggest that the rays make smaller angles with the surface beneath the Pacific than with the surface of continents, a feature that would be explained if the Pacific segment lacked a granitic layer (Fig. 18-16). The curvature of the rays in passing through the continental granitic layer insures their striking the surface at a higher angle than they would in the absence of such a layer (Fig.

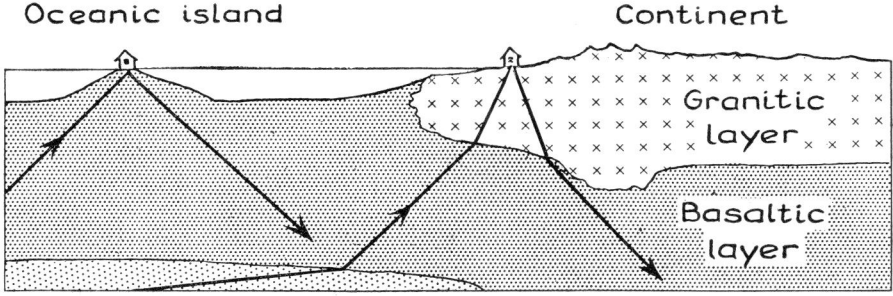

FIGURE 18-16. *Angles of incidence of earthquake waves arriving beneath ocean basins and continents. (After B. Gutenburg,* Internal Constitution of the Earth, *courtesy Dover Publications.)*

18-16). Hence the larger energy carried by waves reflected from a continental surface suggests the absence of such a layer beneath the Pacific floor.

The surface (L) waves tell a similar story. We know that surface waves differ from body waves in that their speeds vary with their wave lengths. This means that we can tell something about how the elastic properties of the material below the surface vary with depth, for a surface wave of a given wave length affects a layer about as deep as the wave is long. A wave with a length of 3 km. travels at a rate determined by the elastic properties of the material within 3 km. of the surface; one with a wave length of 200 km. travels at a rate affected by the elastic properties of all the rocks from the surface to a depth of about 200 km.

Surface waves of short wave lengths travel at different speeds in different parts of the crust. Their speeds across segments of the Pacific floor are higher than across other areas, but longer surface waves, whose speed is more dependent on the properties of deeper material, travel at nearly the same speed, whether across continent or ocean basin.

The American seismologist Beno Gutenberg has shown that these facts are consistent with the absence of a granitic layer from the floor of the Pacific. Possibly rock like that deep within the intermediate layer of the continents floors the Pacific, but perhaps even this is absent, and the lower layer is present immediately beneath the deep-sea sediments. It also seems likely that the granitic layer is absent or very thin and patchy beneath the Atlantic and Indian oceans, so that the intermediate layer of the continental segments directly underlies the oceanic sediments. Indeed, the speeds suggest that the lower layer may be present just below the sediments over much of the North Atlantic.

Short surface waves lose considerable energy in crossing the boundary of the Pacific Basin. Such losses do not occur at other oceanic borders, nor are they found with waves of very long period even at the Pacific boundary. Hence, although at a depth of 60 km. (37 miles) or more the elastic properties of the rocks are practically identical over all parts of the earth, there is considerable difference in the rocks at shallower depths beneath continents and oceans. This contrast is most pronounced between the Pacific Basin and the continents, though it also exists between the continents and the other oceanic floors.

The Deep Interior

The fact that the speed of body waves increases with depth indicates that the rigidity and incompressibility of the deeper rocks increase more or less steadily downward. The increased speed cannot be due to lower density of the deeper layers, for we know from physical measurements that the density of the earth as a whole (5.52) is far higher than that of surface rocks (2.2 to about 3.2). Furthermore, the approach of large surface blocks to isostatic balance implies that they are essentially floating on a substratum that must have a higher density.

This relation of elasticity increasing faster than density with depth continues for a long way below the discontinuity at the base of the crust, and then changes abruptly. *P* waves emerging at about 102° to 104° epicentral distance have followed a curved path whose deepest portion lies about 2,900 km. (1,740 miles) beneath the surface. Within a short distance beyond this range, however, both *P* and *S* waves suddenly fade out. The S wave cannot be detected beyond this zone (except those that are reflected from discontinuities), and the *P* wave is found only by sensitive seismographs. But at epicentral distances of about 143°, a new and very powerful *P* wave reappears. The travel time for this wave is much slower, however, than would be found from time-distance curves of the ordinary *P* waves (Fig. 18-17).

If this deep-penetrating *P* wave had traveled through the center of the earth at the speed it had at a depth of 2,900 km. it should arrive on the opposite side of the earth about 16 minutes after the earthquake. Instead, it arrives 4 minutes later. This delay can only mean that the speed through the central core of the earth is far less than it is above the core. Also, if the speed in the core is slower than in the upper part of the earth, we can see a

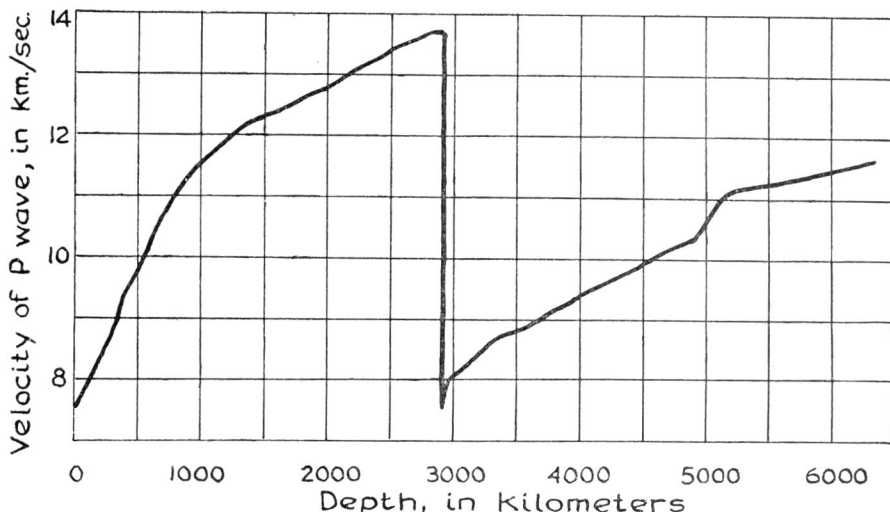

FIGURE 18-17. *Graph of velocity of the P wave in the interior of the earth, as related to the depth of its penetration.* (*After B. Gutenburg,* Internal Constitution of the Earth, *courtesy Dover Publications.*)

reason for the weak records between epicentral distances of 102° and 143° —the so-called *shadow zone.*

A medium in which speed is low will bend the wave paths toward it, just as a reading glass bends the light rays to a focus in line with its thickest portion (the speed of light is less in glass than in air). Thus the central part of the earth acts like a huge converging lens. This is the basis for the interpretation indicated in Figure 18-18, which shows the earth with a central *core,* whose radius is about 3,400 km. (2,100 miles), surrounded by a *mantle* that extends from the core boundary to the Mohorovičić discontinuity at the base of the *crust.*

Though the speed of the *P* wave through the core is twice the speed of sound in steel, it travels much slower than in the deeper parts of the mantle. The diagram of Figure 18-18 shows how these facts explain both the shadow

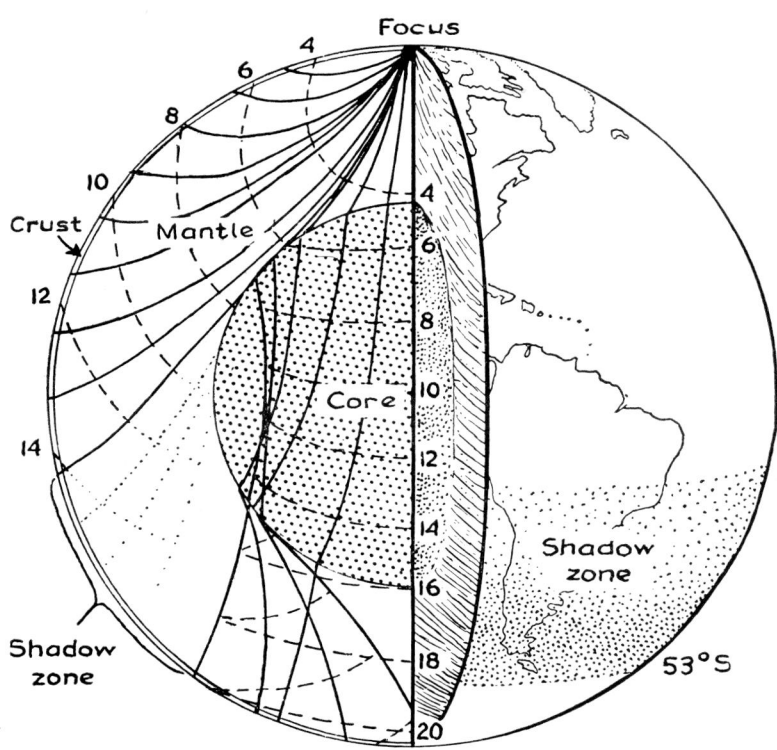

FIGURE 18-18.   *The shadow zone of an earthquake originating near the North Pole. The cutout shows the effect of the core on the pattern of wave paths and wave fronts. (After B. Gutenburg, 1928.)*

zone and the extraordinary strength of the wave at 143°, for here are focused waves that impinge upon a considerable segment of the core. The abruptness of the core boundary is strikingly shown by Canadian records of a South Pacific earthquake. At Toronto (epicentral distance 141°) only a very faint record was made, but at Ottawa, only 2° farther from the epicenter, the P wave was very strong.

Beyond an epicentral distance of 103°, the S waves, if present at all, are very faint. According to some seismologists, the waves may not be absent but merely weak and obscured by reflected waves of various kinds, so that they cannot be identified. But the concensus of experts is that the S waves are not transmitted through the core. This means that the core has practically no rigidity. (See footnote, p. 479). As the absence of rigidity is characteristic of all fluids, these seismologists commonly refer to the core as liquid. There is some evidence to suggest, however, that the very central part of the core, extending from a depth of about 5,000 km. to the earth's center at 6,350 km., has different properties from the outer core. This "inner core" may have considerable rigidity and thus be solid rather than liquid, but the evidence is far from conclusive.

Summary

Study of seismologic records has thus led to a model of the earth that seems to fit nearly all the known data. This model has a central core, perhaps solid, with a radius of about 1,300 km. This is surrounded by an outer, presumably liquid, core extending to a radial distance of 3,400 km. where it has a sharp boundary against the mantle. The mantle may be interrupted by some discontinuities but none is considered important. It is bounded at its top by the Mohorovičić discontinuity. This discontinuity is *the base of the crust*, one of the fundamental features of earth structure. The crust apparently consists of different materials beneath continent and ocean. The continental crust, about 30 to 40 km. (18 to 25 miles) thick on the average, consists of an intermediate layer 20 to 40 km. (12 to 25 miles) thick and averaging perhaps 25 km. (15.5 miles), that is probably of basaltic composition, overlain by a granitic layer 10 to 30 km. (6 to 20 miles) thick and averaging about 15 km. (10 miles), which in turn is blanketed by sedimentary rocks to an average depth of about 2 km. (1.5 miles). The oceanic crust either lacks the granitic layer entirely or it is thin and patchy. The intermediate layer may also be lacking or else very thin over most of the Pacific Ocean, and from part of the Atlantic and Arctic oceans.

Other Clues to the Earth's Interior

What other information can we glean about the mysterious depths of the earth? The evidence from study of earthquake waves confirms our early inference that the continents stand high above the ocean basins because they are predominantly composed of rocks less dense than those that underlie the oceans. They are huge rafts of granitic material, underlain by basalt; the oceans are floored by rocks probably like those of the lower part of the basaltic layer of the continents. Both these layers are "floating" on a denser "lower layer," a part of the very thick mantle of the earth which lies beneath the Mohorovičić discontinuity. We can infer from the speeds of seismic waves that the upper part of this mantle consists of a rock much like peridotite. But of what materials are the lower mantle and the core composed? Can we make reasonable deductions about them despite their inaccessibility, and, if so, what reasonable distribution of composition and density can be inferred?

The increase in earthquake speed downward throughout the mantle seems fairly regular. As we go deeper in the earth the pressure increases tremendously and any reasonable estimates of density distribution indicate pressures of many millions of pounds per square inch at the boundary of the core. In the laboratory, tests of rock at higher and higher confining pressures usually reveal increasing rigidity and incompressibility. Perhaps the entire mantle is composed of something near peridotite in its determinable properties, and the greater elasticity at depth is merely due to the great confining pressure. But this does not account for the sharp discontinuity at the core boundary. It seems clear for many reasons that the material forming the core boundary must be of different composition from that above. What is it?

We of course do not know. From analogy with meteorites, which many astronomers suppose represent fragments of other planets, the most likely material seems to be an alloy of metallic nickel and iron, essentially in a molten state. Though the pressures involved at these depths are far beyond attainment in the laboratory, physicists have dealt with the question theoretically and concluded that a molten alloy of nickel and iron might be expected to give the measured speed of the P wave. It has been suggested, too, that the abundance of iron silicates in the mantle implies that they should be underlain by metallic iron, just as slags rich in iron silicates float on molten iron in a crucible.

How do these deductions check with what we know about the distribution of density in the earth's interior? The total mass of the earth is 5.98×10^{19} metric tons. From this and its volume, the average density has been computed as 5.52. Now, the average density of visible rocks is about 2.8, and only a few ores have densities as high as the *average* for the earth as a whole. It is obvious that the density must increase downward and that the deeper parts of the earth must be composed of material extremely dense compared to any we know as natural substances.

Inasmuch as the earth shells seem to be concentric, as shown by the smoothness and lack of scattering of time-distance curves, any assumptions we make about the changes in density with depth include assumptions as to the angular momentum, or rotational inertia, of the earth. Each particle of matter in a rotating body contributes to the angular momentum in proportion to its mass and its distance from the axis of rotation. Thus when we assume a particular density for material at a distance of, say, 3,000 km. from the center of the earth, the portion of the earth's angular momentum contributed by a shell of that density and radius can be computed. Thus estimates of the changes in density with depth in the earth can be checked against the astronomically measured value of the earth's angular momentum.

The Australian geophysicist Bullen has made perhaps the most careful study of this matter, taking into account both seismic data and the angular momentum. His results are shown graphically in Figure 18-19. Here it can be seen that the density in the center of the earth can hardly be less than 11.5 (which is higher than that of lead at the surface) and may be as high as 17.2. At the outer boundary of the core the density can hardly be less than 9.4 or

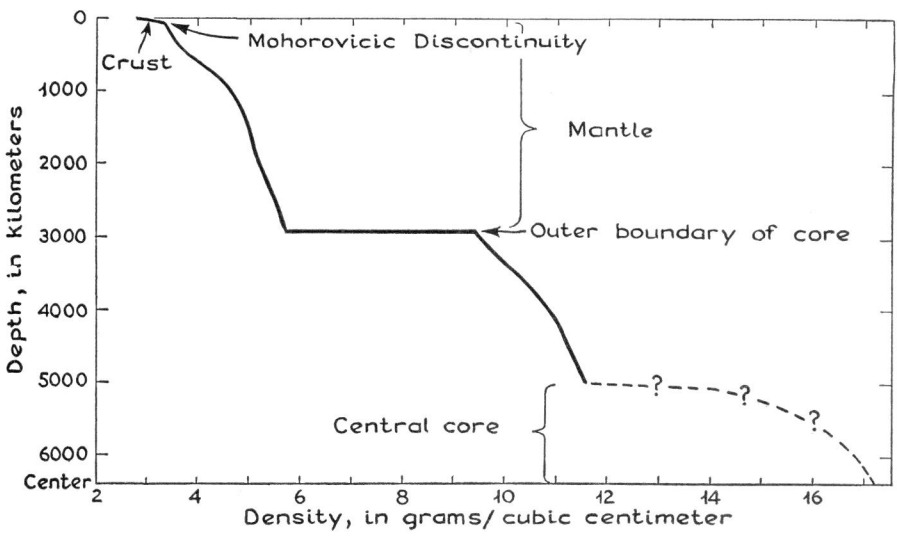

FIGURE 18-19. *Graph showing the variation of density with depth.* (*Data from K. E. Bullen, 1940 and 1942.*)

greater than 9.6, both values higher than those of iron (7.9) or nickel (8.6) at the earth's surface. He concluded that his values for the mantle are considerably more precise than these ranges would suggest for the uncertainties within the core.

Such density distribution is consistent with the earth's mantle being a peridotite with about the same composition as that of stony meteorites. It is also consistent with a core composed of molten iron and nickel, similar to some metallic meteorites. The high densities could be due entirely to the very high pressures that must prevail deep within the earth. Although the core boundary appears to be a fundamental discontinuity requiring a marked change in composition, it may be that some iron-nickel alloy is included in the lower part of the mantle, but in the solid rather than liquid state. If the amount of such metallic material diminishes gradually upward within the mantle, so that the upper part is near a peridotite in composition and the peridotite then changes abruptly to basalt above the Mohorovičić discontinuity, we would have a reasonable accordance between the facts of geology relating to igneous rocks (Chap. 17) and the geophysical measurements.

Temperature Within the Earth

Every well or mine shaft sunk into the rocks proves that the temperature of the earth increases downward. The rate of increase of temperature with depth is highly variable from place to place. In areas of hot springs and geysers, such as Yellowstone Park, temperatures of several hundred degrees Centigrade have been reached in drill holes only a few hundred feet deep. On the other hand, the temperature increase in the gold mines of the Transvaal, South Africa, is so slow that mining has been carried to depths of more than 9,000 feet. The increase here is at the rate of about 1° C. for 300 feet depth. Although the rise in temperature with depth is thus greatly variable from place to place, there can be only one physical interpretation: The earth is losing heat to outer space, for heat flows only from bodies of higher temperature to those of lower.

The actual rate of heat loss depends on two factors: the rate of temperature change with depth (the so-called *temperature gradient*), and the heat conductivity of the rocks. There have, unfortunately, been but few measurements of both these factors at the same localities. But the few that have been made indicate that, despite the wide range of the rate of temperature increase, the heat losses are closely similar. The reason is that the conductivity varies with gradient, and is high where the gradient is low, and low where the gradient is high. The product of gradient and conductivity, which determines the actual heat transferred, is thus not far from the same, whether measured in England, Scotland, South Africa, or Colorado, though there are small, but probably significant, differences.

The actual heat loss is very small. As the sun supplies about 8,000 times as much per square foot or other unit of surface area as is lost, it has been computed that the surface temperature would fall only about 0.01° C. if the internal heat were cut off entirely.

Before the discovery of radioactivity it was believed that since the earth is losing heat, it must necessarily be cooling off. In fact, as was mentioned in Chapter 8, this assumption was the basis of Lord Kelvin's statement that the earth could not be more than 400 million years old. When it was discovered that certain atoms spontaneously disintegrate to yield others, and that this change is accompanied by the evolution of large amounts of energy, it became necessary to reconsider the question of whether the earth is growing colder.

Because they are by far the most abundant, uranium, thorium, and potassium (K_{40}) are the three radioactive elements whose disintegration contributes most to the earth's heat budget. Sampling of the wide variety of rocks exposed at the earth's surface shows that these elements are far from uniformly distributed. All three radioactive elements are more abundant in granite and granodiorite than in gabbro or basalt, and more abundant in basalt than in peridotite. The content in olivine from peridotites is even lower than in peridotite as a whole. It has been computed that a layer of granite 13 km. thick, and having the average radioactivity found in granites exposed at the earth's surface, would supply all the heat loss. The average basalt contains about one-third as much radioactivity as the average granite, the average olivine from peridotite about 1 per cent as much as granite.

If the radioactive content within the crust were the same as that of equivalent rocks at the earth's surface, we would reach the interesting conclusion that the earth is not cooling at all, but is, instead, gradually becoming hotter. For, as we have seen earlier in the chapter, there is an average thickness of about 15 km. of granite and about 20 km. of basaltic material above the Mohorovičić discontinuity. The granite alone would supply more heat than is escaping at the surface, and the basaltic layer would add another 40 per cent to this. Even though the contribution of a unit volume of peridotite from the deeper mantle would be only 1 per cent of that of granite, the great volume of the peridotite in the mantle would also ensure a heating earth.

Inasmuch as no laboratory experiments have shown the slightest tendency for high pressures to slow down radioactive disintegration, there seem to be only two possibilities: Either the surface rocks are relatively enriched in radioactivity with respect to the similar rocks at depth, or else the earth is actually becoming warmer. But the heating is not so great as to liquefy the mantle—at least in very large volumes—for the transmitted shear waves show that it is a rigid solid. Much more careful sampling and measurement of radioactivity is needed before this question can be settled, but at present it appears likely that the temperature within the earth does not melt the rocks in large volumes until the core boundary is reached. What the temperature here may be is pure conjecture, though it seems probable that it is less than 6,000° C. (the temperature of the sun's surface) and perhaps about 3,000°.

The volcanoes, of course, attest to pockets of molten rock in the earth, either within or just below the crust. Perhaps these pockets result either from localized radioactivity (which would explain the relatively high radioactive content of surficial lavas) or from release of pressure along faults (which would lower the melting point of the rocks and permit them to melt, despite the fact that under full confining pressure at these depths they are solid). Such melting would, in turn, concentrate radioactive elements within the liquid because radioactive minerals are associated selectively with the low-melting constituents. Thus by some such mechanism we see how magma bodies may form in the crust, and how the rocks of the crust are higher in radioactive content than the rocks deeper in the earth.

Facts, Concepts, Terms

FAULTS AND EARTHQUAKES
　　Elastic rebound theory

P WAVES, S WAVES, L WAVES

EPICENTER, FOCUS, TIME-DISTANCE CURVES
　　Isoseismals, intensity scales
　　Earthquake magnitude
　　Deep-focus earthquakes
　　Earthquake distribution
　　The crust of the earth
　　Continental layering
　　The crust beneath the oceans
　　The core and mantle

THE EARTH'S THERMAL GRADIENT

SOURCES OF INTERNAL HEAT

Questions

1. In many Central American earthquakes well-built masonry buildings have been destroyed while bamboo huts nearby were undamaged. Can you suggest a reason?
2. The Bouguer anomalies for the Alps (see Chap. 3) suggest a greater thickness of light rocks than beneath the lowlands to the north. What would this suggest regarding the heat flow to be expected in the two areas?
3. During the San Francisco earthquake the porch was sheared off a house and moved more than 10 feet by movement on the rift, without knocking over the brick chimney, whereas well-built structures 4 miles from the rift were demolished. What factors can you suggest to account for this?
4. Why is the Mohorovičić discontinuity of greater importance than those within the crust?
5. What does the elastic rebound theory suggest concerning the distribution of changes of elevation near a fault line while strain is accumulating preparatory to a *vertical* displacement?
6. Assume that a mountain range with an average height of 10,000 feet is buoyed up isostatically by a "mountain root." The average density of the range and root is 2.7, that of the substratum 3.3. If this mountain range lost an average thickness of 1,500 feet by erosion, how high would it have to rise isostatically to restore isostatic equilibrium?

Suggested Readings

1. Jeffreys, Harold, *Earthquakes and Mountains*, Methuen & Co., London, 1935.
2. Daly, R. A., *Our Mobile Earth*, Scribner's Sons, New York, 1926.

19. *Mountains*

Geosynclines

THE GREAT AMERICAN geologist James Hall made a useful observation in 1859. The early Paleozoic rocks of the Interior Lowland of North America are everywhere only a few hundred or at most a few thousand feet thick. But in the Appalachian Mountains, rocks of the same age are many times this thickness. Furthermore, the rocks of the Interior Lowlands are largely limestone and dolomite, with subordinate clastics, whereas clastic rocks or metamorphic rocks derived from them predominate in the Appalachians. Thus the rocks of the mountain belt are both much thicker and more generally clastic than those of the same age in the continental plate to the northwest.

Geologists have since found comparable facies changes associated with almost all the other ranges: the Alps, Himalayas, Urals, Andes, and the Cordilleran System of North America. In many of these, thicknesses of as much as 6 or even 8 miles of strata have been measured, whereas the average thickness over the adjacent continental plates is not more than a few thousand feet.

The strata comprising all of these great chains are chiefly volcanics and sediments or their metamorphic derivatives. Some preserve rainprints, mudcracks, salt-crystal casts, and other indications that they were deposited in shallow water, and some, like the coals of Pennsylvania, are even land-laid beds. Fossils found in them are nearly all of shallow-water organisms. Though a few strata in the Alps, West Indies, and Timor have been interpreted as deep-sea oozes, their associations make this doubtful, and, in any case, their volume is trivial. These great piles of strata do not, then, represent the filling of an ocean deep, even though they are thick enough to fill any existing deep to overflowing, for shallow-water deposits recur time after time through sections as thick as 40,000 feet. Only one conclusion seems possible: The crust in these areas was being depressed at a rate roughly equal to that of sediment accumulation; the surface of the sediments was never far above or below sea level during all the time of their accumulation.

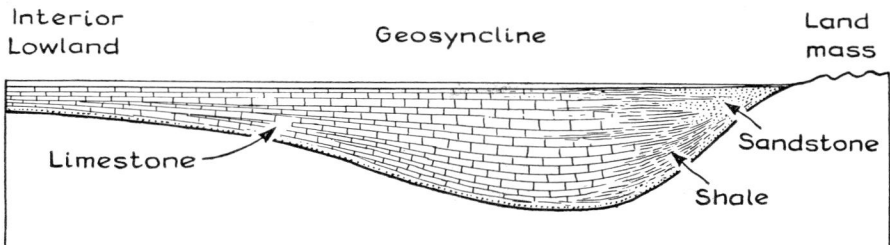

Interior
Lowland

Geosyncline

Land
mass

Limestone

Sandstone

Shale

FIGURE 19-1. *The thickening of strata from the Interior Lowland into the Appalachian geosyncline. (After A. W. Grabau, 1924.)*

The basement beneath these accumulations must have warped slowly downward; the trough thus formed has been called a *geosyncline,* and the pile of strata filling it a *geosynclinal prism* (Fig. 19-1).

The association Hall recognized, of mountain chains with geosynclines, has since been found generally, though not quite universally. Mountains thus differ from continental plates in a fundamental way:

Their sedimentary thicknesses are far greater.

The thickening and coarsening of sediments from the stable continental plate into the Appalachians suggest that most of them were derived from a land southeast of the present mountains—either the Piedmont, the Coastal Plain, or, perhaps, an area now covered by the Atlantic. Similar relations have been found for other geosynclines.

Inferences as to the exact area furnishing the sediment cannot be certain, however, within wide limits, as a glance at the map of Burma will show. The Irrawaddy dumps its huge load 450 miles from the delta of the Brahmaputra—a distance as great as that from Pittsburgh, Pennsylvania, to Birmingham, Alabama—yet both rivers derive much of their load from within a few miles of each other in the eastern Himalayas. Many modern examples of this sort show how hazardous is precise inference as to the source of ancient sediments, even though the point where they entered the ancient sea can be quite closely determined.

Though the exact sources of sediments are uncertain, there is less ambiguity about the general position of the shoreline from which they entered the seas. Thus three elements have been recognized in common association: the *geosyncline* itself, with its thick pile of strata; the *foreland,* a part of the continental plate covered only by relatively thin sediments; and, on the opposite side of the geosyncline, the *hinterland,* which supplied the bulk of the sediments but of which we know little. Obviously, strata equivalent to those of the geosyncline must be lacking or sparse in the hinterland, for this was undergoing erosion. Also, many of the postulated hinterlands are now veneered with younger rocks or drowned beneath the sea.

Though the association of geosynclines with most mountain ranges is unmistakable, the picture given in the preceding paragraph is too broad to apply to all ranges. For example, the Yellow Sea has many features suggestive of a geosyncline. This shallow sea is the resting place for the great load of detritus brought from western China by the swollen torrents of the Yellow River and the Yangtze. The deltas of these streams are like the ancient Devonian delta that has been recognized in the rocks of the Catskills and Appalachians of New York and Pennsylvania, and which was built by streams entering the Appalachian geosyncline from the vanished hinterland to the east. But the "foreland" of the Yellow Sea is deep sea fringed by islands, not part of the stable continental plate of Asia. Similarly, the accumulation of Cretaceous, Tertiary, and Quaternary sediments along the coast of the Gulf of Mexico is clearly approaching geosynclinal dimensions, but the "foreland" of the geosynclinal prism is the deep water of the Gulf.

These modern analogs of the ancient geosynclines do not fit simply into a pattern like the Paleozoic Appalachian geosyncline or the Mesozoic and Tertiary Alpine geosyncline. Yet, the similarities in abrupt thickening of strata to prodigious amounts along a fairly simple belt warrant calling them geosynclinal. Quite possibly, in the distant geologic future a great mountain range will rise along the now swampy coast of Louisiana and Mississippi, but present knowledge does not fully warrant such a prophecy. For example, a thick series of pre-Cambrian strata in western Montana, although now a part of the towering Rockies, lay essentially undisturbed in their depositional attitude through all of Paleozoic and most of Mesozoic time—roughly 400 million years—despite the fact that they are more than 30,000 feet thick and cover an area half as large as the Alps. Also, considerable mountain masses, such as Ruwenzori in Africa, the highlands of Labrador, and the Adirondack Mountains, are not built from deformed geosynclinal sediments.

In summary, we can say that the association of mountain ranges with geosynclinal prisms is far too common to be accidental—there must be some genetic connection between them—but the quiescence of geosynclinal accumulations as thick as those of most mountain ranges through hundreds of millions of years, and the occurrence of mountains not made of thick clastic sediments, show that geosynclines are not essential prerequisites of mountain uplift. Factors other than excessively thick sediment accumulation are needed for mountain-making.

Mountain Structure

We have mentioned that most mountains differ from the continental plates in their much more deformed rocks. A study of the structural details of some of the most completely investigated ranges permits some inferences as to the forces that deformed them.

The Appalachians

The Appalachians rise from beneath the coastal plain of Alabama and extend northeastward in a sinuous belt 1,500 miles long to Nova Scotia and New Brunswick, where their deformed rocks are cut off by the Atlantic shore. In general, individual faults and folds lie about parallel to the trend of the belt and persist for many miles, so that geologic cross sections spaced several miles apart generally resemble their neighbors closely. In the central Appalachians the folds toward the northwest are open, but, as one proceeds southeastward, the folds become tighter, with steeper limbs. The northwesterly of these tighter folds are merely asymmetrical, with the northwest limbs steeper than those on the southeast, but farther southeast the folds are

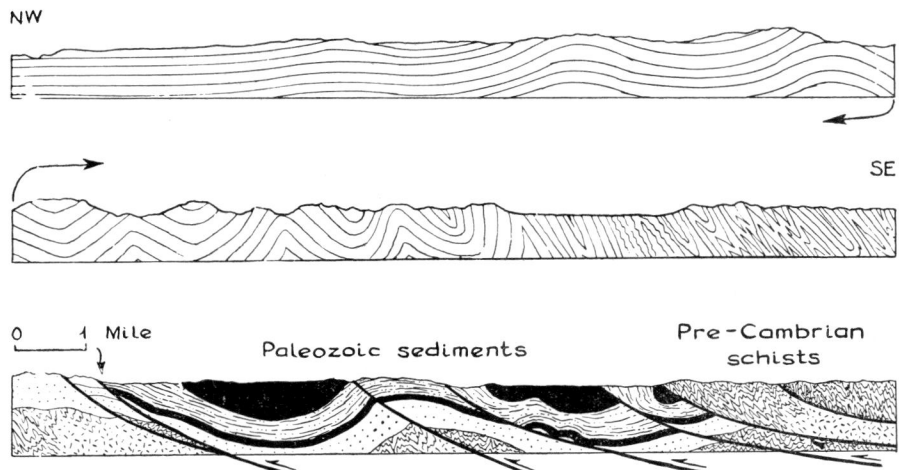

FIGURE 19-2. *The two sketches at the top are the left and right halves of an idealized section across the Appalachians, showing the increase in deformation toward the southeast. (After W. B. Rogers, 1843.) The lower section shows repetition of beds by thrust faulting in the Appalachian Mountains of southeast Pennsylvania. (After Geologic Map of Pennsylvania, 1931.)*

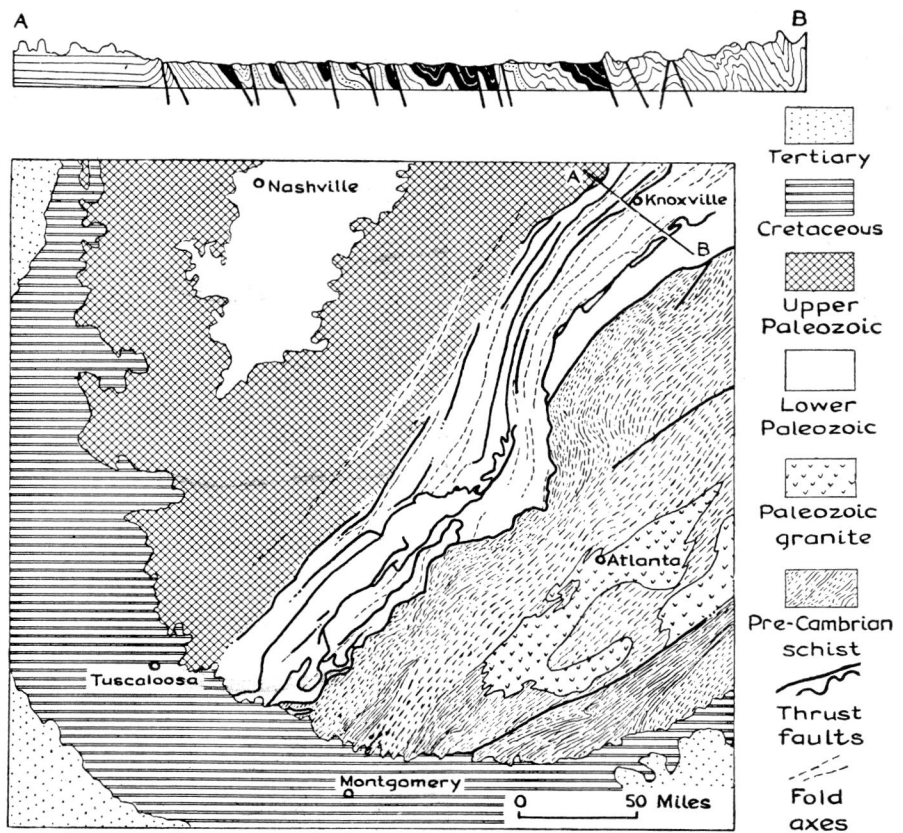

FIGURE 19-3. *Geologic map of the Southern Appalachian region. The cross section* (top) *shows how the Ordovician rocks near Knoxville have been repeated eight times by thrust faults. The vertical scale of the section is greatly exaggerated, hence the low dip of the thrusts has been steepened. (After Geologic Map of the United States, 1933.)*

overturned, so that both limbs dip southeast (Fig. 19-2). Locally, at the southeast side, thrust faults have repeated certain beds of Cambrian age as many as five times (Fig. 19-2). Furthermore, at the northwest the rocks are unmetamorphosed, but as they are traced southeastward they become more altered, and the southeast side of the mountain belt is underlain by slates, schists, and gneisses, with some unmetamorphosed granite plutons.

In West Virginia and to the south, the main mass of the folded belt is locally thrust several miles northwestward over the rocks of the foreland. These great thrusts, some of which have been folded, were first recognized from the pattern of the outcrops; their existence has been conclusively confirmed by oil wells drilled through rocks of the upper plate into younger strata beneath the fault. In the latitude of Knoxville as many as six large thrust faults have been recognized (Fig. 19-3). Great thrust faults also mark the northwest border of the folded belt in the Hudson-Champlain Valley in northern New York and Vermont. Some of these faults have rocks of the same age both above and below them. But the rocks of the two plates are of wholly different facies: One is largely carbonate rocks and the other is metamorphosed clastic rocks. In other words, rocks of the same age, but deposited so far apart that their composition is wholly different, have been brought in contact by thrusting.

Crustal Shortening. A series of symmetrical anticlines and synclines might be pictured as forming in either of two ways: by differential uplift and sinking of the basement on which the folded strata lie or, perhaps, by a force which pushes the edges of the strata horizontally so that they wrinkle and warp like a pile of blankets on a table when shoved together. But systematically asymmetric folds that all lean to the northwest and thrust faults that have brought rocks northwestward for many miles as we see them in the Appalachians cannot be explained by vertical movements. Here the rocks must have been shoved together horizontally. When the folds of the least-deformed part of the Appalachians—in Pennsylvania—are straightened out in imagination, by measuring the length of beds following all the convolutions shown in geologic sections and then comparing this with the width of the folded belt, it has been found that points originally 81 miles apart have been brought nearer together by 15 miles. In other words, the superficial crust of the earth has locally been shortened 18 per cent. This measurement takes no account of the shortening by thrust faults farther to the southeast. In the southern Appalachians, where half a dozen thrust faults in a single cross section each required several miles of overriding, the shortening must have been far greater. Similar studies show comparable or greater shortening in many other ranges. As an example we will cite the best-studied range on earth, the Alps.

The Alps

The Alps lie in a great arc looped northward through Switzerland to Vienna, where they die out beneath the Hungarian plains. They are part of the great mountain system that extends from the Spanish Pyrenees across southern Europe and Asia through the Himalayas and the Malay Peninsula.

Like the Appalachians, the Alps rise on the site of a great geosyncline, but the strata that filled this trough are of Mesozoic and early Cenozoic age rather than Paleozoic. The foreland lies to the north, where relatively thin

Mesozoic rocks rest unconformably on Paleozoic and older rocks. Within the Alps Mesozoic and Tertiary rocks of wholly different facies and far greater thicknesses are found.

Despite the marvelous cliff exposures, Alpine structure is so complex and the facies changes are so great that the arduous labor of hundreds of geologists during more than a century of intensive study still leaves many problems unsolved, but the broader features are well established. A much simplified map of the Alps is shown in Figure 19-4.

Divisions of the Alps.   As in the Appalachians, the structure and history of the different segments along the trend of the range differ. The major division is between the western and eastern Alps, for there is a profound difference both in facies and structure along an irregular line trending south from Lake Constance along the Rhine just west of the Silvretta Alps. The western Alps lie chiefly in Switzerland, though their western arc extends into France and the southern parts are in Italy. The eastern Alps are chiefly in Austria, with their northern border in Bavaria and their southern in Italy.

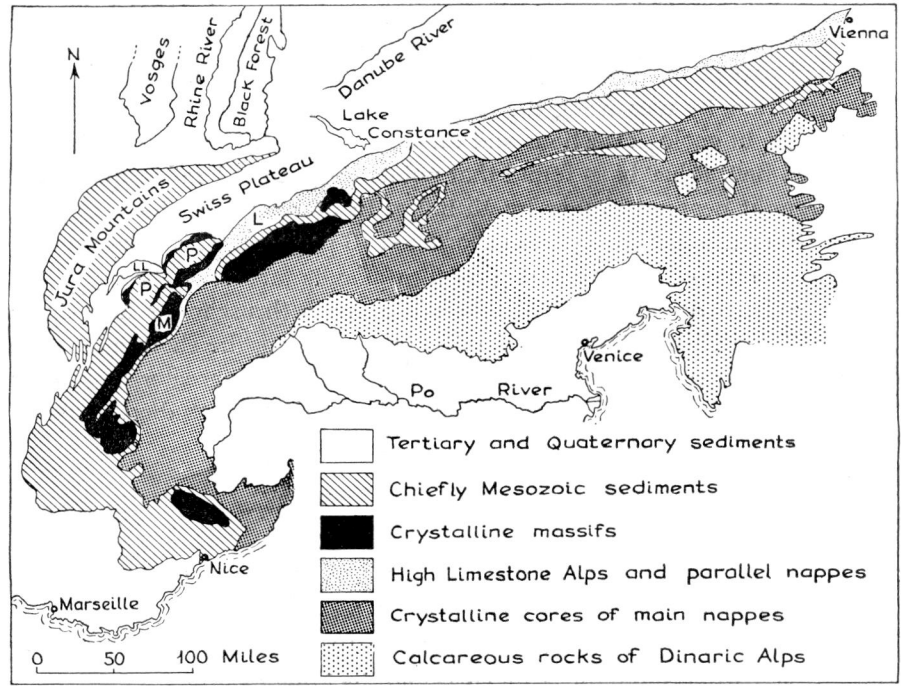

FIGURE 19-4.   *Generalized map of the Alps.* **P,** *Prealps;* **LL,** *Lake Geneva.* (*After R. Staub, redrawn from L. W. Collet,* Structure of the Alps, *Edward Arnold and Co., 1927.*)

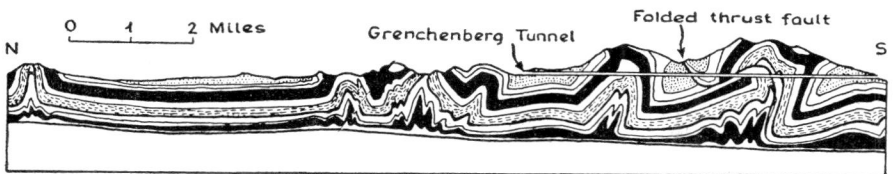

FIGURE 19-5. *Cross section of part of the Swiss Jura. (After Buxtorf, redrawn from E. B. Bailey, 1935.)*

We will discuss only the western Alps, as a sample of this magnificent range, outlining its geology from north to south.

The Jura Mountains   The Jura Mountains (Fig. 10-18) consist of fore-land rocks of Mesozoic age folded into anticlines separated by almost flat synclines. The amplitude of the anticlines increases southward, and some of the southern ones are broken by small thrust faults (Fig. 19-5). Significantly, even the largest anticlines expose no rocks older than Middle Triassic. This fact, together with exposures in railroad tunnels and in areas north and west of the mountains, suggests that the Jura folds were formed by the crumpling of a thin sheet of sedimentary rocks that has been torn loose from its foundation and pushed northward (Fig. 19-5). The geometry of the folds does not permit them to extend downward into the Lower Triassic, Permian, and older rocks that are known to underlie the mountains. Clearly, the surficial rocks were sheared from the nearly flat beds beneath, much as we can crumple a tablecloth by sliding it over the table. This is merely the outer-most and least of many Alpine examples of large-scale horizontal movement of the rocks, yet the surface rocks in the southern Jura have moved several miles from the site of their deposition.

The Swiss Plateau   Between the Jura and the Alps lies the Swiss plateau with the beautiful lakes Geneva, Neuchatel, Zurich, and Constance. Most of the rocks are Tertiary, chiefly coarse conglomerates derived from the Alps to the south. This is shown both by their southward coarsening and by the pebbles, which can commonly be matched with their parent formations, though, as we shall see, not always with their present neighbors. Though broadly synclinal, this area is locally strongly folded, and along its southern border is overridden by great thrust sheets.

The Prealps   Between the Swiss plateau and the main limestone Alps is a chain of lower but still impressive mountains that range from about 3,500 to 6,500 feet above sea level. In places they are separated from the main range to the south only by a zone of saddles in their high projecting spurs. These are the Prealps. Their magnificent cliffs expose chiefly Mesozoic rocks that rest in thrust contact on the crumpled Tertiary of the Swiss plateau.

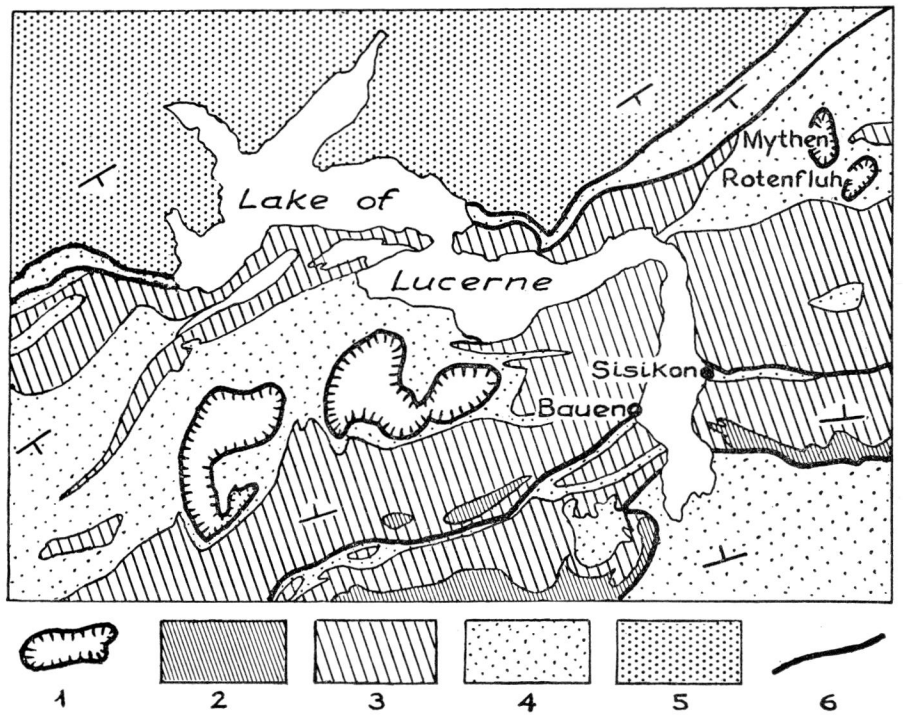

FIGURE 19-6.   *Geologic map showing remnants of three of the great thrust sheets of the Prealps. The symbols are: (1) Klippes of far-traveled Mesozoic rocks, (2) Jurassic, (3) Cretaceous, (4) chiefly Eocene, (5) chiefly Miocene, (6) thrust faults. (After E. B. Bailey,* Tectonic Essays, *Clarendon Press, Oxford, 1935.)*

Thus they are thrust sheets whose rocks are completely foreign to the basement on which they lie (see map, Fig. 19-6, and sketch, Fig. 19-7). Such thrust masses that have been carved by erosion from a formerly continuous sheet are called *klippes* (German: "cliffs"). Similar masses have been recognized in many other mountain chains.

Figure 19-6 portrays the kind of evidence used in interpreting the complex geology of the Prealps. A brief study of this map shows that the klippes of the Mythen and Rotenfluh are merely the highest of at least three thrust sheets piled one on top of the other. At the south end of Lake Lucerne the Eocene rocks clearly dip beneath Jurassic rocks which are overlain by Cretaceous and Eocene. These rocks pass at Sisikon and Bauen beneath the Cretaceous and Eocene on which rest the Triassic of the Rotenfluh and Mythen.

The High Limestone Alps and the Crystalline Massifs   The lower thrust sheets beneath the klippes of the Prealps continue into the High Limestone

FIGURE 19-7.   *Sketch of the Mythen klippe. The steep peaks of white Mesozoic limestone rest on Eocene shale. (After L. W. Collet,* Structure of the Alps, *Edward Arnold and Co., 1927.)*

Alps. They are largely the sheared-off limbs of great anticlinal folds that have been overturned so far that they lie nearly flat. Such sheets are called by the Swiss geologists *nappes* (French) or *Decken* (German). South of them, where the plunge of the folds allows them to be seen, lie the highly sheared anticlinal cores from which the sedimentary mantles have been nearly completely severed and shoved northward to make the nappes of the Limestone Alps.

These cores consist of granite, gneiss, schist, and other metamorphic rocks such as are seen in the magnificent tower of Mont Blanc. A few highly distorted plant fossils have been found in anthracite coal associated with these metamorphic rocks; from them we know that at least part of the core is of Carboniferous age. The overlying sedimentary rocks have yielded some Permian and Triassic fossils and contain pebbles of the core metamorphics. We thus have in the cores of the folds from which the nappes of the High Limestone Alps have been sheared off and thrust northward a rock succession much like that of the foreland north of the Jura. In both places, Permian and Triassic rocks rest with great unconformity on steeply tilted and locally highly metamorphosed Carboniferous rocks associated with pre-Permian intrusive masses. The relation of these "crystalline (metamorphosed) massifs" to the mantles that have been stripped from them to form the High Limestone Alps, and of these rocks, in turn, to the thrust sheets of the Prealps, can be seen in the map and sections of Figure 19-8, much simplified from the detailed original mapping of the great Swiss geologist Argand.

The Swiss geologists consider the Mont Blanc massif and others like it to

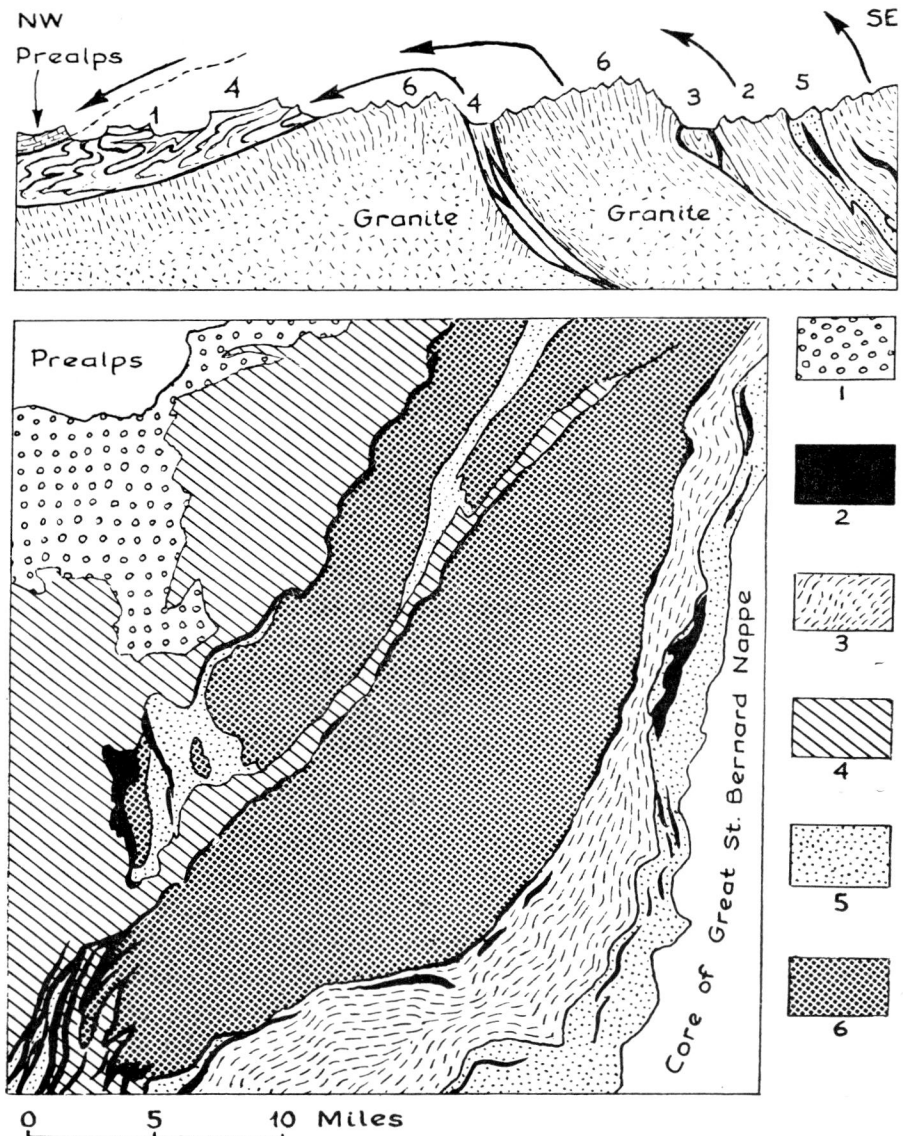

FIGURE 19-8.   *Map and section showing the crystalline cores of the nappes that form the High Limestone Alps. The arrows of the section indicate the movement from the region of the cores to the nappes. (1) Tertiary, (2) chiefly Jurassic schists, (3) Triassic, (4) High Limestone nappe, (5) Carboniferous, (6) crystalline massifs. (After L. W. Collet,* Structure of the Alps, *Edward Arnold and Co., 1927.)*

have moved only a few kilometers. By contrast, the great thrust sheet of the High Limestone Alps has been torn loose and overfolded northward like cloth slid over a table and crumpled into a flat fold. This sheet can be followed foot by foot for more than ten miles to the northwest, where it disappears beneath the higher thrust sheet of the Prealps. This is evidence of considerable horizontal movement in the earth's crust. What, then, are we to think when, several miles farther northwest, identical rocks reappear from beneath the Prealpine sheet and rest in fault contact on the Tertiary of the Swiss plateau?

One strongly overfolded crystalline core after another is found in the southern Alps, each separated from the other by highly deformed schists. Some of these schists have yielded Mesozoic or early Tertiary fossils. The geometrical arrangement of the great thrust masses demands tremendous shortening of the visible crust—a shortening that cannot, of course, be accurately measured, but which certainly must amount to many scores of miles. Thus the crust beneath northern Italy must lie scores, if not, indeed, a hundred or more miles, closer to the Swiss plateau than it did in Eocene time!

The evidence is not only geometric. The facies of the rocks composing the thrust masses demand great travel also. For example, Arnold Heim mapped the Cretaceous of the nappes of the High Limestone Alps in eastern Switzerland. Their pattern justified his drawing the cross section (Fig. 19-9) reproduced, much simplified, at the top of the diagram. The distortion of the rocks along both faults suggested that the overlying rocks moved northward

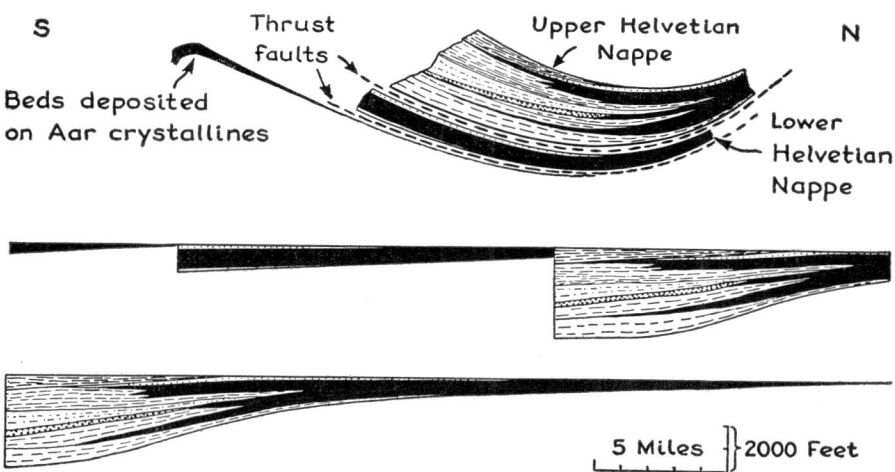

FIGURE 19-9. *Hypothetical reconstructions of the relative positions of the several thrust sheets north of the Aar region of eastern Switzerland. (After A. Heim.)*

with respect to those beneath, on both thrusts. The facies of the Cretaceous rocks in the nappes practically demand this type of movement. For example, if one wished to regard the upper thrust sheet (now exposed to the north of the lower) as having always been north of it, and, similarly, that the Cretaceous covering the Aar massif has always been south of the lower nappe as it now is, one must accept the unlikely arrangement of sedimentary facies shown in the second diagram. If, however, we overrule our "common sense" reaction, that such great travel is "unreasonable," and accept the geometrical evidence at its face value, we can reconstruct the logical arrangement of sedimentary facies shown in the lowest diagram. Obviously, this facies arrangement is not only consistent with the geometrical evidence, but it offers almost conclusive proof that both lower and upper nappes traveled northward, and that the upper sheet has come from farther south than the lower.

Summary of Alpine Structure.   Although the few examples cited can only hint at the almost incredibly complex structure of the Alps, they suffice to show that prodigious horizontal forces must have operated here, just as they did in the Appalachians. The crust has been shoved together as though by the jaws of a great vise, and the material caught between the jaws has been squeezed out and shoved northward relative to the underlying rocks for distances measured in tens of miles (Fig. 19-10). In the process, the rocks have been dynamically metamorphosed—all transitions can be found in a particular rock mass between almost unaltered shale and highly metamorphosed schist. Granitic rocks caught in the deformation have been partly or wholly converted into gneiss, with all transitional stages apparent. These are significant observations, for, as we shall see, both geometrical patterns and metamorphic rocks identical with those now found in the towering Alps are repeated again and again in the great pre-Cambrian shield areas of the earth.

Time of Deformation.   The Alpine folding took place in many stages, but most of it was in Oligocene and early Miocene time, as rocks of these ages are involved in the thrusting, whereas the Upper Miocene and Pliocene rocks nearby are much less deformed. The Appalachian folding occurred

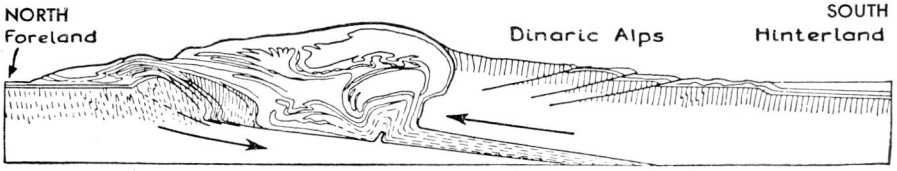

FIGURE 19-10.   *Idealized reconstruction of the structure of the Alps, neglecting erosion. (After Emile Argand, 1916.)*

through a long period, as recorded by angular unconformities within the Ordovician of New York and Quebec, between Ordovician and late Silurian in the Hudson Valley, within the Devonian of the Canadian Maritime Provinces, and, greatest of all, between Carboniferous and Triassic. Abrupt coarsening of sediments at several other times during this long period suggests other pulses of folding that have not been proved by unconformities. The folds and faults of the Appalachians as we now see them may be said to date chiefly from the late Paleozoic. In Scotland and Scandinavia all these features can be matched in mountains whose deformation dates from Silurian times. Examples could be given from many other parts of the world. It is tempting to conclude that the linear ranges were all formed by the crumpling and piling together of geosynclinal sediments by horizontal compressive forces operating at fairly shallow levels in the earth's crust. Certainly, most ranges do show features that demand crustal shortening. But it would be premature to assert that all do.

The Basin Ranges

Most of Nevada, western Utah, and large parts of Oregon, Idaho, Arizona, New Mexico, west Texas, and eastern California constitute a geologic province characterized by isolated mountain ranges separated by desert plains and basins. Some of these ranges are relatively low, but others exceed 12,000 feet in altitude and 100 miles in length. Indeed, the westernmost of these ranges, the Sierra Nevada, is about the size of the Alps. As Dutton long ago said, on a relief map of North America the ranges resemble an army of caterpillars crawling toward Mexico. The geology of this large area is far less well known than that of the Appalachians or the Alps; only small segments have been mapped in detail. Yet our information shows that these mountains differ notably in structure and history. One fact is very impressive: These mountains were formed from parts of the crust that in themselves show widely differing earlier histories, and not from a uniform geosyncline.

Much of the *Basin Range* province was the site of an early Paleozoic geosyncline; Cambrian and Ordovician rocks attain thicknesses of many thousands of feet in eastern California and southern Nevada. In later Paleozoic time, however, the region was occupied by two smaller subsiding basins separated by a highland, for there are many thousands of feet of Devonian and Carboniferous rocks in western Utah and eastern Nevada, and again in western Nevada, whereas in parts of central Nevada the Permian rests directly on Ordovician. Whether a true mountain range was formed at this time, and its possible extent if it were, are questions that await further detailed mapping. This point is almost irrelevant to our present purpose, however, for whatever the size and trend of the emergent area, the Mesozoic history of the different parts of the province was very different. Western Nevada had a huge geosynclinal accumulation of Triassic rocks; the Triassic is thin and continental in southern Nevada and absent from much of western Utah and southern Arizona. Similarly, the Jurassic is geosynclinal in western Nevada and continental in southern Nevada. Cretaceous rocks are unknown in western Nevada and over much of western Utah; they are thin and continental in southern Nevada, but form a thick geosynclinal accumulation in southern Arizona and New Mexico. The Nevada Jurassic geosyncline underwent strong folding in early Jurassic and again in late Jurassic time. Following the second of these episodes the great plutons of the Sierra Nevada were emplaced in the folded belt. Long erosion uncovered these plutonic rocks, and they were afterward flooded by Tertiary volcanic fields. As a result of these various histories, the ranges differ greatly in structure and internal composition, although their gross forms and relations to the lowlands are closely similar. For example, the Sierra Nevada (largely plutonic, with local patches of Tertiary volcanics on its crest and western slope), Steens Mountain, Oregon (almost wholly of Tertiary lava flows), the Wasatch Range (pre-Cambrian, Paleozoic, and Mesozoic folded rocks cut by early Tertiary plutons), and the Little Hatchet Mountains of southern New Mexico (chiefly Lower Cretaceous sedimentary rocks with some Tertiary volcanics) have had such different early histories that we recognize significant variations from the folded ranges discussed before.

Basin Range Structure.   What is the feature common to the Basin Ranges? Though not all the ranges have been mapped even in reconnaissance surveys, it appears significant that all those in which alluvium does not mask the critical features give evidence of uplift along normal faults (Fig. 19-11). The observed faults dip at angles ranging between 40° and the vertical, averaging nearly 60°. Where the body of the range consists of folded rocks or of thrust masses, these normal faults cut across the older folds and faults. Where the intervening basins are not masked by alluvium

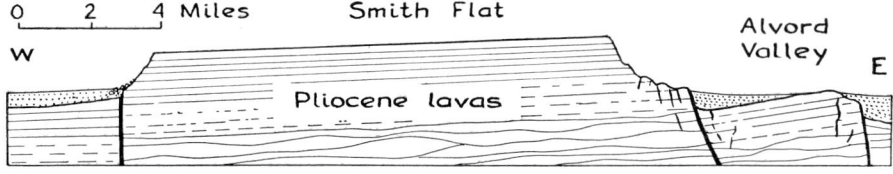

FIGURE 19-11.   *Cross section of a relatively uneroded fault block mountain range, Central Steens Mountains, Oregon.*

it can be shown that, as in Pleasant Valley west of the Sonoma Range (Chap. 9), they are downdropped with respect to the ranges.

In other words, the Basin Ranges, unlike those described above, owe their existence directly to differential vertical movements of adjacent blocks in the earth's crust. No lateral compression of the surface rocks is necessary.

Summary of Mountain Structure

Let us recapitulate the features in which mountains differ from the continental plates:

1. *The sial beneath the mountains is far thicker than that beneath the continental plates.* This generalization is supported by gravity measurements in many areas, by seismic data for many others. In the Colorado Rockies it is less directly shown by a heat flow sufficiently larger than average to suggest the existence there of an unusually thick prism of highly radioactive, and hence sialic, rocks. We have insufficient data to support such a statement with regard to the Basin Ranges as a whole, and in view of the narrowness of many of these ranges, it seems improbable that the blocks composing them contain sial notably thicker than that beneath the adjacent basins.

2. *The rocks of the mountains are more deformed than those of the continental plates.* This generalization seems clearly valid for nearly all ranges, though again we must note that as far as the Basin Ranges are concerned it does not apply to those composed largely of Tertiary volcanic rocks. Nor does it apply in an obvious cause-and-effect relation to the nonvolcanic ranges of this province, for the complex structures they display are far older than the uplift of the present ranges and have not been intensified during uplift.

3. *Granitic plutons are confined to mountains (and to shield areas).* This general statement seems valid, though plutons younger than the folding are only of trivial extent in the Alps and are absent from many ranges such as the California Coast Ranges.

4. *Sedimentary thicknesses are greater in the mountain belts than in the continental plates.* This generalization is well supported by data from many ranges. But it does not apply as a distinguishing feature of the mountains and basins of the Basin Ranges. Nor, as was noted in connection with the pre-Cambrian of Montana, does unusual sedimentary thickness suffice to cause mountain-making. And it fails to apply on a local scale, for it is extremely doubtful whether the Colorado Rockies ever had eroded from them Cretaceous deposits as thick as those found in the nearby intermont basins, which locally contain as much as 20,000 feet of rocks of this age.

5. *In many of the great mountain chains the rock structures are such as to demand large horizontal shortening of the surficial crust.* Again, the Basin Ranges offer a conspicuous exception.

What is the meaning of these generalizations in terms of mountain origin? We find no specific features that we can identify as the "jaws" of a vise that has squeezed together wide segments of the crust into a narrow belt. How has this come about? We will return to this enigma after we consider the subsequent history of such a compressed belt, however it was formed.

The Subsequent History of Folded Belts

Whatever the mechanism by which tightly folded belts are made, isostasy demands that the average level of the base of the folded sedimentary prism must sink (see Fig. 3-6). Local masses may be sustained by the strength of the crust, but excess loads of regional extent cannot be. The sial composing the outer crust has a density of about 2.7. We do not know the precise density of the underlying sima, but it is certainly greater or isostasy could not prevail. Estimates from seismic data range between 3.0 and 3.3 for the density just below the Mohorovičić discontinuity. Let us take the higher of these figures. If, by shortening of a belt of the outer crust, it is thickened 1 km., in order to maintain isostatic balance the basement must sink far enough to displace a mass of sima equal to that of the added sial. This means that the bottom would sink $\frac{2.7}{3.3} \times 1$ km., or more than 800 meters. Thus if we imagine an overthrust sheet 1 km. thick thrust over a wide expanse of country, the surface elevation would increase only about 200 meters. Such an estimate as this cannot be taken literally, as the quantitative values to be assigned are too indefinitely known. But there can be little doubt of the general validity of the reasoning, for it agrees with two significant facts about folded mountains: the long persistence of uplift along them and their association with granitic plutons. It also conforms with some leading facts regarding the geology of the great oceanic deeps.

Several features testify to some uplift and concurrent erosion of mountain ranges during their folding. Among these are: unconformities within the geosynclinal prism, the general presence of coarse clastic deposits in the outer gently folded parts of mountain chains, and the common presence in them of pebbles that could have come only from rocks of far greater age exposed by deep erosion. Apparently, however, much of the uplift of all ranges takes place by vertical bowing upward, after folding has ceased. Perhaps the most compelling evidence of this are the unconformities. At their southwest end, the Appalachians fall off in altitude and disappear as a topographic unit beneath the Cretaceous beds of the Gulf Coast. The map of Figure 19-3 shows that the surface relief formed during the folding of the southern Appalachians had been entirely erased by erosion before the Cretaceous deposits were laid down. Yet the present Appalachians attain their greatest height at Mount Mitchell, about 200 miles along the strike to the northeast. Obviously, Mount Mitchell must have attained most of its present elevation of 6,710 feet in post-Cretaceous time.

Less convincing but still strong evidence of long-delayed uplift of many ranges is found in the erosion surfaces carved across them. The profiles of present streams have characteristically high gradients in their source areas (Chap. 12). In many mountain ranges, however, the streams have their sources in summit uplands of relatively low relief, whose gentle sloping surfaces cut smoothly across rocks that vary greatly in composition and structure, without reflecting these differences in hill and valley forms (cf. Fig. 12-28). The area must once have been a low plain on which streams of low gradient were working. Yet these uplands are now dissected by deep trenches and canyons, with the foaming torrents of modern streams tumbling down them. The contrast in gradient strongly suggests that the summit upland was formed when the streams all had low gradients such as could have existed only when the area stood much nearer to sea level than at present. The evidence is very strong that the mountain mass had been nearly or almost erased as a topographic feature before it was again uplifted.

When we analyze this situation, our idealized picture of the thickened prism of sial beneath the mountain belt may suggest part of the answer. If

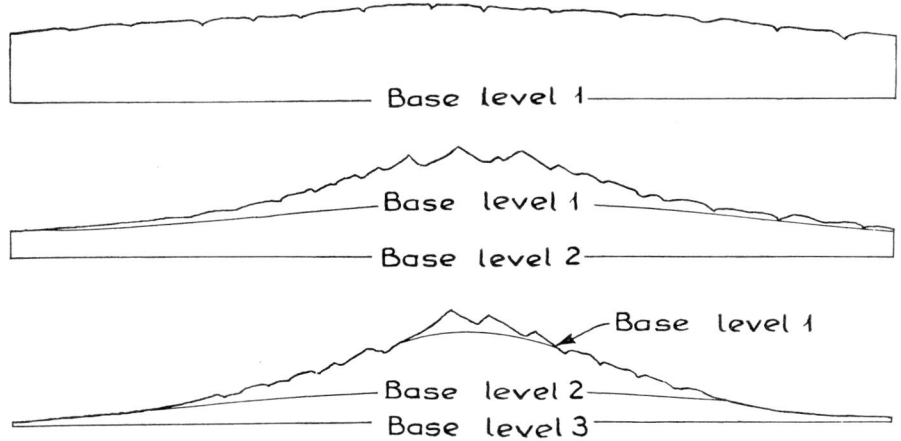

FIGURE 19-12. *Diagram to show how summit elevation may be increased by isostatic uplift even though the average level of the mountain block is being lowered.*

the range is to remain in isostatic balance, far more than 1 km. of material must be eroded from the surface in order to lower the summit of the range 1 km. The situation is the reverse of that prevailing at the time of thrusting. In fact, the erosion of 1 km. of material would reduce the average elevation by only about 200 meters under the simplified assumptions of our illustration. And since erosion by streams with concave side-slopes removes a very large proportion of the mass before the divides are greatly lowered, it is entirely possible that the elevation of the divides will actually increase through isostatic uplift at the same time that the average elevation of the block is decreasing by valley cutting (Fig. 19-12).

The Mountain Root

So far we have considered the thickened sial as maintaining all its properties unchanged, but this is certainly not true. All rocks whose structural relations show evidence of deep burial and deformation are now metamorphosed. Clearly, this requires high temperatures. Mere burial beneath other rocks will raise the temperature of the geosynclinal sediments by blanketing in the earth's internal heat; in addition, the rocks contain radioactive elements whose disintegration contributes to the heating. If a thick column of sialic rock is heated, it may expand enough to raise the surface very notably. Inasmuch as heating is slow, a long time may elapse before the surface of the prism rises because of this effect. Perhaps this is an important factor in the long-delayed uplift of many ranges.

But if the burial is to very great depths, such as the 60-km. root of the Sierra Nevada, heat flow is extremely slow, and bowed-down sial may come in close contact with hot sima. Under these conditions it may gradually heat up to the point of fusion, forming a mass of granitic magma. The abundance of migmatites in deeply eroded mountain chains suggests such refusion, and so, too, does the practical confinement of granitic plutons to mountain chains. But a root so softened by partial melting can no longer maintain itself as a root; it must gradually spread sideways and flatten out. In so doing, it would thicken the sial alongside the range, perhaps accounting in this way for the high-standing plateaus or high plains that border many ranges, for example the Rocky Mountains and Himalayas.

The Shield Areas

Each of the continents contains large areas over which pre-Cambrian rocks are at the surface, among them most of eastern Canada, Finland and Sweden, northeastern Siberia, India, much of South Africa, western Australia, and eastern Brazil. They are called the *pre-Cambrian shields*. At their borders most of them are overlain by Cambrian rocks, resting upon an unconformity of relatively low relief.

Most of the shield areas are only a few hundred feet above sea level, and the character of the surrounding sediments of younger age suggests that this has been their condition through most of later time. However, not all the shield areas stand low: the Adirondacks of New York (an outlier of the Canadian shield), much of Labrador, the mountainous areas near Lake Baikal on the Siberian shield, and large parts of the South African and Brazilian shields stand a few thousand feet above the sea. They are as high, for example, as much of the Urals and Appalachians (which were folded near the end of the Paleozoic), or the Scottish Highlands (where the folding was near the middle of the Paleozoic). Yet these are exceptional parts of the shields, most of them lie close to sea level.

Owing to their lack of fossils, it is impossible to correlate pre-Cambrian rocks over wide areas. Accordingly, in the shield areas we cannot reconstruct the geology with anything like as much confidence as for such ranges as the Appalachians. Nevertheless, despite this handicap, the structures in much of the shield can be deciphered well enough to prove the presence of old mountain chains.

The evidence lies in the structural patterns which are brought out by detailed mapping. Figure 17-18 shows the geology of part of the Baltic shield. When the attitude and relations of the rocks are compared with those depicted on Figures 19-3 or 19-10, the similarity in complexity and general patterns is apparent. On the average, the shield areas have considerably more granite. Furthermore, migmatites, which are rare or absent from granitic borders in most of the younger ranges), are abundant. They make up about a quarter of the exposed rocks in Finland. It seems reasonable to infer that in these belts we have the deeply eroded roots of ancient mountains. The cores of many Paleozoic and younger ranges contain more intensely deformed rocks than their external zones, and the extent of the metamorphic rocks is proportionately greater in older ranges (Appalachians, late Paleozoic; and Sierra Nevada, mid-Mesozoic) than in mid-Tertiary ranges like the Alps. Such metamorphic cores are absent from most late Tertiary and Quaternary ranges. The contorted and metamorphosed zones of the pre-Cambrian shields thus seem what we should expect at lower and lower erosion levels in a folded mountain chain.

Perhaps the generally low topography and structural stability of the shield areas—for it must be remembered that the shields project below the flat-lying strata of the continental plates and thus underlie most of the continents at a depth averaging less than a mile—are due to long-continued erosion of their surfaces, combined with lateral spreading of the former mountain roots until, in the long course of geologic time, the sial beneath them is just thick enough to maintain them in isostatic balance near sea level. Further erosion is very slow. They may be further depressed by sedimentation, but for the most part they are stable. Why some relatively small parts of them may, after long standing in this low position, rise to mountainous heights, as the Adirondacks, is a problem for which no generally satisfactory answer has yet been found.

Ocean Deeps and the Belts of Negative Anomalies

In Chapter 2 we pointed out the close association of most of the great oceanic deeps with arcuate chains of islands. The many earthquakes that occur close to the deeps imply crustal unrest in these areas, as does the common, though not invariable, association of active volcanoes with the island arcs. As the deeps are the maximum departures of the earth's surface from sea level, and many of them immediately adjoin large islands (Mindanao, Cuba, Porto Rico, Sumatra, Java) or even the continent of South America (Atacama Deep, Leeward Trough), the deeps must be fairly young geologically or they would have been filled with sediment. We have noticed (Chap. 16) that the sediment dredged from the Philippine Deep east of Mindanao is obviously derived from that island, and had both deep and island existed for a geologically long time, the deep would surely have been partly filled, if not destroyed.

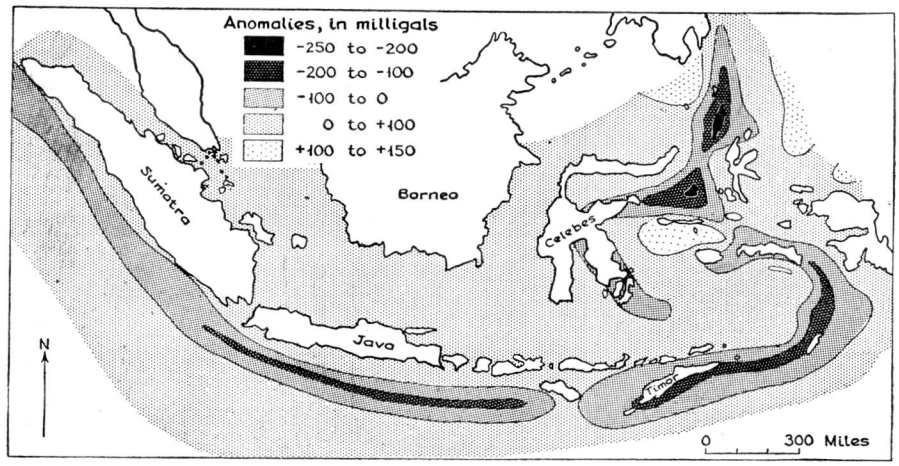

FIGURE 19-13. *Map of the East Indies, showing gravity anomalies. (After F. A. Vening-Meinesz, 1934.)*

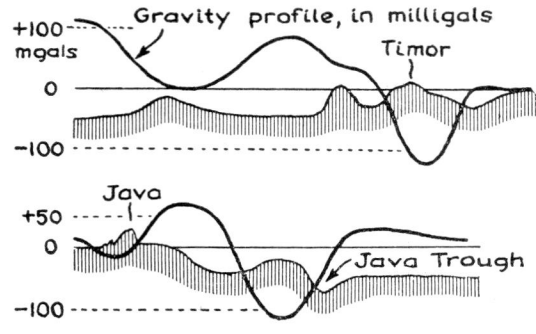

FIGURE 19-14. *Cross sections in the East Indies, showing the relations of gravity anomalies and topography. The upper section is NW-SE through the south end of Timor; the lower is oriented north-south, through the east end of Java. (After F. A. Vening-Meinesz, 1934.)*

One of the greatest geological discoveries of the last generation was that there are remarkable deviations from isostasy associated with some of the island arcs and ocean deeps. This discovery we owe to the distinguished Dutch geodesist F. A. Vening-Meinesz, who, between 1923 and 1932, determined the value of gravity at many points at sea. His later work developed the information summarized in Figure 19-13.

As we noted in Chapter 3, if the earth's surface were level and the crust homogeneous, the attraction of gravity would be determined entirely by the latitude at which the observation is made. The anomalies shown on this chart were determined by subtracting this theoretical value of gravity from the actual value, after certain corrections have been made. The corrections include allowances for the depth of the submarine beneath the surface, the topography of the land and sea floor for long distances around the station, the fact that sea water differs greatly in density from the rocks underlying land stations, and, finally, an allowance for "regional compensation" (equivalent to assuming that a very large crustal block centering at the station is in isostatic balance with the rest of the crust). These actual computations are laborious and far beyond the scope of this book, but the meaning of them is not.

The great belt of negative anomalies can mean only that the belt is underlain by rocks whose density is far less than is usual at equivalent depths elsewhere—in other words, by light rocks. The belt is so narrow—only 30 to 100 miles in the strongly anomalous part—that the deficiency in gravity demands a very deep septum (30–40 miles) of light rocks to account for it.

The belt is about 3,000 miles long. For much of its course it lies close to the axis of a deep oceanic trough, though some sections lie on either side, and others overlie a submarine ridge that divides the deep in two. In Timor, Tanimber, Kei, and Ceram Islands, this ridge comes to the surface, but the negative strip continues. On both sides of the negative strip the gravity anomalies become positive over broad, not obviously linear areas. The positive anomalies are considered to mean that the denser rocks deep in the crust are closer to the surface than they normally are, and hence exert greater than normal attraction (Fig. 19-14).

To summarize, along the negative strip light crustal rocks are greatly thickened and extend to unusual depths, forming a long narrow wedge in the denser rocks on either side.

Now these very large deviations from isostatic equilibrium are not at all common. The uplift of Scandinavia after removal of the glacial ice indicates that even in the shield areas (long unaffected by intense crustal movements or igneous activity since before the Paleozoic, and which might, therefore, be considered unusually strong) the crust has responded to far smaller forces than these great density differences indicate in the East Indies. Both the numerous earthquakes and widespread active volcanoes indicate that the crust is probably unusually weak along the East Indian arc, as compared, for example, with one of the shield areas. Vening-Meinesz pointed out that these facts can only be reconciled by assuming that the rocks of the negative strip are being *currently held down so that they are prevented from floating at the level appropriate to their density.* In other words, some selective downward force of very great magnitude is now operating here. The rocks within these belts must be under very great shearing stress, as they are being dragged down against their natural tendency to float upward. Clearly, such conditions might produce folding and thrusting of the type we associate with mountain chains. Conforming with this deduction, the islands along the belt are, indeed, composed of highly deformed rocks and, as we saw in Chapter 9, are locally capped by warped and uplifted coral reefs. Do we have here a mountain chain in process of formation?

Speculations Regarding Mountain-Making

Although the student will recognize that the mere statement of a geologic "fact" almost of necessity carries with it an element of inference, the facts so far stated in this chapter are generally regarded by geologists as well established. However, the objectives of science include not merely the description of natural phenomena, but also the coordination of data into broader generalizations, and the fitting of these into a comprehensive theory. Yet, though most geologists agree on the facts and even on most of the generalizations we have discussed, hardly a one would state that any satisfactory theory has been advanced to explain the making of mountains. Yet theories and speculations are valuable in the progress of science, for they suggest tests to be made which may modify or confirm some of the basic generalizations. What are some of the speculations suggested to explain the great enigma of mountain-making?

Theory of a Shrinking Earth

One speculation favored by many geologists in the early days of geology (and supported today by some) was that the earth is a shrinking body. In shrinking, the more rigid rocks of the crust must accommodate themselves to a smaller inner core by crumpling, just as the skin of an apple crumples and wrinkles when it dries. It was assumed that the deformation of rocks along mountain chains has been localized by the unusual thickness of relatively weak sediments in the geosynclines. The earth is losing heat at the surface, as is proved by the increase of temperature measured in mines and oil wells. Before radioactivity was discovered, it seemed certain, therefore, that the earth must be cooling off and, hence, contracting.

Although many geological objections have been made to this theory, perhaps the principal one is that shrinking should shorten every great circle equally. The present distribution of folded mountains is far from that to be expected on this theory. A second argument is less easily tested, but seems valid: The Alpine-Himalayan chain is essentially the only considerable east-west chain on earth. Considering scale and strength, it seems unlikely that the crust is strong enough to transmit the great forces half-way around the earth so as to concentrate all the shortening into this single belt. Also, the distances between fold belts of other trends seem far too great. The most serious objection, however, and one that most geologists regard as insuperable, is that, from reasonable estimates of the distribution of radioactive materials in the earth, it seems probable that all the heat now escaping from the surface is of radioactive origin: The earth as a whole is probably not becoming appreciably cooler. Indeed, it may be heating up, for computations of heat transfer show that even in 2 billion years the earth cannot have lost any appreciable heat by conduction from beneath a depth of 700 km. Until these objections are overcome, this once popular theory seems discredited.

Theory of Continental Drift

Probably every schoolboy who has studied a globe has independently discovered that if the Americas were pushed eastward they would almost fit the coast of Africa and Europe. This rough "fit" is even more striking when we notice that the Mid-Atlantic Ridge swings sharply eastward at about the right place to fit the Guinea Coast of Africa. Nearly a century ago, Antonio Snider published the speculation that the continents east and west of the Atlantic had formerly been joined and have since drifted apart. This idea seemed so fantastic to geologists of that time that it was lost sight of until about 40 years ago, when it was again independently suggested by the American geologist F. B. Taylor and the German Alfred Wegener. If whole continents can move so far, the shortening of the crust by a few scores of miles in the mountain belts becomes insignificant. Any force that could move the Americas westward two or three thousand miles could furnish the Rockies and Andes as a minor by-product. Although a mechanism adequate for bodily displacing the continents is unknown, this theory has, nevertheless, won favor from many geologists.

Wegener assumed that the continents float about on the denser substratum almost like rafts on a millpond, and that the ocean floors and the deep layers beneath are so weak that they yield almost like a fluid to very small forces. Two forces tend to move the continents. The first is centrifugal: The sial masses float higher than the sima and so are further from the earth's axis. Hence the centrifugal force acting on them is greater, and they tend to drift toward the equator. The second is the tidal attraction of the sun and moon which, as the earth rotates from west to east, tends always to pull the continents westward. Wegener thought the Alpine-Himalayan chain was formed when Eurasia collided with Africa and India in response to the first force, and that the Andes and Rockies were piled up by friction as the Americas were dragged westward by the tidal pull.

Both these forces exist, but computations show that they are millions of times too small to have folded the rocks. If so, one may wonder why the theory is still seriously considered. But many geological phenomena are well established—continental glaciation, for one—that have so far defied explanation. If geologists were convinced that the continents were formerly more closely grouped, they would accept the fact, even though no adequate mechanism has yet been suggested. Let us briefly consider the geological evidence advanced in support of such a grouping.

Obviously, more than the apparent "fit" of the Atlantic shores is needed. Indeed, this "fit" at a depth of 10,000 feet is far less impressive, and there were notable differences in the continental outlines in Cretaceous or even in mid-Tertiary time. The main arguments advanced are the similarities in the trends of mountain belts, in strata, and in fossils of land-dwelling plants on opposite sides of the Atlantic.

In New England and Newfoundland, the fold axes of mountain belts formed during the middle and late Paleozoic are cut off by the shore. Mountain structures of similar, but not identical, ages in the British Isles and western Europe also end at the Atlantic shore (Fig. 19-15). Likewise, pre-Devonian fold axes in Brazil and Uruguay end at the Atlantic, as do some folds of about the same age in South Africa. The apparent interruption of

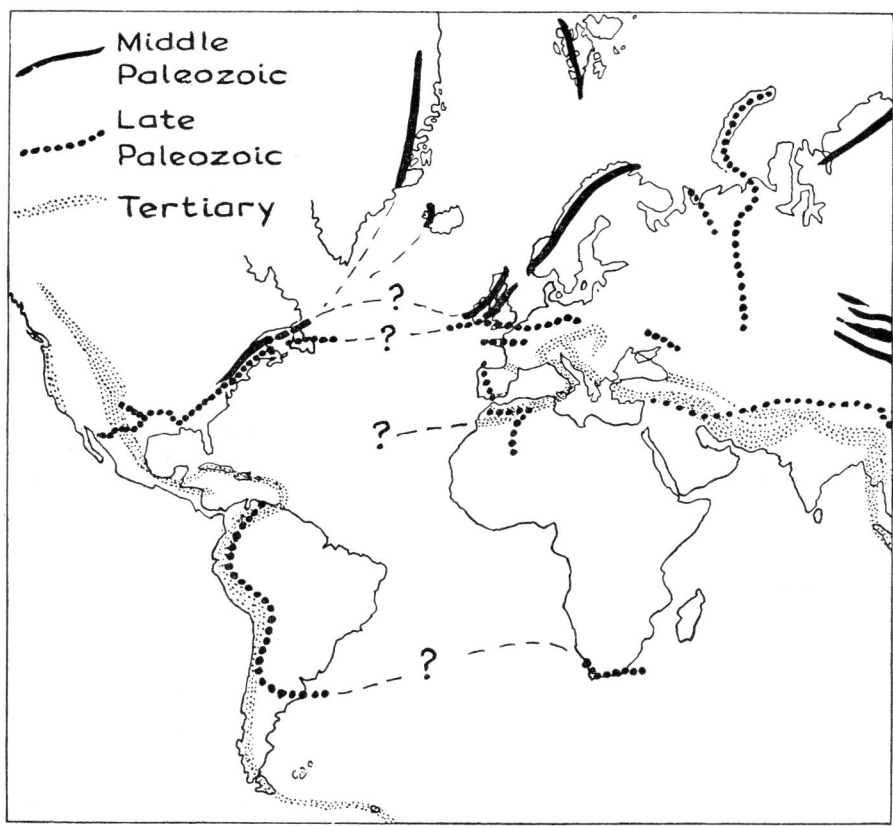

FIGURE 19-15. *Map showing trend lines of folded belts on both sides of the Atlantic. (Data chiefly from J. H. F. Umbgrove, 1947.)*

these fold lines by the ocean suggests, but does not prove, that they were formerly joined and have since drifted apart. The hazard in such an assumption is shown by the fact that the mid-Tertiary chains of the Atlas and Pyrenees ranges also end at the shore, and some geologists have tried to find their western continuation in the Greater Antilles. It is clear from the pattern of the gravity anomalies and island arcs, however, that the Antilles structures never extended eastward to join them, but curve southward through the nearly drowned arc of the Lesser Antilles to join the Venezuelan ranges. It is as likely that the mid-Paleozoic folds of Newfoundland curved slightly to connect with those of east Greenland as that they joined those of the British Isles.

More suggestive, though still open to question, are similarities in the strata of South America and Africa. In both regions the oldest fossil-bearing rocks are Upper Silurian or Devonian. The fossils resemble those of the Falkland Islands but differ from those of northern lands. In both regions similar continental rocks overlie these marine beds, and both areas contain late Carboniferous tillites. Tillite of about the same age is found also in South Australia, Madagascar, and India (Fig. 19-16). Many fossil plants are associated with this tillite, and coal beds overlie it. In all these widely separated areas two genera of fernlike plants, *Gangamopteris* and *Glossopteris*, occur, but they are found in the northern hemisphere only in Russia, northeastern Siberia, and India. The *Glossopteris* flora has been cited as "proving" a complete separation of the southern lands and India from the rest of the northern hemisphere, and the name "Gondwana Land" has been given this hypothetical continent that included all the areas with these plants. However, such vagaries in plant distribution are not rare. Many Permian plants of Texas and Arizona are not known to occur in the more easterly parts of North America, nor in Europe, but are found in China. These are never cited as evidence of a former union between Asia and North America.

Perhaps the strongest argument for a drastically different pattern of continents in late Paleozoic time lies in the glacial pattern. As Figure 19-16 shows, glaciers invaded South America from the east and the Falkland Islands from the north, apparently radiating from a point now in the South Atlantic. Glaciers entered Australia from the south as far as latitude 28° S. An ice cap centered near 15° S in Africa, leaving its northern moraines on the equator. The Indian glacier centered in the Aravalli Mountains, north of Bombay, in latitude 20° N. Characteristic rocks from here are found in the

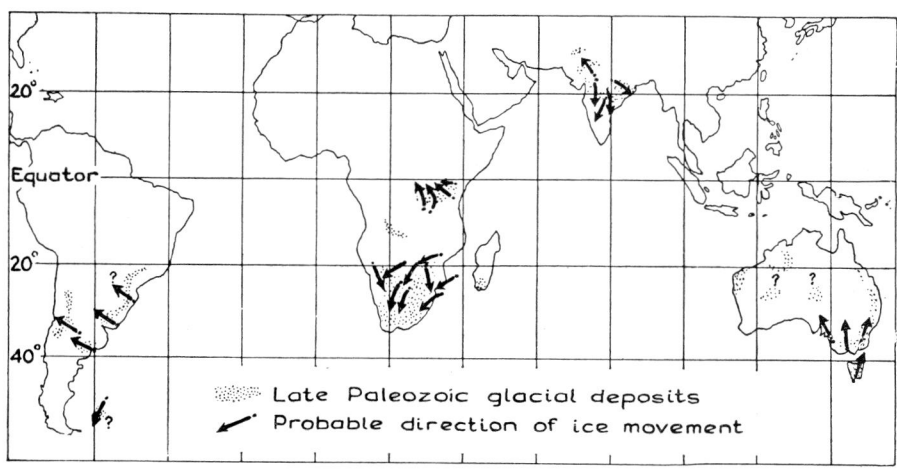

FIGURE 19-16. *Map showing areas of Late Paleozoic continental glaciation.* (*See also Figure 19-17.*) (*Data from A. L. Du Toit, 1937, and D. N. Wadia.*)

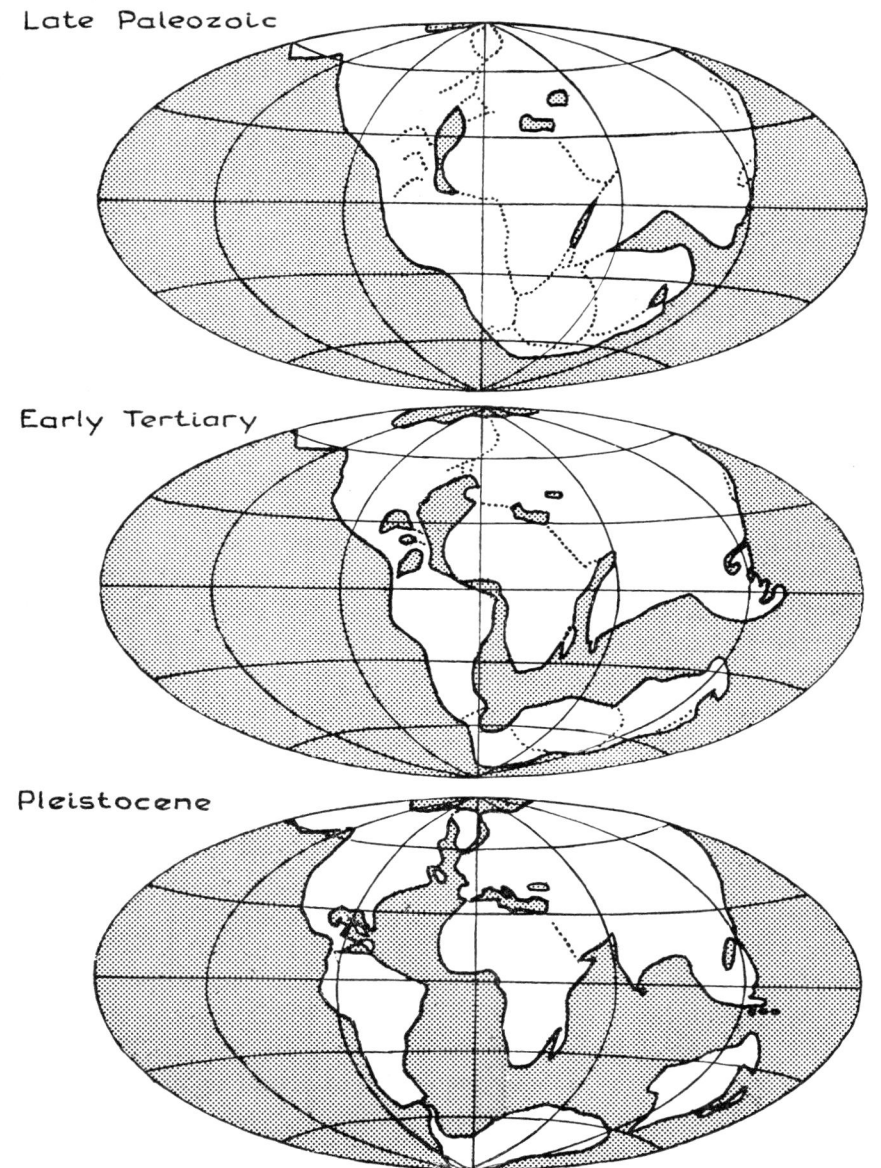

Late Paleozoic

Early Tertiary

Pleistocene

FIGURE 19-17. *Wegener's conception of the drifting of continents. (After A. Wegener, 1912.)*

tillite hundreds of miles away, both to the southeast and northwest. How are we to account for continental ice sheets in the present tropics?

One suggestion is that the earth's poles have shifted in position, and another, sometimes linked with it, that the now-scattered parts of Gondwana Land were formerly a single mass near the South Pole (Fig. 19-17).

Mere shifting of the earth's axis without moving the land masses would not avoid the embarrassment of continental ice sheets in the tropics, for the tillites are now so placed that no possible shift would fail to leave one or another within 20° of the equator. Wegener thought, therefore, that in the late Paleozoic the Gondwana fragments were gathered into a single mass, with the South Pole at about the common junction of Africa, Australia, Antarctica, India, and Madagascar. His sketch of assumed relations is shown in Figure 19-17. He thought the American and European coals (of about the same age as the Gondwana glaciation) represented tropical peat bogs, whereas the *Glossopteris* coals were formed in peat bogs like those of Ireland or Alaska. The tillite near Boston, within a few miles of the Rhode Island coals, is thus interpreted as that of a mountain glacier. But the Russian and Siberian *Glossopteris* localities must have been in the tropics.

Others besides Wegener have tried to explain the climates of the late Paleozoic by different groupings of the continents. All require special explanations for nearly as many phenomena as they explain. All the plants of the late Paleozoic are extinct; few paleobotanists have much faith in climatic inferences from them, for there are many examples of later adaptations of plants to environments that differ widely from those in which they arose.

The earth's angular momentum is great. Like a gyroscope, it strongly resists change in its axis of rotation, and astronomers do not believe it possible that the polar axis shifts in other than very small amounts.

Another good reason for doubting that the drifting of continents causes mountain folding is that, even if we grant it for the Tertiary mountain chains, we are left with no satisfactory mechanism for all the far older ones. Though the theory is a brilliant tour-de-force, its support does not seem substantial. Nevertheless, it has focused attention on one of the most difficult problems of geology: If the continents are, indeed, light masses floating on a denser substratum, how is it possible for an area once continental later to become a part of the ocean floor?

Evidence for Deep Submergence of Continental Margins

Though facies changes of many ancient formations have suggested that land areas have locally been drowned to considerable depths, the recent work of Maurice Ewing, an American geophysicist, has placed the matter beyond reasonable argument for the Atlantic coastal plain off the United States. With instruments for recording seismic waves set up in the rocks by artificial explosions, Ewing was able to trace several strata with distinctive elastic properties from their exposures on land (and in deep wells) far out to sea, in fact, to the edge of the continental shelf (Fig. 19-18). In this way he determined that the rocks which lie unconformably upon the folded Appalachians form a huge embankment, in places more than 3 miles thick. The basement of this embankment has the same elastic properties as the metamorphosed rocks exposed in the Appalachian Piedmont, and it is practically certain that the unconformity separating this from the younger series has been identified for about 100 miles offshore. In other words, the eroded land surface beneath the Mesozoic sediments that we find at many places in the Piedmont, has been traced almost to the edge of the continental shelf, where it is more than 2 miles below the sea.

Figure 19-18 is interesting for two reasons. First, in the present context, it shows that former continental areas may become submerged to the depth of the ocean floor near the continental border; second, it suggests the probability that this continental shelf, above the crystalline floor, is a sedimentary embankment built of Mesozoic and Tertiary rocks, and not in any large degree a cut surface planed off by marine erosion, although there is no doubt that marine erosion has modified it. Although features such as these pose difficult problems, in the light of the earth's general isostatic condition, they furnish strong reason to doubt that continental drifting is necessary to bring about the present wide separation of terrestrial plant and animal fossils that we thought required former land connections like those assumed for Gondwana Land.

Another theory that may hold promise of reconciling these many diverse factors has recently come under consideration. This is the theory of mountain-building by convection currents within the body of the earth.

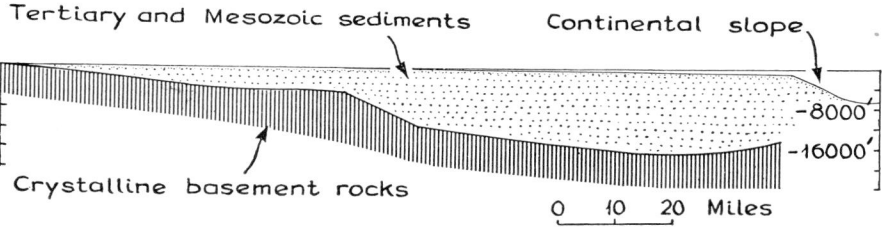

FIGURE 19-18.   *The structure of the Atlantic Coastal shelf off Cape May. (Modified from Maurice Ewing, 1950.)*

The Convection Theory

Any fluid heated from below tends to lose heat in two ways: by conduction and by convection. In large masses conduction is ineffective because the amount of heat transferred in a given time decreases as the square of the thickness of the mass through which it must be transmitted. The effectiveness of convection depends on the rate of overturn of the liquid. Convection is favored by low viscosity, low conductivity (or large size, which has the same effect), and by a large increase in temperature with depth.

Now it may seem absurd to think of the subcrustal part of the earth as like a liquid subject to convective overturn, for the speed of transverse seismic waves shows that it is a rigid solid to the core boundary at 2,900-km. depth. Nevertheless, we know from the folded structures of sedimentary rocks in mountain ranges and from the contortion of many metamorphic rocks that these, though not molten, have been plastic and have flowed. And isostasy shows that over broad regions the subcrustal material must act much like a dense liquid in which the lighter crust is floating. The viscosity of the subcrustal material has been computed from the rate of uplift of the Scandinavian areas in response to the unloading of glacial ice (Chap. 13). It is very high, but, since the temperature increases downward, it is possible that the viscosity is less near the core, despite the tremendous pressure there. If, then, the deeper parts of the mantle contain radioactive material in amounts equal to those of meteorites, heating by radioactive disintegration will make the mantle unstable, and may start a slow convective overturn.

The theory has been analyzed by the Israeli physicist Pekeris on the idealized scheme of Figure 19-19, which, for convenience in computation, assumes two polar continents and an equatorial ocean. He found that convective currents moving more than 3 cm. per year may be present beneath such idealized continents, and that their drag on a rigid crust would be strong enough to crush the crustal rocks even if they were three times as strong as tests indicate. Other theoretical work on the problem, assuming different convective depths, yield comparable results. These computations are all based on a constant rate of convective overturn, such as would take place in liquids with no shear strength but with appropriate viscosities.

The American geophysicist Griggs, however, has pointed out that the subcrust is not a true liquid but has a certain strength, even though this is small under the high temperatures deep within the earth. He assumed, therefore, that in any particular convection cell overturn is not at a constant

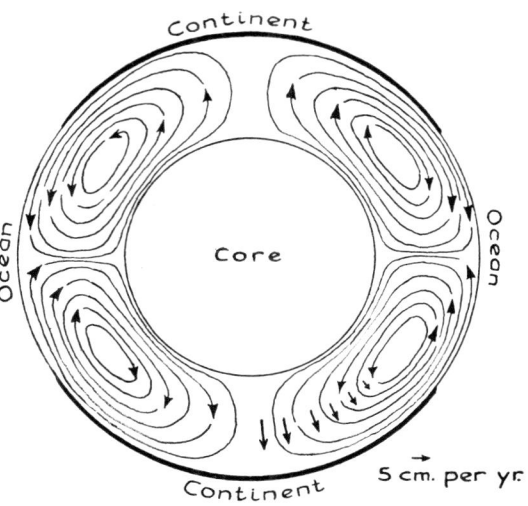

FIGURE 19-19. *A scheme of convection currents in an idealized earth with polar continents and an equatorial ocean. The rate of movement in centimeters per year is indicated by the straight arrows. (After C. L. Pekeris, redrawn from D. T. Griggs, 1939.)*

rate, but that a certain threshold strength must be overcome before it can begin. Furthermore, when the driving force sinks below a certain amount, convection will cease rather abruptly, because the rock strength must always be exceeded if motion is to take place. Thereby he derived a cyclic scheme of convection, starting slowly, speeding up to a maximum, and then slowing down and stopping until the cooler material has again been heated by conduction from the core and by radioactivity (Fig. 19-20). Making cer-

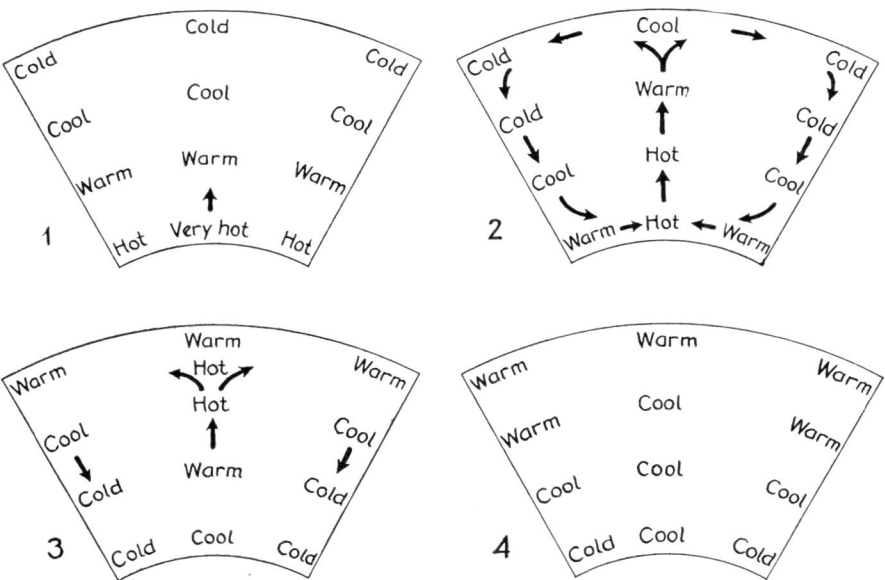

FIGURE 19-20. *Cross sections through the mantle illustrating four phases of Griggs' convection theory. (After D. T. Griggs, 1939.)*

tain assumptions he thinks reasonable, the several phases would be about as follows:

First phase: slowly accelerating currents—25 million years
Second phase: rapid currents—5 to 10 million years
Third phase: decelerating currents—25 million years
Fourth phase: quiescence—500 million years

It seems probable that during the quiescent period some other convection cell may be active, so that it is impossible to put a definite time limit to the crustal drag by currents at one or another place.

Kuenen of Holland and Griggs have both constructed models carefully scaled down to simulate earth conditions. These models suggest that, if convection does take place in the mantle, all the features of folding and shortening of the surface rocks, overthrusting, and the crowding together of roots of light crustal material as they are dragged downward into the substratum may be formed in this way (Fig. 19-21). In the model the subcrustal currents are made by slowly rotating drums. The subcrustal material is simulated by glycerine, and the crust by cylinder oil mixed with fine sawdust. (Physical analysis indicates that these substances have about the right properties to emulate the action of the rocks in their natural dimensions.)

Properly to evaluate these experiments and the theoretical basis on which they rest demands more background in science than can be given in this book. Nevertheless, we can see that a surprisingly large number of the phenomena connected with mountains are coordinated by this hypothesis.

Perhaps the weakest aspect of the convection theory is the formation of the geosyncline: The period assigned to slowly accelerating currents whose downward drag develops the crustal sag to be filled by sediments is much too short for most geosynclines. But, if this weakness is overlooked, we can see that overthrusting and intense deformation of the passive sedimentary cover, the dragging down of the crust to form "negative belts," and the later uplift of the mountain belt after the current stops are all nicely accounted for. When the currents stop, the root is no longer held down and will rise isostatically. Furthermore, in being dragged down the root enters a region of higher temperature, and the highly sheared and deformed crustal rocks of which it is composed would recrystallize to metamorphic rocks and be injected by magma. The magma may be produced by liquefaction of subcrustal material which is already near the melting point and ready to "flash" into liquid in zones of fracture where pressure suddenly falls. Some of the root may only partly remelt to form migmatites, and part may flow upward as granitic batholiths to congeal at higher levels in the mountain

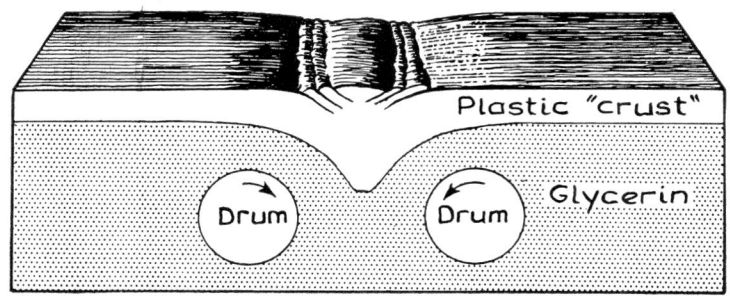

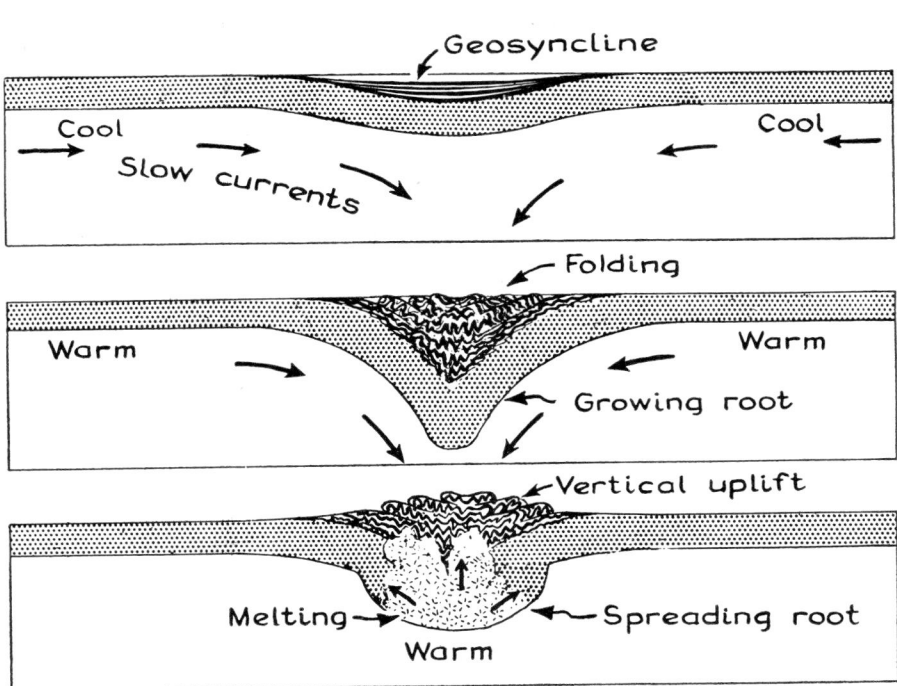

FIGURE 19-21. Top, *model showing how convective currents can drag a plastic crust together and form a "root" in the subcrustal layer.* Bottom, *cross sections suggesting conditions during three stages of mountain building, according to the convection theory. (Modified from D. T. Griggs, 1939.)*

structure. Further, it is reasonable to believe that a softened and partly melted root may flow laterally as well as upward, thereby accounting for the common widespread uplift that affects not only the folded belt itself, but also the country alongside, the surface rocks of which are unfolded.

On such an hypothesis, the belts of negative anomaly in the East Indies and the comparable belts in the West Indies and along some of the island arcs of the Pacific, such as the Marianas and Japanese arcs, would be interpreted as due to the dragging down of the crust by existing subcrustal currents. The fact that deep-focus earthquakes seem definitely alined in planes that dip away from the ocean deeps supports the idea that such currents are actually dragging the oceanic crust beneath the land masses alongside. The fault plane marked by the earthquake epicenters may be thought of as a shear zone plunging down with the descending currents.

Obviously, it is far too soon to accept the convection theory, at least in its present form, as a satisfactory explanation of the complexities of mountain-making. Until further knowledge is gained as to conditions deep within the earth—and such knowledge is necessarily indirect and difficult to evaluate—we cannot even be sure that the necessary driving force can be furnished. The dearth of conclusive evidence for the other theories suggested gives no support for this one. Yet it is possible, since so many phenomena can be naturally fitted into this scheme, that it holds seeds for an ultimately satisfactory explanation of the fascinating enigma that confronts us: an earth at least two billion years old but still, beneath its passive crust, endowed with the energy literally to move mountains! Indeed, if evidence for continental displacement ever becomes so strong as to be compelling, it may be that subcrustal convection currents will offer the only mechanism by which to bring these, too, about.

Facts, Concepts, Terms

GEOSYNCLINE; FORELAND; HINTERLAND

EVIDENCE OF CRUSTAL SHORTENING

NAPPE; KLIPPE

ASSOCIATION OF PLUTONS WITH HIGHLY DEFORMED AREAS

DATING MOUNTAIN FOLDING BY UNCONFORMITIES

CRUSTAL THICKENING FAR EXCEEDS MOUNTAIN ELEVATION

VERTICAL UPLIFT LONG AFTER FOLDING AND DEEP EROSION

MIGMATITES AND MOUNTAIN ROOTS

BELTS OF STRONG NEGATIVE GRAVITY ANOMALY

THE THEORY OF CONTINENTAL DRIFT
    Geographic and geologic evidence; available forces

Questions

1. In both the East and West Indies, volcanoes are arranged on an inner arc parallel to the arc of the ocean deeps and associated negative anomalies. The surface on which the deep-focus earthquakes occur slopes downward toward this inner arc (Fig. 18-13). What does this suggest as to the source of the magmas?

2. The Mohorovičić discontinuity beneath the Atlantic Piedmont lies at about 10 miles depth. Beneath the folded Appalachians it is at a depth of about 30 miles. What qualitative difference would you expect to find in the average heat flow to the surface in the two areas?

3. Many of the highest ridges of the Appalachians are synclinal in structure. In the light of Chapters 9 and 12 can you offer a suggestion for the reasons?

4. What inferences can you make regarding the forces causing crustal deformation from such features as the Great Glen fault (Chap. 10) and the San Andreas rift (Chap. 18)?

5. Many of the buildings of western England are roofed with slate, whereas those near London are chiefly tiled. Can you draw from this any inferences as to the regional geology of England?

6. Many deep wells drilled in the search for oil have shown the existence of a long, rather narrow mass of granite beneath the Carboniferous strata of eastern Kansas and southeastern Nebraska. What features of the rocks brought up by the drill would enable you to decide whether this mass is unconformably buried by the sedimentary rocks or whether it invaded the sediments after they were deposited?

7. The Triassic rocks of the Atlantic slope of North America are commonly thought to have been formed under conditions similar to those of the present Great Basin. What features would you expect them to have from which such an origin was inferred?

8. In western Nevada a fossil-rich Permian limestone lies nearly horizontally across upturned slates beneath. A few fossils identifiable as Ordovician in age have been found in the slate. What history is recorded by these relationships?

9. What features would enable you to distinguish a sill of granitic rock (whose lower contact only is exposed) from a block of granite thrust over flat-lying sedimentary rocks?

10. How would you distinguish a klippe (an erosional remnant of a thrust sheet) from the erosional remnant of a resistant bed in an undeformed sedimentary

series? Assume in both cases that the remnant is of Carboniferous rocks resting upon flat-lying Devonian.

Suggested Readings

1. Bucher, W. H., *The Deformation of the Earth's Crust*, Princeton University Press, Princeton, 1933.
2. Jeffreys, Harold, *Earthquakes and Mountains*, Methuen and Co., London, 1935.
3. Umbgrove, J. H. F., *The Pulse of the Earth*, Martinus Nijhoff, The Hague, 1947. (Especially Chapter 2, pp. 26-39.)
4. Daly, R. A., *Architecture of the Earth*, D. Appleton-Century Co., New York, 1938.
5. Bailey, E. B., *Tectonic Essays, Mainly Alpine*, Clarendon Press, Oxford, 1935.

20. *Mineral Resources*

The Industrial Revolution

OUR WORLD differs more from that of the Founding Fathers than did theirs from the world of Alexander the Great. In 1800, nearly four-fifths of the population of Great Britain and Italy was rural, and more than nine-tenths in the rest of Europe. Land transport was by wagons on roads hardly better than those Caesar used in Gaul; and Napoleon's crossing of the Alps was little less of a feat than Hannibal's 2,000 years before. Today a Zulu miner travels third-class to his labor compound in the Rand gold field in greater comfort than Louis XIV could command between Versailles and Paris. By our standards, the Zulu's lot is hard and his pay pitiably small; yet he is better clothed and fed than most of the people of George the Third's England. He is fortunate, indeed, when compared to a slave in the mines of Laurium, whose life expectancy was four years, and whose labors beneath the Athenian lash won the silver that sustained the Golden Age of Greece.

Most of us now reject the slave-holding philosophy of Plato and Pericles, but it is not primarily our ethics that account for these differences. Material goods are perhaps as unevenly distributed today as under most of the cultures of the past, but the standard of living is higher.

The change began with two events of the Eighteenth Century. Neither induced as much notice at the time as the intrigues of Bonnie Prince Charlie or the campaigns of Frederick the Great. But about 1730 a Shropshire Quaker, Abraham Darby, discovered how to use coke in smelting iron; and in 1768 James Watt invented the steam engine. These men made possible cheap iron, steel, mechanized power—and the industrial age. Without machinery, population would long since have outstripped food supply the world over, as Malthus had predicted, and as it indeed has in China and India, where the Industrial Revolution has only feeble roots.

Now, as always, agriculture is the basic industry. But a wholly agricultural economy imposes sharp limits on a division of labor and the increased productivity this allows. As transport improved, first with iron rails and then locomotives and steam-driven ships, a specialization formerly unknown made possible tremendous savings in manpower. Now a twelve-year-old girl operating a machine loom in a Lancashire mill could turn out 35 yards of calico daily, and thus, in a year, clothe about 1,200 persons.

This little girl's own existence was doubtless as dismal as any Norman serf's, and even today the "better life for all" is an ideal that is still far from governing our world. Yet this very ideal would be pathetically ludicrous if there had been no industrial revolution. Even now the food supply of much of the world is below that required at the subsistence level; it would be a pitifully smaller fraction of the need if we were to revert to the economy of 1800.

These things are commonplace and generally accepted. But what is not so widely understood is that all these changes in living standards ultimately depend upon the world's diminishing nonreplenishable assets—its mineral resources.

The Mineral Basis of Civilization

Throughout history mineral resources have played a greater role than is usually recognized. Today this role is second only to that of agriculture. Even in ancient times the impact of mineral wealth on national power can be clearly traced, though most historians ignore it. The Greeks who turned back the Persian hosts at Marathon were armed with bronze swords and shields, while many of the enemy had only leathern shields and stone weapons; the Greek fleet at Salamis was built by the Athenian profits from the silver-lead mines of Laurium, discovered only a few years earlier. In fact, these profits paid the mercenaries who fought Athens' battle of the Peleponnesian Wars, and with the exhaustion of the mines came the end of Athens as a military power. Philip burst from the wild Macedonian Mountains, and his son, Alexander the Great, swept over the world, financed by the flush production of gold—roughly a billion dollars in modern equivalent—from the new mines on Mount Pangaeus. When Scipio drove the Carthaginians from Spain and won for Rome the gold, iron, copper, silver, and mercury of the Peninsula, he sealed the fate of Carthage.

These are but a few examples from preindustrial days. Today the impact of mineral resources on national power and well-being is even greater. Gold and silver could hire mercenaries and influence campaigns, but useful goods could not be created from them; they merely gave control of the few goods available to one group rather than another. They still possess this conventional value, but living standards and national power depend only incidentally on them. The useful goods of the world depend on the mineral fuels and the industrial metals—iron, copper, aluminum, lead, and others. It was no accident that Britain maintained the *Pax Britannica* through the Nineteenth Century; her industrial and military supremacy came from the happy fortune that her "tight little island" held the greatest known mineral wealth per acre of any similar area of the world, together with a population intelligent and aggressive enough to exploit it.

At one time or another in the Nineteenth Century, Great Britain was the world's largest producer of iron, coal, lead, copper, and tin. From these came her machines, mills, and the great cities founded on them. Before 1875 she built more miles of railroad than any of the larger continental countries. Her flourishing internal markets and her products, carried to all the world by the British merchant marine, brought her the greatest wealth any country in history had ever enjoyed. True, the cheap foodstuffs she received eventually ruined the island's agriculture, but her favorable trade balance enabled British capital to take hold of Malayan tin, Spanish iron, many of the mines and oil fields of Mexico, Chile, Iran, Australia, Burma and the United States. These holdings saw her through one world war and kept her credit through a second. When her flag followed her mining investments into South Africa and the Boers were defeated, she gained control of more than half the world's production of new gold and ultimately of great deposits of copper, chromite, diamonds, asbestos, and manganese.

Nowhere better than in the United States can be seen the close dependence of living standards and national power upon minerals. Before 1840 manufacturing was inconsequential and required heavy subsidies and tariffs to compete with the advanced British industries. The small and scattered iron deposits along the eastern seaboard did indeed supply enough of the local demand to influence the British Parliament in 1750 to outlaw their exploitation; after Independence there was slow growth, but as late as 1850 the iron production was only about half a million tons. In 1855 the "Soo Canal" brought the rich Lake Superior iron deposits within economic reach of Pennsylvania coal; by 1860 iron production had trebled and by 1880 it had passed that of Great Britain.

It was the greater productivity of Northern industry and the weight of armament, supplies, and equipment flowing over its superior railway net that were decisive in the Civil War. The Tredegar iron works at Richmond was the only one worth mentioning in the Confederacy. Alone it could not compete with the overwhelming output of the Pennsylvania furnaces.

Our huge internal market, our prodigious endowment in all the minerals basic to manufacturing, and our favorable agricultural heritage all contributed to make our country the most powerful one in the world at present. During the Battle of the Bulge, our troops hurled at the Germans more metal than was available in all the world in Napoleon's time. That cynical remark of a general, "God is on the side of the most cannon," has been repeatedly justified. Cannons, plowshares, tanks, and tractors all are made of metals, and all are transported by the mineral fuels.

Salient Features of Mineral Resources

That mineral resources are concentrated in relatively small areas, and that they are exhaustible and irreplaceable, are facts which have social and political implications often—one might almost say, generally—overlooked by those not familiar with the mineral industry. Their effect on society is so profound that no student of geology should fail to recognize them and the geological factors that determine them.

Sporadic Distribution. As we shall see in more detail later in this chapter, mineral deposits of all kinds are essentially "freaks of nature." An abnormal pituitary may make a man a giant, although his physiological processes are otherwise normal. Similarly, mineral deposits result from normal geological processes but under exceptional conditions. Only a few geological environments favor the formation of mineral deposits.

These favored spots are by no means evenly distributed over the earth. Nearly nine-tenths of the world's nickel comes from less than a score of mines in the Sudbury district, Ontario. A single mine at Climax, Colorado, produces about as large a share of the world's molybdenum from an area of far less than one square mile. Nearly 30 per cent of the copper produced in the United States since 1880 has come from an area of less than four square miles at Butte, Montana. The Rand gold field of South Africa, unusual for its great extent compared with most gold districts, produces half the new gold of the world from an area about 50 miles long and 20 wide.

Though the mineral fuels are far less localized, they underlie only a trivial part of the continents. Less than 25 per cent of Pennsylvania, a leading coal state, is underlain by coal. The East Texas oil field, the greatest thus far found in the United States, which for several years yielded about a quarter of the nation's oil, covers an area about 10 by 40 miles—a mere dot on the vast expanse of Texas.

The economic implications of this unequal distribution of the mineral fuels are decisive. Modern chemistry may be able to make a rayon purse (perhaps better than a silk one) out of a sow's ear, but only with the expenditure of energy. Today this means mineral fuels or, more rarely, water power. Without such adequate sources of cheap fuel the dreams of many countries of emulating the United States in manufacturing are foredoomed to failure. Other countries, such as the Scandinavian, have higher educational standards, or, like Argentina, a higher agricultural output per capita than we. But none is so fortunate as the United States in the combination of high average education and hence skilled population, great agricultural productivity, and the most nearly balanced mineral resources of any free-trade area in the world.

Exhaustibility. All mineral deposits are limited in extent. They represent unusual associations of geologic factors that have permitted their concentrations. Once the valuable materials are extracted by mining, all that is left are holes in the ground. This is the fate of all mines, even of the greatest.

It is true that the mines of Almaden, Spain, have yielded mercury since the days of the Carthaginians and still hold the richest known reserves of this metal. But they are almost unique among metal deposits. The mines of Cornwall—the "Cassiterides" or Tin Islands—furnished tin to the Phoenicians and a trickle of this metal throughout history until within a generation. In the Nineteenth Century they led the world in production. Now the mines are worked out, and the Cornish miners have dispersed throughout the world to carry their traditional mining lore far from Britain. The greatest single oil well, the Cerro Azul No. 4, in the Tampico Field, Mexico, after yielding nearly 60 million barrels of petroleum, suddenly gave forth only salt water. Neither the old Cornish tin mines nor the Cerro Azul will ever yield a new crop.

The valley of the Nile has been the granary of the Mediterranean through most of recorded history, and huge areas of China and India have been farmed nearly or quite as long. The forests of Norway that built the Viking ships still produce lumber. But the mines of Freiberg, where Werner's Mining Academy flourished, and where even now a mining school persists, have long been abandoned. Belgium and Wales, with their cheap coal and metallurgical traditions, are still centers of smelting, but the local metal mines on which industries were founded have been closed for generations. Potosi, which supplied silver in tons to the viceroyalty of Peru; the fabulous Comstock Lode of Nevada; the Broken Hill zinc deposits of New South Wales; and the copper deposits of Michigan are not quite dead, but they are pale shadows, indeed, of their former greatness. *There is no second crop of minerals!* And, lest the meaning of this fact, economically, politically, and socially, be overlooked, let it be noted that *more metal has been mined since 1910 than in all preceding history.*

It seems inevitable that in a generation hence the world supply of minerals will be won only at greatly increased costs of energy. In the long sweep of history we are in the period of flush production of mineral resources. It is apparent that the maintenance of our present standard of living is by no means a foregone conclusion; unless tremendous technological improvements can be made, the inevitable increase in energy requirements to win a pound of iron, a gallon of oil, or a ton of coal must be reflected in falling standards of living. Such are the inevitable results of the geological conditions that determine the localization of mineral resources in the earth's crust.

What the exhaustibility of mineral resources means socially can be seen in the long roll of our own Western "ghost towns," where a few families remain in place of thousands. More dramatically, it is seen in the "distressed" coal mining towns of England, where the mines are not, strictly speaking, exhausted, but where rising costs due to the necessity for deeper mining, more pumping of ground water, and longer hauls from coal face to portal have weakened the competitive position of the mine in world trade. Unemployment, wage cuts, and lower living standards have followed. Only by drastic technologic changes can costs be kept down.

Although geological conditions compel the long-term trend in mineral procurement to be more expensive in terms of energy costs, it does not necessarily follow that economic costs on an industry-wide basis need rise. Many times, technical improvements have made it feasible to rework a deposit once thoroughly exploited by an outmoded method. Gold dredges now operate at a profit on placer deposits that contain only a few cents worth of gold per cubic yard—deposits that were exhausted from the standpoint of the Alaskan "sourdough" working with sluice boxes. Similarly, most of the world's copper is taken today from deposits that were impossible to exploit by methods of 50 years ago. There are limits to the development of low-grade deposits, however, and it is important to realize that these technological limitations, as well as the sporadic distribution and exhaustibility of mineral resources, place upon the mineral industries restrictions which are different *in kind* from those affecting most other economic activities.

Cycles of Mineral Production. The mineral industry, like others, is constantly shifting. Newly discovered deposits and new technologies may increase prosperity for a time, but then the higher costs of deep mining and exhaustion of the deposits eventually bring harder times. The American economic geologist, D. F. Hewett, has analyzed the history of the mineral industry in many countries and finds a surprisingly consistent sequence of stages. Briefly put, these are as follows:

1. Period of mine development: exploration, new discoveries; boom towns; many small mines and a few large deposits recognized; rapid increase in output.

2. Period of smelter development: few new deposits found; small mines

being worked out; greater output from larger mines; many smelters competing for ore.

3. Period of industrial development: lower costs, higher living standard; rapid increase in internal and external markets and wealth; approaching height of commercial power.

4. Period of rapid depletion of *cheap* domestic raw materials: higher costs of mining and of recovered metals; more power required; bitter commercial rivalry with competitive sources of raw materials; trade balance unfavorable, with gradual loss of home trade to foreign competitors.

5. Period of decreasing domestic and foreign trade: higher costs of manufacture because of using foreign raw materials; lower living standards and declining commercial power; stiff competition for cheap foreign materials, often leading to international friction and wars.

Of course, Hewett did not maintain that deviations from this pattern are unknown nor that all phases are inevitable, but the sequence has enough generality on the basis of past history to be worth serious attention. The stages in the mining industry itself may be recognized even in countries like Bolivia or Malaya, where lack of power prevents any noteworthy manufacturing activity. Roughly speaking, stage 1 may be represented by Rhodesia; stage 2 by Canada; early stage 3 by the U.S.S.R.; late stage 3 and early stage 4 by the United States (note our heavy investments in petroleum in the Caribbean and Near East, in copper in Chile and Rhodesia, in iron in Brazil, Venezuela, and Cuba); stage 4 by Germany; and stage 5 by Great Britain, which passed through the first three stages during the Eighteenth and Nineteenth centuries.

The Economic Importance of Mineral Resources

More than half the value of mineral production of the United States is supplied by the energy sources: the *mineral fuels,* petroleum, gas, and coal. Mineral resources other than fuels include the *metalliferous* deposits, the source of our metals; and the *nonmetallic* deposits, such as building stone, cement rock, clay, sand, and gravel. From 1940 to 1950 the total annual value of mineral products in the United States has ranged between 6 and 10 billion dollars. Although mineral resources are only about 5 per cent of the total national production, they have played a crucial part in industry, for all heavy manufacturing depends on them.

The Mineral Fuels

The mineral fuels are the most important mineral resources. They are essential for heat and power and for metal refining. They are also sources of many

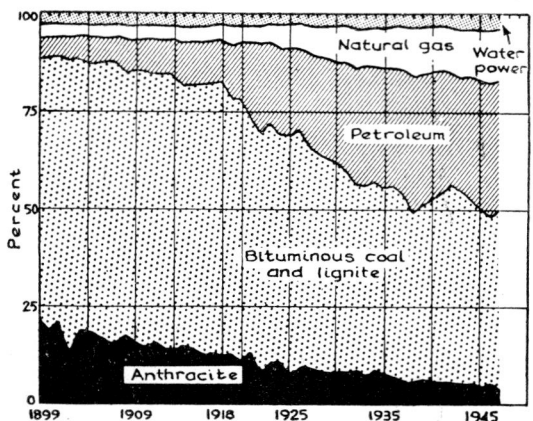

FIGURE 20-1. *Percentages of total energy from fuels used in the United States. (After U. S. Bureau of Mines,* Minerals Yearbook, 1949.)

chemicals and nitrogen fertilizers. They are especially important to the geological profession. The search for oil is the principal business of more than half of all geologists.

The Industrial Revolution was based on coal. It is still the basic fuel, though petroleum is displacing it in the field of transport. Coal still furnishes the world more energy than do oil and gas, despite the rapid increase of petroleum production in the United States and elsewhere. Even in the United States half of the energy is still supplied by coal (Fig. 20-1). Of the other half, two-thirds is from oil, and most of the rest from natural gas, with water power much the smallest contributor.

World production of coal in 1947 was 1,750 million tons, mostly in Europe and the United States, with a probable value at the mines of about six billion dollars. World production of petroleum was three billion barrels (about 550 million tons) in 1947, mostly in the United States, Venezuela, Russia, and the Persian Gulf region. The value at the wells was almost six billion dollars (Fig. 20-2), nearly equal to that of the coal, though its tonnage was less than one-third as great. Per heat unit, oil and gas bring much higher

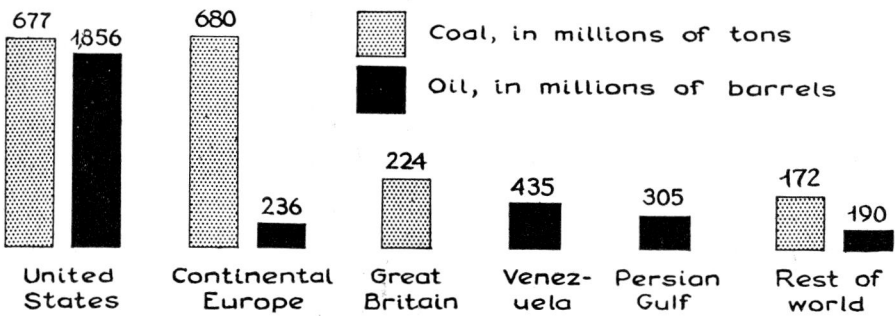

FIGURE 20-2. *World coal and oil production during 1947. (Data from U. S. Bureau of Mines and Bituminous Coal Institute, 1948.)*

prices than coal because of their greater convenience, absence of ash, and ready transportation through pipes.

Coal

Coal is a brownish-black to black combustible rock (see Chap. 5). It forms beds that range from a fraction of an inch to many feet in thickness, interstratified with shale, sandstone, and other sedimentary rocks (Fig. 20-3). A single sequence of strata may include several coal beds. In West Virginia 117 different beds have been named. They are distributed through several thousand feet of strata, which, taken together, are called the *coal measures.* These and many other coal-bearing strata include many alternations of marine and nonmarine beds. The coal beds are in the nonmarine parts of the section and themselves contain evidence of nonmarine origin. They are composed chiefly of flattened, compressed, and more-or-less altered remains of land-dwelling plants: wood, bark, roots (some in place), leaves, spores, and seeds.

Coal Rank. Coals appear to have been formed chiefly from plant residues that accumulated in swamps. There is a continuous series from brown *peat,* obviously made up of slightly modified plant residues, to a hard, black, glistening type of coal without recognizable plant remains. The principal members of the series are *peat, lignite, subbituminous coal, bituminous coal,*

FIGURE 20-3. *Coal beds on Lignite Creek, Yukon region, Alaska.* (*Photo by C. A. Hickcox, U. S. Geological Survey.*)

and *anthracite coal.* Except for peat, which is not considered as coal, their more obvious features are shown in Table 20-1.

When coal is heated in the absence of air, water vapor and organic gases are given off. These are called the volatile matter. The woody and other plant components in peat are complex compounds of carbon, oxygen, and

TABLE 20-1. *Distinctive Features of Coal of Various Ranks*

KIND OF COAL	PHYSICAL APPEARANCE	CHARACTERISTICS
Lignite	Brown to brownish black	Poorly to moderately consolidated; weathers rapidly; plant residues apparent
Subbituminous coal	Black, dull or waxy luster	Weathers easily; plant residues faintly shown
Bituminous coal	Black, dense, brittle	Does not weather easily; plant structures visible with microscope; burns with short blue flame
Anthracite coal	Black, hard	

hydrogen. In the air they oxidize and rot away, yielding chiefly carbon dioxide and water, but if air is excluded by geologic burial they slowly alter into many solid products as well as some gases. Among the solid products is finely divided black elemental carbon, a substance whose presence distinguishes coal from peat. The higher the proportion of elemental ("fixed") carbon and the lower that of volatile matter, the higher the rank of the coal is in the series from peat to anthracite. Clay or sand washed into the swamp where the coal accumulated are left as ash when the coal is burned. This lessens the heating value and increases waste, and thus detracts from coal value.

Most bituminous and anthracite coals are of Carboniferous age. Such coals are largely concentrated in Europe and eastern North America. Most high-rank coal in Europe is in a belt extending from Britain across Belgium, Luxembourg, and Germany to southern Russia. This is the industrial heart of Europe.

Coal Reserves. Coal is generally so abundant that only the thicker, more accessible, and higher rank deposits are now being mined. An estimate of the coal reserves thus means little unless limits of thickness, depth, and quality (including both rank and ash percentage) are stated in the estimate. Officials in different countries rarely use the same limits, so that estimates for different areas are rarely comparable. The older estimates for the United

States and Canada included beds as thin as 14 inches. Coal cannot be mined from these thin beds by the mechanical methods now used for more than 90 per cent of United States' underground production. The thin beds are no more desirable for surface stripping, which in 1946 yielded more than one-fifth of the country's bituminous coal. Accordingly, these older estimates are not economically significant. Now (1950) the U. S. Geological Survey is in the midst of a program that may, after several years, yield a satisfactory modern estimate of the nation's coal reserves. Such a program involves much more than statistics. Especially in the western United States, low- and medium-rank reserves are large, but there has been relatively little mining. Much field work must be done before estimates can be made. Coal outcrops are traced and recorded on maps, thicknesses measured, drill cores studied, folds and faults worked out, and overburden calculated in the manner outlined in Chapter 10. The coal tonnages must then be calculated in beds of various thicknesses and to various depths.

Using the old 1922 figures as the best available currently for the United States and a set of German estimates (1938) as best for the rest of the world, the total coal reserves of the world, including all ages and ranks, reach the enormous total of more that 7 trillion tons.

Of this, North America contains 58.8 per cent (Fig. 20-4). Coal fields cover about one-ninth of the United States (Fig. 20-5). The distribution of the chief ranks of coal by states is shown in Figure 20-6. Though Wyoming and North Dakota lead in amount, the former has little high-rank coal and the latter none. In 1947 the overwhelmingly greatest output was of bituminous

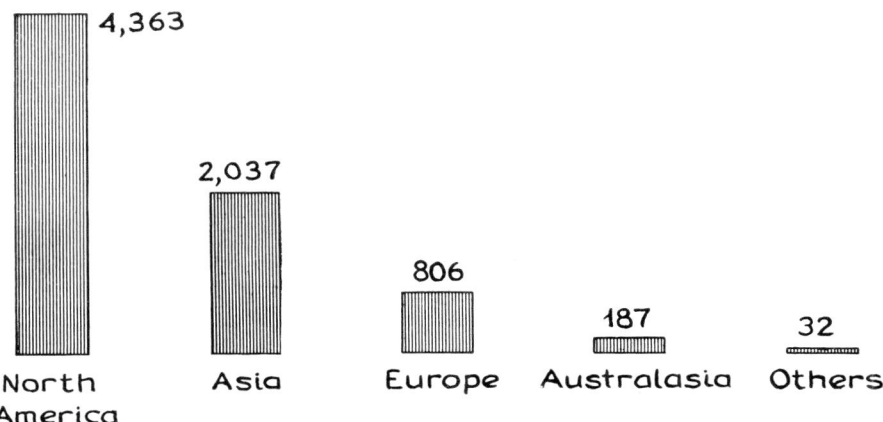

FIGURE 20-4. *The coal reserves of the world, in billions of tons. (Data from Bituminous Coal Institute, 1948.)*

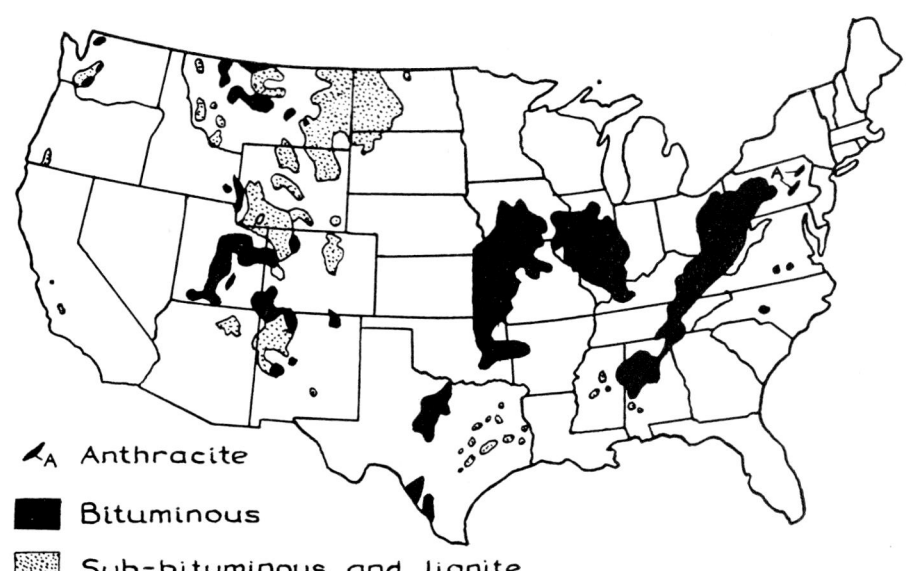

FIGURE 20-5.   *Map showing the coal fields of the United States. (After Bituminous Coal Institute, 1948.)*

and anthracite coal, produced by West Virginia, Pennsylvania, Kentucky, Illinois, and Ohio, in that order.

Oil and Gas

Earth oil (petroleum) and natural gas are found in similar environments and usually together. Commercial accumulations are almost entirely limited to sedimentary rocks and to special circumstances. Producing areas are generally called "pools," although the fluids fill pore spaces in rocks (like ground water) rather than open caverns.

The requisites for an oil pool appear to be four: (1) a source rock, (2) a permeable reservoir rock which will yield the oil rapidly enough to make drilling worth while, (3) an impermeable cap rock, and (4) a favorable structure, so that the cap rock can retain the oil below ground.

The essential feature of a *reservoir rock* is the presence of connected pores or cavities through which a liquid can move. Permeability (Chap. 14) thus is its prime characteristic, though the percentage of pore space, and hence the storage capacity of a given volume of the rock, is also obviously significant. Permeability is commonly measured in darcys (Chap. 14) and millidarcys (thousandths of a darcy). Even the tavern loafers of an oil town now speak enthusiastically of a permeability of 500 millidarcys and deprecatingly of a mere 50 millidarcys. The majority of reservoir rocks are sandstones, though some are limestones or dolomites, either granular, with interstitial pore spaces, or jointed and cavernous. Shattered brittle rocks like chert, granite, and schist are permeable, productive reservoir rocks in a few fields.

A *cap rock* must be practically impermeable to oil or gas. Most are shale, but nonporous limestone seals some reservoirs, and others are sealed by the asphalt left near the surface where oil has escaped and evaporated. Such spots where oil escapes to the surface are called oil *seeps* or, in Spanish countries, *breas*. Drilling beneath them has frequently led to the development of commercial oil accumulations.

Oil and gas are much rarer pore fillings in rocks than water, and their commercial accumulations require *favorable structure*. As water fills most spaces below the water table, and oil is lighter than water and hence floats on it, the favorable structural positions are high in the reservoir rock, directly beneath the cap rock. Gas, which is lighter than oil, rises to the top of the

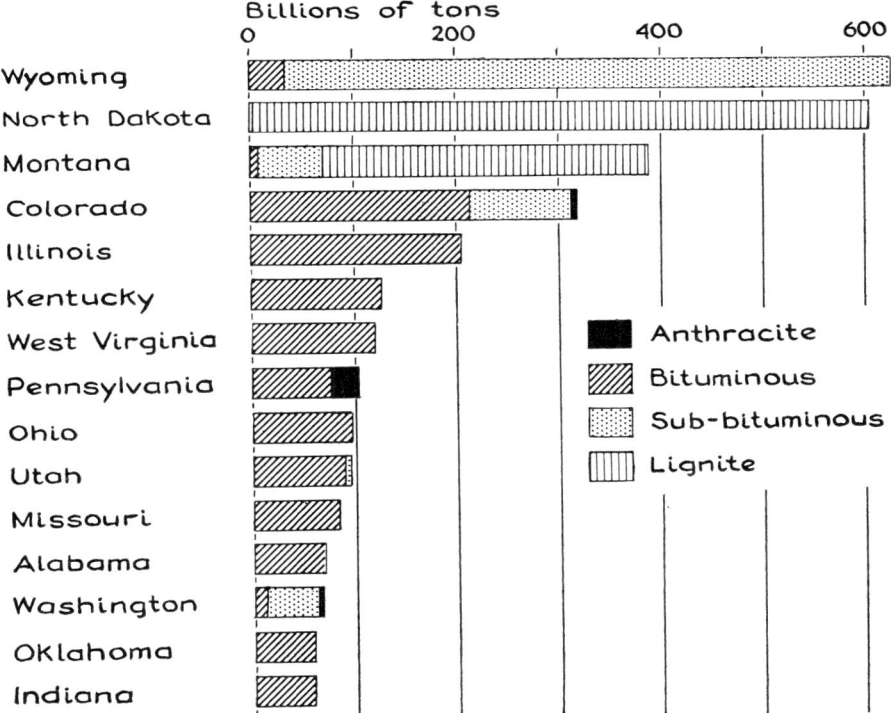

FIGURE 20-6. *The original coal resources of the United States. Other states than those shown had a total of about 200 billion tons. (After Bituminous Coal Institute,* Bituminous Coal, *1948.)*

oil column, though, when greatly compressed, much gas is dissolved in the oil underground and is released only when the pressure is decreased. Some favorable structures are diagrammed in Figure 20-7.

Most oil fields are along the crests of elongate anticlinal folds, often called domes. Typical are the Salt Creek Dome, Wyoming; the Kettleman Hills North Dome, California; and the huge oil field of Bahrein Island in the Persian Gulf. Second in number are the "salt dome" fields, formed alongside plugs of massive salt that have been injected from beneath, tilting the strata (Fig. 20-7). There are many of these along the Gulf of Mexico in Texas and Louisiana, of which the famous Spindletop Field, one of the earliest, is representative. Similar accumulations have been found along the Caucasus and in Rumania in the Carpathian foothills. Finally, there are the stratigraphic traps, with the reservoir capped generally unconformably by an overlying blanket of impermeable sediment. The greatest of all American oil fields, East Texas, is a stratigraphic trap on a gently dipping sand-

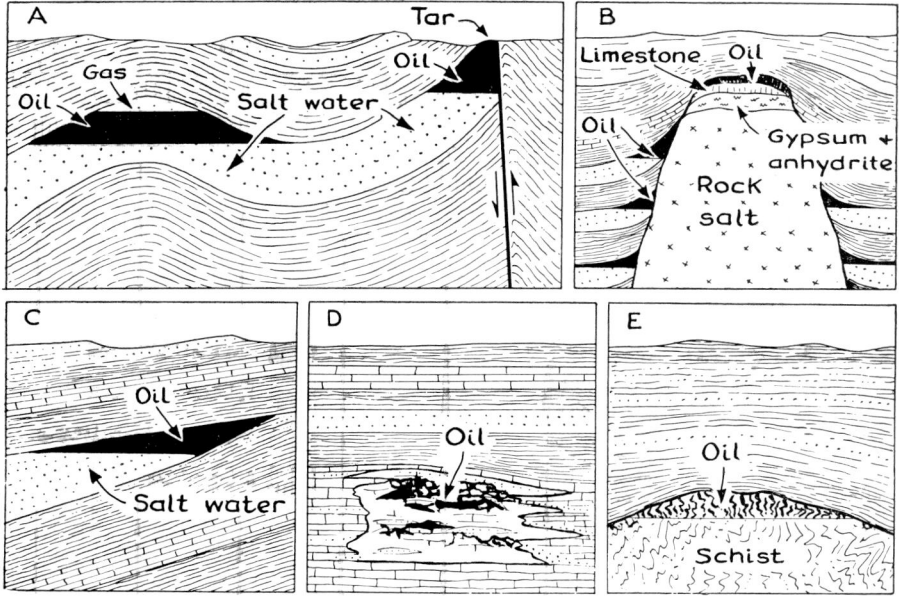

FIGURE 20-7.   Structures favorable to the commercial accumulation of oil and gas. A, anticlinal fold, with reservoir sand underlain by shale (possible source rock) and overlain by shale (cap rock). Note the reservoir sand capped by tar at outcrop, making second trap for oil. B, salt dome, with oil at crest and on flanks. C, trap in sandstone, formed where unconformity is overlain by shale. D, porous limestone reef reservoir, in impermeable limestone and shale. E, fractured schist reservoir, beneath domed shale.

stone bed whose former outcrop is covered by younger strata. Some traps contain gas but no oil.

Source Rocks of Petroleum. The conclusions as to reservoir rock, cap rock, and favorable structure are well-established geological generalizations inductively based on repeated observations. Oil is found in permeable rocks, confined by nonpermeable, in structural positions determined by hydraulic laws. But the sources of earth oil are matters of less certain inference. Some very important conclusions seem well established, but further advances are needed.

Practically all geologists, though not all chemists, believe that petroleum originates exclusively in sediments. Nearly all oil pools are in sedimentary rocks. The rare exceptions, mostly in California, are in shattered schists or gneisses that have been raised by earth movements to positions where they could have received their oil by migration upward from nearby sandstones that are oil bearing (Fig. 20-7). Most pools are separated from the nearest igneous or metamorphic rock by thousands of feet of barren sediments that do not have a trace of oil.

Oil fields, too, are in or near thick accumulations of marine or deltaic sediments that contain large volumes, measured in cubic miles, of more or less organic shales. These shales contain up to one or two per cent compounds of carbon manufactured by plants or animals; hence the name organic. Study of the black organic shales from many oil-producing regions suggests that the source material may have been largely marine ooze. The original synthesis of carbon dioxide and water must have been by plants, perhaps in large part by the very abundant diatoms. The organic compounds of carbon, hydrogen, and oxygen thus made may have been repeatedly worked over in the digestive tracts of many kinds of animals in or on the floor of the sea before final burial beneath accumulating ooze. Even after burial they may have been further altered by bacteria, with the elimination of oxygen. Some oozes may have accumulated in stagnant basins where the sea water contains insufficient oxygen to permit transforming the hydrogen-carbon residues into water and carbon dioxide again. However, the final transformation into liquid oil is hidden in mystery. It must have awaited the accumulation of a cover of younger sediments.

The tentative prehistory of petroleum just outlined is the product of much chemical and geological research. Three additional generalizations seem justified, though not all are universally accepted. First, little oil has been found in areas containing only fresh-water sediments, despite the vast quantities of "oil shale" (containing no free oil, but from which liquid oil can be distilled) in the Eocene lake deposits of Utah, Colorado, and Wyoming. Second, oil and coal are not the liquid and solid products, respectively, of the alteration of peat, even though both oil and coal are found in the same sedimentary sequences in the mid-continent region of the United States. Third, as no oil has been found in process of formation or in Pleistocene sediments—formed during the last million years or so—the oil-forming process seems very slow. The hope, expressed by some chemists, that earth oil may now be forming at somewhere near the rate of consumption seems quite unjustified geologically.

Oil Map of the World. Because of the economic importance of oil, a knowledge of petroleum geology is essential to domestic or foreign statesmanship. Where is oil now being produced? What reserves are still in the ground? Where may additional supplies be found in the future?

First, we may mark off the areas where igneous and metamorphic rocks crop out, calling these *impossible* or *wholly unimportant* because of the absence of marine organic sediments. We can add to them as *unfavorable* the areas where only thin or nonmarine or impervious sediments overlie the crystalline rocks (Fig. 20-8). Then, we may outline the areas where oil is being produced and add to them those where thick marine sediments have yielded evidences of oil, such as seepages. These are obviously the *most favorable* areas. Finally, we may distinguish an intermediate *possible* group

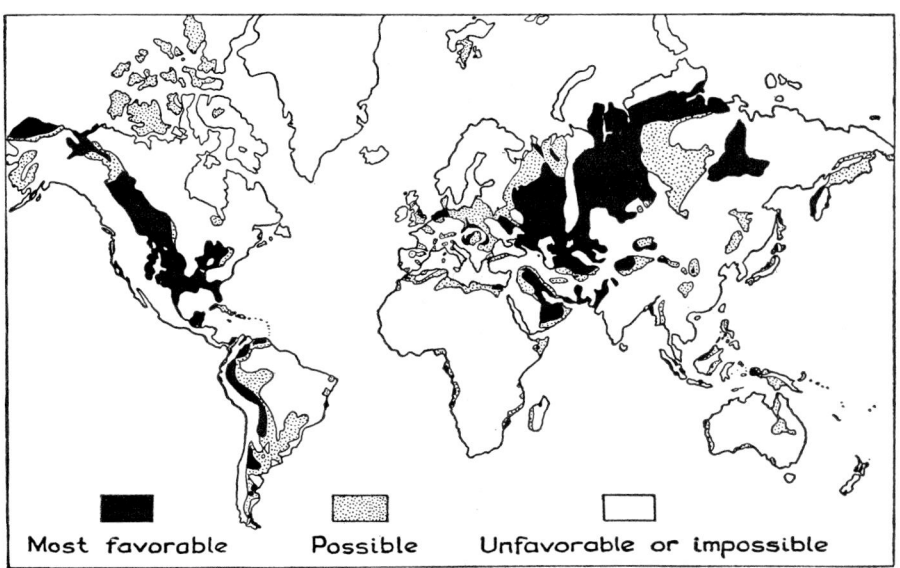

Most favorable　　　　Possible　　　Unfavorable or impossible

FIGURE 20-8. *Oil map of the world, based on the probability of finding oil in any one region.* (*After Arabian American Oil Company,* Middle East Oil Developments, *1948.*)

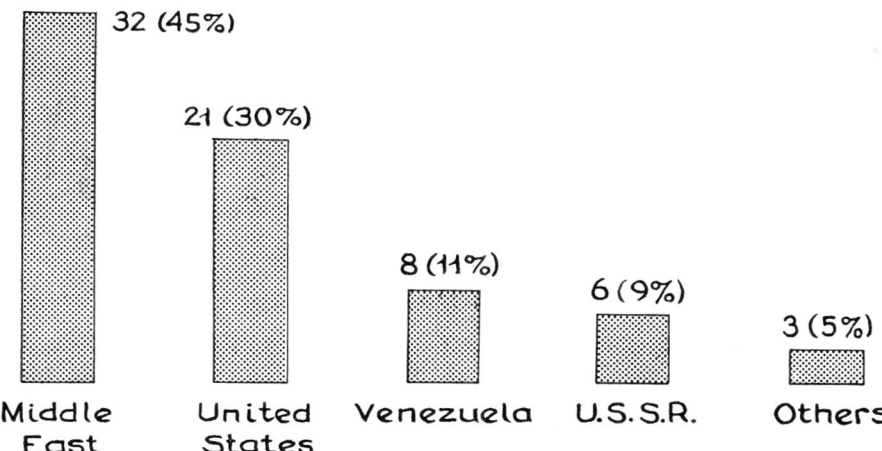

FIGURE 20-9. *Oil reserves of the world (figures indicate billions of barrels). (Data from Arabian American Oil Company, 1948.)*

of areas, containing thick marine sediments but no positive indications of petroleum.

Such a map is valuable but leaves many questions unanswered. Which areas have produced the most oil? Which have the most left? To date the United States had produced more than half of the world's oil. But what about proved reserves? The United States still has much oil, as shown by Figure 20-9. However, supremacy has passed to the region about the Persian Gulf. This area, which includes Iraq, Iran, Kuwait, and eastern Arabia, is one of the two great petroliferous regions of the world. The other is the border of the Caribbean and Gulf of Mexico, including Venezuela, eastern Mexico, and our Gulf Coast. Of these two regions, the Persian Gulf, with a much smaller past production, appears to have a far more promising future. The fields of Iran, Arabia, and Kuwait are intrinsically so rich and are being managed so much better than were the American fields in the days of wasteful competition that the net values are almost unimaginably great. Never in human history have there been such prizes of concentrated wealth. Even now, at the beginning of development, more than a million barrels flow daily from a few hundred wells. Compare this with America's hundreds of thousands of wells, with an average production of 11 barrels per day, and remember the vast wealth that has come from even these wells. Obviously, much of the history of the immediate future will be connected with the oil of the Middle East.

No inventory of potential oil provinces would be complete without mention of the continental shelves. These are repositories of huge volumes of sedimentary rocks, a large part of which is undoubtedly marine. Though the technological difficulties of exploration and exploitation are tremendous, there can be little doubt that vast amounts of petroleum are contained in them. Whether these can be economically extracted in the face of the high costs of offshore work and the probable competition from synthetic oil from coal or oil shale is an important problem for the future.

Oil Finding.   The first earth oil put to human use oozed from oil seeps. Noah's Ark may have been caulked with asphalt from a Mesopotamian seepage as are the present native boats of that country. The first successful well, in Pennsylvania (1859), was drilled beside a seepage, as were many later discovery wells in all parts of the world. Most of these have been small producers, but Cerro Azul No. 4, the greatest single producer in oil history, was an exception. This well, near Tampico, Mexico, blew in February 10, 1916, for 260,000 barrels a day, the column of oil rising 598 feet into the air. This well produced almost 60 million barrels of oil before suddenly yielding only salt water. It was drilled in limestone and apparently penetrated a real pool of oil and gas floating on salt water in the caverns.

Seepages are obvious clues to the presence of oil underground, but absence of seepages does not deter exploration. By 1883 random drilling in Pennsylvania and West Virginia had shown that the productive oil and gas wells were grouped along anticlinal crests. I. C. White announced the anticlinal theory of accumulation which gradually came to be generally accepted as a guide to exploration. Between 1900 and 1918 most oil companies built up geological staffs and gradually transformed their exploratory programs from random drilling to systematic search guided by surface geology. By 1928 the drill had tested most anticlines recognizable from surface geology in the petroliferous parts of the United States and had discovered hundreds of oil fields in this way.

Nevertheless large areas are covered by surface beds that are known to be unconformable upon older ones. Structures in the surficial rocks do not reveal all the anticlines beneath the unconformity and, of course, give little clue to stratigraphic traps. Systematic exploration demanded some means of determining geologic structure at depth.

Geophysical Methods.   The first geophysical method applied was the *gravity survey*. A few domes along the Texas coast had yielded much oil from the flanks or crests of huge salt plugs (Fig. 20-7). As salt is lighter than most other rocks, it was reasoned that measurements of the relative value of gravity over the flat coastal country should reveal similar buried salt plugs by the smaller attraction such masses should exert (see Chap. 3). Systematic surveys did, in fact, reveal many such anomalously low values of gravity, and by 1930 the coast of Texas and Louisiana was dotted with salt-dome oil fields located by drilling these "gravity lows." The application of gravity methods, however, is greatly handicapped in most regions because of slight contrasts in density of the rocks and the difficulties introduced by rough local topography. Accordingly, gravity surveys have been useful mainly in flat terrain and in seeking salt domes which contrast sharply in density with adjacent rocks.

Seismic methods of geophysical exploration, first applied about 1924, have proved more generally useful than gravity methods. Of the several methods used, "reflection shooting" has been most successful. This involves the firing of small charges of dynamite in shallow holes to produce artificial earthquakes of small intensity. The travel times of the elastic waves are recorded on small field seismographs with an accuracy of tenths of a second. Though the interpretation of such records is a highly technical problem and not always free from ambiguity, it is possible to identify reflections from strata of different elastic properties and, by comparing travel times to different points on the surface, to deduce the shape of the buried strata.

By this principle many oil fields in structure not apparent from surface geology have been found. For example, the Louden field in Illinois was so accurately located by seismic work that the oil company was able to lease in advance of drilling nearly all of the 20,000 acres that have since been proved productive. This is a major field with an ultimate recovery estimated at about 200 million barrels. Similarly, many Texas fields, and those hidden by the alluvium in the San Joaquin Valley of California, have been discovered by the seismograph.

Subsurface Methods   By the late twenties, so many wells had been drilled in and near oil fields that geology at depth could often be determined almost independently of the surface rocks. This is obviously extremely important where unconformities, lensing of beds, or facies changes are present. The major problem is that of correlating beds between wells. If this can be done, favorable structures may be further explored and unfavorable ones avoided.

Three principal methods, *lithologic, paleontologic* and *electrical,* are employed in correlation. Lithologic correlations are based on study of well cuttings or cores from the drill holes. An illustration is that in Figure 20-10.

Paleontologic methods are needed where the rocks are not so readily distinguished as the anhydrite and red shale of the Yates field. As the larger fossils that may be present are generally ground to bits in drilling, the principal paleontological materials are "microfossils," chiefly *Foraminifera* (Fig. 16-25). These tiny fossils are extremely abundant in many strata and are small enough to survive the grinding in well-drilling. Some zones 10 or 20

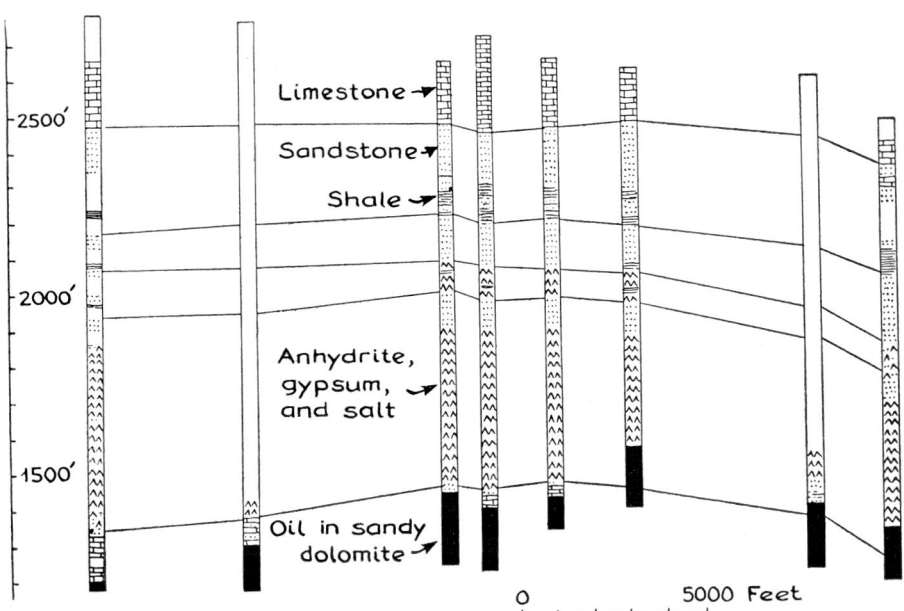

FIGURE 20-10. *SW-NE section across the Yates oil field, west Texas, showing rock correlation from well to well. (After G. C. Gester and H. J. Hawley, Structure of Typical American Oil Fields, American Association of Petroleum Geologists, 1929.)*

feet thick can be distinguished by their microfossils and even traced from one oil field to another nearby. Hundreds of micropaleontologists are now working in the oil industry to make correlations by means of these fossils.

Many methods of correlation by physical properties have been tried. Of these the most widely used is the *electric log*. Electrodes lowered into the uncased well measure differences in electrical characteristics of the beds penetrated. These characteristics vary markedly because of differences in the rocks themselves, their porosity, and the kind of fluid, whether oil, salt water, or fresh water, that occupies the pores. Figure 20-11 illustrates the structure of a flat anticline in Texas worked out from electric logs. The results verified the structure indicated by an earlier seismic survey.

Summary of Methods of Oil Exploration This brief review of the occurrence of oil and of some of the many technical and scientific approaches to oil exploration shows that, although the presence of oil in a particular place at a particular depth cannot be foretold in advance of drilling, the search is not a blind one. Oil is localized in response to definite physical laws, in structures or stratigraphic traps that are geologically determined. The

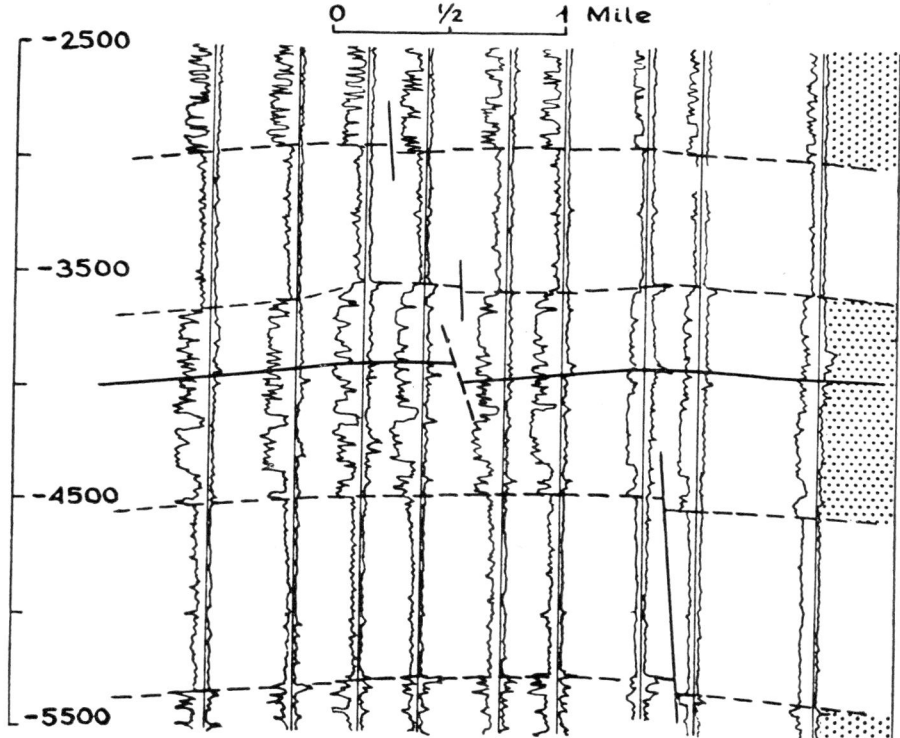

FIGURE 20-11. *SW-NE section across Odem dome, south Texas, showing correlation (dashed lines) from electric logs. The irregular lines to the right of the well location lines record the resistance of the rocks to an electric current; the lines to the left measure the amount of natural current produced by the rocks themselves. Note that the latter line effectively marks the position of the three sandstone units dotted at the right. (After Society of Exploration Geophysicists, Geophysical Case Histories, 1949.)*

problem of finding it is a geological one, the solution of which depends on many methods of geophysics, physics, petrography, paleontology, and chemistry. Systematic and economical exploitation of fields already found, and the cutting down of waste involved in random drilling have resulted from such efforts; hence, elaborate geological departments are maintained by all the large oil companies.

Ore Deposits

Ore deposits are rock masses from which metals are obtained commercially. Every ore body results from the selective concentration of one or more elements in which the rock is greatly enriched as compared to the average of the earth's crust. How great this enrichment must be can be realized from the fact that more than 99 per cent of the earth's crust is made up of only ten elements, and of these ten only aluminum, iron, and magnesium are industrial metals. Most of the other industrial metals—copper, zinc, lead, tin, and others—are found in the earth's crust only in very small amounts (Table 20-2).

TABLE 20-2. *Abundance of the Metals in Igneous Rocks*

ELEMENT	PERCENTAGE	ELEMENT	PERCENTAGE
Aluminum	8.13	Cobalt	0.0023
Iron	5.0	Uranium	0.0004
Magnesium	2.09	Lead	0.00016
Titanium	0.44	Tungsten	0.00015
Manganese	0.1	Antimony	0.0001
Chromium	0.02	Mercury	0.00005
Vanadium	0.015	Silver	0.00001
Zinc	0.0132	Gold	0.0000005
Nickel	0.008	Platinum	0.0000005
Copper	0.007	Radium	0.0000000013
Tin	0.004		

Thus even an aluminum ore near the lowest usable grade (about 25% metal) has three times as much of the metal per pound as the average igneous rock; an iron ore with 50 per cent iron has 10 times as much as the average; a copper ore with 1 per cent copper more than 100 times the average. Most gold mines operate on less than 1/3 oz. of gold per ton of ore, but even this is 20,000 times the content of the average igneous rock. Since the "average rock" may be thought of as forming under "average" conditions, these concentrations must reflect either very unusual geologic conditions or the carrying of ordinary geologic processes to unusual perfection. As we shall see, examples of both aberrations from "normal" are found.

Few ore deposits contain metals as such. Most are rock bodies containing one or more minerals in which the metallic element is combined with other elements (Table 20-3). This combination must be such that the useful metal may be economically recovered. Such valuable minerals are called *ore minerals*. The separation of the ore minerals from the associated useless minerals—the *gangue* minerals—and the extraction of the valuable metals from the ore mineral itself is called *metallurgy*.

The definition of ore is purely economic: It is a rock mass that can be worked commercially for the extraction of a useful metal. There is no stipulation as to the exact percentage of the metal in the ore—only that it be enough to work at a profit. A particular rock may pass from subore to ore

TABLE 20-3. *The Common Ore Minerals*

METAL	MINERAL	ELEMENTS CONTAINED	PERCENTAGE OF METAL IN ORE MINERAL	CHEMICAL COMPOSITION
Gold	Native gold	Gold	50 to 100 (alloyed with silver)	Au and Ag
Silver	Native silver	Silver	100	Ag
	Argentite	Silver, sulfur	87.1	Ag_2S
Copper	Native copper	Copper	100	Cu
	Chalcopyrite	Copper, iron, sulfur	34.6	$CuFeS_2$
	Chalcocite	Copper, sulfur	79.8	Cu_2S
	Enargite	Copper, arsenic, sulfur	48.3	$3Cu_2S \cdot As_2S_5$
Lead	Galena	Lead, sulfur	86.6	PbS
Zinc	Sphalerite	Zinc, sulfur	67	ZnS
	Franklinite	Zinc, iron, oxygen, manganese	About 12	$(Fe, Mn, Zn)O \cdot (Fe, Mn)_2O_3$
Iron	Hematite	Iron, oxygen	70	Fe_2O_3
	Magnetite	Iron, oxygen	72.4	Fe_3O_4
Aluminum	Bauxite (actually a mixture of several minerals)	Aluminum, oxygen, hydrogen	35 to 40	$Al_2O_3, 2H_2O$ (varies)

with increased metal prices (as happened with many mercury deposits during both world wars), with improved metallurgical techniques (as did many deposits of copper, zinc, and lead when the "flotation process" of separating ore minerals from gangue and from each other was developed), or with subsidies (as did the very low-grade Rhineland iron deposits under the Nazi "self-sufficiency" program). Conversely, with price decline or higher mining and metallurgical costs a material may change from a valuable ore into worthless rock.

Many factors other than metal prices determine whether a particular material is or is not ore. Among these are:

1. The size, shape, and depth of the deposit. (All these greatly affect the cost of mining.)
2. The amenability of the ore to metallurgical treatment. (Fine-grained mineral aggregates must be ground to very small grain size for clean separation from the gangue minerals; this is more costly than coarse grinding, which, in turn, is more costly than direct smelting of the ore.)
3. The distance to metallurgical centers or to market. (Brazilian iron ores, though richer than the Lake Superior ores, have lagged in exploitation because of distance from fields of coking coal.)

A dramatic example of the influence of these factors on mining is the following: The great copper mine at Bingham Canyon, Utah, can mine ore containing as little as 0.85 per cent (17 pounds of copper to the ton), yet 40 years ago masses of pure copper up to several tons in weight that were occasionally found in the Michigan mines were not ore, as they could not be effectively blasted, and it costs too much to chisel them out. A generation ago rock containing 50 per cent iron could not be mined on the Minnesota "Iron Ranges" because blast furnaces were designed to use only higher-grade ores. But with the exhaustion of the high-grade ores, furnace practice has been changed so that these lower-grade materials can be used. The happy combination of abundant local coal and limestone flux that makes simple the metallurgy permits the use of ores with as little as 35 per cent iron at Birmingham, Alabama, though such material would be useless in Montana, California, or even in Minnesota, our greatest iron ore state.

Ore Formation

Processes of Concentration. The processes that concentrate minerals into economic deposits have been both mechanical and chemical, either acting alone or in combination. These processes do not differ from those we have already noticed: weathering, solution, transportation and sedimentation at the surface, and volcanism and flow of solutions below the surface. Ore bodies are *rocks,* and at one place or another nearly every rock-forming process has produced a valuable mineral deposit. This is indicated in the brief tabulation of geologic types of mineral deposits, to which many others could be added (Table 20-4). These few are selected because they are economically important and illustrate different ways in which particular

TABLE 20-4. *Types of Mineral Deposits*

TYPE	MANNER OF FORMATION	REPRESENTATIVE DEPOSIT
Magmatic segregation	By settling of early-formed minerals to the floor of a magma chamber during consolidation;	Layers of magnetite and chromite in the Bushveld lopolith, South Africa.
	by settling of late-crystallizing but dense metalliferous parts of the magma, there to crystallize in the interstices of older silicate minerals;	Copper-nickel deposits of Norway and parts of those of the Sudbury district, Canada.
	or to be later injected along faults and fissures of the wall rocks.	Injected bodies of magnetite in Sweden (greatest in Europe) and in New York.
	By direct magmatic crystallization.	Diamond deposits of South Africa.
Contact-metamorphic	By replacement of the wall rocks of an intrusive by minerals whose components were derived from the magma.	Magnetite deposits of Iron Springs, Utah.
Hydrothermal de-deposits (deposits from hot watery solutions)	By filling fissures in and replacing both wall rocks and the consolidated outer part of a pluton by minerals whose components were derived from a cooling magma. These differ from contact-metamorphic deposits in having fewer silicate minerals and by more obvious fissure control.	Copper deposits of Butte, Mont., and Bingham, Utah; lead deposits of Idaho; zinc deposits of Missouri, Oklahoma, and Kansas; silver deposits of the Comstock Lode; gold of the Mother Lode, Calif., and Cripple Creek, Colo.
Sedimentary beds	Sedimentary deposits under conditions which lead to deposition of relatively pure minerals;	Salt and potash deposits of Stassfurt, Germany, and of New Mexico.
	or of rocks unusually rich in particular elements;	Iron deposits of Lorraine, France.
	or of rocks in which the detrital grains of valuable minerals are concentrated because of superior hardness or density.	Placer gold deposits of Australia, California, and probably of the Rand gold field, South Africa; of the beach placers of Nome, Alaska, and Travancore, India.
Residual deposits	By weathering, which causes leaching out (dissolving) of valueless minerals, thereby concentrating valuable materials originally of too low grade into workable deposits;	Iron ores of Minnesota, Cuba, and Bilbao, Spain. Barite deposits of Missouri.
	or by this concentration together with further leaching and enrichment of the valuable mineral itself.	Bauxite (aluminum) ores of Arkansas, France, Hungary, and British Guiana.

FIGURE 20-12. *Banded chromite, Seldovia district, Alaska. (Photo by P. W. Guild, U. S. Geological Survey.)*

elements have been selectively concentrated to many times their normal proportions.

Magmatic Segregations As we noted in Chapter 17, many of the larger floored intrusive bodies show a density stratification, a more-or-less regular banding with the dense, early crystallizing minerals near the base. In the Palisade sill the mineral so concentrated by sinking was olivine, of no economic value. But in some places economic minerals have accumulated in similar ways to form ore deposits. Among these are the chromite ores of the Bushveld, South Africa, and of the Stillwater region, Montana. A representative specimen of such segregated chromite is illustrated in Figure 20-12.

Floored intrusions of a variety of gabbro called norite, with concentrations of nickel sulfides and copper sulfides near their bases, are found in localities as widely scattered as Norway, Canada, and South Africa. The great Norwegian geologist Vogt (1893) pointed out that the bulk composition of these intrusives resembles that of smelter charges of sulfide ores, that is, a small percentage of sulfide in a large preponderance of silicates of aluminum, magnesium, iron, and calcium. When such a sulfide ore is smelted, the molten sulfides (the *matte*) sink to the bottom of the crucible, while the *slag* of silicates floats to the top. Industrially, the sulfides are separated by draining them off from the bottom. Vogt suggested that the same mechanism might have operated in nature to produce the observed concentrations.

Among such deposits are the greatest nickel deposits of the world, those of Sudbury, Ontario, which occur along the base of the lopolith shown in Figure 17-15. Here nickel sulfides are commonly molded on the silicate minerals, suggesting strongly that they were molten after the silicates had crystallized and that they accommodated themselves to the intergrain spaces. Certain details of the Sudbury deposits suggest that the final distribution of the sulfides may have been affected by other factors as well as by gravity; but the settling process is generally accepted as an important step in the formation of the ore.

Contact-Metamorphic Deposits In Chapter 5 it was pointed out that alongside some intrusives the wall rocks contain quite different minerals than they do at a distance. In many places beds of limestone can be traced right up to the intrusive contact, and features such as bedding and even fossils in the rock are still identifiable, although the limestone near the contact may be completely transformed to garnet, pyroxene, amphibole, and epidote. Such mineral changes in an already existing rock at contacts with an intrusive prove that material has been transferred from magma to wall rock during the cooling of the intrusive. If the new minerals were merely concentrated from the limestone by solution and removal of other constituents, there would have been a change in volume, but this is denied by the preservation of the distinctive structures mentioned. The new material must have been added from the magma. Supporting this idea is the observed deposition of iron-rich minerals such as magnetite and hematite from gases escaping through fissures at Vesuvius and Katmai.

Only gases or relatively dilute solutions of low viscosity could so intimately permeate the rocks as to bring about these changes without disturbing the finer textural features of the rock. Equally evident is the fact that part of the material formerly comprising the rock must have been removed in the same solutions. The *replacement* of calcite by garnet or pyroxene must have taken place volume by volume, the atoms of calcite not used being removed at the same moment that the garnet crystal structure was being built in that particular space. This volume-for-volume replacement is one of the most widespread phenomena of geology. It takes place not only at intrusive contacts, but in many other environments where circulating ground water or other fluids can transfer material. The contact-metamorphic deposits are distinctive, however, because of the high-temperature minerals they contain.

There are innumerable examples of contact-metamorphic deposits yielding many useful products: garnet (for sandpaper); corundum (Al_2O_3, for emery wheels, and as rubies and sapphires for gem use); copper; zinc; iron; and lead. The iron deposit of Iron Springs, Utah, is representative (Fig. 20-13). The deposit is pod-shaped, having replaced a limestone, locally to its full thickness, but elsewhere only in part. The ore is a mixture of magnetite and hematite, with a little apatite and minor amounts of garnet, pyroxene, and quartz. Of these minerals, only the quartz is found in the unaltered limestone. The rest have been introduced from the magma, con-

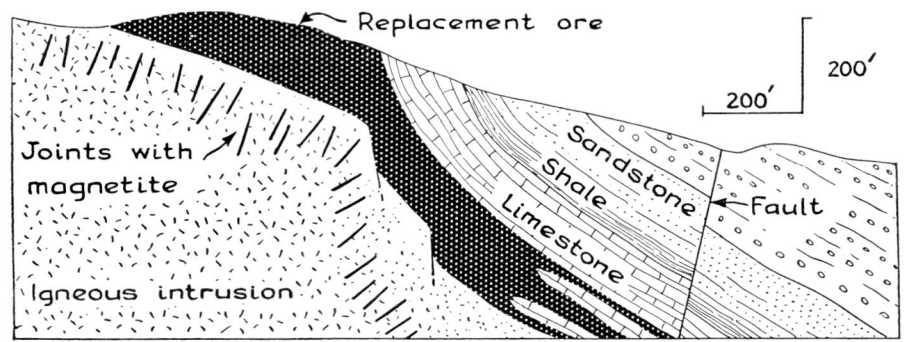

FIGURE 20-13. *Contact-metamorphic deposit at Iron Springs, Utah. (After J. H. Mackin,* Guidebook to the Geology of Utah, *Utah Geol. Survey, 1947.)*

currently with the removal of calcium and carbon dioxide from the limestone.

Hydrothermal Deposits   Most ores of copper, lead, zinc, mercury, silver, and many of those of gold and tungsten are classed geologically as hydrothermal deposits; that is, they were formed by deposition from hot watery solutions, as proved by the following facts:

1. Closely similar deposits form in hot springs and fumaroles.

2. Many deposits are localized along faults and fissures that cut pre-existing rocks, therefore they must have crystallized from fluids that penetrated cracks in the host rock.

3. The minerals of many are identical, though the deposit is followed from one kind of wall rock to another, and so must have formed in an environment independent of that prevailing during the formation of the wall rock.

4. As the adjacent wall rocks often show drastic mineral changes, though delicate structures are preserved in them inherited from the time of their own origin, the alterations must have been brought about by solutions so fluid that their passage did not mechanically disturb the rocks.

5. The well-developed crystal faces found on many of the minerals imply growth from solution, for such features can be readily produced in the laboratory and can be seen forming from solutions in nature.

That the solutions inferred from these features were of *magmatic* origin cannot always be proved, but is strongly implied by:

1. The close association of most of the deposits with intrusive masses.

2. The clustering of deposits about particular intrusives.

3. The especial abundance of these deposits near the upper parts of plutons.

4. The identity of the minerals of some of these deposits and in the wall rock alongside with those that have been seen forming in volcanic areas; and also the identity of some of the minerals with minerals found in contact-metamorphic deposits. As there is an essentially unbroken chain of deposits of intermediate characteristics, it seems probable that these, like the extreme varieties, are of magmatic origin.

The characteristics of fumaroles, geysers, and hot springs have been described in Chapter 17, where the evidence for the magmatic derivation of some of the water and of several metallic minerals was presented. Among those that have been seen forming are magnetite, galena, sphalerite, cinnabar (HgS, the principal source of mercury), and others. In a few places, such as the sulfur mines of Sicily and the mercury deposit at Sulphur Bank, California, economic mineral deposits are directly associated with hot springs, and their valuable minerals were obviously deposited by the ascending hot waters. The similarities of ore minerals and of wall-rock alteration in these deposits to those of other deposits where hot springs are no longer active are so close that most geologists agree that the so-called hydrothermal deposits have been formed from magmatic solutions. Along the walls of metallic deposits, just as along the walls of hot springs, the rocks have been altered to clay minerals, chlorite, and other hydrous minerals, doubtless by the action of long-vanished hot solutions.

Paradoxically, one of the strongest evidences that hydrothermal deposits are of magmatic origin, and not formed from circulating ground water merely set in motion by magmatic heat, lies in the fact that they do *not* accompany many igneous bodies. Many intrusives are barren of ore deposits, though others in identical country rock nearby have great suites of them. An outstanding characteristic of ore deposits is their habit of being grouped into clusters. The argument runs as follows: We know that ground water fills the pores of all rocks to great depths. Any intrusive must stimulate the ground-water circulation by heating the wall rocks. It would therefore be possible to maintain, as was done by some geologists for many years, that the metals of ore deposits were leached from the country rocks and reconcentrated by heated ground water. Even if the original metal content of the rock were very low, one might expect ore deposits to form if the solutions gathered from large volumes of rock were constrained to flow through small channels.

If this were true, there should be no favoring of one kind of intrusive rock over another. The presence or absence of an ore deposit near a particular pluton would not depend on the magma's composition. But in many areas this is obviously not the case. For example, nearly all the great copper deposits of Arizona and Utah are associated with or lie within intrusives that are very similar in composition, although other intrusives in the same region,

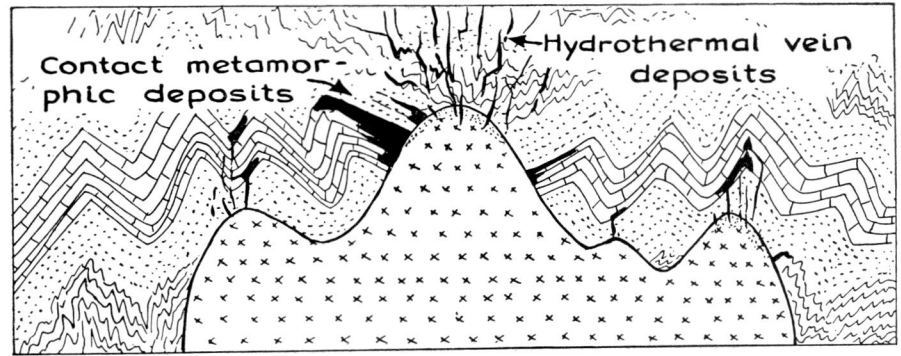

FIGURE 20-14. *Concentration of ore deposits in and near the apices of a large intrusion.*

having no associated copper ores, are comparable in size, country rock, and structural setting. This strongly supports the idea that the parental magma of some intrusives was somehow richer in copper than that of others. Just as with the nickel and copper associated with many norites, we are completely at a loss to explain this pristine enrichment of the magma in a few elements. But the association is so well established as to leave little doubt that the bulk of the metals was derived from the magma and not from the wall rocks.

Even those intrusives closely associated with hydrothermal deposits are not uniformly ore-bearing. Where erosion has been deep enough to disclose the form of the pluton, the ore deposits appear grouped about the higher parts—the apex or "cupolas." Few deposits lie above the low parts of the roof or along the steeper sides. Further, in plutons so deeply eroded as to destroy all remnants of the roof, ore bodies are rare. These relations suggest that the ore-depositing solutions were residues from crystallization of the magma, and that they were concentrated in the higher part by the upward convergence of the walls (Fig. 20-14).

Form and Relations of Hydrothermal Ore Bodies   The commonest form of hydrothermal ore body is the *vein*. Unlike the cylindrical veins of animals and plants, mineral veins are generally tabular—hundreds or even thousands of times as long and wide as they are thick. They may lie at any angle from vertical to horizontal; they "pinch" and "swell," branch and swerve. Many occupy faults, as shown by offsets of geologic contacts along them. Some follow dikes or bedding planes, and still others occupy joints. The so-called "true fissure veins" are veins that differ sharply in mineral content and structure from their wall rocks and generally break away from them cleanly. These are so common in some districts that many of the mining laws of the

United States are based on the fallacious idea that *all* ore deposits are "true fissure veins."

Many veins, particularly in the volcanic rocks (where the geologic relations indicate they were formed at depths of only a few hundred feet), are filled with quartz; calcite, and other carbonate minerals; a little feldspar; sulfides; and, perhaps, native gold. These minerals are arranged in bands (Fig. 20-15) in definite sequence from the walls toward the center of the vein. Unfilled cavities, lined with well-formed crystals, testify to the fact, already clear from the crustlike arrangement of the minerals, that the vein fills a formerly open channel, and that its minerals were deposited from solutions.

Other veins are not sharply separable from their walls, but the vein matter blends into the walls. Microscopic study shows that the vein matter has replaced the wall rock without disturbing it, just as in the contact-metamorphic deposits. These are *replacement veins.* Parts of a vein may show evidence of replacement, while other parts have features pointing to mineral growth in open spaces.

Lodes are unusually thick veins or groups of veins. Some are scores or even hundreds of feet thick. Rocks are not strong enough to hold open a space hundreds by thousands of feet along strike and dip—the size of many lodes. Yet features such as chunks of wall rock lying along the lower side of the vein and the crustified arrangement of well-formed crystals in the vein (Fig. 20-16) prove filling of open spaces. Careful study of many large veins and lodes shows that early-formed quartz has been broken and recemented by later quartz. This recognition of several generations of mineral filling and

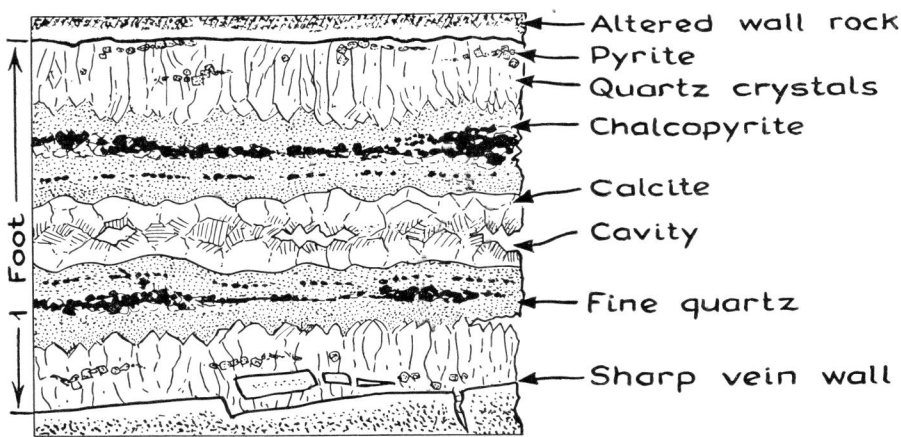

FIGURE 20-15. *A section across a banded vein.*

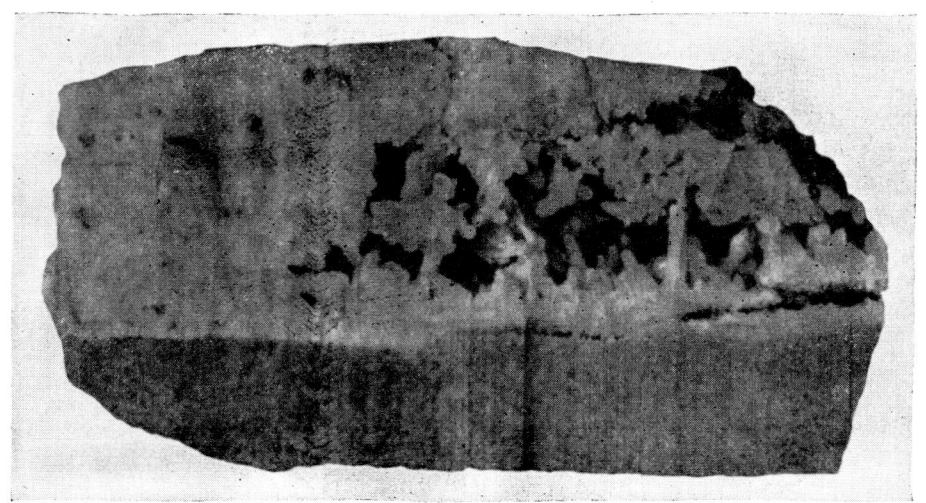

FIGURE 20-16. *Part of crustified quartz vein from Grass Valley, California. (Photo by W. D. Johnston, U. S. Geological Survey.)*

rupture suggests that the fault localizing the vein or lode was recurrently active a long time. Movement along an irregular fault would necessarily bring bends into contact and leave openings between (Fig. 20-17). Renewed faulting after such spaces were filled would produce new openings to be filled in their turn, and so on.

Disseminated Deposits   Other hydrothermal deposits are very irregularly shaped. Notable among these are the great so-called "porphyry copper"

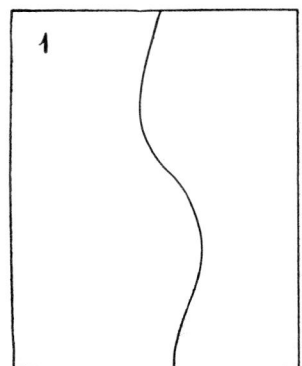 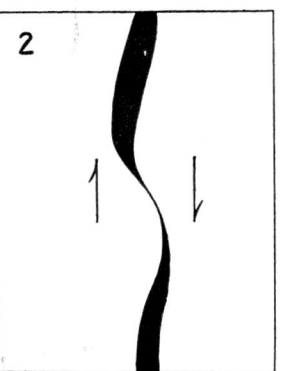 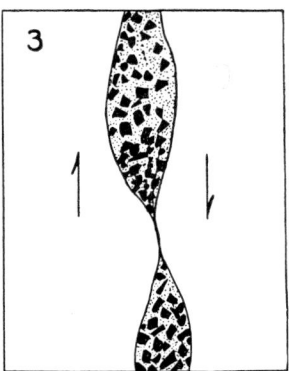

FIGURE 20-17. *Vein deposits along an irregular fault. (1) Fracture, before movement; (2) ore filling cavities, after movement; (3) renewed movement with fracturing of old ore and emplacement of new.*

deposits such as those of Bingham, Utah; Ely, Nevada; Morenci, Arizona; and the greatest of all copper deposits, that of Chuquicamata, Chile. Most of the ore in these districts is disseminated along small cracks in porphyritic intrusive rocks, impregnating and replacing them with copper and iron sulfides. The rock surrounding and within the ore body is generally highly altered to clay and fine-grained mica, with a little epidote and chlorite. The rock is so intimately veined by sulfides and quartz along narrow cracks that hardly a piece as big as a tennis ball can be found without them even through volumes of thousands of cubic yards. The cause of the intimate shattering of these great rock masses to permit such thorough impregnation with solutions is a mystery. It has been noticed that many of these deposits occupy the higher parts of plutons. They have, therefore, been attributed to shattering which is connected with the streaming of volatiles to the top of the intrusive mass as it slowly congealed.

A different kind of disseminated deposit is illustrated by the lead ores of southeastern Missouri and the zinc-lead deposits of the "Tri-State District" (Oklahoma, Missouri, and Kansas). These lie along irregular and indefinite "runs" in limestone (Fig. 20-18). The ores are sulfides that replace the carbonate wall rocks, and, although the pattern of the ore bodies shows the control of faults or other fractures, there are no well-marked veins. The sulfides are accompanied by dolomite and calcite, often well crystallized. It seems that these minerals were introduced in solutions that permeated the entire rock but deposited their ores selectively in certain beds rather than others, perhaps because of some obscure differences in porosity or chemical composition.

Summary   The hydrothermal deposits are rocks that blend indistinguishably into contact-metamorphic deposits, on the one hand, and into deposits in fumarole vents and hot springs on the other. Observations at fumaroles and hot springs show that the minerals so formed change with the tempera-

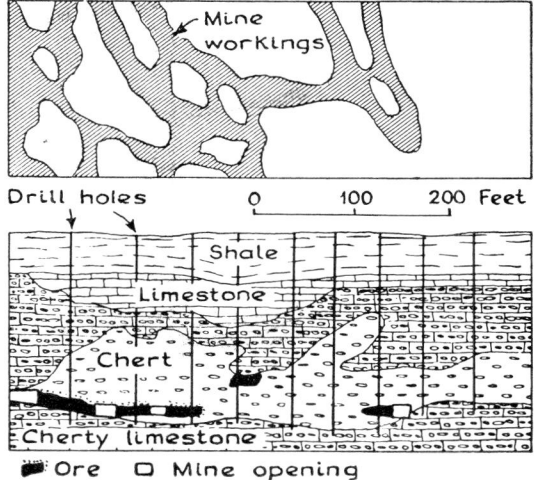

FIGURE 20-18. *Map (top) and section showing the irregular distribution of ore and silicification in limestone of a Missouri zinc-lead mine. (Redrawn by permission from W. Lindgren, Ore Deposits, McGraw-Hill Book Co.)*

ture—as temperature falls, the earlier minerals become unstable and are dissolved, and new minerals form. Although we see nowhere in a single district a complete series of ore deposits filling the range from hot springs to contact-metamorphic, parts and gradations in such a series are apparent in many districts. By observing the transitions of these partial series in many parts of the world, it appears possible to arrange ore deposits in a reasonably systematic sequence. It is probable that this series reflects the changes in composition, temperature, and pressure of magmatic solutions as they pass through and react with the minerals of different rocks and cool as they approach the surface. Where conditions favorable to the deposition of one or more ore minerals have been sufficiently constant during the passage of a large volume of solution, ore bodies have resulted. Where conditions changed too rapidly, or the volume of solutions was too small, no ore was formed, though small quantities of minerals identical with those of ore bodies were deposited. "Gold is where you find it"—but it has been proved many times that there are identifying geologic factors which have controlled its deposition, factors which enable us to locate additional ore bodies or hidden extensions of those already known.

Of paramount importance is the structural setting that guided the ore-forming solutions. Channels for the "feeding" solutions, whether faults, fissures, schistosity, or bedding planes, have all been geologically determined. By detailed mapping it has often been possible to recognize them and to point to new possibilities where similar structural settings exist. Second to structure as a localizing factor is the existence of particular strata readily susceptible to replacement by ore minerals. If the ore deposits confined within a particular bed are unusually abundant or rich, it is obvious that places where this bed intersects a channelway for solutions are good prospecting grounds. There are many technical ways in which mining is facilitated by geologic mapping and other studies. In the aggregate these constitute one of the most important contributions of geology to society. With the depletion of our richest and most readily found ore deposits, it is inevitable that still more intensive geologic studies will be needed to supply the industrial demands of the world.

Sedimentary Deposits   Many economic deposits have come about through sedimentation. Among these, the clay used in ceramics and cement; limestone for building stone or mortar and cement; dolomite; rock salt; gypsum; and potassium salts have already been discussed (Chaps. 5 and 16). The aggregate value of these deposits is very large; it depends primarily on the kind and amount of impurities in the sedimentary rock, on the cost of mining, and on the distance to market.

Metallic ores also form sedimentary strata, often of wide extent. Among these are the iron ores of Alabama, Newfoundland, northeastern France, and some of those of England. The ore beds in the Birmingham district of Alabama are of Silurian age, as shown by the associated fossils. Similar deposits of the same age are widespread from New York southward in the Appalachians, but the only considerable mines are near Birmingham, where the iron-rich beds are thickest, locally attaining 20 feet. The ore beds are generally sharply bounded from the adjoining beds of sandstone and shale, but in places are graded into sandstone. In these ores hematite cements the rock and also coats and replaces some of the fossils. The sand grains and pebbles are also surrounded by rounded pellets of hematite. These show, by the abrasion of the hematite-covered surfaces, that the iron mineral was formed before the agitation of the grains was stopped by their burial. Many of the hematite granules are flattened in such a way as to suggest that they were soft and were squeezed during the consolidation of the rock. Fossils composed of calcite found here are undeformed. These facts convince most students of these deposits that the present hematite granules were formerly soft, jellylike aggregates of iron rust on the sea floor. The iron oxide was deposited in the sea on nuclei of clastic grains. Presumably, the material was a hydrous iron oxide, from which the water was driven off by heat and pressure after the rock was buried. These deposits are relatively low in iron. Their workability depends on their ease of smelting, because the associated calcite acts as a flux in the blast furnace, and on their location close to large sources of coking coal.

The present seas apparently contain no such concentrations of iron-rich sediments as must have been present during deposition of the Birmingham ores, but some iron is now being precipitated on the sea floor. We know from analyses of river waters that iron is leached from the rocks in great quantities and that it is less abundant in sea than in river water. The Silurian deposit is thought to have formed under peculiar conditions whereby iron was carried to the sea in great abundance. In other words, they resulted from an unusual concentration in time and place, but from a normal geologic process.

Placer Deposits We saw in Chapter 6 that during normal weathering the feldspars and other silicate minerals tend to decompose to clays, so fine grained and so weakly coherent as to break up with the slightest transportation. Quartz is normally quite stable and hence is concentrated in the stream beds of a granitic region where it makes up most of the sand, even though it comprises only 10 or 20 per cent of the bedrock. In the same way, minerals such as magnetite, chromite, diamond, gold, and cassiterite (SnO_2, the chief source of tin) are chemically stable under most climatic environments. These minerals are also heavier than the average. Accordingly, they tend to fall to the bottom of the stream and to accumulate on the riffles while quartz of the same grain size is carried on. Where streams from bedrock regions that furnish such minerals reach the sea, the heavy minerals are also concentrated by the waves. This is the origin of the "black sands" of the Oregon coast (chromite), of the raised beaches of Nome, Alaska (gold), of the monazite (thorium) sands of the Malabar Coast of India, of the diamond beaches of Southwest Africa. In all these localities, minerals that form only very minor parts of the rocks inland have been concentrated because of their density and stability under weathering.

Facts, Concepts, Terms

DEPENDENCE OF INDUSTRY ON MINERALS

SPORADIC DISTRIBUTION OF MINERAL RESOURCES

EXHAUSTIBILITY OF MINERAL RESOURCES

MINERAL FUELS THE PRINCIPAL ENERGY SOURCES

THE SEQUENCE OF COAL ALTERATION
   Peat to anthracite

ECONOMIC FACTORS IN COAL MINING
   Quality; thickness and structure of beds; overburden; ground-water conditions; marketing conditions

REQUISITES FOR THE CONCENTRATION OF PETROLEUM
   Reservoir rock; cap rock; source rocks; favorable structure

STRUCTURE OF OIL FIELDS
   Anticlines; salt domes; stratigraphic traps

OIL PROVINCES

GEOPHYSICAL EXPLORATION
   Gravity surveys; seismic prospecting

SUBSURFACE GEOLOGY
   Correlation by foraminifera; by lithology; by electrical properties

ORE DEPOSITS

DEFINITION OF ORE

ORE MINERALS; GANGUE MINERALS

FACTORS IN EXPLOITATION OF A MINERAL DEPOSIT
   Size; shape; depth; grade of ore; amenability to cheap metallurgical processes; distance to market; price

MECHANICAL PROCESSES OF ORE CONCENTRATION
Weathering; solution; transportation; sedimentation

CHEMICAL PROCESSES OF CONCENTRATION
Magmatic segregation; metamorphic transfer; deposition from hot waters

Questions

1. Why did Holland and Denmark, which suffered greatly during World War II, oppose the suggestion to abolish the commercial production of coal from the Ruhr area in Germany despite their fear of renewed German aggression?
2. Why, in view of the great mineral endowment of the United States, did Congress authorize "stock-piling" of certain minerals as a defense measure?
3. From the description of placer deposits, explain how they are used as guides to ore in bedrock. Why does a prospector carry a "pan"?
4. In view of the nonreplenishable nature of mineral deposits, what are the advantages and disadvantages of a tariff on minerals?
5. Although they were not described in the text, draw a cross section of an oil field which is a "fault trap" in a gently dipping sequence of sediments.
6. Why has the Persian Gulf oil region been developed more economically than the principal American oil areas?
7. Determination of the depth of relatively recent, unconsolidated river-deposited sediments in the estuary of the Congo River is important for the solution of the problem of submarine canyons (Chap. 15). What methods now routine in oil finding might be used to determine this depth at enough points to give the form of the bedrock surface? Which method would be quickest?
8. Sandstones used for making glass must consist of almost pure quartz. How is such purity attained by natural processes?
9. Why are placer deposits usually the first to be discovered in a gold mining region?
10. Pyrite is common in many deposits of copper, lead, zinc, and gold. Why is a conspicuously brown-stained outcrop often a guide to a buried ore deposit?

Suggested Readings

1. *Bituminous Coal Facts and Figures,* Bituminous Coal Institute, Washington, D. C., 1948.
2. Lalicker, Cecil G., *Principles of Petroleum Geology,* Appleton-Century-Crofts Co., New York, 1949.
3. Lovering, T. S., *Minerals in World Affairs,* Prentice-Hall, New York, 1943.
4. *World Atlas of Petroleum,* American Geographical Society, New York, 1950.

5. Rickard, T. A., *Man and Metals*, McGraw-Hill Book Co., New York, 1932.

6. *The Mineral Resources of the United States*, by the staffs of the U. S. Geological Survey and Bureau of Mines, Public Affairs Press, Washington, D. C., 1948.

APPENDIX I

Techniques of Topographic Mapping

To USE a topographic map requires some knowledge of the way it is made. There are many different methods and techniques; we will outline here only some of the more important steps. Because of its simplicity we have chosen the "Plane Table Method" as an example. This method is widely used in making geologic maps, but most topographic maps are now made from airplane photographs.

The first step in the preparation of a topographic map by any method is to acquire both *horizontal* and *vertical control* for the measurements. To attain this double control, we must first establish the position of a point on the earth's surface (a) by its latitude and longitude and (b) by its altitude with respect to sea level. The more points thus determined, the better our control. If we knew the elevation, latitude, and longitude of every point on the surface, we could quickly construct the map; but it is impossible within a given time to determine them for every point in a large area.

Once the horizontal and vertical control and the direction of the North-South line through one point have been established, we can quickly determine many other points by a process called *triangulation.*

The first step in triangulation is the selection and measurement of a *base line.* The base line is a straight line from the point for which we have established control to another point. Each end of the base line is marked by a stake holding a flag (Fig. I-1), or by some other suitable marker, and the distance between them is measured carefully with a steel tape.

The accuracy of the whole map depends on the base line; therefore, as a check, the measurement is generally repeated. After the base line has been measured, it must also be carefully plotted on the *plane table sheet* (paper on which the map is to be constructed) in accordance with the reduced scale chosen for the map. Actually, the plotted length determines the scale, for if there are errors either in the measurement or in the plotting of the base line, they will be carried throughout the map.

Once the base line has been plotted, a *plane table* (essentially a drawing board mounted on a tripod) is set up over one end of the base line. Then the edge of an *alidade* (a telescope attached to a ruler base) is placed along the line drawn on the plane table sheet to represent the base line (Fig. I-1). The plane table, carrying the alidade, is rotated until the telescope points directly at the flag on the other end of the base line, and then clamped firmly in position. It is now correctly *oriented,* since the base line on the ground and the plotted base line on the plane table sheet have exactly the same trend (azimuth) with respect to true north.

With the table still clamped in this position, the telescope is pointed successively toward each of the other flags, such as A, B, C in Figure I-1, or

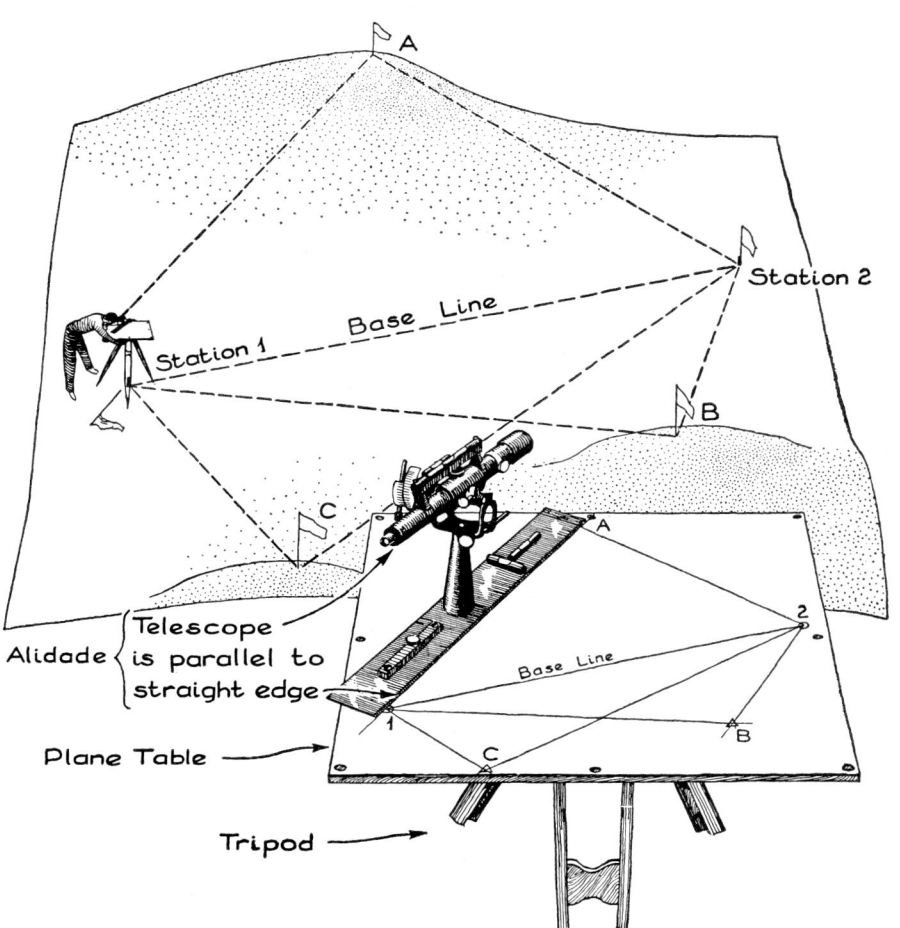

FIGURE I-1.   *Mapping with the plane table and alidade.*

other marked points that are visible, and lines are drawn along the edge of the alidade to indicate the directions of these lines of sight. The plane table is then taken to the other end of the base line, oriented in the same manner by sighting back to the first flag, and the process of sighting upon and drawing lines toward each of the other visible points is repeated. The point of intersection of the two lines of sight toward an object, one line of sight having been drawn from the first end of the base line and the other from the second end, marks the true position of that object on the reduced-scale map (Fig. I-1). This point and the two ends of the base line form the apexes of a triangle. The new point can then be used, just as if it were one end of the base line, to determine the position of additional points, extending the system of triangles (triangulation net) within a certain area. The more points thus determined, the better the control.

A surveyor's *transit* or *theodolite* can be used in triangulation instead of a plane table and alidade. With a transit, the horizontal angle from the base line to the point to be determined is carefully measured at each end of the base line. Knowing the two angles, and having measured one side (the base line) of the triangle, we can compute the position of the apex of the triangle by trigonometry and plot it on the map.

Vertical control is established in the same way, and much of it at the same time, as the triangulation. We establish the elevation of one end of our base line by running levels to it from some point of known elevation. For this, we use a telescope equipped with a level, and sight at a rod graduated in hundredths of a foot. With the elevation of one point in our triangulation system thus established, we can readily compute the elevation of a second station. We already know the distance between the two stations. The only other measurement we need for the computation is the angle between the horizon and the line of sight to the station. The alidade is equipped with a vertical arc (Fig. I-1) for making this measurement. Thus, the vertical control is extended throughout the triangulation net, and the elevations of many points are determined.

By these and other auxiliary techniques, not described here, for obtaining horizontal and vertical control, we locate enough additional points to allow the topographer to sketch the contours in their proper relation to the positions and elevations of the many points thus located and plotted. The map that results from this procedure is, of course, a generalization.

Many factors other than the already mentioned base-line determinations affect its accuracy—for example, the number and spacing of the control points, trees that obscure the ground forms, the skill of the topographer, and the amount of time he has at his disposal for study of the shape of the land surface. Topographical engineers rate a map as excellent if, on testing it, not more than 10 per cent of the elevations are in error by more than one-half the vertical distance between two successive contours.

Recently, great strides have been made in preparing topographic maps from photographs taken from airplanes. The basic principles are, nonetheless, the same as those employed in making maps on the ground and, for the basic horizontal and vertical control, a preliminary ground survey is still necessary.

Good topographic maps are available for relatively few parts of the entire land surface of the earth. Less than half the area of the United States has been mapped on a scale that permits drawing contours with intervals as small as one hundred feet. Poland and several of the "backward Balkans" have much better map coverage. For large parts of the earth's surface, we have only crude maps.

Fortunately, explorers have always been interested in the height of mountains, so that the altitudes of the principal peaks in all the mountain ranges of the earth have been determined. Many scattered observations have also been made at other points on the earth's surface not covered by topographic maps. Though still far from satisfactory, data are available from which we can make a reasonably accurate estimate of the relief of the land surface and of its average elevation.

Construction of Hydrographic Maps

In the construction of hydrographic maps of the sea floor, soundings were formerly made by measuring the length of a rope or wire paid out until it reached bottom. As a measurement in the deep part of the ocean would require several hours, it is not surprising that few wire soundings were made, except at shallow depth near the coasts.

In recent years, a new method, *sonic sounding*, has superseded measurement of depth by wire or rope. In sonic sounding the time required for a sound signal to travel from a ship to the sea bottom and rebound is measured. The depth is calculated from the speed of sound in sea water. By sonic sounding it is now an easy matter for a ship to chart a continuous record of the depths traversed while it is under way. The position of the ship is determined to within a distance of a few hundred feet by radio signals from shore stations.

Although sonic sounding has greatly increased our knowledge of the ocean floor, the vastness of the sea, the lack of interest of many navigators in obtaining detailed information of this kind from little-traveled sea lanes, and the cost of operating a vessel for surveying purposes alone, still conspire to prevent more than a mere sampling of the form of the ocean floor.

The hydrographer, compared with the topographer, is severely handicapped, for he is unable to see the sea bottom and therefore cannot choose the most suitable points to use for control in mapping. A series of points of equal depth can be connected by a contour line in several ways, but obviously only one such contour line represents the actual form of the sea floor. On land the topographer can see the topographic forms and sketch between his points accordingly; the hydrographer must get additional control or else make an interpretation which will probably be inaccurate in minor details.

APPENDIX II*

International Atomic Weights. 1949

	SYM-BOL	ATOMIC NUMBER	ATOMIC WEIGHT†		SYM-BOL	ATOMIC NUMBER	ATOMIC WEIGHT†
Actinium	Ac	89	227	Dysprosium	Dy	66	162.46
Aluminum	Al	13	26.97	Erbium	Er	68	167.2
Americium	Am	95	[241]	Europium	Eu	63	152.0
Antimony	Sb	51	121.76	Fluorine	F	9	19.00
Argon	A	18	39.944	Francium	Fr	87	[223]
Arsenic	As	33	74.91	Gadolinium	Gd	64	156.9
Astatine	At	85	[211]	Gallium	Ga	31	69.72
Barium	Ba	56	137.36	Germanium	Ge	32	72.60
Beryllium	Be	4	9.013	Gold	Au	79	197.2
Bismuth	Bi	83	209.00	Hafnium	Hf	72	178.6
Boron	B	5	10.82	Helium	He	2	4.003
Bromine	Br	35	79.916	Holmium	Ho	67	164.94
Cadmium	Cd	48	112.41	Hydrogen	H	1	1.0080
Calcium	Ca	20	40.08	Indium	In	49	114.76
Carbon	C	6	12.010	Iodine	I	53	126.92
Cerium	Ce	58	140.13	Iridium	Ir	77	193.1
Cesium	Cs	55	132.91	Iron	Fe	26	55.85
Chlorine	Cl	17	35.457	Krypton	Kr	36	83.7
Chromium	Cr	24	52.01	Lanthanum	La	57	138.92
Cobalt	Co	27	58.94	Lead	Pb	82	207.21
Columbium; see Niobium‡				Lithium	Li	3	6.940
Copper	Cu	29	63.54	Lutetium	Lu	71	174.99
Curium	Cm	96	[242]	Magnesium	Mg	12	24.32

(Continued)

* From Pauling, *General Chemistry*, W. H. Freeman and Co., San Francisco, 1949.
† A value given in brackets is the mass number of the most stable known isotope.
‡ The English names of these elements have been changed recently, by action of the International Union of Pure and Applied Chemistry.

International Atomic Weights. 1949 (Continued)

	SYM-BOL	ATOMIC NUMBER	ATOMIC WEIGHT†		SYM-BOL	ATOMIC NUMBER	ATOMIC WEIGHT†
Manganese	Mn	25	54.93	Samarium	Sm	62	150.43
Mercury	Hg	80	200.61	Scandium	Sc	21	45.10
Molybdenum	Mo	42	95.95	Selenium	Se	34	78.96
Neodymium	Nd	60	144.27	Silicon	Si	14	28.06
Neptunium	Np	93	[237]	Silver	Ag	47	107.880
Neon	Ne	10	20.183	Sodium	Na	11	22.997
Nickel	Ni	28	58.69	Strontium	Sr	38	87.63
Niobium	Nb	41	92.91	Sulfur	S	16	32.066
Nitrogen	N	7	14.008	Tantalum	Ta	73	180.88
Osmium	Os	76	190.2	Technetium	Tc	43	[99]
Oxygen	O	8	16.0000	Tellurium	Te	52	127.61
Palladium	Pd	46	106.7	Terbium	Tb	65	159.2
Phosphorus	P	15	30.98	Thallium	Tl	81	204.39
Platinum	Pt	78	195.23	Thorium	Th	90	232.12
Plutonium	Pu	94	[239]	Thulium	Tm	69	169.4
Polonium	Po	84	210	Tin	Sn	50	118.70
Potassium	K	19	39.096	Titanium	Ti	22	47.90
Praseodymium	Pr	59	140.92	Tungsten; see Wolfram‡			
Promethium	Pm	61	[147]	Uranium	U	92	238.07
Protactinium	Pa	91	231	Vanadium	V	23	50.95
Radium	Ra	88	226.05	Wolfram	W	74	183.92
Radon	Rn	86	222	Xenon	Xe	54	131.3
Rhenium	Re	75	186.31	Ytterbium	Yb	70	173.04
Rhodium	Rh	45	102.91	Yttrium	Y	39	88.92
Rubidium	Rb	37	85.48	Zinc	Zn	30	65.38
Ruthenium	Ru	44	101.7	Zirconium	Zr	40	91.22

APPENDIX III

Identification of Minerals

THE LABORATORY techniques in most common use today for the identification of minerals are the following:

Petrographic Analysis

Petrographic analysis is the most frequently used method for the identification of both minerals and rocks. In one method a small piece of the substance to be identified is ground with abrasives on a revolving plate until it is 0.03 millimeters (about 0.001 inch) in thickness—much thinner than a sheet of paper. This is called a *thin section*. It is mounted between thin glass slides, and is examined under the petrographic microscope. In a thin section most minerals become transparent, or nearly so, and the optical properties that serve to distinguish different minerals can be readily measured with the microscope.

An alternative petrographic method is to crush the mineral to powder, place the powder in a drop of liquid of known optical properties on a glass slide, cover with a thin glass plate, and examine the fragments immersed in the liquid under the petrographic microscope.

X-Ray Analysis

As previously explained (Chapter IV), it is possible by means of X-rays to work out the internal structure, or geometric arrangement of the ions or atoms in a mineral. Since the internal structure is the most diagnostic character of a mineral, X-ray analysis is one of the most fundamental methods of mineral identification.

Chemical Analysis

A chemical analysis, or even a qualitative chemical test for some particular element, will sometimes help to identify an unknown mineral; however, even a complete chemical analysis may fail to establish the identity of a mineral. Some distinct minerals such as diamond and graphite have, as we pointed out, identical chemical compositions and cannot be distinguished chemically. Furthermore, most minerals are highly insoluble silicates, difficult to treat by standard chemical procedures which require dissolving the substance to be analyzed. Most minerals are also "solid solutions" whose compositions vary widely. For these reasons standard chemical procedures are little used in ordinary mineral identification, though they may be employed in special kinds of research on minerals.

As supplements to petrographic and other methods, however, a few special chemical techniques have proved useful in mineral identification. Many minerals are too opaque to be readily identified by ordinary petrographic methods. In such cases simple chemical tests can be made on the surface of the thin section or on a polished piece of the mineral while it is being examined under the microscope.

The spectroscope is widely used to detect elements that may be present in small amounts in a mineral. Its use requires merely that the mineral be heated in an arc until it vaporizes.

Determination by Physical Properties

The common rock-forming minerals, as well as many of the rarer minerals of economic value, can usually be identified without special instruments by a careful study of their physical properties. This method suffices for recognition of the 30 minerals listed at the end of this appendix. The more important physical properties are:

Cleavage. Many minerals *cleave* (break) *along smooth planes controlled by the internal structure of the crystal* (Figs. 4-6 and III-1). The

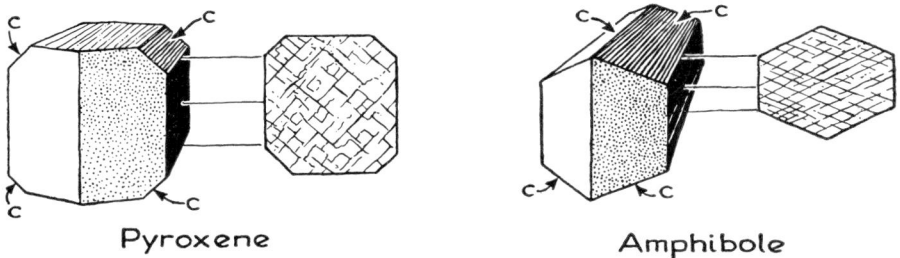

Pyroxene   Amphibole

FIGURE III-1. *The relation between crystal form and cleavage in pyroxene and amphibole. The cleavage planes parallel the crystal faces,* c, *and are shown in a sectional cut to the right of each crystal.*

broken fragments have characteristic shapes because the number of cleavage planes and the angles between the cleavages are characteristic for a particular mineral.

Fracture.   Many minerals do not cleave along smooth planes but fracture irregularly. Such rough fragments are less readily identified than cleavage fragments, but some minerals, of which quartz is an example, usually break with characteristic curved surfaces (*conchoidal fracture*). Others have a splintery or *fibrous fracture* that helps to distinguish them.

Form.   As mentioned previously, *minerals tend to crystallize into definite, characteristically shaped crystals, bounded by smooth planes called crystal faces.* When crystal faces are present, their shapes and interfacial angles are diagnostic (Figs. 4-2 and III-2), but many minerals occur in shapeless, granular forms, or in crystals so small that the crystal faces are not visible. In some minerals, crystal faces are parallel to cleavage surfaces; in others, they are not. For this reason, they should, of course, be carefully distinguished from cleavage surfaces. The distinction is not difficult if one remembers that crystal faces appear only on the outside of the crystal, whereas cleavage surfaces appear only on the broken or cracked fragments of a crystal.

Color.   All specimens of some minerals, such as magnetite and galena, have a constant or uniform color; but others, such as quartz and calcite, are variable in color because of pigments that may be present as impurities. Even in minerals with constant intrinsic color, alteration of the surface

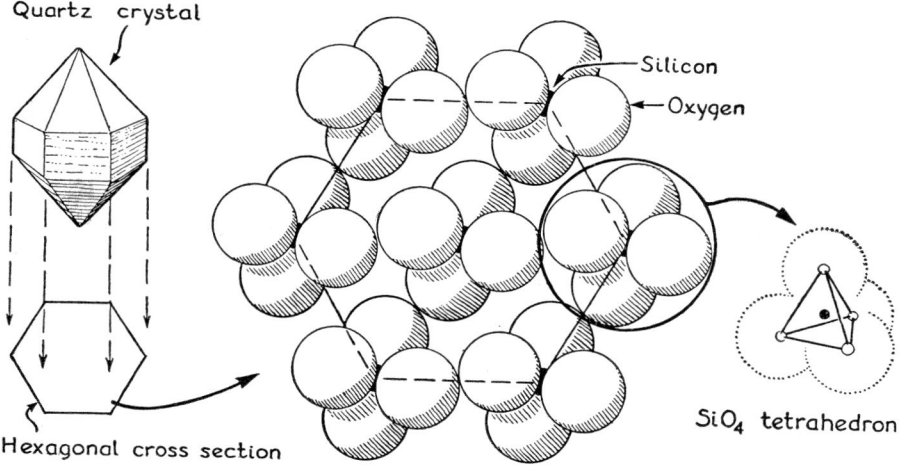

FIGURE III-2.   *The crystal form of quartz and its internal hexagonal structure. Each oxygen is linked to another silicon above or below the layer of tetrahedra shown in the drawing, so that the overall ratio of Si:O is 1:2.*

through exposure to air and moisture may change the surface color. Nevertheless, the color of a freshly broken surface is often diagnostic, and even the color of the surface film of alteration aids in identifying some minerals.

Streak. The color of the powder of a mineral—which is called the "streak" because it is easily obtained from the mark made by rubbing the mineral against a piece of unglazed porcelain—is more constant and, in some minerals, more helpful in identification than the color of the mineral in a larger mass. The streak of a mineral may be similar to, or it may be entirely different from, the color of the mineral itself: Silvery gray galena gives a silvery gray streak; but both the black and red varieties of hematite show a characteristic red streak that is very helpful in identifying the mineral.

Luster. The luster of a mineral refers to the way ordinary light is reflected from its surfaces. *Metallic luster* is like that of polished metals; *vitreous luster* is like that of glass; *adamantine* like that of diamond. Other self-explanatory terms used to describe luster are *resinous, silky, pearly,* and *dull* or *earthy.*

Hardness. The relative hardness of two different minerals can be determined by pushing a pointed corner of one firmly across the flat surface of the other. If the mineral with the point is harder, it will scratch or cut the other. Laboratory tests of the hardness of minerals are usually recorded in terms of a *scale of hardness* ranging from 1 to 10. The numbers refer to the hardness of 10 minerals, arranged in order of increasing hardness:

1. talc	6. orthoclase
2. gypsum	7. quartz
3. calcite	8. topaz
4. fluorite	9. corundum
5. apatite	10. diamond

When specimens to make up this series are not available, it is convenient to know that the steel of a pocketknife is about 5½ in this scale, a copper penny 3½, and the thumbnail about 2½.

Specific Gravity. The specific gravity or density of a mineral is given by the formula:

$$\text{Sp. Gr.} = \frac{(\text{weight of mineral in air})}{(\text{weight of mineral in air}) - (\text{weight of mineral in water})}$$

Specific gravity, therefore, is stated as a number indicating the *ratio of the weight of the substance to that of an equal volume of water.* Specific gravity can be determined by several different instruments ranging from an ordinary spring scale for large specimens to sensitive microbalances for smaller ones. The specific gravity of different minerals ranges widely.

With a little practice, fairly small differences in specific gravity can be detected from the "heft" of a moderate-sized specimen held in the hand. Quartz, with a specific gravity of 2.65, may be used as the standard of comparison. Gypsum (Sp. gr. 2.2–2.4) would then be called light; olivine (3.2–3.6) heavy; magnetite (5.0–5.2) very heavy.

Other Properties.   Many other physical properties are useful in identifying minerals. Some minerals are attracted by a magnet, others are not; some conduct electricity better than others; some "fluoresce" or glow in beautiful color when ultraviolet light is played on them; some are characterized by fine striations or "twinning lines" on certain cleavage surfaces; others have different striations—"growth striations"—on certain crystal faces. Minerals also differ in fusibility; solubility; in their reactions to simple chemicals, such as bubbling when dilute hydrochloric acid is applied to them; and in many other ways, but the major physical properties already discussed will suffice to identify the more common kinds.

List of Minerals

The grouping of the minerals in the list below is not alphabetical but is based on similarities in properties, associations, or uses. The first twenty minerals are common rock-forming minerals. The remaining 10 are comparatively rare but are important as ore minerals. The properties most useful in sight identification are italicized. Some minerals have more than one name; the less common names are given in parentheses. The chemical name and chemical formula follow the mineral name.

MINERAL	FORM	CLEAVAGE	HARD-NESS	SP. GR.	OTHER PROPERTIES
CALCITE. Calcium carbonate, $CaCO_3$	"Dog-tooth" or flat crystals showing excellent cleavages; granular, showing cleavages; also masses too fine grained to show cleavages distinctly.	*Highly perfect, in 3 directions at oblique angles,* yielding rhomb-faced fragments (Fig. 4-3).	3	2.72	Commonly colorless, white or yellow but may be any color owing to impurities. Transparent to opaque, transparent varieties showing *strong, double refraction* (e.g., 1 dot seen through calcite appears as 2). Vitreous to dull luster. *Effervesces* readily *in cold dilute hydrochloric acid.*
DOLOMITE. Calcium magnesium carbonate, $CaMg(CO_3)_2$	*Rhomb-faced crystals showing good cleavage;* also in fine-grained masses.	*Perfect cleavage* in 3 directions at oblique angles, as in calcite.	3.5–4	2.9	Variable in color, but commonly white. Transparent to translucent. Vitreous to pearly luster. *Powder will effervesce slowly in cold dilute hydrochloric acid, but coarse crystals will not.*
GYPSUM. Hydrous calcium sulfate, $CaSO_4 \cdot 2H_2O$	Tabular crystals, and cleavable, granular, fibrous, or earthy masses.	*One perfect cleavage, yielding thin, flexible folia;* 2 other much-less-perfect cleavages.	2	2.2–2.4	Colorless or white, but may be other colors when impure. Transparent to opaque. Luster vitreous to pearly or silky. Cleavage flakes flexible but *not elastic* like those of mica.
HALITE. (Rock salt) Sodium chloride, $NaCl$	Cubic crystals (Fig. 4-5) in granular masses.	Excellent cubic cleavage (3 directions mutually at right angles).	2–2.5	2.1	*Colorless to white,* but of other colors when impure. The color may be unevenly distributed through the crystal. Transparent to translucent. Vitreous luster. *Salty taste.*
KAOLINITE. Hydrous aluminum silicate, $H_4Al_2Si_2O_9$. Representative of the 3 or 4 similar minerals common in clays.	Commonly in soft, compact, *earthy masses.*	Crystals always so small that cleavage is invisible without microscope.	1–2	2.2–2.6	White color, but may be stained by impurities. *Greasy feel. Adheres to the tongue,* and *becomes plastic when moistened.* "Clay-like" odor when breathed upon.

(Continued)

MINERAL	FORM	CLEAVAGE	HARD-NESS	SP. GR.	OTHER PROPERTIES
OPAL. Hydrous silica, with 3% to 12% water, SiO_2nH_2O. Does not have a definite geometric internal structure, hence is a mineraloid, not a true mineral.	*Amorphous.* Commonly in veins or irregular masses showing a banded structure. May be earthy.	None; *conchoidal fracture.*	5.0–6.5°	2.1–2.3	*Color highly variable, often in wavy or banded patterns.* Translucent or opaque. *Somewhat waxy luster.*
CHALCEDONY. (Cryptocrystalline quartz.) Silicon dioxide, SiO_2.	Crystals too fine to be visible; sometimes conspicuously banded, or in masses.	None; *conchoidal fracture*	6–6.5	2.6	Color commonly white or light gray, but may be any color owing to impurities. Distinguished from opal by *dull* or *clouded luster.*
QUARTZ. (Rock crystal.) Silicon dioxide, SiO_2.	*Six-sided prismatic crystals,* terminated by 6-sided triangular faces; also massive. (Fig. III-2).	None or very poor; *conchoidal fracture.*	7	2.65	Commonly *colorless* or white, but may be yellow, pink, amethyst, smoky-translucent, brown, or even black. *Transparent* to opaque. *Vitreous* to *greasy luster.*
ORTHOCLASE. (Potash feldspar.) Potassium aluminum silicate, $KAlSi_3O_8$.	Boxlike crystals (Fig. 4-2b); massive, with excellent cleavage.	One perfect and 1 good cleavage, making an angle of 90°.	6	2.5–2.6	Commonly *white, gray, pink* or pale yellow; rarely colorless. Commonly opaque but may be transparent in volcanic rocks. Vitreous. Pearly luster on better cleavage. *Distinguished from plagioclase by absence of striations on the better cleavage.*
PLAGIOCLASE. (Soda-lime feldspars.) An isomorphous group of sodium calcium aluminum silicates, $NaAlSi_3O_8$ to $CaAl_2Si_2O_8$.	In well-formed crystals and in cleavable or granular masses.	*Good cleavage in 2 directions, nearly at right angles* (86°). May be poor in some volcanic rocks.	6–6.5	2.6–2.7	Commonly *white* or *gray,* but may be other colors. Some gray varieties show a play of colors called *opalescence.* Transparent in some volcanic rocks. Vitreous to pearly luster. Distinguished from orthoclase by the presence on the *better cleavage surface* of *fine parallel lines* or *striations.*

MINERAL	FORM	CLEAVAGE	HARD-NESS	SP. GR.	OTHER PROPERTIES
MUSCOVITE. (White mica; isinglass.) A complex potassium aluminum silicate, $KAl_3Si_3O_{10}(OH)_2$ approximately, but varying.	Thin, scalelike crystals and scaly, foliated aggregates.	*Perfect in 1 direction, yielding very thin, transparent, flexible leaves.*	2–3	2.8–3.1	*Colorless,* but may be gray, green, or light brown in thick pieces. *Transparent to translucent.* Pearly to vitreous luster.
BIOTITE. (Black mica.) A complex silicate of potassium, iron, aluminum, and magnesium, variable in composition but approximately $K(Mg,Fe)_3 AlSi_3O_{10}(OH)_2$.	Thin, scalelike crystals, commonly 6 sided, and in scaly, foliated masses.	*Perfect in 1 direction, yielding thin, flexible scales.*	2.5–3	2.7–3.2	Black to dark brown. Translucent to opaque. Pearly to vitreous luster. White to greenish streak.
PYROXENE. A solid-solution group of silicates. Chiefly silicates of calcium, magnesium, and iron, with varying amounts of other elements.	Commonly in short, 8-sided, prismatic crystals; *the angle between alternate faces nearly 90°.* Also as compact masses and disseminated grains.	*In 2 directions at nearly 90°* (Fig. III-1). Cleavage not always well developed; in some specimens, conchoidal or uneven fracture.	5–6	3.2–3.6	Commonly greenish to black in color. Vitreous to dull luster. Gray-green streak. Distinguished from amphibole by the *right-angle cleavage, 8-sided crystals,* and by the fact that most crystals are short and stout, rather than long, thin prisms, as in amphibole (Fig. III-1).
AMPHIBOLE. A group of complex, solid-solution silicates, chiefly of calcium, magnesium, iron, and aluminum. Similar to pyroxene in composition, but containing a little hydroxyl (OH^-) ion. The commonest of the many varieties of amphibole is *hornblende.*	*Long, prismatic, 6-sided crystals;* also in fibrous or irregular masses of interlocking crystals and in disseminated grains.	*Two good cleavages meeting at angles of 56° and 124°* (Fig. III-1).	5–6	2.9–3.2	Color black to light green; or even colorless. Opaque. *Highly vitreous luster on cleavage surfaces.* Distinguished from pyroxene by the *difference in cleavage angle* and in crystal form. Amphibole also has much better cleavage and higher luster than pyroxene.

(Continued)

MINERAL	FORM	CLEAVAGE	HARD-NESS	SP. GR.	OTHER PROPERTIES
OLIVINE. Magnesium iron silicate, $(Fe,Mg)_2SiO_4$.	Commonly in small, *glassy grains* and granular aggregates.	So poor that it is rarely seen; *conchoidal fracture*	6.5–7	3.2–3.6	*Various shades* of *green,* also yellowish; opalescent and brownish when slightly altered. *Transparent* to translucent. *Vitreous luster.* Resembles quartz in small fragments but has characteristic greenish color.
GARNET. A group of isomorphous silicates having a general formula with variable proportions of different metallic elements. The most common variety contains calcium, iron, and aluminum, but garnet may contain many other elements.	Commonly in well-formed *equidimensional crystals* (Fig. 4-2c) but also massive and granular.	None; *conchoidal or uneven fracture.*	6.5–7.5	3.4–4.3	Commonly *red, brown,* or *yellow,* but may be other colors. Translucent to opaque. *Resinous* to *vitreous luster.*
ICE. Hydrogen oxide, H_2O.	Irregular grains; lacelike flakes with hexagonal symmetry; massive.	None; conchoidal fracture.	1.5	0.9	Colorless, white or blue. Vitreous luster. *Melts at 0°C.,* so is liquid at room temperature.
CHLORITE. A complex group of hydrous magnesium aluminum silicates containing iron and other elements in small amounts.	Commonly in *foliated* or *scaly masses;* may occur in tabular, 6-sided crystals resembling mica.	*Perfect in 1 direction,* yielding thin, flexible, but inelastic, scales.	1–2.5	2.6–3.0	*Grass-green* to *blackish-green* color. Translucent to opaque. Greenish streak. Vitreous luster. *Very easily disintegrated.*
SERPENTINE. A complex group of hydrous magnesium silicates, roughly $H_4Mg_3Si_2O_9$.	Foliated or fibrous, usually massive.	Commonly in 1 plane, sometimes in prisms. Fracture usually conchoidal or splintery.	2.5–4 °	2.5–2.65	*Feel smooth, sometimes greasy.* Color *leek-green* to *blackish-green* but varying to brownish-red, yellow, etc. *Luster resinous* to *greasy.* Translucent to opaque. Streak white.

MINERAL	FORM	CLEAVAGE	HARD-NESS	SP. GR.	OTHER PROPERTIES
EPIDOTE. A complex group of calcium, iron, aluminum silicates, $Ca_2(Al,Fe)_3(SiO_4)_3(OH)$.	*Short, 6-sided crystals or radiate crystal groups* (Fig. 4-2a) and in granular or compact masses.	One good cleavage; in some specimens, a second poorer cleavage at an angle of 115° with the first.	6–7	3.4	Characteristic *yellowish-green* (*pistachio green*) color. Vitreous luster.
MAGNETITE. A combination of ferric and ferrous oxides, Fe_3O_4.	Well-formed, 8-faced crystals, more commonly in compact aggregates, disseminated grains or loose grains in sand.	None; conchoidal or uneven fractures; may show a rough parting resembling cleavage.	5.5–6.5	5.0–5.2	*Black.* Opaque. Metallic to submetallic luster. *Black streak. Strongly attracted by a magnet.* Magnetite is an important iron ore.
HEMATITE. Ferric iron oxide, Fe_2O_3.	Highly varied, compact, granular, fibrous, or earthy, micaceous; rarely in well-formed crystals.	None, but fibrous or micaceous specimens may show parting resembling cleavage; splintery to uneven fracture.	5–6.5†	4.9–5.3	Steel-gray, reddish-brown, red, or iron-black in color. Metallic to earthy luster. *Characteristic red streak.* Hematite is the most important iron ore.
"LIMONITE." Microscopic study shows that the material called limonite is not a single mineral. Most "limonite" is a very finely crystalline variety of the mineral **GOETHITE** containing absorbed water. Hydrous ferric oxide with minor amounts of other elements, roughly $Fe_2O_3 \cdot H_2O$.	Compact or earthy masses; may show radially fibrous structure.	None; conchoidal or earthy fracture.	1–5.5	3.4–4.0	*Yellow*, brown, or black in color. Dull earthy luster, which distinguishes it from hematite. *Characteristic yellow-brown streak.* A common iron ore.

† but earthy varieties appear much softer

(Continued)

MINERAL	FORM	CLEAVAGE	HARD-NESS	SP. GR.	OTHER PROPERTIES
PYRITE. ("Fool's gold.") Iron sulfide, FeS₂.	*Well-formed crystals,* commonly cubic, with striated faces (Fig. 4-2d); also granular masses.	None; uneven fracture.	6–6.5	4.9–5.2	Pale *brassy-yellow color;* may tarnish brown. Opaque. Metallic luster. Greenish-black streak. Brittle. Not a source of iron, but used in the manufacture of sulfuric acid. Commonly associated with ores of several different metals.
CHALCOPYRITE. Copper iron sulfide, CuFeS₂.	Compact or disseminated masses, rarely in wedge-shaped crystals.	None; uneven fracture.	3.5–4	4.1–4.3	Brassy to *golden-yellow. Tarnishes* to blue, purple, and reddish iridescent films. Greenish-black streak. Distinguished from pyrite by deeper yellow color and softness. A common copper ore.
CASSITERITE. Tin dioxide, SnO₂.	Well-formed, 4-sided prismatic crystals terminated by pyramids; 2 crystals may be intergrown to form knee-shaped twins; also as rounded pebbles in stream gravels.	None; curved to irregular fracture.	6–7	7	Brown to *black. Adamantine* luster. White to pale yellow streak. Chief ore of tin.
SPHALERITE. Zinc sulfide (nearly always containing a little iron), ZnS.	Crystals common, but chiefly in fine to coarse-granular masses.	*Highly perfect, in 6 directions at 60°.*	3.5–4	3.9–4.2	Color ranges from white to black but commonly *yellowish brown. Translucent* to opaque. *Resinous* to *adamantine luster.* Streak white, pale yellow or brown. Most important ore or zinc.
GALENA. Lead sulfide, PbS.	Cubic crystals common, but mostly in coarse to fine granular masses.	*Perfect cubic cleavage* (three directions mutually at right angles).	2.5	7.3–7.6	*Silvery gray color.* Metallic luster. Silvery gray to grayish black streak. Chief ore of lead.

MINERAL	FORM	CLEAVAGE	HARD-NESS	SP. GR.	OTHER PROPERTIES
URANINITE. (Pitch-blende.) Uranium oxide, UO_2 to U_3O_8.	Regular 8-sided or cubic crystals; massive.	None; fracture uneven to conchoidal.	5–6	6.5–10	Color black to brownish black. Luster submetallic, pitchlike, or dull. Chief mineral source of uranium, radium, etc.
CARNOTITE. Potassium uranyl vanadate, $K_2(UO_2)_2(VO_4)_2 \cdot 8H_2O$.	Earthy powder.	Not apparent.	very soft	4.1 approx.	*Brilliant canary-yellow color.* An ore of vanadium and uranium.

APPENDIX IV

Identification of Rocks

To USE the tables and lists below as guides in identifying unknown rock specimens, the student must have clearly in mind the basic distinctions between sedimentary, igneous, and metamorphic rocks, and the range in texture shown by each of the three major classes. This fundamental material is given in Chapter 5, and is summarized for the sedimentary rocks on p. 74, for the igneous rocks on pp. 85-87, and for the metamorphic rocks on pp. 94-95.

After examining a specimen of rock, but before referring to the tables and lists in this appendix, the student should ask himself the following three basic questions, and, if in doubt about the answers, should refer back to the material in Chapter 5 and Appendix III:

1. Is the specimen an igneous, sedimentary, or metamorphic rock? (Chapter 5)
2. What is its texture? (Chapter 5)
3. Of what minerals is it composed? (Appendix III)

List of Common Sedimentary Rocks

Conglomerate. Conglomerate is cemented gravel. Gravel is an unconsolidated deposit composed chiefly of rounded pebbles. The pebbles may be of any kind of rock or mineral and of all sizes. Most conglomerates, especially those deposited by streams, have much sand and other fine material filling the spaces between the pebbles. Some cleanly washed beach conglomerates contain little sand.

Breccia. Sedimentary breccias resemble conglomerate except that most of their fragments are angular instead of rounded. They accumulate in the same way as conglomerates into which they commonly grade. Since their constituent fragments have been little worn, however, it is apparent that the components of breccia underwent relatively less transportation and wear before they were deposited.

There are many other kinds of breccias besides sedimentary breccias. Volcanic breccias, as well as sedimentary breccias, are described in this appendix; glacial breccias in Chapter 13; and fault breccias in Chapter 10.

Sandstone.   Sandstone is cemented sand. It may grade into either shale or conglomerate. Sand, by definition, consists of particles from 2 to 1/16 mm. in diameter. The grains in most sands are chiefly quartz, but many other minerals and even fragments of other rocks may be found. In some sandstones, feldspar grains or rock fragments may be more abundant than quartz. In many sandstones the grains are all about the same size, but in others they are far from uniform.

Sand accumulates in many different environments. Some sand is deposited

---

TABLE IV-1. *Sedimentary Rocks*

I. Clastic Sedimentary Rocks

CONSOLIDATED ROCK	ORIGINAL UNCONSOLIDATED DEBRIS	DIAMETER OF FRAGMENTS
Conglomerate	Gravel (rounded pebbles)	More than 2 mm.
Breccia	Rubble (angular fragments)	More than 2 mm.
Sandstone	Sand	2 to 1/16 mm.
Shale	Mud, clay, and silt	Less than 1/16 mm.

II. Organic and Chemical Sedimentary Rocks

CONSOLIDATED ROCK	ORIGINAL NATURE OF MATERIAL	CHEMICAL COMPOSITION OF DOMINANT MATERIAL
Limestone	Shells; chemical and organic precipitates	$CaCO_3$
Dolomite	Limestone, or unconsolidated calcareous ooze, altered by solutions	$CaMg(CO_3)_2$
Chert	Siliceous organic material and precipitates	SiO_2 and SiO_2nH_2O
Peat and **Coal**	Plants	C, plus compounds of C, H, O
Salt deposits	Evaporation residues from the ocean or saline lakes	Varied, but *rock gypsum* ($CaSO_42H_2O$) and *rock salt* (NaCl) are most common

by streams; some is heaped up in dunes by the wind; some is spread out by waves and currents along beaches or on the sea floor.

Shale.   Shale is hardened mud. Mud is a complex mixture of very small mineral particles less than 1/256 mm. in diameter (chiefly *clay*), and coarser grains (called *silt*) from 1/256 to 1/16 mm. in diameter. Shale frequently contains small bits of organic matter.

The predominant minerals of shale are the hydrous aluminum silicates called clay minerals, but most shales also contain appreciable amounts of mica, quartz, and other minerals. Shale splits readily along closely spaced planes, nearly or quite parallel to the stratification. Some rocks of similar grain size and composition show little layering and break into small angular blocks. These are more correctly called *mudstone* instead of shale.

Shales accumulate in many different environments. As the main load brought down to the sea by great rivers is mud, it is not surprising that shale is the most abundant marine sedimentary rock. Mud deposited in deltas, on lake bottoms, and on plains along sluggish rivers may also harden into shale.

Many shales are black, perhaps because of large amounts of carbon-rich organic matter in various stages of decomposition or due to the precipitation of black iron sulfide (FeS_x), by sulfur bacteria. The iron sulfide may later crystallize into pyrite (FeS_2), forming small, brass-colored crystals sprinkled through the rock. Many black, dark gray, or green mudstones owe their color to decomposed volcanic material.

Limestone.   Limestone is composed almost entirely of the mineral calcite. Some limestones, such as the reef-limestones described in Chapter 5, contain abundant shells, many from microscopic organisms. Others are composed of microscopic crystals of calcite, perhaps precipitated out of sea water as by-products of certain chemical reactions accompanying the life processes of organisms such as algae and bacteria. Limestone is also forming today by direct precipitation of calcium carbonate in shallow warm seas, in hot springs, and in saline lakes.

Fine-grained, white, *calacareous ooze* (calcareous means calcite-rich) is abundant in parts of the southwest Pacific and on the shallow Bahama Banks of the Atlantic. Some of these oozes consist largely of microscopic shells, but much of that in the Bahamas appears to have been precipitated by chemical or biochemical agencies. Although shallow, warm seas appear to be the chief locale for the accumulation of limestone, calcareous oozes also cover much of the deep ocean basins. Limestone also forms locally in freshwater lakes, as deposits around hot springs, and as deposits from the saline waters of desert basins.

Limestones differ greatly in texture and color depending on the size of the shells or crystals and the impurities they contain. Some black limestones contain organic matter, as shown by a strong, fetid odor when freshly broken. Limestone deposited from hot springs commonly is full of small irregular holes and may be stained yellow or red by iron oxides. Most limestones, however, are light colored, and contain many fossils.

Dolomite. Dolomite rock is composed chiefly of dolomite, the mineral of the same name. Dolomite resembles limestone, and it also grades, by almost imperceptible variation in the amounts of the minerals calcite and dolomite, into limestone. Chemical and microscopic tests are generally necessary to distinguish dolomite definitely from limestone and to determine the relative amounts of the minerals calcite and dolomite in the rock.

Most dolomite appears to result from alteration of limestone or its parental calcareous ooze by magnesia-bearing solutions. In many, the alteration is thought to have taken place during slow deposition, by the action of the magnesium ion in sea water on the calcareous ooze. Some limestone, however, changed to dolomite long after it was deposited and consolidated.

Dolomite has rarely, and perhaps never, been deposited directly as a precipitated sediment.

Chert, and Other Fine-Grained Siliceous Rocks. Rocks composed almost entirely of fine-grained silica are common, but they rarely form large masses. Many different kinds of *siliceous* (siliceous means silica-rich) sedimentary rocks have been described and named, but the most common is *chert,* a hard rock with grains so fine that a broken surface appears uniform and lustrous.

Chert nodules, resembling a knobby potato in shape and size, are common in limestone and dolomite. Dark-colored chert nodules are often called *flint.* Chert also appears as distinct beds and as thin, wedge-like, discontinuous layers. Beds of chert are commonly associated with volcanic deposits.

The microscope shows that some cherts are made up largely of spines or lace-like shells of silica (opal) secreted by microscopic animals and plants. In other cherts, siliceous fossils are rare or absent, but siliceous shells may have been partially dissolved and reprecipitated as structureless silica during cementation. Abundant undissolved siliceous shells usually make the rock porous and light in weight. An example is *diatomite,* a white rock composed almost entirely of the siliceous shells of microscopic plants called *diatoms.*

Not all siliceous rocks are of organic origin. Some are believed to have precipitated around silica-bearing submarine hot springs. Many fine-grained siliceous rocks have been formed by the replacement of wood, limestone, shale, or other materials by silica-bearing solutions. *Petrified wood* is a familiar example.

Peat and Coal.   Peat and coal are not plentiful sedimentary rocks but their economic importance justifies their mention here.

Peat is an aggregate of slightly decomposed plant remains. It can be seen in process of accumulation in swamps and shallow lakes in temperate climates and even on steep hillsides in wet semi-arctic regions. Coal is the result of compression and more thorough decomposition of the plant material in ancient peat bogs which were buried under later sediments. Coals grade from *lignite,* which differs little from peat, through *bituminous* to *anthracite,* which may be 90 per cent or more of carbon. From evidence obtained in mines and by geologic mapping, we infer that the grade of the coal depends largely on the depth to which it has been buried (i.e., the pressure and heat to which it has been subjected).

Salt Deposits.   Salt deposits vary greatly in mineral composition and texture. They are now being formed by the evaporation of land-locked masses of sea water, as at the Runn of Kutch in northwest India and in saline lakes like Great Salt Lake. If we evaporate sea water completely to dryness in the laboratory, we see many different salts precipitated from it, but *rock salt* (halite, $NaCl$) being the most abundant residue. In nature, however, calcium sulfate, which occurs both as a hydrated form, *rock gypsum* ($CaSO_4 2H_2O$), and as the anhydrous mineral called *anhydrite* ($CaSO_4$) is much more common than rock salt. Gypsum separates out early in the process of evaporation and will, therefore, accumulate in quantity from water bodies that are not saline enough to precipitate halite. Rock gypsum, accompanied by little or no rock salt, is abundant in the Paris Basin of France, in the Dakotas, and elsewhere. Thick beds of rock salt, accompanied by gypsum and anhydrite, are found in Texas, New Mexico, Germany, Iran, India and many other areas.

In a few places where relatively complete evaporation of sea water has occurred, deposits of potassium salts and other valuable, late-crystallizing minerals are found. Many rare and useful mineral products such as potash, salsoda, borax, nitrates, sodium sulfate, and epsom salts are recovered from salt deposits formed by the evaporation of ancient desert lakes.

Notes on the Igneous Rock Table

In Table IV-2 *textures are listed at the left;* the *major differences in mineral composition along the base.* Names of major rock groups are in **boldface type.** Thus, the texture of any rock group appears to the left of its name, and the predominant minerals it contains below it. *Rocks high in silica are found to the left of the table,* and *the silica content decreases gradually to the right.* This difference is reflected in the color of the rocks. Those at the left of the table are commonly light colored because of the abundance of high-silica minerals such as quartz and feldspar. Those at the right of the table are dark colored because they are rich in ferromagnesian minerals. Glassy rocks, of course, contain few, if any, recognizable minerals and the columns showing predominant mineral composition do not strictly apply. Here, as with the aphanitic rocks, we depend, in field classification, on the minerals represented among the phenocrysts. If no phenocrysts are present, the rock can only be classified after microscopic work or chemical analysis.

TABLE IV-2. *Igneous Rocks**

	Feldspar and quartz predominate	Feldspar predominates (no quartz)	Ferromagnesian minerals predominate, but feldspar also abundant (no quartz)	Ferromagnesian minerals predominate (no feldspar or quartz)
Pyroclastic	**Volcanic tuff** (fragments up to 4 mm. in diameter) **Volcanic breccia** (fragments more than 4 mm. in diameter)			Rocks of the texture and composition represented by this part of the table are rare or unknown.
Glassy	**Obsidian** (if massive glass) **Pumice** (if a glass froth)		**Basalt glass**	
Aphanitic (generally porphyritic-aphanitic)	**Rhyolite†** and **Dacite‡** (contains phenocrysts of quartz and feldspar)	**Andesite**	**Basalt**	
Granular	**Granite†** and **Granodiorite‡**	**Diorite**	**Dolerite** (if fine-grained) **Gabbro**	**Peridotite** (with both olivine and pyroxene) **Pyroxenite** (with pyroxene only) **Serpentine** (olivine and pyroxene altered)

* Note: **Porphyry**, omitted from this table, is described in the "List of Common Igneous Rocks."
† **with** dominant orthoclase.
‡ **with** dominant plagioclase.

List of Common Igneous Rocks

Volcanic Tuff. Volcanic tuff is a fine-grained pyroclastic deposit composed of fragments less than 4 mm. in diameter. Most of the fragments are volcanic glass, either in microscopic slivers or in frothy bits of pumice (see p. 607). Other common constituents are broken phenocrysts and chunks of solidified lava. Fragments of the basement rock on which the volcano rests may also be present.

Pumice and other kinds of glass fragments have been seen to form by the explosive disruption of sticky lava highly charged with gases. Evidently the gas pressure increases until it exceeds the containing pressure on the magma; then the pent-up gases separate into bubbles, causing the lava to expand tremendously and to froth. Upon breaking out to the surface, the froth disrupts further into a great explosion cloud of glass fragments and pumice.

The fragments of a volcanic tuff may be cemented together in the same way as the fragments of a sedimentary rock. The component fragments of some tuffs, however, appear under the microscope to have been *welded*— partly melted and stuck together. The heat necessary for welding is thought to have been supplied by reactions of the hot gases in the eruption cloud with one another and with air. Such *welded tuffs* are common. Before C. N. Fenner, an American geologist, recognized their true mode of origin by observations of the products of the 1912 eruption of Mont Katmai in Alaska, welded tuffs were thought to be lava flows. Even today, they are often confused with rhyolite and dacite lavas because of the close similarity of the welded fragmental matrix to the aphanitic texture of lava flows.

Volcanic Breccia. Volcanic breccia is composed dominantly of fragments more than 4 mm. in diameter. In general, fragments of lava are more abundant than in tuff; glass slivers and pumice may be scarce. *Scoria* (see p. 77) is abundant in some breccias.

Some volcanic breccias are formed like the tuffs, but many are products of volcanic mudflows. Heavy rains falling on the steep slopes of a volcanic cone have been seen to set great avalanche-like slides of unconsolidated pyroclastic debris in motion. Other mudflows are formed by eruption clouds falling into rivers or by explosive eruptions through crater lakes. The water-soaked volcanic debris may travel for many miles down stream valleys.

Obsidian. Obsidian is natural glass, formed chiefly from magmas of rhyolitic, dacitic, or andesitic composition. It is lustrous and breaks with a curved fracture. Most obsidians are black with sparsely disseminated grains of magnetite and ferromagnesian minerals, but may be red or brown from the oxidation of the iron by hot magmatic gases. In thin pieces, obsidian is almost transparent.

Obsidian forms lava flows and rounded domes above volcanic vents. It also is found as thin selvages along the edges of intrusions, and, rarely, makes small intrusive masses. Most intrusive obsidians have a dull, pitch-like luster, and are called *pitchstone*.

Pumice.   Pumice is obsidian froth (see Volcanic Tuff, above). Its light-gray to white color and abundant tiny bubbles are characteristic. The bubbles are so numerous that pumice will float on water. Pumice is abundant as fragments in tuffs and breccias. It also may form distinct flows, or more commonly, it caps flows of obsidian or rhyolite, and grades downward into the unfrothed lava beneath.

Basalt Glass.   Basalt glass is a jet-black natural glass formed by chilling of basaltic magma. Unlike obsidian, it is not noticeably transparent on thin edges. Basalt glass has never been found in large flows like obsidian; on this fact is based the inference that basalt magma crystallizes much more readily than rhyolite. Basalt glass forms thin crusts on the surfaces of lava flows, small fragments in volcanic breccia, and thin contact selvages in volcanic necks and dikes. Breccias of basalt glass form in abundance when basalt magma is extruded into water and quickly quenched.

Rhyolite.   Rhyolite has an aphanitic groundmass generally peppered with phenocrysts of quartz and orthoclase. The color of rhyolite ranges widely, but generally is white or light shades of yellow, brown, or red. Most rhyolites are flow banded; that is, they show streaky irregular layers as in taffy, formed by the flowing of the sticky, almost congealed magma.

Dacite is like rhyolite except that it contains predominant plagioclase instead of orthoclase. It bears the same relation to rhyolite that granodiorite does to granite (see below).

Rhyolite and dacite are found in lava flows and as small intrusions.

Andesite.   Andesite is an aphanitic rock, generally porphyritic, that resembles dacite but contains no quartz. Plagioclase feldspar is the most common phenocryst, but pyroxene, amphibole, or biotite may appear. Most andesites are flow banded, though not so conspicuously as rhyolites. Andesites range from white to black, but many are dark gray or greenish gray.

Andesite is abundant as lava flows and as fragments in volcanic breccia, particularly in the high, volcano-capped mountain ranges such as the Andes (from which it is named), the Cascades, and the Carpathians. Andesite also forms small intrusive masses.

Basalt.   Basalt is a black to medium-gray aphanitic rock. Many basalts are nonporphyritic, but some contain phenocrysts of plagioclase, olivine, or other ferromagnesian minerals.

Basalt is the world's most abundant lava and is very widespread, forming great lava plateaus that cover thousands of square miles in the northwestern United States, India, and elsewhere. It is the chief constituent of the isolated oceanic islands. Although it typically forms lava flows, basalt is also common in small intrusive masses.

Granite.   Granite, characterized by a granular texture, has as its two most abundant minerals, feldspar and quartz. They make most granites light colored. Biotite and/or hornblende are also present in most granites.

Technically, the term *granite* is reserved for those granular quartz-bearing igneous rocks with potash feldspar (orthoclase) as the chief feldspar. Those with dominant plagioclase are called *granodiorite*. (Compare rhyolite and dacite above.) Granodiorite can usually be distinguished from granite by the fine striations that characterize one cleavage surface of plagioclase.

Geologic mapping shows that granite and granodiorite are present in great bulk in the earth's crust. They form large intrusive masses along the cores of many mountain ranges and in areas where deep erosion has occurred such as northeastern Canada, the Scandinavian region, and eastern Brazil. They are typical continental rocks and have never been found on isolated oceanic islands such as the Hawaiian or Caroline groups.

Some granites are of metamorphic instead of igneous origin (Chaps. 17 and 19).

Diorite.   Diorite is a granular rock composed of plagioclase and lesser amounts of ferromagnesian minerals. The most common ferromagnesian minerals are hornblende, biotite, and pyroxene.

In general, diorite masses are much smaller than those of granite or granodiorite.

Gabbro.   Gabbro is a granular rock composed chiefly of pyroxene and plagioclase, commonly with small amounts of other ferromagnesian minerals, especially olivine. If ferromagnesian minerals predominate over the plagioclase so that the rock is dark-colored, it is generally correct to call it gabbro, though the microscopic distinction from diorite rests on the composition of the plagioclase, a character not determinable with the unaided eye.

Gabbro is widely distributed in both large and small masses. Dikes and thin sills of fine-grained gabbro are especially common. In these small intrusions, the mineral grains are commonly so small that they are barely recognizable without the aid of the microscope. Such gabbros, intermediate in grain size between basalt and normal gabbro, are called *dolerite.**

Peridotite, Pyroxenite, and Serpentine.   Granular rocks composed almost entirely of ferromagnesian minerals and without feldspar are common in some areas. If the rock contains olivine as a conspicuous constituent, it is called *peridotite;* if it is made up almost wholly of pyroxenes, it is called *pyroxenite.*

---

* Some geologists prefer the name *diabase* in place of dolerite.

Olivine is a very unstable mineral easily altered to a mixture of greenish hydrous minerals. Some varieties of pyroxene are also easily altered. It will be shown (Chap. 17) that these alterations probably occur soon after consolidation of the magma and are caused by the hot gases and solutions escaping from the crystallizing peridotite or perhaps from nearby granite or gabbro masses. Such altered peridotites and pyroxenites are called *serpentine*. Because serpentine is composed almost entirely of secondary minerals which did not solidify directly out of the magma, it is often classed as a metamorphic rock instead of an igneous rock. Nearly all plutonic igneous rocks, however, show some features that suggest alteration and "working over" by hot gases during the last stages of crystallization, although they are not commonly modified as much as serpentine.

Serpentine forms sills, dikes, and other small intrusive masses.

Porphyry.   The ancient term *porphyry* is used rather indefinitely. It is commonly applied to porphyritic-textured, fine-grained intrusive igneous rocks in which phenocrysts make up 25 per cent or more in volume. The groundmass may be either coarse-grained aphanitic, or fine-grained granular. The name of the rock whose composition and texture fit the groundmass part of the rock is usually prefixed to the word porphyry. Thus *diorite porphyry* has a fine-grained granular groundmass and contains abundant phenocrysts of plagioclase and perhaps some ferromagnesian mineral. *Andesite porphyry* is similar except that the groundmass is aphanitic.

The noun "porphyry," as distinguished from the adjective "porphyritic," should not be applied to porphyritic rocks with a coarse-granular groundmass or to porphyritic lava flows containing a few scattered phenocrysts. The former should be called *porphyritic diorite* and the latter *porphyritic andesite* if they have the same composition as diorite and andesite.

Granite porphyry, granodiorite porphyry, and diorite porphyry are widespread as dikes near granite and granodiorite masses. Rhyolite porphyry, dacite porphyry, and andesite porphyry are common in volcanic plugs and other small intrusive masses.

List of Metamorphic Rocks

Hornfels (German: "horny rock").   Hard, unfoliated, very fine-grained rock which breaks into sharp angular pieces. In many hornfels traces of original structures such as stratification, flow banding, or slaty cleavage can be seen; but the rock will not break along them. The mineral composition is highly variable, and grains are, in general, too small to be recognizable without the microscope.

Hornfels is formed by the partial or complete recrystallization, near an igneous intrusion, of fine-grained rocks such as shale, shaly limestone, slate, chlorite schist, tuff, and lavas.

Quartzite. Very hard, sugary-textured granoblastic rock, composed al-

TABLE IV-3. *Metamorphic Rocks*

	NAME	TEXTURE	COMMONLY DERIVED FROM	CHIEF MINERALS
UNFOLIATED OR FAINTLY FOLIATED	Hornfels	Hornfelsic	Any fine-grained rock	Highly variable
	Quartzite	Granoblastic, fine grained	Sandstone	Quartz
	Marble	Granoblastic	Limestone, dolomite	Calcite, magnesium and calcium silicates
	Tactite	Granoblastic, but coarse and variable	Limestone or dolomite plus magmatic emanations	Varied; chiefly silicates of iron, calcium, and magnesium, such as garnet, epidote, pyroxene, amphibole
	Amphibolite	Granoblastic	Basalt, gabbro, tuff	Hornblende and plagioclase, minor garnet and quartz
FOLIATED	Slate (and Phyllite)	Slaty	Shale, tuff	Mica, quartz
	Chlorite schist	Schistose to slaty	Basalt, andesite, tuff	Chlorite, plagioclase, epidote
	Mica schist	Schistose	Shale, tuff, rhyolite	Muscovite, quartz, biotite
	Amphibole schist	Schistose	Basalt, andesite, gabbro, tuff	Amphibole, plagioclase
	Gneiss	Gneissose	Granite, shale, diorite, mica schist, rhyolite, etc.	Feldspar, quartz, mica, amphibole, garnet, etc.
	Migmatite	Coarsely banded, highly variable	Mixtures of igneous and metamorphic rocks	Feldspar, amphibole, quartz, biotite

most entirely of interlocking quartz grains. Quartzite breaks across the grains, not around them as in most sandstones. Colors range from white through pale buff to pink, red, brown, and black; but most quartzite is light colored.

Quartzite is formed by the metamorphism of quartz sandstone. It is a widely distributed metamorphic rock.

Sandstone with a cement of silica (sedimentary "quartzite") is difficult to tell from metamorphic quartzite since both break across the grains. The distinction is usually not difficult with the petrographic microscope, for the cement can be distinguished readily from the original sand grains. Metamorphic quartzite can also be distinguished from silica-cemented sandstone by the rocks associated with it in the field, for true quartzite is associated with other metamorphic rocks, and sandstone with other sedimentary rocks.

Marble. Granoblastic, fine to coarse-grained rock composed chiefly of calcite and/or dolomite. Many marbles show a streaky alternation of light and dark patches; others show brecciated structures healed by veinlets of calcite.

Marble is formed by the metamorphism of limestone and dolomite; if from dolomite, it commonly contains magnesium-bearing silicates such as pyroxene, amphibole, and serpentine.

Tactite ("to touch"). Granoblastic, but variable in texture, grain size, and mineral composition. Tactite is rich in silicates of calcium, iron, and magnesium—such as amphibole, pyroxene, garnet, and epidote. It occurs usually where limestone or dolomite has been invaded by granite or granodiorite. From this it is inferred that fluids escaping from the congealing magma have carried into the limestone large quantities of silica, iron, and other substances which combined with the calcite and dolomite to form new minerals. Ores of iron, copper, tungsten, and other minerals are often associated with these rocks.

Amphibolite. Granoblastic-textured, commonly coarse-grained rock consisting chiefly of amphibole and plagioclase. Garnet, quartz, and epidote may be present in small quantities.

Amphibolites have been formed by the metamorphism of basalt, gabbro, and rocks of similar composition; some are derived from impure dolomite.

Slate (and Phyllite). Very fine-grained, exceptionally well-foliated rocks. Because of the excellent foliation, they split into thin sheets. Mineral grains are too small to be identified without the microscope or X-rays. *Slate* is dull on cleavage surfaces; *phyllite* is shiny and coarser grained, containing some mineral grains large enough to be identified by the eye. Slate and, to a lesser extent, phyllite commonly show remnants of sedimentary features such as stratification, pebbles, and fossils.

Slate and phyllite are abundant. Most were formed by the metamorphism of shale, but, locally, from tuffs or other fine-grained rocks.

Chlorite Schist. Green, very fine-grained schistose to slaty rock. It is commonly soft, greasy, and easily pulverized with the fingernail. Composed of chlorite, plagioclase, and epidote, but all except chlorite may be in grains too small to identify. Remnants of original volcanic structures such as phenocrysts and scoria may be present.

Chlorite schists are abundant. They are often called *green schist* or, if poorly foliated, *greenstone,* from the color of the abundant chlorite. Most have been formed by the metamorphism of basalt or andesite and their corresponding tuffs, but some have been derived from dolomitic shale, gabbro, and other ferromagnesian rocks.

Mica Schist.   Schistose rock composed chiefly of muscovite, quartz, and biotite in varying proportions, any one of which may predominate.

Mica schist is one of the most abundant metamorphic rocks. Like slate, it has commonly been formed from shales and tuffs, less commonly from feldspathic sandstone, shaly sandstone, rhyolite, and other rocks. It represents more intense metamorphism than slate.

Amphibole Schist.   Schistose rock, composed chiefly of amphibole and plagioclase, with more or less garnet, quartz, or biotite.

It is a common metamorphic derivative of basalt, gabbro, chlorite schist, and related rocks.

Gneiss.   Coarse-grained, gneissose rock with distinct bands or lenses of different minerals. Mineral composition variable, but feldspar especially abundant. Other common minerals are quartz, amphibole, garnet, and mica.

Gneisses are among the most abundant metamorphic rocks. They may be derived from many different rocks—granite, granodiorite, shale, rhyolite, diorite, slate, and schist, among others.

Migmatite.   Migmatites are highly complex rocks (described in Chaps. 17 and 19). In general, they are intimate small-scale mixtures of igneous and metamorphic rocks, characterized by a roughly banded or veined appearance. They are common and widespread, especially near large granite masses. The mineral composition of migmatites is complex and highly variable, but most contain abundant feldspar, quartz, biotite, and amphibole.

Index

Glacial floors, 307
  floor, 124, 301
  ice, layering in, 290
Glacial loading and isostasy, 323-324
Glacial polishing, 307
  steps, 308
  theory, development of, 311
Glaciation, astronomical influences in, 321
  causes of, 319-323
  cosmic theories for, 321
  drainage changes contemporaneous with, 314-319
  effect on sea level, 266
  fluctuation in carbon dioxide as a cause of, 320
  fluctuations in solar radiation as possible cause of, 321
  geographic factors in, 320
  in the geologic past, 309, 319
  in Great Britain, 311
  in New England, 311
  in Swiss plain, 311
  late Paleozoic, 319
  Pleistocene, 310
  pre-Cambrian, 319
  pre-Pleistocene, 319
Glacier, loading of, 294
Glacier motion, 289-290
  nature of, 290
  proof of, 289
Glaciers, defined, 289
  effect on topography, 307
  erosion and deposition by, 294
  influence of precipitation on, 286
  kinds of, 291
  recrystallization in, 290
  rock, 226
  thickness of, 323
Glacio-fluvial deposits, 304
Glacio-lacustrine deposits, 306
Glassy texture, 86
Glossopteris flora, 532-534
Gold, mineral, 565
  placer deposits of, 577-578
Gondwana Land, 532
Graded stream, 263
Gradient, stream, defined, 248
  temperature, 502-503
Grain size, downstream decrease in, 273
Grand Canyon of the Colorado, evidence of erosion, 138-140
  importance of downslope movements in forming, 240-241
Grand Coulee, Wash., glacial origin of, 317
Grand Coulee Dam, Washington, landslide threat, 244
Granite, abundance of, 83
  as an oceanic precipitate (Werner), 83
  as the "original crust," 83

Granite, intrusive nature of, 83-84, 455-458
Granites, with sedimentary structures, 85, 457-458
Granite contacts, gradational, 85, 456-458
Granitic domes formed by weathering, 110
Granitic injections, criteria of, 83-84, 456-459
  intermittent character of, 84
Granodiorite, weathering of, 108
Granular texture, 86
Gravitation, Law of, 29
Gravity anomaly, 45
  Bougeur, 45
Gravity measurements and isostasy, 44-46
Gravity at sea, 34, 526
Gravity surveys in oil exploration, 560
Great Altai Mts., 18
Great Barrier Reef, Australia, 424
Great Glen fault, Scotland, 208
Great Lakes, glaciation in region of, 315-317
Great Salt Lake, Utah, 354
Great Serpentine Dike, Rhodesia, 450
Greenland, glaciers in, 293
Grenville gneiss, 169
Griggs, David T., 536, 538
Ground Moraine, 304
Ground water, alkaline, 339
  cementation by, 342
  composition of, 338
  defined, 327
  discharging into the sea, 336
  in carbonate rocks, 339
  in volcanic rocks, 342
  legal aspects of, 348
  movement of, 333
  movement, rate of, 335
  recharge of, 345, 347
  solution by, 339-343
  source of, 327
  supplies of in United States, 343
Groynes, effect on sediment transport, 400
Guglielmini, contribution to crystal study, 58
Gulf Coast, similar to ancient geosynclines, 508
Gulf of Mexico, difference in level from Atlantic, 384
Gulf Stream System, 383-384
Gully development in deserts, 355
Gumbo, 312
Gumbotil, 312
Gutenberg, Beno, 496

H

Halemaumau, Hawaii, 437-438
Half-life of a radioactive element, 168

Hall, James, 506
Halley, Edmund, 127
Hangchow River, China, tidal bore in, 398
Hanging glaciers, 291
  valleys, 309
"Hardpan," 112
Hard water, 338
"Harrisburg surface," 282
Hawaiian Islands, 24
  ground water in, 337
  olivine sands on beaches, 410
Headward erosion, 276
Heat budget of the earth, 100
Heat flow, to surface, 502-504
Heim, Arnold, 517
Herculaneum, destruction of, 75
Herodotus, 267
Hewett, D. F., 548
High Limestone Alps, structure of, 514-515
Himalayas, 18
  a part of Alpine Mountain System, 511
Hindu Kush Mts., 18
Hinterland, 507
Horizontal control of maps, 581
Horns, glacial, 309
Hot springs, 464-467
  magmatic water in, 465
Hudson-Champlain Valley, facies changes in, 511
Hudson submarine canyon, 429
Humus, formation of, 105-106
Hutton, James, 68
Huygens, Christian, discoverer of "double refraction," 59
Hydraulic gradient, 333
  effect of water supply on, 333
Hydraulic jump, 254
Hydraulic pressure, of waves, 391
Hydrographic maps, construction of, 584-585
Hyrology, 126
Hydrothermal ore deposits, 570
  form of, 572-576
  origin of, 570-572
  transitions between, 575-576

dependence of grain size on rate of cooling, 85-86
textures of, 85-87
Inclusions, significance of, 81
  in plutons, 459-460
Incompressibility, rock, 479, 483
India, glaciation in, 532
Indian Ocean, floor of, 23
Industrial revolution, 543-544
Inertia, 33
Inert gas structure, 56
Infiltration capacity, 130
Influent streams, 333
Injection gneiss, 457
Inselbergs, 361
Interior basins, groundwater in, 344
Interior drainage in deserts, 352
Interior Plateau, British Columbia, 19
Intermediate layer, 492, 493
Intrusive igneous rocks, defined, 79
Intrusive masses, deep, 453-458
  igneous, 449-460
Ion, definition of, 56
Iran, sand dunes in, 372
  petroleum reserves of, 559
Iranian plateau, 19
Iraq, petroleum reserves of, 559
  soils in, 354
Iron deposits, sedimentary, 577
  minerals, 565
Iron Springs, Utah, iron deposit, 569-570
Irrawaddy River, 507
Island arcs, Pacific, 24
  earthquakes associated with, 490, 492
"Island mountains," 361
Isobaric surfaces in the sea, 382
Isomorphism, definition of, 63
  importance of ionic diameter in, 64-65
  mechanism of, 64-65
Isoseismals, 485-487
Isostasy, 35-36
  perfection of, 46
Isostatic compensation, 36
Isostatic evidence from glaciation, 324
Isostatic relations in mountain belts, 522-528

I

Ice caps, 294
Ice loading, isostatic effects of, 178-179
Ideal profile, value of concept, 274
Igneous activity, role of volatiles in, 463-464
  contacts, recrystallization near, 91-92
Igneous rocks, classification of, 85
  correlation between chemical and mineralogical composition, 85
  definition of, 69, 74

J

Japan Sea, 24
Japan trench, 24
Japanese arc, 24
Japanese Islands, 19
Joints, defined, 206
Joly, John, 166
Jura Mountains, 162, 200, 513
  structure of, 513
Jurassic System, 162

(2)

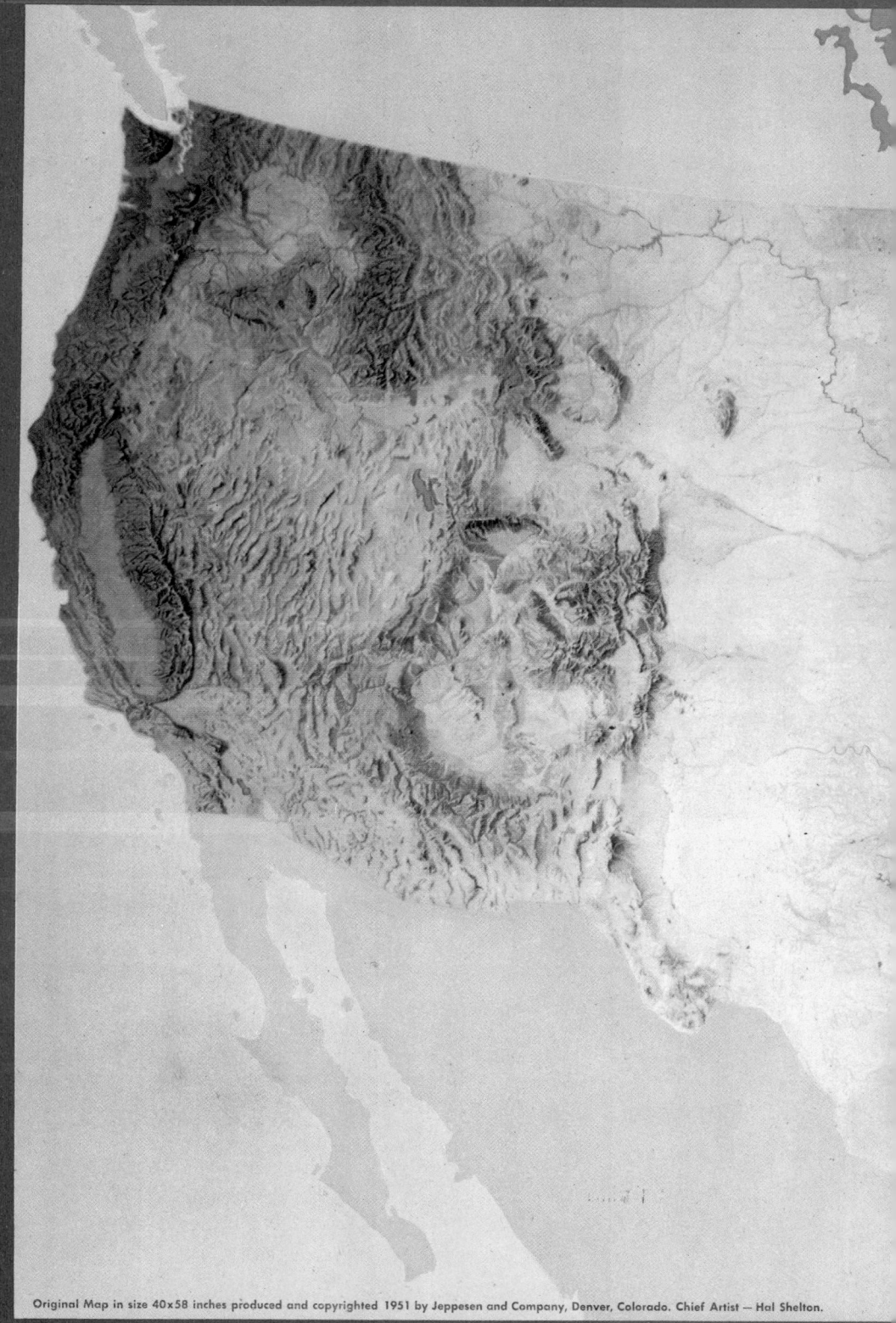

Original Map in size 40x58 inches produced and copyrighted 1951 by Jeppesen and Company, Denver, Colorado. Chief Artist — Hal Shelton.